INTERNATIONAL SERIES I.

NATURAL PHILOSOPHY

General Editor: D. TER HAAR

VOLUME 30

Problems in Quantum Mechanics

Other Pergamon Titles of Interest

Books

DAVYDOV
Quantum Mechanics, 2nd Edition
Biology and Quantum Mechanics

HEINE
Group Theory in Quantum Mechanics

LANDAU & LIFSHITZ
Course of Theoretical Physics, Vol 3 — Quantum Mechanics, 3rd Edition

TAYLOR
Mechanics: Classical and Quantum

*Journals**

Progress in Quantum Electronics

Progress in Surface Science

Reports on Mathematical Physics

**Free specimen copies available on request*

A full list of titles in the International Series in Natural Philosophy follows the index.

PROBLEMS IN QUANTUM MECHANICS

BY

F. CONSTANTINESCU

A.v. Humboldt Fellow, University of Munich, and University of Cluj, Rumania

AND

E. MAGYARI

University of Craiova, Rumania

TRANSLATED BY

V. V. GRECU

EDITED BY

J. A. SPIERS

PERGAMON PRESS

OXFORD · NEW YORK · TORONTO · SYDNEY · PARIS · FRANKFURT

U.K.	Pergamon Press Ltd., Headington Hill Hall, Oxford OX3 0BW, England
U.S.A.	Pergamon Press Inc., Maxwell House, Fairview Park, Elmsford, New York 10523, U.S.A.
CANADA	Pergamon Press Canada Ltd., Suite 104, 150 Consumers Rd., Willowdale, Ontario M2J 1P9, Canada
AUSTRALIA	Pergamon Press (Aust.) Pty. Ltd., P.O. Box 544, Potts Point, N.S.W. 2011, Australia
FRANCE	Pergamon Press SARL, 24 rue des Ecoles, 75240 Paris, Cedex 05, France
FEDERAL REPUBLIC OF GERMANY	Pergamon Press GmbH, 6242 Kronberg-Taunus, Hammerweg 6, Federal Republic of Germany

First edition 1971
Reprinted (with corrections) 1976
Reprinted 1978, 1982

Library of Congress Catalog Card No. 73-116573

Printed in Great Britain by A. Wheaton & Co. Ltd., Exeter

ISBN 0-08-019008-1 (Flexicover)

Contents

Contents

Foreword

WHEN Dr. Constantinescu and Dr. Magyari approached me about the possibility of the publication of an English edition of their collection of problems on quantum mechanics I was very pleased to arrange this. Although there exist a few good collections of this kind, none of those has the same extensive coverage which the present volume gives, and I felt that it would be a valuable addition to the existing literature. Quantum mechanics courses are getting more and more advanced, and topics which only a few years ago were deemed to be fit only for graduate courses now appear in undergraduate courses. It is thus very much to be welcomed to have a problems book which covers such topics as the Dirac equation, second quantization, many-body problems, and the density matrix. I hope and expect that quantum mechanics teachers as well as students will find the present book a great help, and I wish it every success.

Magdalen College, Oxford D. TER HAAR

A Note on the Layout of this Book

EACH chapter is essentially self-contained, and is in three parts. In the first part, basic propositions relating to a particular topic are given and some consequential theorems are stated without proof. In the second part, a set of graded problems leads the reader to prove the theorems already stated, and to derive further theorems and applications. Detailed solutions of the problems are given in the third part. Equations in the first part are labelled with the number of the chapter (in roman numerals) followed by a serial number; those in the second and third parts with the number of the problem followed, respectively, by a serial *letter* or a serial *number*. The letter A followed by a number is a reference to an epuation in the Appendix.

CHAPTER I

The Mathematical Formalism of Quantum Mechanics

1. Hilbert Spaces

A set of entities E is said to constitute a linear (complex) space if, on the elements of the set, operations of commutative addition, and of multiplication by a complex number, are defined and are such that the results of these operations are also element of E. The term "vector" is often used to denote the elements of such a space; these vectors should not be confused with the vectors of ordinary geometric space.

Let H be a linear (complex) space such that, to every pair of elements x, y of H, there can be made to correspond a complex number, called the *scalar product* (x, y) of the elements x and y, and having the following properties:

(a) $(x, y) = (y, x)^*$,

(b) $(x_1+x_2, y) = (x_1, y)+(x_2, y)$,

(c) $(\lambda x, y) = \lambda^*(x, y); \qquad (x, \lambda y) = \lambda(x, y)$,

(d) $(x, x) \geqslant 0$, the equality holding only for the "null element" $x = 0$. The "norm" of an element is defined to be $\|x\| = +\sqrt{(x, x)}$.

H is then said to be a *Hilbert space* provided it is complete.†

In any Hilbert space a set of elements can be found, such that any element of the space can be written as a linear combination of these. Such a set is called a basis.

In quantum mechanics, every state of a physical system is associated with a vector, sometimes called a "ket", and denoted by the general symbol $|\ \rangle$, into which may be inserted symbols denoting eigenvalues, quantum numbers, etc., which specify the state in question. From the principle of superposition of states, it follows that the ket vectors of a system together form a linear space.

† An ordered set $\{x_n\}$ of elements of H is said to converge to the limit x_0, if $\|x_n - x_0\| \to 0$ as $n \to \infty$. If any ordered set $\{x_n\}$ of H, which is such that $\|x_n - x_m\| \to 0$ as $n, m \to \infty$, converges to a limit which is an element of H, then H is said to be "complete".

The scalar product of any two ket vectors $|u\rangle$ and $|v\rangle$ of a system is a complex number, written as $\langle v|u\rangle$. The symbol $\langle v|$ is sometimes called the "bra" vector corresponding to the ket $|v\rangle$, since a "bra" and a "ket" together form a "bra(c)ket". This somewhat whimsical terminology is due to Dirac.

The set of all ket vectors of a system form a Hilbert space.

2. Operators in Hilbert Space

Any procedure A whereby each vector $|u\rangle$ of a Hilbert space H is related to one and only one vector $|v\rangle = A|u\rangle$ of that space is called an *operator* of that space.

An operator A is said to be "linear" if, for any vectors $|u\rangle$ and $|v\rangle$, and any complex numbers λ and μ,

$$A(\lambda|u\rangle+\mu|v\rangle) = \lambda A|u\rangle+\mu A|v\rangle. \tag{I.1}$$

Two operators A and A^+ are said to be Hermitian adjoints if, for any vectors $|v\rangle$ and $u\rangle$ of H,

$$\langle v|A^+|u\rangle = \langle u|A|v\rangle^*. \tag{I.2}$$

An operator A is said to be "Hermitian" (or "self-adjoint") if $A = A^+$.

An operator U is said to be "unitary" if $UU^+ = U^+U = 1$.

If A is an operator and $|u\rangle$ is such that

$$A|u\rangle = a|u\rangle, \tag{I.3}$$

where a is a real or complex number, then $|u\rangle$ is said to be an "eigenvector" of A with the "eigenvalue" a. The eigenvalues of Hermitian operators are always real numbers. There may exist more than one linearly independent eigenvector of A with the same eigenvalue a. Any linear combination of these is then also an eigenvector of A with the same eigenvalue a. The number of such linearly independent eigenvectors (if greater than one) is called the "degeneracy" of the eigenvalue a. The set of all eigenvalues of A is called the "spectrum" of A.

The spectrum of an operator may be discrete, or may be continuous, or may have a discrete part and a continuous part. Let us denote by $|nr\rangle$ an eigenvector of A with the eigenvalue a_n belonging to the discrete part, and by $|\nu\varrho\rangle$ an eigenvector of A with the eigenvalue a_ν belonging to the continuous part of the spectrum. (Here r and ϱ are indices which distinguish between degenerate states having the same eigenvalue.)

For the eigenvectors of Hermitian operators, the following orthonormalization relations are satisfied for all values of the indices (provided that for degenerate eigenvalues a suitable choice of eigenvectors is made):

$$\begin{aligned}\langle nr|n'r'\rangle &= \delta_{nn'}\delta_{rr'},\\ \langle nr|\nu'\varrho'\rangle &= 0,\\ \langle \nu\varrho|\nu'\varrho'\rangle &= \delta(\nu-\nu')\,\delta(\varrho-\varrho'),\end{aligned} \tag{I.4}$$

where $\delta_{nn'}$, is the Kronecker delta and $\delta(\nu-\nu')$ is the Dirac delta function.

If the set of all eigenvectors $|nr\rangle$ and $|\nu\varrho\rangle$ of the Hermitian operator A spans (by linear combination) the entire space H, this set is said to be complete, and the operator A is said to be an "observable" (the operator A is then capable of having a physical interpretation).

The operator $[A, B] = AB - BA$ is called the "commutator" of the operators A and B. If $[A, B] = 0$, A and B are said to commute. If two observables A and B commute, they have a complete set of eigenvectors in common, and conversely.

3. The Matrix Representation of Vectors and of Operators

For any given vector $|u\rangle$, the number $\langle n|u\rangle$, where $|n\rangle$ is an eigenvector of some observable A, can be regarded as an element having "row number" n of a one-column matrix (u); the whole matrix is then said to be a "representation" of the ket $|u\rangle$. The bra vector $\langle u|$ is represented by the Hermitian adjoint matrix whose elements are $\langle u|n\rangle$.

For a given operator Ω, the eigenvectors $|m\rangle$ and $|n\rangle$ of A enable one to define the number $\Omega_{mn} = \langle m|\Omega|n\rangle$, which can be regarded as the element with "row number" m and "column number" n of a square matrix (Ω). The whole matrix is then said to be a representation of the operator Ω. Note that, in this representation, A itself is represented by a diagonal matrix.

Properties of representations:

(a) The Hermitian adjoints of operators are represented by the Hermitian adjoints of the matrix representations of the original operators.

(b) Algebraic relations between vectors and operators lead to equivalent algebraic relations between their representative matrices (thus, in particular, the equations which define the eigenvalues and eigenvectors of operators become matrix eigenvalue equations).

(c) The trace $\mathrm{Tr}(\Omega)$ of any Hermitian operator Ω, defined as $\sum_m \Omega_{mm}$, is independent of the representation used to define it.

Let us denote by $|n\rangle$, and by $|\omega\rangle$, eigenvectors of the observables A and B, respectively. The two sets of eigenvectors define two possible representations, $\{A\}$ and $\{B\}$ say. The transformation from the first representation (of vectors and operators) to the second can be made with the help of the unitary matrix $S = (\langle\omega|n\rangle)$, thus

$$(u)_B = S(u)_A \tag{I.5}$$

$$(\Omega)_B = S(\Omega)_A S^+. \tag{I.5'}$$

If there is a one-to-one correspondence between the vectors $|n\rangle$ and $|\omega\rangle$, the matrix S is square and corresponds to a unitary operator U such that $|n\rangle = U|\omega\rangle$. U is then said to define a unitary transformation of vectors and operators. Under a unitary transformation of vectors and operators, all scalar products remain unchanged and all Hermitian adjoints transform into Hermitian adjoints.

If $|\phi\rangle$ is the state vector of some physical system at a given time t, then the "matrix" $\langle\mathbf{r}|\phi\rangle = \phi(\mathbf{r})$, say, with "row number" $\mathbf{r}$ = the position vector(s) of the particle(s) constituting the system, is called the "wavefunction of the system in coordinate representation". Note that the scalar product of two state vectors can be written, in terms of wavefunctions, as follows:

$$\langle\psi|\phi\rangle = \int \langle\psi|\mathbf{r}\rangle\langle\mathbf{r}|\phi\rangle\, d\mathbf{r} = \int \psi^*(\mathbf{r})\,\phi(\mathbf{r})\, d\mathbf{r}$$

Problems

1. Derive the Schwartz inequality

$$|\langle u|v\rangle| \leqslant \sqrt{\langle u|u\rangle}\sqrt{\langle v|v\rangle},$$

in which $|u\rangle$ and $|v\rangle$ are any two vectors of a Hilbert space H.

2. Derive the triangle inequality†

$$\sqrt{\langle u+v|u+v\rangle} \leqslant \sqrt{\langle u|u\rangle} + \sqrt{\langle v|v\rangle}.$$

3. Show that the validity of the relation

$$\langle\alpha|\beta\rangle = \sum_\nu \langle\alpha|u_\nu\rangle\langle u_\nu|\beta\rangle, \tag{3a}$$

for any arbitrary vectors $|\alpha\rangle$ and $|\beta\rangle$, is a necessary and sufficient condition for the system of orthonormal vectors $|u_1\rangle, |u_2\rangle, \ldots, |u_\nu\rangle, \ldots,$ to be complete.

4. Let S_1 be a subspace of a Hilbert space H and S_2 its complementary orthogonal subspace. Any vector $|u\rangle$ can be written as a sum of its projections in the two subspaces, $|u\rangle = |u_{S_1}\rangle + |u_{S_2}\rangle$. Show that the projection operator P_{S_1}, which is such that $|u_{S_1}\rangle = P_{S_1}|u\rangle$, is Hermitian, and satisfies the equation $P_{S_1}^2 = P_{S_1}$.

5. Show that if the subspace S_1 of the preceding problem is taken to be the subspace spanned by a single normalized vector $|a\rangle$, then the corresponding projection operator is given by

$$P_a = |a\rangle\langle a|.$$

6. Show that if $|nr\rangle$ denotes the eigenvectors of an observable, the following "closure relation" is valid

$$\sum_{n,r} |nr\rangle\langle nr| = 1.$$

7. A projection operator P_M is said to be greater than or equal to another projection operator P_N, i.e. $P_M \geqslant P_N$, if the subspace N is contained in M. Show that

(a) The relation $P_M \geqslant P_N$ satisfies the axioms required of any relation of inequality.
(b) $[P_M, P_N] = 0$,

† To simplify the writing, the notation $|\lambda_1 u+\lambda_2 v\rangle$ is used instead of $\lambda_1|u\rangle+\lambda_2|v\rangle$, λ_1 and λ_2 being complex numbers. Then $\langle\lambda_1 u+\lambda_2 v| = \lambda_1^*\langle u| + \lambda_2^*\langle v|$.

(c) The relation $P_M \geqslant P_N$ is equivalent to the statement that $\langle u | P_M | u \rangle \geqslant \langle u | P_N | u \rangle$ for any vector $|u\rangle$ of the Hilbert space.

8. Consider a set of vectors $|\nu\rangle$, in which ν is a continuous index which can take all values in the interval (ν_1, ν_2). Show that if the vectors $|\nu\rangle$ are orthonormal in the sense that $\langle \nu' | \nu \rangle = \delta(\nu' - \nu)$, then the operator

$$P = \int_{\nu_1}^{\nu_2} |\nu\rangle \, d\nu \langle \nu|$$

is the projection operator of the subspace spanned by the set of vectors $|\nu\rangle$.

9. Show that if a unitary operator U can be written in the form $U = 1 + i\varepsilon F$, where ε is a real infinitesimally small number, then the operator F is Hermitian.

10. A Hermitian operator A is said to be positive-definite if, for any vector $|u\rangle$, $\langle u | A | u \rangle \geqslant 0$. Show that the operator $A = |a\rangle\langle a|$ is Hermitian and positive-definite.

11. If A is a Hermitian positive-definite operator (see problem 10), then

$$|\langle u | A | v \rangle| \leqslant \sqrt{\langle u | A | u \rangle} \sqrt{\langle v | A | v \rangle}. \tag{11a}$$

Show that Tr $(A) \geqslant 0$, and that the equality holds if and only if $A = 0$.

12. Show that the operator F defined by the relation

$$F = \iint |s\rangle \, k(s, t) \langle t| \, dt \, ds,$$

in which the kernel $k(s, t)$ is real, is a linear Hermitian operator.

13. Show that the differential operator

$$p = \frac{\hbar}{i} \frac{d}{dx},$$

is linear and Hermitian in the space of all differentiable wavefunctions $\langle x | \phi \rangle = \phi(x)$, say, which vanish at both ends of an interval (a, b).

14. The translation operator $\Omega(a)$ is defined to be such that

$$\Omega(a) \, \phi(x) = \phi(x+a).$$

Show that:

(a) $\Omega(a)$ may be expressed in terms of the operator $p = \dfrac{\hbar}{i} \dfrac{d}{dx}$,

(b) $\Omega(a)$ is unitary.

15. Given three operators A, B and C, express the commutator $[AB, C]$ in terms of the commutators $[A, C]$ and $[B, C]$.

16. Let $H(\mathbf{r})$ be an operator acting on the wavefunctions $\psi(\mathbf{r})$, and 0 a coordinate transformation operator which acts on the wavefunctions in such a way that $0\psi(\mathbf{r}) = \psi(\mathbf{r}')$. Show that if $H(\mathbf{r})$ is invariant under the coordinate transformation 0, i.e. $H(\mathbf{r}') \equiv H(\mathbf{r})$, then $[H, 0] = 0$.

17. Find the expansion of the operator $(A-\lambda B)^{-1}$ in a power series in λ, assuming that the inverse A^{-1} of A exists.

18. Show that if A and B are two operators satisfying the relation $[[A, B], A] = 0$, then the relation $[A^m, B] = mA^{m-1}[A, B]$ holds for all positive integers m.

19. Show that

$$[p, x] = -i\hbar, \quad [p, x^n] = -ni\hbar x^{n-1}, \quad n > 1, \tag{19a}$$

$$[p, A] = -i\hbar \frac{dA}{dx} \tag{19b}$$

where $p = \dfrac{\hbar}{i}\dfrac{d}{dx}$ and $A = A(x)$ is a differentiable function of x.

20. If the characteristic equation $f(\lambda) = 0$ for an observable A is given, show that $f(A) = 0$.

21. Let $|u\rangle$ and $|v\rangle$ be two vectors of finite norm. Show that

$$\mathrm{Tr}\,(|u\rangle\langle v|) = \langle v|u\rangle.$$

22. If A is any linear operator, show that A^+A is a positive-definite Hermitian operator whose trace is equal to the sum of the square moduli of the matrix elements of A in any arbitrary representation. Deduce that $\mathrm{Tr}\,(A^+A) = 0$ is true if and only if $A = 0$.

23. Show that if A and B are two positive-definite observables, then $\mathrm{Tr}\,(AB) \geqslant 0$.

24. Diagonalize the Hermitian matrix

$$\varrho = \begin{Bmatrix} \varrho_{11} & \varrho_{12} \\ \varrho_{21} & \varrho_{22} \end{Bmatrix}, \quad \mathrm{Tr}\,(\varrho) = 1,$$

by means of a unitary matrix of the form

$$U = e^{(1/2)i\phi\sigma_2}e^{(1/2)i\psi\sigma_3},$$

where

$$\sigma_2 = \begin{Bmatrix} 0 & -i \\ i & 0 \end{Bmatrix}, \quad \sigma_3 = \begin{Bmatrix} 1 & 0 \\ 0 & -1 \end{Bmatrix}.$$

25. The derivative of an operator $A(\lambda)$ which depends explicitly on a parameter λ is defined to be

$$\frac{dA(\lambda)}{d\lambda} = \lim_{\varepsilon \to 0} \frac{A(\lambda+\varepsilon)-A(\lambda)}{\varepsilon}.$$

Show that

$$\frac{d}{d\lambda}(AB) = \frac{dA}{d\lambda}B + A\frac{dB}{d\lambda}, \tag{25a}$$

$$\frac{d}{d\lambda}(A^{-1}) = -A^{-1}\frac{dA}{d\lambda}A^{-1}. \tag{25b}$$

26. Show that the operator $B(t)$ defined by the expression

$$B(t) = e^{iAt}B_0e^{-iAt},$$

where A and B_0 are operators independent of the parameter t, is a solution of the integral equation

$$B(t) = B_0 + i\left[A, \int_0^t B(\tau)\,d\tau\right].$$

27. Show that, for any two operators A and L,

$$e^LAe^{-L} = A + [L, A] + \frac{1}{2!}[L, [L, A]] + \frac{1}{3!}[L, [L, [L, A]]] + \dots \tag{27a}$$

28. Show that $e^Ae^B = e^{A+B}e^{-(1/2)[A, B]}$ if $[[A, B], A][[A, B], B] = 0$. (28a)

29. Verify Kubo's indentity

$$[A, e^{-\beta H}] = e^{-\beta H}\int_0^\beta e^{\lambda H}[A, H]\,e^{-\lambda H}\,d\lambda, \tag{29a}$$

where A and H are any two operators.

30. Show that a necessary and sufficient condition for two linear operators A and B to be equal (to within a phase factor), i.e. for $A = Be^{i\alpha}$, is that

$$|\langle u|A|v\rangle| = |\langle u|B|v\rangle| \tag{30a}$$

should hold for any pair of linearly independent kets $|u\rangle$ and $|v\rangle$.

31. Show that a necessary and sufficient condition for a linear operator U to be unitary is that the matrix elements $\langle i|U|k\rangle$ of this operator in a given representation should obey the following equations:

$$\begin{aligned}&\sum_i |\langle i|U|k\rangle|^2 = 1,\\ &\sum_i \langle i|U|h\rangle\langle i|U|k\rangle^* = 0, \qquad h \neq k,\end{aligned} \tag{31a}$$

The convergence of the sums may be assumed.

Solutions

1. Let us write for a start† $\langle u|u\rangle+\langle v|v\rangle-2\,\mathrm{Re}\,\langle u|v\rangle = \langle u|u\rangle+\langle v|v\rangle-\langle u|v\rangle-\langle v|u\rangle = \langle u-v|u-v\rangle \geqslant 0$. Hence

$$\mathrm{Re}\,\langle u|v\rangle \leqslant \tfrac{1}{2}(\langle u|u\rangle+\langle v|v\rangle). \tag{1.1}$$

If we substitute $c\,|u\rangle$ and $(1/c)\,|v\rangle$ for $|u\rangle$ and $|v\rangle$ (c being a positive real number) then the left-hand side remains unchanged and the right-hand side becomes $\frac{1}{2}(c^2\langle u|u\rangle+1/c^2\langle v|v\rangle)$. Since this expression is greater than $\mathrm{Re}\,\langle u|v\rangle$, its minimum also satisfies the inequality. This minimum is attained when $c^2 = \sqrt{\langle v|v\rangle/\langle u|u\rangle}$. Hence

$$\mathrm{Re}\,\langle u|v\rangle \leqslant \sqrt{\langle u|u\rangle}\,\sqrt{\langle v|v\rangle}. \tag{1.2}$$

Let us now subsitute $e^{-i\alpha}\,|u\rangle$ for $|u\rangle$, where α is a real number. The right-hand side of (1.2) is unchanged, while the left-hand side becomes

$$\mathrm{Re}\,(e^{i\alpha}\langle u|v\rangle) = \cos\alpha\ \mathrm{Re}\,\langle u|v\rangle-\sin\alpha\ \mathrm{Im}\,\langle u|v\rangle.$$

This expression has a maximum when

$$\sqrt{(\mathrm{Re}\,\langle u|v\rangle)^2+(\mathrm{Im}\,\langle u|v\rangle)^2} = |\langle u|v\rangle|,$$

and thus the relation to be proved is obtained, viz.,

$$|\langle u|v\rangle| \leqslant \sqrt{\langle u|u\rangle}\,\sqrt{\langle v|v\rangle}. \tag{1.3}$$

From the proof it can be seen that the equality holds only if $|u\rangle$ or $|v\rangle$ is zero, or $|u\rangle$ and $|v\rangle$ are linearly related.

2. We have that

$$\langle u+v|u+v\rangle = \langle u|u\rangle+\langle u|v\rangle+\langle v|u\rangle+\langle v|v\rangle = \langle u|u\rangle+\langle v|v\rangle+2\,\mathrm{Re}\,\langle u|v\rangle.$$

Using (1.2) we obtain

$$\langle u+v|u+v\rangle \leqslant \langle u|u\rangle+\langle v|v\rangle+2\sqrt{\langle u|u\rangle}\,\sqrt{\langle v|v\rangle} = (\sqrt{\langle u|u\rangle}+\sqrt{\langle v|v\rangle})^2$$

and thus

$$\sqrt{\langle u+v|u+v\rangle} \leqslant \sqrt{\langle u|u\rangle}+\sqrt{\langle v|v\rangle}. \tag{2.1}$$

The equality holds only if $|u\rangle$ or $|v\rangle$ is zero, $|u\rangle$ and $|v\rangle$ or are linearly related.

3. Let us show first that the condition is sufficient. For this we have to prove that no vector $|u\rangle$ different from zero and orthogonal to all vectors of the system can be found. If

† See the footnote on page 4.

we suppose that there exists such a vector, we can choose $|\alpha\rangle = |\beta\rangle = |u\rangle$ and we have $\langle u|u\rangle = 0$, whence $|u\rangle = 0$ which is contrary to the supposition we have made.

From the completeness of the set $|u_1\rangle, |u_2\rangle, \ldots, |u_\nu\rangle$, it follows that

$$|\alpha\rangle = \sum_{\nu=1}^{\infty} a_\nu |u_\nu\rangle, \quad a_\nu = \langle u_\nu|\alpha\rangle$$
$$|\beta\rangle = \sum_{\nu=1}^{\infty} b_\nu |u_\nu\rangle, \quad b_\nu = \langle u_\nu|\beta\rangle \tag{3.1}$$

and thus that

$$\langle\alpha|\beta\rangle = \sum_{\mu,\nu=1}^{\infty} \langle\alpha|u_\mu\rangle\langle u_\nu|\beta\rangle\langle u_\mu|u_\nu\rangle = \sum_{\nu=1}^{\infty}\langle\alpha|u_\nu\rangle\langle u_\nu|\beta\rangle. \tag{3.2}$$

Remarks: Taking $|\beta\rangle = |\alpha\rangle$, we have

$$\langle\alpha|\alpha\rangle = \sum_{\nu=1}^{\infty} |\langle\alpha|u_\nu\rangle|^2. \tag{3.3}$$

The equality (3.3) is called Parseval's relation. It is valid for any complete set $|u_1\rangle, \ldots, |u_\nu\rangle, \ldots$ If the set is not complete it can be seen that

$$\langle\alpha|\alpha\rangle \leqslant \sum_{\nu=1}^{\infty} |\langle\alpha|u_\nu\rangle|^2. \tag{3.4}$$

4. Take the scalar product with $|v\rangle$ of both sides of the equation

$$P_{S_1}|u\rangle = |u_{S_1}\rangle \tag{4.1}$$

so that

$$\langle v|P_{S_1}|u\rangle = \langle v|u_{S_1}\rangle.$$

Similarly

$$\langle u|P_{S_1}|v\rangle^* = \langle u|v_{S_1}\rangle^* = \langle u_{S_1}|v_{S_1}\rangle^* = \langle v_{S_1}|u_{S_1}\rangle = \langle v|u_{S_1}\rangle = \langle v|P_{S_1}|u\rangle$$

for any $|u\rangle$ and $|v\rangle$, i.e. P_{S_1} is Hermitian.

Applying the operator P_{S_1} to both sides of (4.1) we have that

$$P_{S_1}^2|u\rangle = P_{S_1}|u_{S_1}\rangle = |u_{S_1}\rangle = P_{S_1}|u\rangle,$$

for any $|u\rangle$, hence $P_{S_1}^2 = P_{S_1}$.

5. In this case we can write $|u\rangle = |u_a\rangle + |u_{S_2}\rangle$. By hypothesis $\langle a|u_{S_2}\rangle = 0$ and $|u_a\rangle = c|a\rangle$, c being a complex number. Taking the scalar product of $|u\rangle$ and $|a\rangle$ we have $c = \langle a|u\rangle$, i.e. $|u_a\rangle = |a\rangle\langle a|u\rangle$, whence the relation to be proved follows.

6. Let P_A be the projection operator on the subspace spanned by the vectors $|nr\rangle$. Since A is an observable, we have $P_A = 1$. On the other hand the projection on a subspace is equal to the sum of all projections onto vectors forming a base in this subspace. In particular,

$P_A = \sum_{n,r} P_{nr}$ where P_{nr} is the projection operator on the vector $|nr\rangle$. Thus we have that $\sum_{n,r} P_{nr} = 1$. But, from the preceding problem, $P_{nr} = |nr\rangle\langle nr|$, and hence

$$\sum_{n,r} |nr\rangle\langle nr| = 1. \tag{6.1}$$

Remarks: From the solutions of problems 3 and 6 it follows directly that, for any orthonormal set of vectors, the closure relation and the completeness condition are equivalent.

7. (a) From the definition it follows immediately that, if $P_M \geqslant P_N$ and $P_N \geqslant P_M$, then $P_M = P_N$, and that, if $P_M \geqslant P_N$ and $P_N \geqslant P_Q$, then $P_M \geqslant P_Q$.

(b) the inequality $P_M \geqslant P_N$ implies that $P_M P_N = P_N$ and $P_N P_M = P_N$, and hence that $[P_M, P_N] = 0$.

(c) In general, for the projection operator P_{S_1}, we have that

$$P_{S_1}|u\rangle = |u_{S_1}\rangle, \quad |u\rangle = |u_{S_1}\rangle + |u_{S_2}\rangle.$$

Consider the operator $1-P_{S_1}$, then

$$(1-P_{S_1})|u\rangle = |u\rangle - P_{S_1}|u\rangle = |u\rangle - |u_{S_1}\rangle = |u_{S_2}\rangle = P_{S_2}|u\rangle,$$

and consequently $1-P_{S_1}$ is also a projection operator. Now

$$\langle u|P_{S_1}|u\rangle + \langle u|P_{S_2}|u\rangle = \langle u|(P_{S_1}+P_{S_2})|u\rangle = \langle u|u\rangle.$$

But $\langle u|P_{S_2}|u\rangle = \langle u|P_{S_2}^2|u\rangle$ is non-negative, and

$$\langle u|P_{S_1}|u\rangle \leqslant \langle u|u\rangle. \tag{7.1}$$

We can therefore write $\langle u|P_N|u\rangle = \langle u|P_M P_N P_M|u\rangle \leqslant \langle u|P_M|u\rangle$.

Conversely, if $|u\rangle$ is a vector of the complementary space of M we have $P_M|u\rangle = 0$, and, in accordance with the relation $\langle u|P_M|u\rangle \geqslant \langle u|P_N|u\rangle$, we have $P_N|u\rangle = 0$. Then, if $|u\rangle$ belongs to N, $(1-P_M)|u\rangle = 0$ and $P_N(1-P_M)|u\rangle = 0$. If $|u\rangle$ does not belong to M, $(1-P_M)|u\rangle = |u\rangle$. But $P_N|u\rangle = 0$, and therefore for any $|u\rangle$ we have $P_N(1-P_M)|u\rangle = 0$, and hence $P_N = P_N P_M$, which is equivalent to the inequality $P_M \geqslant P_N$.

8. For any arbitrary vector $|u\rangle$, we have that

$$P|u\rangle = \int_{\nu_1}^{\nu_2} |\nu\rangle\, d\nu \langle \nu|u\rangle,$$

and hence

$$\langle \nu'|(1-P)|u\rangle = \langle \nu'|u\rangle - \int_{\nu_1}^{\nu_2} \langle \nu'|\nu\rangle\, d\nu\langle \nu|u\rangle = \langle \nu'|u\rangle - \int_{\nu_1}^{\nu_2} \delta(\nu'-\nu)\, d\nu\langle \nu|u\rangle = 0.$$

Since $P|u\rangle$ is a linear combination of the $|\nu\rangle$ vectors, and, as we have just seen, $(1-P)|u\rangle$ is orthogonal to all of these vectors, it follows that P is a projection operator onto the subspace spanned by the $|\nu\rangle$ vectors (see problem 4). Note that it is sometimes more convenient to write the orthonormalization condition in the form $\langle\nu|\nu'\rangle = f(\nu)\delta(\nu-\nu')$, where $f(\nu)$ is a real and positive function of ν. The vectors so defined are equal to the normalized vectors previously defined, except that each vector $|\nu\rangle$ is multiplied by a constant of modulus $\sqrt{f(\nu)}$. The projection operator on the subspace spanned by the $|\nu\rangle$ vectors is, in this case, given by

$$P = \int_{\nu_1}^{\nu_2} |\nu\rangle \frac{d\nu}{f(\nu)} \langle\nu|. \tag{8.1}$$

9. Since U is unitary, we have that

$$(1-i\varepsilon F^+)(1+i\varepsilon F) = (1+i\varepsilon F)(1-i\varepsilon F^+)$$

and, retaining only the terms to first order in ε, it follows that $F = F^+$.

10. We have that

$$\langle u|A|v\rangle = \langle u|a\rangle\langle a|v\rangle = \langle v|a\rangle^*\langle a|u\rangle^* = \langle v|A|u\rangle^*$$

and also that

$$\langle u|A|u\rangle = \langle u|a\rangle\langle a|u\rangle = |\langle u|a\rangle|^2 \geqslant 0.$$

11. The inequality to be proved can be deduced from the fact that A is positive-definite, by using the same procedure as was used to prove the Schwartz inequality (see problem 1).

Note that $\text{Tr}\,(A) = \sum_n \langle n|A|n\rangle$, where the $|n\rangle$ form the (complete) set of eigenvectors of an observable. It follows directly that $\text{Tr}\,(A) \geqslant 0$. The equality is true if, and only if, for all $|n\rangle$, $\langle n|A|n\rangle = 0$. If this is the case, then, taking $|v\rangle = |n\rangle$ in (11a), it follows that $\langle u|A|n\rangle = 0$ for any $|u\rangle$, and hence that $A|n\rangle = 0$. Since the $|n\rangle$ vectors form a basis we have that $A\times$any linear combination of the $|n\rangle$s $= 0$, i.e. $A = 0$.

12. To prove the linearity of F is straightforward. To prove that it is Hermitian we write, using the notation $\langle t|\phi\rangle = \phi(t)$, $\langle s|\psi\rangle = \psi(s)$,

$$\langle\psi|F|\phi\rangle = \int \psi^*(s)\left[\int k(s,t)\,\phi(t)\,dt\right] ds = \int k(s,t)\,\psi^*(s)\,\phi(t)\,dt\,ds.$$

Also

$$\langle\phi|F|\psi\rangle^* = \left[\int \phi^*(s)\left(\int k(s,t)\,\psi(t)\,dt\right) ds\right]^* = \int k(s,t)\,\psi^*(s)\,\phi(t)\,dt\,ds,$$

whence it follows that $\langle\psi|F|\phi\rangle = \langle\phi|F|\psi\rangle^*$ for any $|\phi\rangle$ and $|\psi\rangle$, i.e. that F is Hermitian.

13. We have that

$$\langle\psi|p|\phi\rangle = \frac{\hbar}{i}\int_a^b \psi^* \frac{d\phi}{dx}\,dx.$$

Integrating by parts we obtain

$$\langle\psi|p|\phi\rangle = \frac{\hbar}{i}(\phi\psi)_a^b - \frac{\hbar}{i}\int_a^b \frac{d\psi^*}{dx}\phi\, dx = -\frac{\hbar}{i}\int_a^b \frac{d\psi^*}{dx}\phi\, dx = \langle\phi|p|\psi\rangle^*.$$

14. (a) Using Taylor's theorem, we can write

$$\Omega(a)\,\phi(n) = \phi(x+a) = \sum_{n=0}^{\infty}\frac{a^n}{n!}\frac{d^n}{dx^n}\phi(x) = \sum_{n=0}^{\infty}\frac{1}{n!}\left(\frac{iap}{\hbar}\right)^n\phi(x) = e^{\frac{i}{\hbar}ap}\,\phi(x).$$

It follows that $\Omega(a) = \exp(iap/\hbar)$.

(b) $\Omega^+(a)$ is defined to be such that, for all $|\psi\rangle$ and $|\phi\rangle$,

$$\langle\psi|\Omega|\phi\rangle^* = \langle\phi|\Omega^+|\psi\rangle,$$

whence

$$\int \phi^*(x)\,\Omega^+(a)\,\psi(x)\,dx = \int \phi^*(x+a)\,\psi(x)\,dx.$$

By making the change of variable $x+a\to x$ in the second integral we obtain

$$\int \phi^*(x)\,\Omega^+(a)\,\psi(x)\,dx = \int \phi^*(x)\,\psi(x-a)\,dx,$$

for any $\phi(x)$. It follows that $\Omega^+(a)\,\psi(x) = \psi(x-a)$, hence that $\Omega(a)\,\Omega^+(a) = \Omega^+(a)\,\Omega(a) = 1$.

15. We have

$$[AB, C] = (AB)C - C(AB) = ABC - CAB + ACB - ACB = A(BC-CB) + (AC-CA)B,$$

i.e.

$$[AB, C] = A[B, C] + [A, C]B. \tag{15.1}$$

16. From the statement of the problem we can see that for any $\psi(\mathbf{r})$,

$$0[H(\mathbf{r})\,\psi(\mathbf{r})] = H(\mathbf{r}')\,\psi(\mathbf{r}') = H(\mathbf{r})\,\psi(\mathbf{r}') = H(\mathbf{r})\,0\psi(\mathbf{r})$$

whence $[H, 0] = 0$.

17. Let the expression we are seeking be written as a series in powers of λ, thus:

$$(A-\lambda B)^{-1} = \sum_{n=0}^{\infty}\lambda^n L_n, \tag{17.1}$$

in which the operators L_n are to be determined. Multiplying on the left by $A-\lambda B$ one obtains

$$1 = \sum_{n=0}^{\infty}\lambda^n(A-\lambda B)\,L_n = AL_0 + \sum_{n=1}^{\infty}\lambda^n(AL_n - BL_{n-1}).$$

By equating coefficients of powers of λ, it is found that $L_0 = A^{-1}$, $L_n = A^{-1}BL_{n-1}$, $n = 1, 2, \ldots,$ and hence that

$$(A-\lambda B)^{-1} = A^{-1}+\lambda A^{-1}BA^{-1}+\lambda^2 A^{-1}BA^{-1}BA^{-1}+ \ldots \tag{17.2}$$

If $[A^{-1}, B] = 0$, (17.2) becomes

$$(A-\lambda B)^{-1} = A^{-1}+\lambda BA^{-2}+\lambda^2 B^2 A^{-3}+ \ldots \tag{17.3}$$

18. To prove this theorem we shall use the method of induction. Note that for $m = 1$, the relation is obvious. Let us suppose that it is true for an arbitrary m and then deduce its validity for $m+1$. Thus, let us suppose that

$$A^m B - BA^m = mA^{m-1}(AB-BA). \tag{18.1}$$

Multiplying on the left by A, we find that

$$A^{m+1}B - ABA^m = mA^m(AB-BA)$$

and, on adding $A^m(AB-BA)$ to both sides, we obtain

$$A^{m+1}B - ABA^m + A^m(AB-BA) = (m+1)\, A^m(AB-BA).$$

But, according to the conditions of the problem, we have that

$$A^m(AB-BA) = (AB-BA)\, A^m.$$

By making use of this fact, a relation of the form (18.1) is finally obtained with m replaced by $m+1$.

19. The first relation (19a) follows directly by applying the commutator to an arbitrary function of x. To establish the second one we use the identity

$$[A, B^n] = \sum_{k=0}^{n-1} B^k[A, B]\, B^{n-k-1}, \tag{19.1}$$

which can be obtained by repeated use of (15.1). Relation (19b) is then verified if $A(x)$ is a polynomial in x or, more generally, if it is a convergent series in powers of x.

20. We shall show that any vector $|u\rangle$ is transformed by the operator $f(A)$ into the null vector, i.e. that $f(A)|u\rangle = 0$, and hence that $f(A) = 0$ is true as an operator relation.

Let $|n\rangle$ be an eigenvector of the observable A with the eigenvalue λ_n, i.e. $A|n\rangle = \lambda_n|n\rangle$.

By applying the operator A r times to both sides we find that $A^r|n\rangle = (\lambda_n)^r|n\rangle$ and hence, for an arbitrary polynomial $P(\lambda)$, $P(A)|n\rangle = P(\lambda_n)|n\rangle$. Since any function $f(\lambda)$ can be approximated arbitrarily closely by polynomials $P(\lambda)$, we conclude that $f(A)|n\rangle = f(\lambda_n)|n\rangle = 0$, since the λ_n are the solutions of $f(\lambda) = 0$.

Let $|u\rangle$ be an arbitrary vector. This can always be written in the form $|u\rangle = \sum_n u_n|n\rangle$.

Applying the operator $f(A)$ to both sides we have

$$f(A)|u\rangle = \sum_n u_n f(\lambda_n)|n\rangle = 0. \tag{20.1}$$

Remarks: The reader can now solve the first part of problem 14 by using the relation (20.1).

21. Let us represent the operator $\Omega = |u\rangle\langle v|$ by a matrix, by using the eigenvectors of some observable. The elements of this matrix will then be

$$\Omega_{mn} = \langle m|\Omega|n\rangle = \langle m|u\rangle\langle v|n\rangle.$$

Then

$$\mathrm{Tr}(\Omega) = \sum_n \Omega_{nn} = \sum_n \langle n|u\rangle\langle v|n\rangle.$$

Using the identity (3.2), the desired result follows.

22. Using the properties of A^+ we have that

$$\langle u|A^+A|v\rangle = \langle v|A^+A|u\rangle^*, \tag{22.1}$$

which expresses the fact that A^+A is Hermitian.

Taking now $|u\rangle = |v\rangle$, and writing $|\omega\rangle = A|u\rangle$, we have that

$$\langle u|A^+A|u\rangle = \langle\omega|\omega\rangle \geqslant 0,$$

and hence A^+A is positive-definite.

Let $|n\rangle$ be a complete set of eigenvectors of an observable. We have then that

$$\mathrm{Tr}(A^+A) = \sum_n \langle n|A^+A|n\rangle.$$

Using the closure property $\sum_n |n\rangle\langle n| = 1$, we find that

$$\mathrm{Tr}(A^+A) = \sum_n \left\langle n\middle|A^+\sum_m\middle|m\right\rangle\langle m|A|n\rangle = \sum_{m,n}\langle n|A^+|m\rangle\langle m|A|n\rangle = \sum_{m,n}|\langle n|A|m\rangle|^2 \geqslant 0.$$

The equality is true only if all the elements $\langle n|A|m\rangle = 0$, which is equivalent to the operator relation $A = 0$.

23. Let $|n\rangle$ be an eigenvector of the observable B. Then

$$\mathrm{Tr}(AB) = \sum_n \langle n|AB|n\rangle = \sum_n \lambda_n\langle n|A|n\rangle. \tag{23.1}$$

But since the observable A is positive-definite, we have $\langle n|A|n\rangle \geqslant 0$. To study the sign of λ_n we write

$$\langle n|B|n\rangle = \lambda_n\langle n|n\rangle \geqslant 0.$$

But $\langle n|n\rangle > 0$ and hence $\lambda_n \geqslant 0$.

By (23.1) it then follows that

$$\mathrm{Tr}\,(AB) \geqslant 0. \tag{23.2}$$

24. Since the matrix ϱ is Hermitian we can write $\varrho_{12} = A+iB$, $\varrho_{21} = A-iB$ where A and B are real numbers. On expanding the exponentials in power series, the following expression is obtained for U

$$U = \begin{Bmatrix} e^{(i/2)\psi}\cos\dfrac{\phi}{2} & e^{-(i/2)\psi}\sin\dfrac{\phi}{2} \\ -e^{(i/2)\psi}\sin\dfrac{\phi}{2} & e^{-(i/2)\psi}\cos\dfrac{\phi}{2} \end{Bmatrix} \tag{24.1}$$

The transformed matrix $\bar{\varrho} = U\varrho U^{+}$ has the form

$$\bar{\varrho} = \begin{Bmatrix} \begin{array}{l} 1-(1-2\varrho_{11})\cos\phi \\ +2\sin\phi(A\cos\psi-B\sin\psi) \end{array} & \begin{array}{l} (1-2\varrho_{11})\sin\phi \\ +2i(A\sin\psi+B\cos\psi) \\ +2\cos\phi(A\cos\psi-B\sin\psi) \end{array} \\ \begin{array}{l} (1-2\varrho_{11})\sin\phi \\ -2i(A\sin\psi+B\cos\psi) \\ +2\cos\phi(A\cos\psi-B\sin\psi) \end{array} & \begin{array}{l} 1+(1-2\varrho_{11})\cos\phi \\ -2\sin\phi(A\cos\psi-B\sin\psi) \end{array} \end{Bmatrix}. \tag{24.2}$$

This is diagonal if

$$(1-2\varrho_{11})\sin\phi+2\cos\phi(A\cos\psi-B\sin\psi) = 0,$$
$$A\sin\psi+B\cos\psi = 0,$$

i.e. if

$$\tan\psi = -\frac{B}{A}, \quad \tan\phi = -\frac{2\sqrt{A^2+B^2}}{1-2\varrho_{11}}. \tag{24.3}$$

With these results the matrix U is determined and $\bar{\varrho}$ becomes

$$\bar{\varrho} = \frac{1}{2}\begin{Bmatrix} 1-\Lambda & 0 \\ 0 & 1+\Lambda \end{Bmatrix}, \tag{24.4}$$

where $\Lambda = \sqrt{(1-2\varrho_{11})^2+4(A^2+B^2)}$.

25. We can write

$$\frac{A(\lambda+\varepsilon)\,B(\lambda+\varepsilon)-A(\lambda)\,B(\lambda)}{\varepsilon}$$
$$= \frac{A(\lambda+\varepsilon)\,B(\lambda+\varepsilon)-A(\lambda+\varepsilon)\,B(\lambda)+A(\lambda+\varepsilon)\,B(\lambda)-A(\lambda)\,B(\lambda)}{\varepsilon}$$
$$= \frac{A(\lambda+\varepsilon)\,[B(\lambda+\varepsilon)-B(\lambda)]}{\varepsilon}+\frac{[A(\lambda+\varepsilon)-A(\lambda)]\,B(\lambda)}{\varepsilon}.$$

Taking the limit as ε tends to zero, we obtain

$$\frac{d}{d\lambda}(AB) = \frac{dA}{d\lambda}B + A\frac{dB}{d\lambda}. \tag{25.1}$$

For the other relation we start from $AA^{-1} = 1$ and differentiate it with respect to

$$\frac{dA}{d\lambda}A^{-1} + A\frac{dA^{-1}}{d\lambda} = 0,$$

i.e.

$$A\frac{dA^{-1}}{d\lambda} = -\frac{dA}{d\lambda}A^{-1}.$$

Multiplying both sides on the left by A^{-1} we obtain

$$\frac{dA^{-1}}{d\lambda} = -A^{-1}\frac{dA}{d\lambda}A^{-1}. \tag{25.2}$$

26. The integral equation can be written as

$$B(t) = B_0 + i\int_0^t [A, B(\tau)]\, d\tau, \tag{26.1}$$

which is equivalent to the differential equation

$$\frac{dB}{dt} = i[A, B(t)] \tag{26.2}$$

with the initial condition $B(0) = B_0$.

Using the expression for $B(t)$ we obtain

$$\frac{dB}{dt} = iAe^{iAt}B_0e^{-iAt} - ie^{iAt}B_0e^{-iAt}A = iAB(t) - iB(t)A = i[A, B(t)].$$

The initial condition is also satisfied.

27. Consider the operator

$$A(s) = e^{sL}Ae^{-sL}, \tag{27.1}$$

where s is a parameter. We have then that

$$\frac{dA(s)}{ds} = Le^{sL}Ae^{-sL} - e^{sL}Ae^{-sL}L = [L, A(s)].$$

Let us differentiate once more, then

$$\frac{d^2A(s)}{ds^2} = \left[L, \frac{dA(s)}{ds}\right] = [L, [L, A(s)]],$$

and so on. We shall now write the operator $e^L A e^{-L} = A(1)$ as a Taylor series expansion about the origin

$$A(1) = A(0) + \frac{1}{1!}\frac{dA(0)}{ds} + \frac{1}{2!}\frac{d^2A(0)}{ds^2} + \dots .$$

Since $A(0) = A$, the required identity follows directly. It can be obtained also in another way, by using the result of the preceding problem.

28. Consider the operator

$$T(s) = e^{As}e^{Bs} \tag{28.1}$$

and differentiate it with respect to s:

$$\frac{dT}{ds} = Ae^{As}e^{Bs} + e^{As}Be^{Bs} = (A + e^{As}Be^{-As})T(s). \tag{28.2}$$

Since the operators $[B, A]$ and A commute, we find, by using the results of problem 18, that

$$[B, A^n] = nA^{n-1}[B, A],$$

$$[B, e^{-As}] = \sum_n (-1)^n \frac{s^n}{n!}[B, A^n] = \sum_n (-1)^n \frac{s^n}{(n-1)!} A^{n-1}[B, A] = -e^{-As}[B, A]s$$

and hence that

$$e^{As}Be^{-As} = B - [B, A]s. \tag{28.3}$$

The relation (28.3) could have been obtained directly from (27a). From (28.2) and (28.3) we have that

$$\frac{dT(s)}{ds} = (A + B + [A, B]s)T(s), \tag{28.4}$$

and $T(s)$ is thus the solution of this differential equation with the initial condition $T(0) = 1$. Since the operators $A+B$ and $[A, B]$ commute, equation (28.4) can be integrated as if they were merely numbers, to give the solution

$$T(s) = \exp[(A+B)s] \exp\{\tfrac{1}{2}[A, B]s^2\}. \tag{28.5}$$

The identity (28a) follows by putting $s = 1$.

29. Let us denote by $C(\beta)$ the left-hand side of (29a) and by $D(\beta)$ the right-hand side. We have evidently

$$C(0) = D(0) = 0. \tag{29.1}$$

If we can also prove that $C(\beta)$ and $D(\beta)$ satisfy the same first order differential equation, then the identity (29a) is valid. This is in fact the case, since

$$\frac{dC}{d\beta} = -AHe^{-\beta H}+He^{-\beta H}A = H(e^{-\beta H}A-Ae^{-\beta H})-(AH-HA)e^{-\beta H}$$

$$= HC-[A, H]e^{-H\beta}, \tag{29.2}$$

and

$$\frac{dD}{d\beta} = HD-e^{-\beta H}e^{\beta H}[A, H]e^{-\beta H} = HD-[A, H]e^{-\beta H}. \tag{29.3}$$

30. The necessity of the condition is evident. To prove its sufficiency, let us consider a representation in which $A_{ij} = \langle i|A|j\rangle$ and $B_{ij} = \langle i|B|j\rangle$ are the matrix elements of the operators A and B.

By (30a), with $|u\rangle = |i\rangle$ and $|v\rangle = |j\rangle$, we have that

$$|A_{ij}| = |B_{ij}| \tag{30.1}$$

for any $|i\rangle$ and $|j\rangle$. On the other hand, with $|u\rangle = |i\rangle$ and $|v\rangle = x_j|j\rangle+x_l|l\rangle$, where x_j and x are arbitrary complex numbers, from (30a) we find that

$$|A_{ij}x_j+A_{il}x_l| = |B_{ij}x_j+B_{il}x_l|. \tag{30.2}$$

Taking into account (30.1), (30.2) can be written as

$$\text{Re}\,[x_jx_l^*(A_{ij}A_{il}^*-B_{ij}B_{il}^*)] = 0. \tag{30.3}$$

Since the complex number $x_jx_l^*$ is arbitrary, it follows from (30.3) that

$$A_{ij}A_{il}^*-B_{ij}B_{il}^* = 0. \tag{30.4}$$

From (30.1) and (30.4) we then have that

$$\frac{A_{ij}}{B_{ij}} = \frac{A_{il}}{B_{il}}, \tag{30.5}$$

which means that the ratio A_{ij}/B_{ij} does not depend on j. On repeating the same argument after interchanging rows and columns, we find that the ratio A_{ij}/B_{ij} does not depend on i either. Taking into account (30.1), we conclude that

$$\frac{A_{ij}}{B_{ij}} = e^{i\alpha}, \tag{30.6}$$

where α is a real number independent of i and of j; i.e. the two operators A and B are equal to within a constant phase factor.

31. Let us consider first the necessity of the condition. Let U be a unitary operator. Then for the matrices representing U and U^+ we have

$$UU^+ = I \tag{31.1}$$

where I is the unit matrix. Since the matrix U^+ is the complex conjugate transpose of U, the relations (31a) follow directly from (31.1). Conversely, from (31a), valid in a particular representation, (31.1) follows. We have to prove only that (31.1) is valid in any representation. Now the transformation from any representation to another is made by a unitary matrix S in the following way (see I.5′),

$$\bar{U} = SUS^+, \quad \bar{U}^+ = SU^+S^+. \tag{31.2}$$

Hence it follows that

$$\bar{U}\bar{U}^+ = (SUS^+)(SU^+S^+) = I, \tag{31.3}$$

i.e. (31.1) is valid in any representation. This fact ensures the unitarity of the operator U.

CHAPTER II

Simple Quantum Systems

1. The Eigenfunctions and the Energy Spectrum

In quantum mechanics the non-relativistic motion of a particle of mass m, in a potential $V(\mathbf{r}, t)$, is described by the time-dependent Schrödinger equation for the wavefunction $\Psi(\mathbf{r}, t)$, of the particle

$$i\hbar \frac{\partial \Psi(\mathbf{r}, t)}{\partial t} = \left[-\frac{\hbar^2}{2m} \nabla^2 + V(\mathbf{r}, t) \right] \Psi(\mathbf{r}, t). \tag{II.1}$$

For conservative systems, $V = V(\mathbf{r})$, equation (II.1) has solutions of the form

$$\Psi_E(\mathbf{r}, t) = \psi_E(\mathbf{r}) \exp\left(-\frac{i}{\hbar} Et\right), \tag{II.2}$$

which describe dynamical states with a well-defined energy E. Then, because of the linearity of the equation, the most general solution $\Psi(\mathbf{r}, t)$ can be written as

$$\Psi(\mathbf{r}, t) = \sum_E C_E \psi_E(\mathbf{r}) \exp\left(-\frac{i}{\hbar} Et\right), \tag{II.3}$$

where the values of E are the eigenvalues of the time-independent Schrödinger equation

$$\nabla^2 \psi_E(\mathbf{r}) + \frac{2m}{\hbar^2} (E - V(\mathbf{r}))\, \psi_E(\mathbf{r}) = 0, \tag{II.4}$$

and the functions $\psi_E(\mathbf{r})$ are the corresponding eigenfunctions (continuous, differentiable, and bounded at infinity), often called the wavefunctions of the stationary states, or the time-independent wavefunctions, of the system. The coefficients C_E are arbitrary constants. The set of all eigenvalues E is called the *energy spectrum* of the system. The eigenfunctions $\psi_E(\mathbf{r})$ are subject to the orthonormality condition:

$$\int \psi_E^*(\mathbf{r})\, \psi_{E'}(\mathbf{r})\, d\mathbf{r} = \delta(E, E'), \tag{II.5}$$

where $\delta(E, E')$ is the Kronecker delta or the Dirac delta function, according to whether E and E' belong to a discrete ("bound state") or to a continuous ("free state") part of the spectrum. In the former case, but not in the latter, ψ_E is square integrable. If $\Psi(\mathbf{r}, t)$ is square integrable, the quantity $|\Psi(\mathbf{r}, t)|^2 d\mathbf{r}$ then gives the probability of finding the particle described by $\Psi(\mathbf{r}, t)$ in the volume element $d\mathbf{r}$ at time t, provided $\Psi(\mathbf{r}, t)$ is normalized to unity, i.e. that

$$\int |\Psi(\mathbf{r}, t)|^2 \, d\mathbf{r} = 1 \tag{II.6}$$

2. The Transmission of Particles through Potential Barriers

According to classical mechanics, a particle approaching a "potential barrier" (i.e. a region of space in which the potential energy of the forces acting on the particle has a maximum, V_{max}, say) will pass through this region if its total energy E is greater than V_{max}, and will be reflected back if E is less than V_{max}. According to quantum mechanics, on the other hand, such a particle, whatever its energy, has in general a finite probability of passing through the barrier, and a finite probability of being reflected from it.

The probability of transmission (or of reflection) can be conveniently expressed in terms of the transmission coefficient T (or the reflection coefficient R), defined as the ratio of the probability flux of the transmitted (or reflected) wave to the probability flux of the incident wave, thus

$$T = \frac{|\mathbf{j}_T|}{|\mathbf{j}_I|}, \qquad R = \frac{|\mathbf{j}_R|}{|\mathbf{j}_I|}$$

where

$$\mathbf{j}(\mathbf{r}, t) = \operatorname{Re}\left[\frac{\hbar}{im}\psi^*(\mathbf{r}, t) \nabla \psi(\mathbf{r}, t)\right] \tag{II.7}$$

By conservation of particles, $|\mathbf{j}_I| = |\mathbf{j}_T| + |\mathbf{j}_R|$, and thus $T+R = 1$.

3. Motion in a Central Field

If the potential energy has spherical symmetry, i.e. if $V(\mathbf{r}) = V(r)$, it is convenient to change from Cartesian coordinates x, y, z to spherical coordinates r, θ, ϕ. The time-independent Schrödinger equation (II.4) then becomes

$$H\psi(r, \theta, \phi) \equiv \left\{\frac{p_r^2}{2m} + \frac{\mathbf{l}^2}{2mr^2} + V(r)\right\}\psi(r\ \theta, \phi) = E\psi(r, \theta, \phi), \tag{II.8}$$

where p_r is the so-called "radial momentum" operator

$$p_r = -i\hbar \frac{1}{r}\frac{\partial}{\partial r} r, \tag{II.9}$$

and

$$\mathbf{l}^2 = -\frac{\hbar^2}{\sin^2\theta}\left[\sin\theta\frac{\partial}{\partial\theta}\left(\sin\theta\frac{\partial}{\partial\theta}\right)+\frac{\partial^2}{\partial\phi^2}\right] \tag{II.10}$$

is the square of the orbital angular momentum operator, the latter being defined by the relation

$$\mathbf{l} = -i\hbar\mathbf{r}\times\nabla. \tag{II.11}$$

The Hermitian operators H, $\mathbf{l}^2$ and $l_z = -i\hbar(\partial/\partial\phi)$ commute among themselves and form a complete set of observables for the motion of a (spinless) particle in a central field.

The simultaneous eigenfunctions of $\mathbf{l}^2$ and of l_z are the spherical harmonics $Y_l^m(\theta, \phi)$, which satisfy the following eigenvalue equations

$$\mathbf{l}^2 Y_l^m(\theta, \phi) = l(l+1)\,\hbar^2 Y_l^m(\theta, \phi), \tag{II.12}$$

$$l_z Y_l^m(\theta, \phi) = m\hbar Y_l^m(\theta, \phi). \tag{II.13}$$

The orbital quantum number l may have any integral value $l = 0, 1, 2, \ldots$, and, for each value of l, the "magnetic" quantum number m has the possible values $m = 0, \pm 1, \ldots, \pm l$.

The simultaneous eigenfunctions of all three observables H, $\mathbf{l}^2$, l_z are solutions of the Schrödinger equation (II.8), and have the form

$$\psi_{Elm}(r, \theta, \phi) = \frac{R_{El}(r)}{r} Y_l^m(\theta, \phi), \tag{II.14}$$

where the functions $R_{El}(r)$ are the solutions of the "radial equation"

$$\frac{d^2 R_{El}}{dr^2}+\frac{2m}{\hbar^2}\left[E-\left(V(r)+\frac{l(l+1)\hbar^2}{2mr^2}\right)\right]R_{El} = 0, \tag{II.15}$$

which are bounded at infinity and have the boundary condition at the origin that

$$R_{El}(0) = 0. \tag{II.16}$$

Thus, for motion in a central field, the solutions of the three-dimensional Schrödinger equation can be found by solving a one-dimensional problem, with an effective potential

$$V_{\text{eff}}(r) = V(r)+\frac{l(l+1)\hbar^2}{2mr^2}, \tag{II.17}$$

in the range $(0, +\infty)$, with the boundary condition (II.16) at $r = 0$.

Problems

1. Show that if the potential energy $V(\mathbf{r})$ can be written as a sum of functions of a single coordinate, $V(\mathbf{r}) = V_1(x_1)+V_2(x_2)+V_3(x_3)$, then the time-independent Schrödinger equation (II.4) can be decomposed into a set of one-dimensional equations of the form

$$\frac{d^2\psi_i(x_i)}{dx_i^2}+\frac{2m}{\hbar^2}[E_i-V_i(x_i)]\,\psi_i(x_i) = 0, \qquad i = 1, 2, 3,$$

with $\psi(\mathbf{r}) = \psi_1(x_1)\,\psi_2(x_2)\,\psi_3(x_3)$ and $E = E_1+E_2+E_3$.

2. The time-independent wavefunctions, i.e. the solutions of the Schrödinger equation

$$\frac{d^2\psi(x)}{dx^2}+\frac{2m}{\hbar^2}[E-V(x)]\,\psi(x) = 0,$$

correspond to bound or to unbound states according to whether they vanish or are merely bounded at infinity. Supposing that $\lim_{x\to\pm\infty} V(x) = V_\pm$ exists, and that $V_+ < V_-$, determine whether a state of energy E is bound or not, (1) if $E > V_-$; (2) if $V_- > E > V_+$; (3) if $V_+ > E$.

3. Show that, in one-dimensional problems, the energy spectrum of the bound states is always non-degenerate.

4. The well-known "oscillation theorem" states that if the discrete eigenvalues of a one-dimensional Schrödinger equation are placed in order of increasing magnitude, $E_1 < E_2 < \ldots < E \ldots$, say, then the corresponding eigenfunctions will occur in increasing order of the number of their zeros, the nth eigenfunction having $n-1$ zeros.

Show that between any two consecutive zeros of the nth eigenfunction, the $(n+1)$th eigenfunction will have at least one zero.

5. What conclusions can be drawn about the parity of the eigenfunctions of the Hamiltonian

$$H(x) = -\frac{\hbar^2}{2m}\frac{d^2}{dx^2}+V(x)$$

if the potential energy is an even function of x, i.e. if $V(x) = V(-x)$?

6. Show that the first derivative of the time-independent wavefunction is continuous even at points where $V(x)$ has a finite discontinuity.

7. Find the eigenfunctions and the energy spectrum of a particle in the potential well given by $V(x) = 0$ if $|x| < a$ and $V(x) = +\infty$ if $|x| > a$ (Fig. II.1).

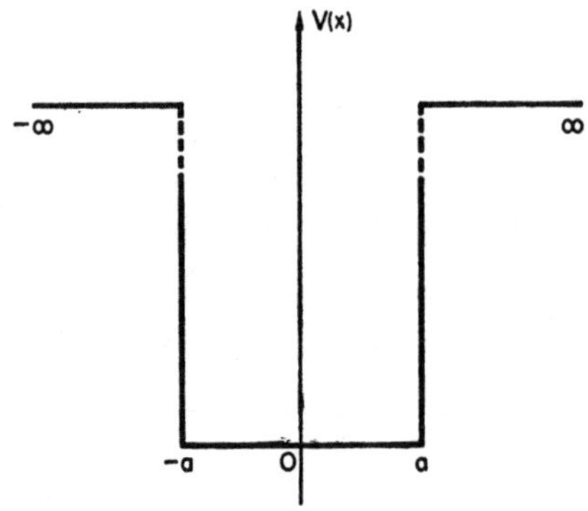

FIG. II.1.

8. A particle is enclosed in a rectangular box with impenetrable walls, inside which it can move freely (Fig. II.2). Find the eigenfunctions and the possible values of the energy. What can be said about the degeneracy, if any, of the eigenfunctions?

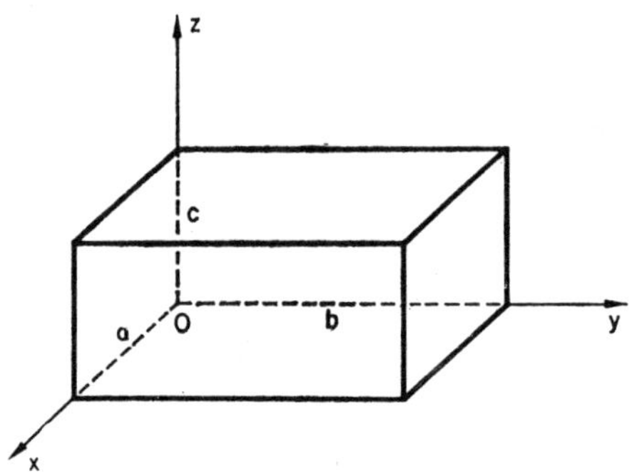

FIG. II.2.

9. Find the energies of the bound states of a particle in the symmetrical potential well given by $V(x) = -V_0$ if $|x| < a$, and $V(x) = 0$ if $|x| > a$ (Fig. II.3), where V_0 is a positive quantity.

10. Find the energies of the bound states of a particle in the potential well given by $V(x) = +\infty$ if $x < 0$, $V(x) = -V_0$ if $0 < x < a$, and $V(x) = 0$ if $x > a$ (Fig. II.4). Compare these values with those of the preceding problem.

11. Find the energy values of a particle in the asymmetric potential well given by $V(x) = V_2$ if $x < 0$, $V(x) = 0$ if $0 < x < a$, $V(x) = V_1$ if $x > a$ (Fig. II.5).

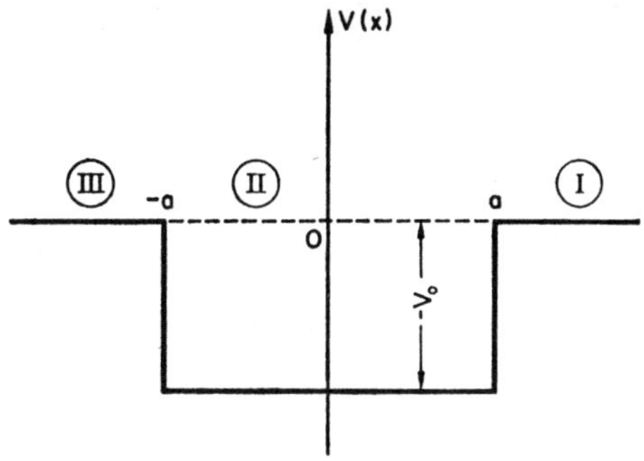

FIG. II.3.

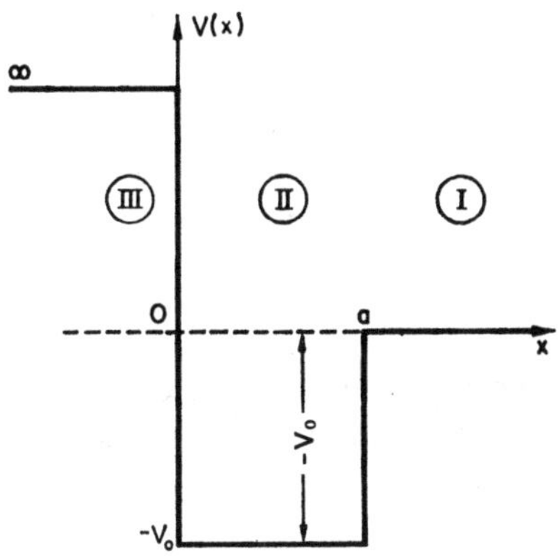

FIG. II.4.

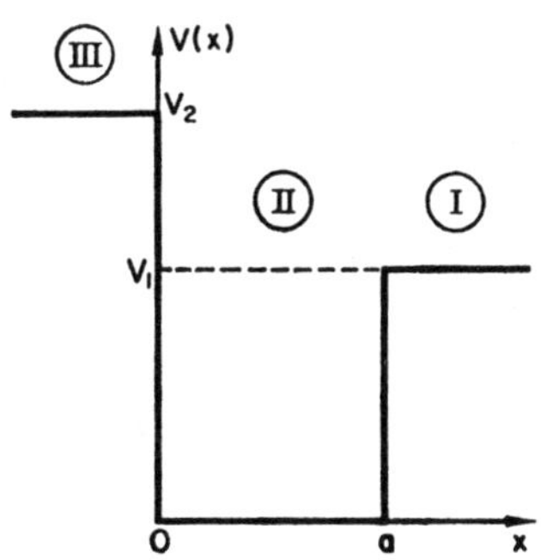

FIG. II.5.

12. Solve the Schrödinger equation for the potential shown in Fig. II.6. Write down the condition which gives the possible energy eigenvalues of a particle in such a potential.

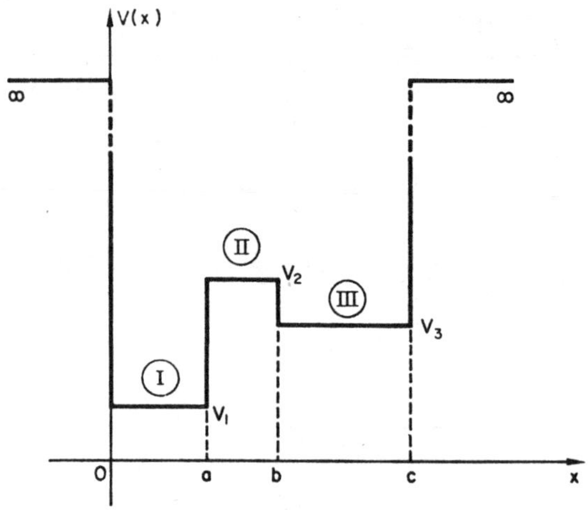

FIG. II.6.

13. Find the normalized energy eigenfunctions of a particle in the potential shown in Fig. II.7. What are the probabilities of finding the particle in the intervals $0 < x < a$ and $a < x < b$ respectively?

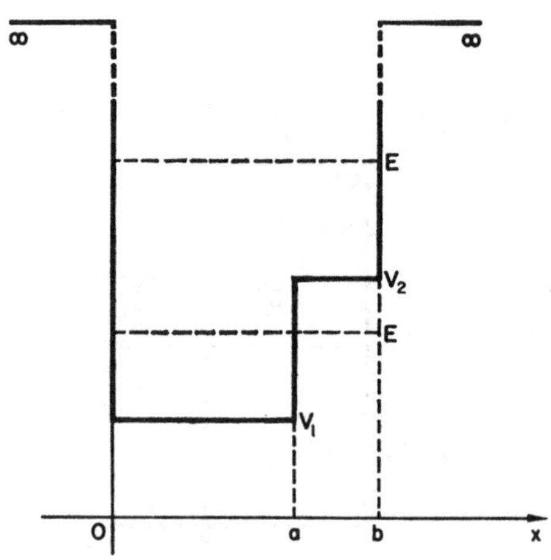

FIG. II.7.

14. Show that the energy spectrum of a particle in the periodic potential of Fig. II.8 has a band structure.

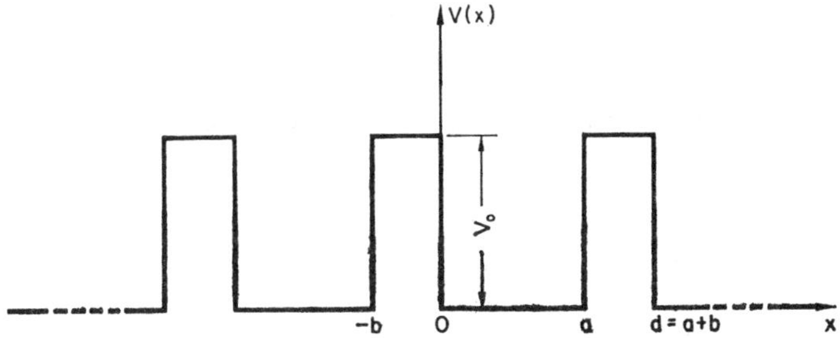

FIG. II.8.

15. Solve the Schrödinger equation for the Pöschl–Teller potential[†]

$$V(x) = \frac{V_0}{\cos^2 \alpha x}, \tag{15a}$$

where V_0 is a positive constant.

16. Solve the Schrödinger equation for the generalized Pöschl–Teller potential

$$V(x) = \frac{V_1}{\sin^2 \alpha x} + \frac{V_2}{\cos^2 \alpha x},$$

in the interval $0 < x < \dfrac{\pi}{2\alpha}$ (V_1 and V_2 are positive constants).

17. In quantum mechanics, by analogy with classical mechanics, a system described by the Hamiltonian

$$H = \frac{p_x^2}{2m} + \frac{m\omega^2}{2} x^2 \tag{17a}$$

is called a "linear harmonic oscillator".

Finds its eigenfunctions and its energy spectrum.

18. In Fig. II.9, the potential energy $V(x)$ and the total energy E_1 of an oscillator in the quantum state with $n = 1$ are represented. Show that the probability of finding the particle in the interval $(x, x+dx)$ is, according to classical mechanics,

$$W_{cl}(x)\, dx = \frac{1}{\pi a}\left(1 - \frac{x^2}{a^2}\right)^{-1/2} dx, \qquad -a \leqslant x \leqslant a.$$

† G. Pöschl and E. Teller, *Z. Physik* **83,** 143 (1933).

Compare this probability with the quantum-mechanical one

$$W_{qu}(x)\,dx = |\psi_1(x)|^2\,dx,$$

where

$$\psi_1(x) = 2(2\pi^{1/2}x_0)^{-1/2}\left(\frac{x}{x_0}\right)e^{-\frac{1}{2}\left(\frac{x}{x_0}\right)^2} \quad \text{and} \quad x_0 = \left(\frac{\hbar}{m\omega}\right)^{1/2}.$$

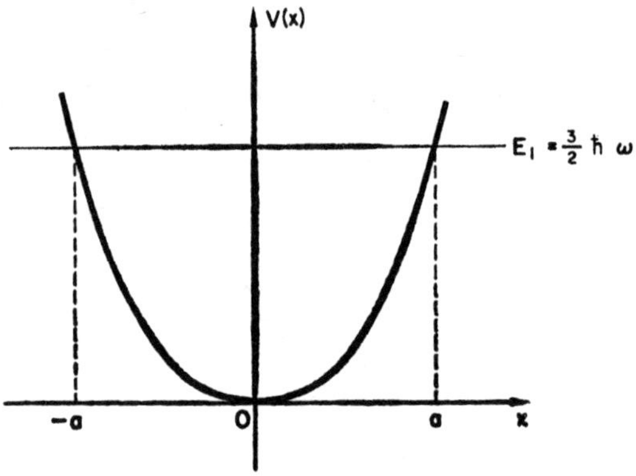

FIG. II.9.

19. Calculate the possible energy values of a particle in the potential given by $V(x) = \infty$ if $x \leqslant 0$, and $V(x) = \frac{m\omega^2}{2}x^2$ if $x > 0$.

20. A system described by the Hamiltonian

$$H = -\frac{\hbar^2}{2m}\nabla^2 + \frac{m}{2}(\omega_1^2 x^2 + \omega_2^2 y^2 + \omega_3^2 z^2) \tag{20a}$$

is called an "anisotropic harmonic oscillator".

Determine the possible energies of this system, and, for the isotropic case ($\omega_1 = \omega_2 = \omega_3 = \omega$), calculate the degeneracy of the level E_n.

21. Find the wavefunction of a particle in the homogeneous field $V(x) = -kx$. What can be said about its energy spectrum?

22. The conduction electrons in metals are held inside the metal by an average potential called the inner potential of the metal. Calculate, for the one-dimensional model given by $V(x) = -V_0$ if $x < 0$ and $V(x) = 0$ if $x > 0$ (Fig. II.10), the probability of reflection and of transmission of a conduction electron approaching the surface of the metal with total energy E, (i) if $E > 0$, and (ii) if $-V_0 < E < 0$.

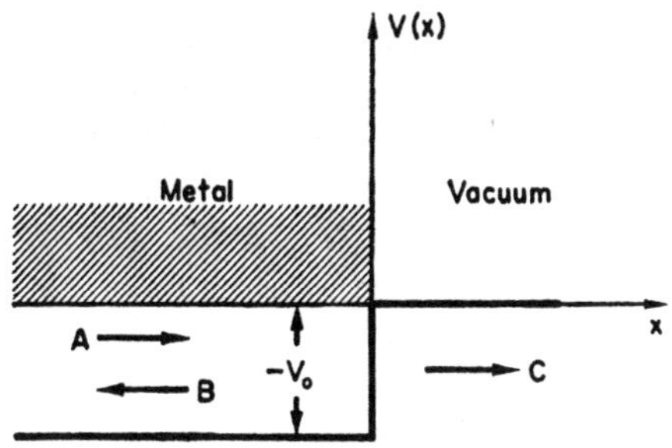

FIG. II.10.

23. A beam of mono-energetic electrons strikes the surface of a metal at normal incidence. Calculate the reflection probability of these electrons if $E = 0{\cdot}1$ eV and $V_0 = 8$ eV.

24. In problem 22 it was supposed that, at the metal–vacuum interface, the potential energy of the electrons jumps from $-V_0$ to 0. Actually this change is continuous over an interval a, whose dimensions are of the order of the interatomic distances in the metal. The potential energy near the surface of the metal can thus be written approximately as

$$V(x) = -\frac{V_0}{1+e^{x/a}} \tag{24a}$$

(Fig. II.11), which, as $a \to 0$, approximates to the previously used discontinuous potential. Calculate the reflection probability of a conduction electron approaching the surface of the metal if (i) $-V_0 < E < 0$, and (ii) $E > 0$.

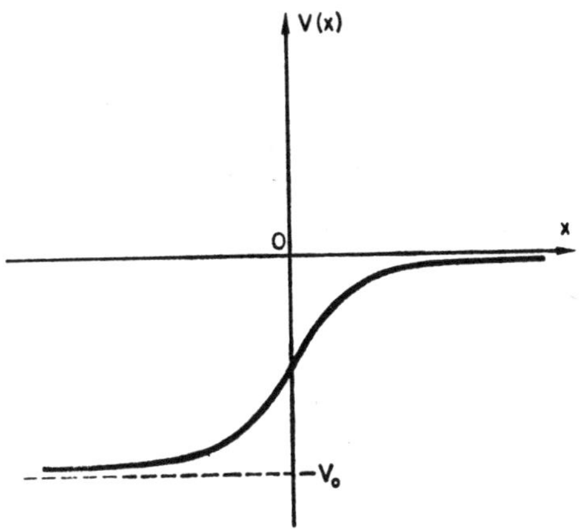

FIG. II.11.

25. Calculate the transmission and the reflection coefficients of a particle having total energy E, at the potential barrier given by $V(x) = 0$ if $x < 0$, $V(x) = V_0$ if $0 < x < a$, $V(x) = 0$ if $x > a$, for the cases $E > V_0$ and $0 < E < V_0$ (Fig. II.12).

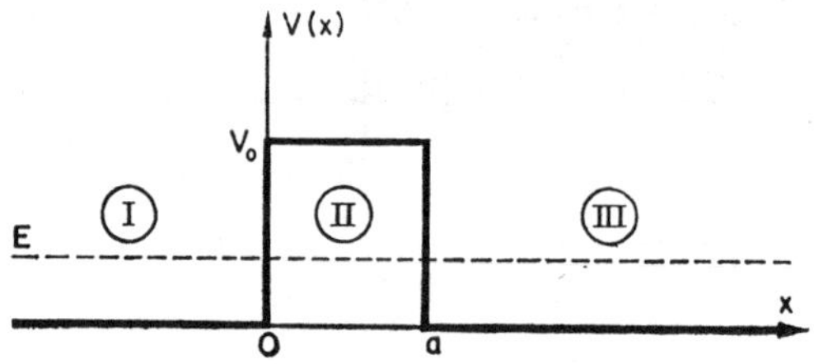

FIG. II.12.

26. A particle of total energy E enters the barrier $V = V(x)$ (Fig. II.13) at the point $x = x_1$ and leaves it at the point $x = x_2$ (the "tunnel effect"). Assuming that the potential energy curve $V(x)$ is sufficiently smooth, let us divide the interval $[x_1, x_2]$ into intervals of length Δx_i, large compared with the relative penetration depth $d_i = \hbar[8m(V(x_i)-E)]^{-1/2}$ of a particle in the rectangular barriers so obtained. Calculate in this approximate way the transmission coefficient T for the whole barrier $V = V(x)$, knowing that $T_i \approx \exp\left[-\frac{1}{\hbar}\sqrt{8m(V(x_i)-E)}\,\Delta x_i\right]$ for the ith rectangular barrier (see problem 25).

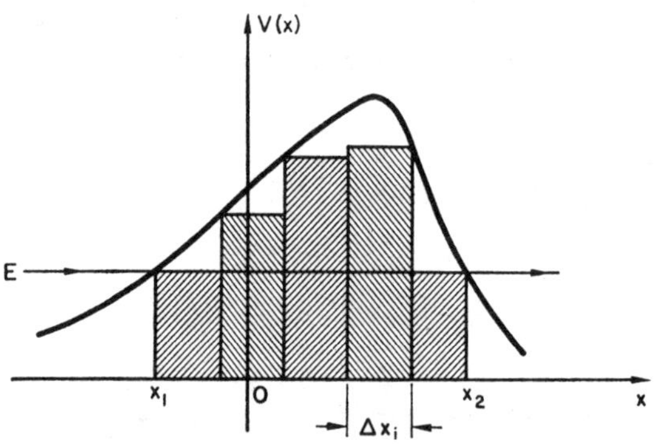

FIG. II.13.

27. It has been shown experimentally that, under the influence of a strong electric field normal to the surface of a metal, there is a flow of conduction electrons out of the metal (the "cold emission" effect). According to classical electrodynamics, those electrons can leave the metal which have enough energy to surmount the potential barrier produced jointly by the electrical image force $-e^2/4x^2$ (which acts upon an electron at a distance x

outside the metal) and by the force $e\mathcal{E}$ due to the external applied field. Although this explanation is qualitatively correct, the quantitative results of classical calculations based upon it are in complete disagreement with the experimental data. This disagreement is resolved when a quantum phenomenon—the tunnel effect—is taken into account. Supposing, for simplicity, that the electric field is homogeneous, determine the dependence of the cold emission current (at a given temperature) on the magnitude of the applied field.

28. The fact that α-particles having energies of a few MeV can leave potential wells with depths of tens of MeV (inside which they find themselves in radioactive nuclei), can be explained by the tunnel effect. Using a simplified model, let $V(r) = -V_0$ if $r < R_0$, $V(r) = \frac{e_1 e_2}{r}$ if $r > R_0$ (Fig. II.14), and calculate Gamow's factor for this barrier, i.e. the transmission probability for α-particles of energy E through the barrier. Express the result in terms of the final velocity of the α-particle, and estimate the mean life of an α-emitting nucleus.

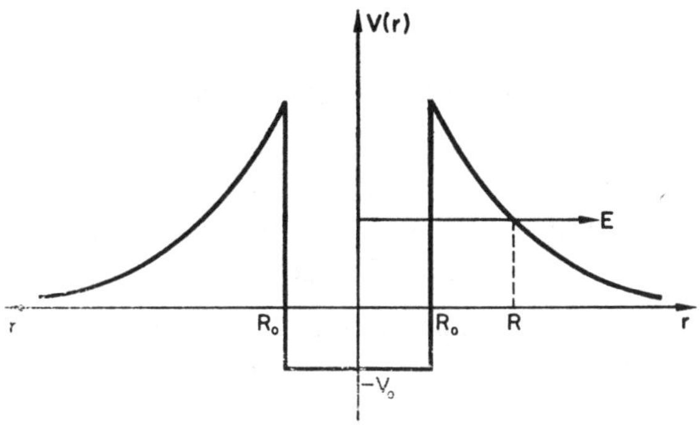

FIG. II.14.

29. Find the eigenfunctions of a free particle, in the limiting case of motion in a central field in which the potential $V(r) \to 0$. Compare these eigenfunctions, based on the complete set of observables H, $\mathbf{l}^2$, l_z, with the "plane wave" eigenfunctions, in which the motion is characterized by the observables p_x, p_y, p_z, and $H = \frac{\mathbf{p}^2}{2m}$, which also form a complete set of observables for a spinless free particle.

30. Find the possible energies of a particle in the spherical potential well given by $V(r) = -V_0$ if $r < a$ and $V(r) = 0$ if $r > a$ (Fig. II.15).

31. Find the energy levels of a particle in the central field

$$V(r) = \frac{A}{r^2} + Br^2, \tag{31a}$$

where A and B are positive constants.

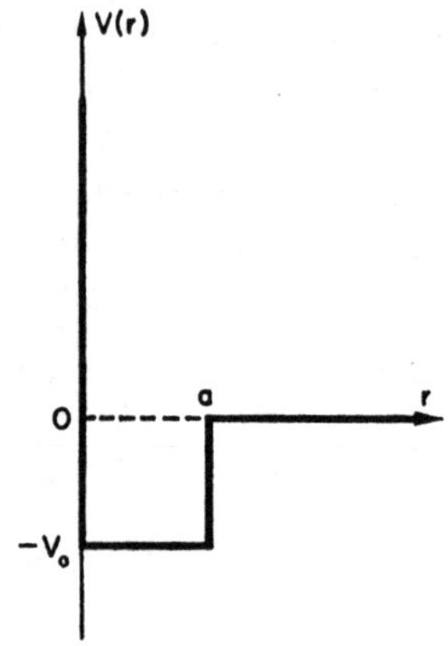

FIG. II.15.

Show that, in the particular case $A = 0$, $B = \frac{1}{2}m\omega^2$, the levels are the same as those found in problem 20 for an isotropic three-dimensional oscillator.

32. Show that in quantum mechanics (as in classical mechanics), the problem of the motion of two interacting particles of masses m_1 and m_2 can be reduced essentially to the problem of the motion of a single particle of effective ("reduced") mass $m = m_1m_2/(m_1+m_2)$ moving in the potential $V(\mathbf{r})$ of the mutual interaction of the two particles.

33. Calculate the energy levels and eigenfunctions of a hydrogen atom. Discuss the degeneracy of these levels.

34. By analogy with classical mechanics, a system described in quantum mechanics by the Hamiltonian

$$H = \frac{1}{2I}\mathbf{l}^2 \tag{34a}$$

is called a "rigid rotator". Here I is the (constant) moment of inertia of the rotator. Determine the corresponding energies and eigenfunctions. What is the degeneracy of the energy levels?

Solutions

1. Substituting the trial function $\psi(\mathbf{r}) = \psi_1(x_1)\,\psi_2(x_2)\,\psi_3(x_3)$ into equation (II.4), and dividing by $\psi_1\psi_2\psi_3$, we obtain

$$\sum_{i=1}^{3}\left(\frac{1}{\psi_i}\frac{d^2\psi_i}{dx_i^2} - \frac{2m}{\hbar^2}V_i\right) = -\frac{2m}{\hbar^2}E.$$

Because the terms in each bracket of the sum contain independent variables, the equality can be valid for all (x_1, x_2, x_3) only if each bracket is a constant, i.e. if

$$\frac{1}{\psi_i}\frac{d^2\psi_i}{dx_i^2}-\frac{2m}{\hbar^2}V_i=-\frac{2m}{\hbar^2}E_i, \qquad i=1,2,3,$$

where the E_i are constants, and $E = E_1+E_2+E_3$.

In particular, if $V(\mathbf{r}) = V_1(x_1)$, the function $\psi_1(x_1)$ is given by the one-dimensional Schrödinger equation

$$\frac{d^2\psi_1}{dx_1^2}+\frac{2m}{\hbar^2}(E_1-V_1)\psi_1=0,$$

and $\psi_2(x_2)$ and $\psi_3(x_3)$ are solutions of the free particle equations

$$\frac{d^2\psi_i}{dx_i^2}+\frac{2m}{\hbar^2}E_i\psi_i=0, \qquad i=2, 3.$$

2. In the first case the difference $E-V(x)$ is positive at both ends of the interval $(-\infty, +\infty)$, and thus the eigenfunctions in these regions oscillate indefinitely between finite bounds and therefore correspond to unbound states. The energy spectrum is continuous and doubly degenerate. In the case $V_- > E > V_+$, the difference $E-V(x)$ is negative in the limit $x \to -\infty$, so that in this asymptotic region only one of the two linearly independent eigenfunctions is bounded (in fact, it has a decreasing exponential). In the other asymptotic region, where $E-V(x)$ is positive, this eigenfunction oscillates indefinitely between finite bounds and in consequence it corresponds to an unbound state. The energy spectrum is continuous and non-degenerate. In the third case the difference $E-V(x)$ is negative in both asymptotic regions. The bound state solution, if it exists, must approach zero exponentially at both the limits $x \to \pm\infty$. It can be shown that such a solution exists only for discrete values of E. The case in which $V_+ > V_-$ is similar to the one in which $V_+ < V_-$, and does not add anything new to the above conclusions. If $V_+ = V_- = V_0$, the eigenfunctions correspond to a bound state if $V_0 > E$, and to an unbound state in a continuous and doubly degenerate energy spectrum if $E > V_0$.

3. For the sake of argument let us suppose that the opposite is true. Let $\psi_1(x)$ and $\psi_2(x)$ then be two linearly independent eigenfunctions with the same energy eigenvalue E. From the equations

$$\psi_1''+\frac{2m}{\hbar^2}(E-V)\psi_1=0, \quad \psi_2''+\frac{2m}{\hbar^2}(E-V)\psi_2=0,$$

we obtain

$$\frac{\psi_1''}{\psi_1}=\frac{\psi_2''}{\psi_2}=\frac{2m}{\hbar^2}(V-E),$$

i.e.

$$\psi_1''\psi_2-\psi_2''\psi_1=(\psi_1'\psi_2)'-(\psi_2'\psi_1)'=0.$$

After integrating this equation we find that

$$\psi_1'\psi_2-\psi_2'\psi_1 = \text{a constant.}$$

Since, at infinity, $\psi_1 = \psi_2 = 0$ (bound states), we must have the constant $= 0$ and hence

$$\frac{\psi_1'}{\psi_1} = \frac{\psi_2'}{\psi_2}.$$

Integrating once more we have $\ln \psi_1 = \ln \psi_2+\ln c$, i.e. $\psi_1 = c\psi_2$, which contradicts the assumed linear independence of the two functions.

4. Consider two eigenfunctions $\psi_n(x)$ and $\psi_{n+1}(x)$ with eigenvalues $E_n < E_{n+1}$. Since the spectrum is supposed to be non-degenerate, all eigenfunctions can be taken to be real, by a suitable choice of phase factors. From the equations

$$\psi_n''+\frac{2m}{\hbar^2}(E_n-V)\psi_n = 0, \quad \psi_{n+1}''+\frac{2m}{\hbar^2}(E_{n+1}-V)\psi_{n+1} = 0,$$

after some simple calculation we obtain

$$\left[\psi_n'\psi_{n+1}-\psi_{n+1}'\psi_n\right]_\alpha^\beta = \frac{2m}{\hbar^2}(E_{n+1}-E_n)\int_\alpha^\beta \psi_n\psi_{n+1}\,dx.$$

Let us take α and β to be two consecutive zeros of ψ_n $(n \geqslant 3)$. Then

$$\left[\psi_n'\psi_{n+1}\right]_\alpha^\beta = \frac{2m}{\hbar^2}(E_{n+1}-E_n)\int_\alpha^\beta \psi_n\psi_{n+1}\,dx.$$

Now, in the interval (α, β), ψ_n does not change sign; suppose, without loss of generality, that $\psi_n > 0$. This means that $\psi_n'(\alpha) > 0$ and $\psi_n'(\beta) < 0$. It follows that $\psi_{n+1}(x)$ must change sign in the interval (α, β), since otherwise the right-hand side of the equality would have the same sign as ψ_{n+1}, while the left-hand side would have the opposite sign. Hence, between two consecutive zeros of ψ_n, ψ_{n+1} has at least one zero. In this connection note also that eigenfunctions which are even (odd) with respect to the reflection $x \to -x$ have an even (odd) number of zeros, and that the ground state is always even.

5. Since $H(x) = H(-x)$, we have

$$H\psi(x) = E\psi(x), \quad H\psi(-x) = E\psi(-x), \tag{5.1}$$

i.e. $\psi(x)$ and $\psi(-x)$ are eigenfunctions of H with the same eigenvalue E. We distinguish two cases:

(1) The level E is non-degenerate.

In this case $\psi(x) = C\psi(-x)$, and hence $\psi(x) = \pm\psi(-x)$, that is, the eigenfunctions corresponding to non-degenerate energy levels are either even or odd. Suppose now that all the energy levels are non-degenerate. Then, if we write the energy eigenvalues in increasing order of magnitude, $E_1 < E_2 < E_3 \ldots$, the corresponding eigenfunctions will occur in increasing order of the number of their zeros, the function corresponding to E_n having $n-1$ zeros (see problem 4). Since the even (odd) functions have an even (odd) number of zeros, it follows that the eigenfunctions will be alternately even and odd, the ground state wavefunction being always even.

(2) The level E is degenerate.

The degeneracy being twofold, the general solution of equation (5.1) can be written as

$$C_1\psi(x)+C_2\psi(-x) = A[\psi(x)+\psi(-x)]+B[\psi(x)-\psi(-x)] = A\psi_e(x)+B\psi_o(x)$$

where $A+B = C_1$, $A-B = C_2$. Thus the two eigenfunctions having the same eigenvalue can be written in the form of a linear combination of two functions of well-defined parity, which are themselves eigenfunctions with the same eigenvalue.

6. Consider a potential $V(x)$ having a finite discontinuity at $x = x_0$ (Fig. II.16). In the interval $(x_0-d,\ x_0+d)$, by replacing $V(x)$ by the line segment shown in the diagram,

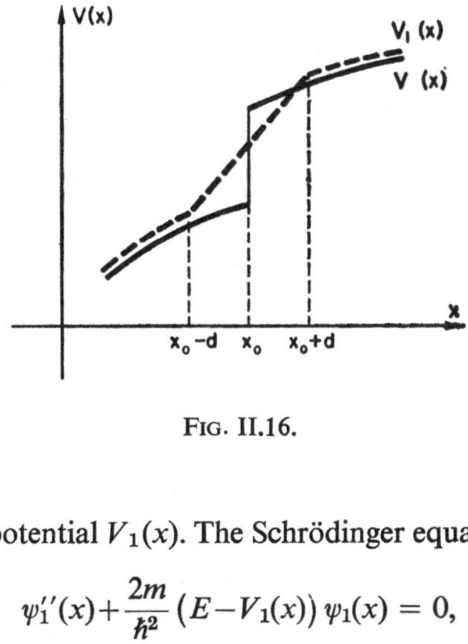

FIG. II.16.

one obtains a continuous potential $V_1(x)$. The Schrödinger equation then becomes

$$\psi_1''(x)+\frac{2m}{\hbar^2}(E-V_1(x))\,\psi_1(x) = 0,$$

whence

$$(\psi_1')_{x_0+d}-(\psi_1')_{x_0-d} = \frac{2m}{\hbar^2}\int_{x_0-d}^{x_0+d}(V_1(x)-E)\,\psi_1(x)\,dx.$$

As $d \to 0$ we have $V_1(x) \to V(x)$, $\psi_1(x) \to \psi(x)$, where $\psi(x)$ is the solution which corresponds to the discontinuous potential $V(x)$ and the same eigenvalue E. Because the integrand is bounded we find that $(\psi')_{x_0+0} = (\psi')_{x_0-0}$ in the limit $d \to 0$. Note that at a point where the potential $V(x)$ has an infinite discontinuity, the derivative wavefunction does not have the above property. The condition of continuity of $\psi(x)$ and of its first derivative is equivalent to the condition of continuity of the "logarithmic derivative"

$$\frac{d}{dx} \ln \psi(x) = \psi'(x)/\psi(x) \quad \text{of} \quad \psi(x).$$

7. For $|x| > a$, the solutions must be identically zero. Inside the well the particle states are bound and the energy spectrum is non-degenerate. Because $V(x) = V(-x)$ the solutions of the equation $\psi''(x)+k^2\psi(x) = 0$, where $k^2 = (2m/\hbar^2)E$, will have well-determined parities, $\psi^{(e)}(x) = A \cos kx$, $\psi^{(0)}(x) = B \sin kx$. The continuity conditions $\psi^{(e)}(\pm a) = 0$ and $\psi^{(0)}(\pm a) = 0$ will be satisfied if $ak = n\pi/2$, where $n = 1, 3, 5, \ldots$, and $n = 2, 4, 6, \ldots$, respectively. Hence

$$\psi_n(x) = \begin{cases} A \cos \dfrac{n\pi}{2a} x = \psi_n^{(e)}, & \text{for } n \text{ odd} \\ B \sin \dfrac{n\pi}{2a} x = \psi_n^{(0)}, & \text{for } n \text{ even.} \end{cases}$$

The possible values of the energy are

$$E_n = \frac{\hbar^2}{2m} k^2 = \frac{\hbar^2 \pi^2 n^2}{8ma^2}, \qquad n = 1, 2, 3, \ldots$$

In this example, the results of problems 4 and 5 can be verified directly. Thus, the eigenfunctions which correspond to the levels $E_1 < E_2 < E_3 < \ldots$, are alternately even and odd, and the eigenfunction corresponding to the energy E_n has $n-1$ zeros in the open interval $(-a, a)$. The orthonormalization relation $\int_{-a}^{a} \psi_m^*(x)\, \psi_n(x)\, dx = \delta_{mn}$ is satisfied if we take $A = B = \sqrt{a}$.

Note the difference between the classical and the quantum behaviour of particles. A particle between two absolutely rigid walls can either perform an oscillatory movement with any specified energy E, or it can remain at rest ($E = 0$), according to classical mechanics. According to quantum mechanics the energy can have only certain values, whose number is, however, infinite. It is significant that the ground state energy E_1 is not zero, i.e. the particle cannot be at rest in the potential well (see problem 12, Chapter III).

8. The potential energy can be written in the form $V(x, y, z) = V(x)+V(y)+V(z)$, where

$$V(x) = V(y) = V(z) = \begin{cases} 0 \text{ if } 0 < x < a, \quad 0 < y < b, \quad 0 < z < c, \\ \infty \text{ outside these intervals.} \end{cases}$$

The Schrödinger equation (II.4) can be decomposed into one-dimensional equations (see problem 1), whose solutions are

$$\begin{aligned} \psi_1(x) &= A \sin k_1x + B \cos k_1x, & 0 < x < a \\ \psi_2(y) &= C \sin k_2y + D \cos k_2y, & 0 < y < b \\ \psi_3(z) &= F \sin k_3z + G \cos k_3z, & 0 < z < c, \end{aligned}$$

where $k_1^2+k_2^2+k_3^2 = \frac{2m}{\hbar^2}E$ and $\psi_1(x)\,\psi_2(y)\,\psi_3(z) = \psi(x, y, z)$.

Because at the walls $\psi = 0$, the eigenfunctions normalized to unity are given by

$$\psi_{n_1n_2n_3}(x, y, z) = \sqrt{\frac{8}{abc}} \sin\left(\frac{n_1\pi}{a}x\right) \sin\left(\frac{n_2\pi}{b}y\right) \sin\left(\frac{n_3\pi}{c}z\right).$$

The energy of the particle is thus quantized, the possible values being given by

$$E_{n_1n_2n_3} = \frac{\pi^2\hbar^2}{2m}\left(\frac{n_1^2}{a^2}+\frac{n_2^2}{b^2}+\frac{n_3^2}{c^2}\right); \qquad n_1, n_2, n_3 = 1, 2, 3, \ldots$$

If the ratio of any two sides is an irrational number, all the energy levels are non-degenerate. Otherwise the energy spectrum is in general degenerate. Thus, e.g., if $a = b = c$, the level for which $n_1^2+n_2^2+n_3^2 = 6$ is threefold degenerate, since three linearly independent eigenfunctions have the same eigenvalue: $E_{121} = E_{112} = E_{211} = \frac{6\pi^2\hbar^2}{2ma^2}$. The ground state E_{111} is, however, always non-degenerate.

9. If bound states exist, the particle's total energy E in such states will be in the range $-V_0 < E < 0$. The eigenvalue equations are then

$$\begin{aligned} &\psi_1''(x) - \frac{2m}{\hbar^2}|E|\,\psi_1(x) = 0, && \text{if} \quad x > a \\ &\psi_2''(x) + \frac{2m}{\hbar^2}(V_0 - |E|)\,\psi_2(x) = 0, && \text{if} \quad |x| < a \\ &\psi_3''(x) - \frac{2m}{\hbar^2}|E|\,\psi_3(x) = 0, && \text{if} \quad x < -a. \end{aligned}$$

Using the notation $\alpha = \left(\frac{2m}{\hbar^2}|E|\right)^{1/2} > 0$ and $\beta = \left\{\frac{2m}{\hbar^2}(V_0 - |E|)\right\}^{1/2} > 0$, the wave-function can be written as

$$\psi(x) = \begin{cases} \psi_1 = Ae^{\alpha x} + Be^{-\alpha x}, & \text{if} \quad x > a \\ \psi_2 = C \sin \beta x + D \cos \beta x, & \text{if} \quad |x| < a \\ \psi_3 = Fe^{\alpha x} + Ge^{-\alpha x}, & \text{if} \quad x < -a. \end{cases}$$

For ψ to describe bound states, we have to take $A = G = 0$ (the vanishing condition at infinity). We can then see a typical quantum-mechanical effect. Since $|\psi_1|^2 = B^2e^{-2\alpha x} \neq 0$, and $|\psi_3|^2 = F^2e^{2\alpha x} \neq 0$, the particle may be found outside the potential well, with a probability which decreases exponentially with distance (Fig. II.17).

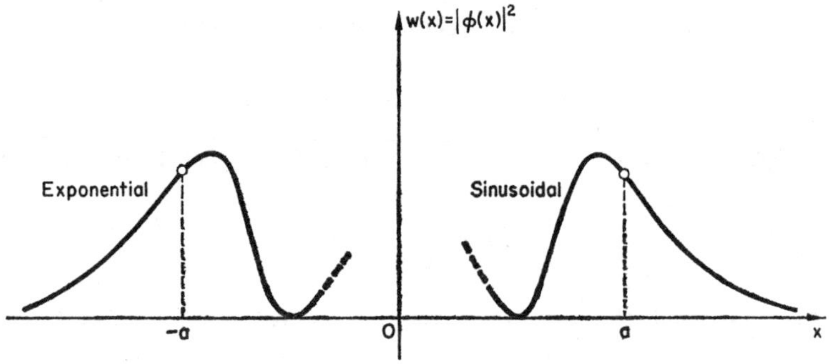

FIG. II.17.

From the continuity conditions at $x = \pm a$ we obtain

$$\left.\begin{aligned} 2C \sin \beta a &= (B-F)e^{-\alpha a} \\ 2\beta C \cos \beta a &= -\alpha(B-F)e^{-\alpha a} \end{aligned}\right\} \tag{9.1}$$

$$\left.\begin{aligned} 2D \cos \beta a &= (B+F)e^{-\alpha a} \\ 2\beta D \sin \beta a &= \alpha(B+F)e^{-\alpha a} \end{aligned}\right\} \tag{9.2}$$

If $C \neq 0$, and thus $B \neq F$, we obtain from (9.1)

$$\beta \cot \beta a = -\alpha. \tag{9.3}$$

If $D \neq 0$, and thus $B \neq -F$, we obtain from (9.2)

$$\beta \tan \beta a = \alpha. \tag{9.4}$$

The relations (9.3) and (9.4) cannot both be satisfied at the same time. We have therefore to distinguish two classes of solution:

(1) $C = 0, \quad B = F \quad$ and $\quad \beta \tan \beta a = \alpha$,
(2) $D = 0, \quad B = -F \quad$ and $\quad \beta \cot \beta a = -\alpha$.

In the first class the eigenfunctions are even and in the second class they are odd, a result which follows directly from problem 5.

The corresponding energy values are given by the solutions of the transcendental equations (9.3) and (9.4), which can be solved graphically as follows. Put $X = \beta a$, $Y = \alpha a$; the energy levels $E = -\dfrac{\hbar^2}{2ma^2} Y^2$ are then obtained from the intersections (if any) of the curves

$$\left.\begin{aligned} X \tan X &= Y \\ X^2+Y^2 &= \frac{2ma^2}{\hbar^2} V_0 \end{aligned}\right\} \quad \text{or} \quad \left.\begin{aligned} X \cot X &= -Y \\ X^2+Y^2 &= \frac{2ma^2}{\hbar^2} V_0 \end{aligned}\right\} \text{ respectively}$$

in the region $X > 0$ and $Y > 0$ (Fig. II.18).

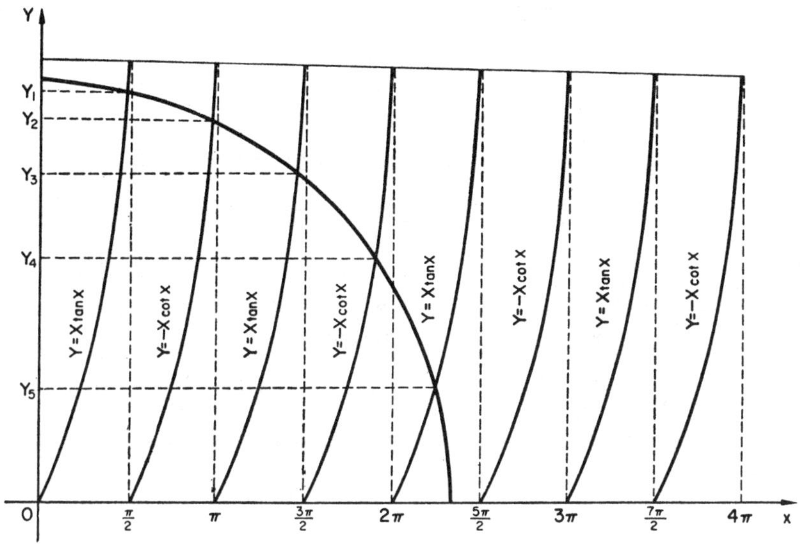

FIG. II.18.

We can see from Fig. II.18 that the number of bound states increases as the product a^2V_0 (the "well parameter") increases, and is finite if a^2V_0 is finite. It can also be seen that if $\dfrac{N}{2}\pi \leqslant R < \left(\dfrac{N+1}{2}\right)\pi$, where $R = \left(\dfrac{2m}{\hbar^2} a^2V_0\right)^{1/2}$ and $N = 0, 1, 2, \ldots,$ then the number of bound states is $N+1$ (for $N = 0$, the condition becomes $0 < R < \pi/2$).

10. The energies of the bound states will lie in the interval $-V_0 < E < 0$, and the eigenfunctions will have the form

$$\psi(x) = \begin{cases} \psi_1 = Be^{-\alpha x}, & \text{if} \quad x > a \\ \psi_2 = C \sin \beta x + D \cos \beta x, & \text{if} \quad 0 < x < a \\ \psi_3 = 0, & \text{if} \quad x < 0. \end{cases}$$

In this case the probability of finding the particle in range III is zero, and in range I it decreases exponentially with x (Fig. II.19).

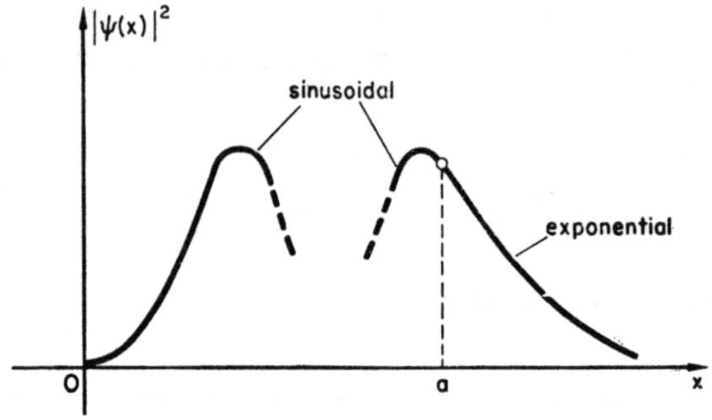

FIG. II.19.

From the continuity conditions at $x = 0$ and at $x = a$ one obtains $D = 0$, and hence

$$\beta \cot \beta a = -\alpha. \tag{10.1}$$

Since this transcendental equation is the same as (9.3), the possible energy values in the present problem will coincide with a part of those found in the previous problem. However, the wavefunctions are now neither even nor odd, since $V(x) \neq V(-x)$.

We observe that, if $R = \left(\frac{2m}{\hbar^2} V_0 a^2\right)^{1/2} < \frac{\pi}{2}$, i.e. if $a^2 V_0 < \hbar^2\pi^2/8m$, then the potential well cannot bind the particle; the condition for the existence of at least one bound state is

$$a^2 V_0 \geqslant \frac{\hbar^2\pi^2}{8m}. \tag{10.2}$$

11. From the behaviour of the wavefunction in the asymptotic regions $x \to \pm\infty$, it follows that the energy spectrum of the particle is either discrete, or continuous and non-degenerate, or continuous and doubly degenerate, according to whether $0 < E < V_1$, or $V_1 < E < V_2$, or $E > V_2$ respectively (see problem 2). The energy of the particle is thus quantized in the range $0 < E < V_1$ and the corresponding eigenfunctions will have the form

$$\psi(x) = \begin{cases} B_1 e^{-\beta_1 x}, & x > a \\ A \sin(kx+\phi), & 0 < x < a \\ B_2 e^{\beta_2 x}, & x < 0, \end{cases}$$

where

$$\beta_{1,2} = \left[\frac{2m}{\hbar^2}(V_{1,2}-E)\right]^{1/2}, \quad k = \left(\frac{2m}{\hbar^2}E\right)^{1/2}.$$

The continuity conditions at $x = 0$ and at $x = a$ give $\beta_2 \tan \phi = k$ and $\beta_1 \tan(ka+\phi) = -k$, or, in other words,

$$\sin \phi = k\hbar(2mV_2)^{-1/2} \quad \text{and} \quad \sin(ka+\phi) = -k\hbar(2mV_1)^{-1/2}.$$

Eliminating ϕ and introducing the notation

$$q = \frac{(2mV_1)^{1/2}}{\hbar}, \quad \xi = \frac{k}{q} = \sqrt{\frac{E}{V_1}}, \quad \cos\gamma = \sqrt{\frac{V_1}{V_2}} \quad \left(0 < \gamma < \frac{\pi}{2}\right),$$

the following transcendental equation is obtained:

$$n\pi - aq\xi = \arcsin\xi + \arcsin(\xi\cos\gamma) \tag{11.1}$$

(here $n = 1, 2, 3, \ldots$, and the values of arc sin are to be taken in the interval $(0, \pi/2)$). The roots of equation (11.1) give the energy levels $E = V_1\xi_1^2$. When E increases from 0 to V_1, ξ increases from 0 to 1, the right-hand side of equation (11.1) increases from 0 to $(\pi/2)+\arcsin(\cos\gamma) = \pi-\gamma$, and the left-hand side decreases from $n\pi$ to $n\pi - aq$. Equation (11.1) can be solved graphically, by determining the abscissae of the intersections of the curve given by $0 < f(\xi) = \arcsin\xi + \arcsin(\xi\cos\gamma) < \pi-\gamma$ with the straight lines $n\pi - aq\xi$ (Fig. II.20).

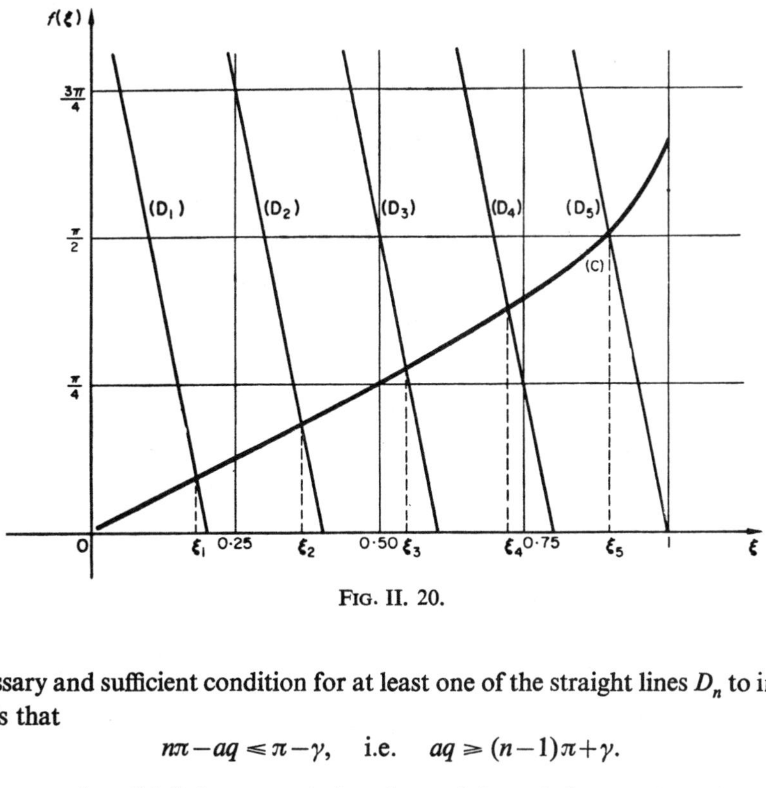

FIG. II. 20.

The necessary and sufficient condition for at least one of the straight lines D_n to intersect the curve C is that

$$n\pi - aq \leqslant \pi - \gamma, \quad \text{i.e.} \quad aq \geqslant (n-1)\pi + \gamma. \tag{11.2}$$

If $aq < \gamma$, equation (11.1) has no solution for real ξ, and thus no bound states exist. If $\gamma \leqslant aq < \pi+\gamma$ there is one eigenvalue E_1, and if $\pi+\gamma \leqslant aq \leqslant 2\pi+\gamma$ there are two eigenvalues E_1 and E_2 $(E_1 < E_2)$, and so on. As can be seen from condition (11.2), the number of energy levels in this potential well is finite. Observe that any bound particle can be found outside the well with a non-vanishing probability. In the particular case $V_1 = V_2$, we have $\cos\gamma = 1$, $\gamma = 0$, and the condition (11.2) is certainly satisfied, at least for $n = 1$, which means that there always exists a bound state in this case (see also problem 9).

12. The solution of the one-dimensional Schrödinger equation satisfying the boundary conditions $\psi(0) = \psi(c) = 0$ can be written in the form

$$\psi(x) = \begin{cases} A_1 \sin k_1 x, & \text{if} \quad 0 < x < a \\ A_2 \sin k_2 x + B_2 \cos k_2 x, & \text{if} \quad a < x < b \\ A_3 (\sin k_3 x - \tan k_3 c \cos k_3 x), & \text{if} \quad b < x < c, \end{cases} \tag{12.1}$$

where

$$k_i = \left[\frac{2m}{\hbar^2}(E - V_i)\right]^{1/2}, \qquad i = 1, 2, 3. \tag{12.2}$$

The continuity conditions at $x = a$ and at $x = b$ give the system of homogeneous equations:

$$\left.\begin{aligned} A_1 \sin k_1 a - A_2 \sin k_2 a - B_2 \cos k_2 a &= 0 \\ A_1 k_1 \cos k_1 a - A_2 k_2 \cos k_2 a + B_2 k_2 \sin k_2 a &= 0 \\ A_2 \sin k_2 b + B_2 \cos k_2 b - A_3 (\sin k_3 b - \tan k_3 c \cos k_3 b) &= 0 \\ A_2 k_2 \cos k_2 b - B_2 k_2 \sin k_2 b - A_3 k_3 (\cos k_3 b + \tan k_3 c \sin k_3 b) &= 0. \end{aligned}\right\} \tag{12.3}$$

Non-trivial solutions for the variables A_1, A_2, A_3, B_2 exist only if the determinant of the coefficients of these variables in (12.3) vanishes, which gives

$$\begin{aligned} &k_3 \cos k_3(c-b)\,[k_2 \sin k_1 a \cos k_2(b-a) + k_1 \cos k_1 a \sin k_2(b-a)] \\ &\quad = k_2 \sin k_3(c-b)\,[k_2 \sin k_1 a \sin k_2(b-a) - k_1 \cos k_1 a \cos k(b-a)]. \end{aligned} \tag{12.4}$$

From (12.2) and (12.4) the possible values of the energy of the particle can be found. We mention that for $E - V_i < 0$, i.e. for imaginary k_i, the trigonometric functions of (12.4) become the corresponding hyperbolic functions.

13. Because this is the particular case $b = c$, $V_2 = V_3$ of problem 12, the energy eigenvalue equation (12.4) becomes

$$k_1 \cot k_1 a + k_2 \cot k_2(b-a) = 0, \tag{13.1}$$

and the corresponding eigenfunctions satisfying the boundary conditions $\psi(0) = \psi(b) = 0$ are

$$\psi(x) = \begin{cases} \psi_1 = A \sin k_1 x, & \text{if} \quad 0 < x < a \\ \psi_2 = A \dfrac{\sin k_1 a}{\sin k_2(b-a)} \sin k_2(b-x), & \text{if} \quad a < x < b. \end{cases} \tag{13.2}$$

From the normalization condition

$$\int_0^b \psi^2\,dx = \int_0^a \psi_1^2\,dx + \int_a^b \psi_2^2\,dx = 1,$$

the following expression for the amplitude A is obtained

$$\frac{1}{A^2} = \frac{a}{2}\left[1-\frac{\sin 2k_1a}{2k_1a}+\frac{2k_2(b-a)-\sin 2k_2(b-a)}{2k_2a\sin^2 k_2(b-a)}\sin^2 k_1a\right]. \tag{13.3}$$

The probabilities W_1 and W_2 of finding the particle in the intervals $0 < x < a$ and $a < x < b$ respectively are given by

$$W_1 = \int_0^a \psi_1^2\,dx = \frac{a}{2}\left(1-\frac{\sin 2k_1a}{2k_1a}\right)A^2$$

and

$$W_2 = 1-W_1 = 1-\frac{a}{2}\left(1-\frac{\sin 2k_1a}{2k_1a}\right)A^2.$$

In the particular case $\frac{2m}{\hbar^2}(V_2-V_1)a^2 = (k_1a)^2-(k_2a)^2 = 1$, $b-a = a$ (i.e. the two intervals have equal widths), equation (13.1) becomes

$$k_2a\cot k_2a = -\sqrt{1+(k_2a)^2}\cot\sqrt{1+(k_2a)^2}, \tag{13.4}$$

and then

$$W_1 = \frac{1-\dfrac{\sin 2k_1a}{2k_1a}}{1-\dfrac{\sin 2k_1a}{2k_1a}+\dfrac{\sin^2 k_1a}{\sin^2 k_2a}\left(1-\dfrac{\sin 2k_2a}{2k_2a}\right)}.$$

The transcendental equation (13.4) can be solved numerically, the smallest eigenvalue being obtained for $k_2a = 1{\cdot}388$, $k_1a = 1{\cdot}710$. In this particular case $W_1 = 0{\cdot}55$ and $W_2 = 0{\cdot}45$.

Note that if $V_1 < E < V_2$, then the eigenfunctions are no longer periodic in the region $a < x < b$. Substituting $k_2 = ik$, we find from (13.1)–(13.3) that

$$k_1\cot k_1a+k\coth k(b-a) = 0,$$

$$\psi(x) = \begin{cases} \psi_1 = A\sin k_1x, & \text{if } 0 < x < a \\ \psi_2 = A\dfrac{\sin k_1a}{\sinh k(b-a)}\sinh k(b-x), & \text{if } a < x < b \end{cases}$$

$$\frac{1}{A^2} = \frac{a}{2}\left[1-\frac{\sin 2k_1a}{2k_1a}-\frac{2k(b-a)-\sinh 2k(b-a)}{2ka\sinh^2 k(b-a)}\sin^2 k_1a\right].$$

14. On account of the periodicity condition $V(x+d) = V(x)$, the solutions of the Schrödinger equation

$$\frac{d^2\psi}{dx^2}+\frac{2m}{\hbar^2}(E-V(x))\psi = 0 \tag{14.1}$$

will also be periodic, in the sense that

$$\psi(x+d) = C\psi(x) \tag{14.2}$$

where C is a constant factor. From (14.2) it follows that $\psi(x+nd) = C^n\psi(x)$. If $|C| \neq 1$, then along one direction of the x axis, $\psi(x)$ increases (decreases) without bound. Solutions having a physical meaning can be obtained only if C is a phase factor, e.g. $C = e^{i\phi}$, where ϕ is real. Then $|C| = 1$ and

$$\psi(x+d) = e^{i\phi}\psi(x). \tag{14.3}$$

In the period $-b < x < a$, let

$$\psi(x) = \begin{cases} Ae^{i\beta x}+Be^{-i\beta x}, & \text{if} \quad -b < x < 0 \\ De^{i\alpha x}+Fe^{-i\alpha x}, & \text{if} \quad 0 < x < a, \end{cases}$$

where

$$\beta = \left[\frac{2m}{\hbar^2}(E-V_0)\right]^{1/2}, \quad \alpha = \left(\frac{2m}{\hbar^2}E\right)^{1/2}.$$

Then, in the following period, $a < x < a+d$,

$$\psi(x) = e^{i\phi}\begin{cases} Ae^{i\beta(x-d)}+Be^{-i\beta(x-d)}, & \text{if} \quad a < x < d \\ De^{i\alpha(x-d)}+Fe^{-i\alpha(x-d)}, & \text{if} \quad d < x < a+d. \end{cases}$$

From the continuity conditions at $x = 0$ and at $x = a$ the following system of equations is obtained

$$\left.\begin{aligned} A+B &= D+F \\ \beta(A-B) &= \alpha(D-F) \\ e^{i\phi}(Ae^{-i\beta b}+Be^{i\beta b}) &= De^{i\alpha a}+Fe^{-i\alpha a} \\ \beta e^{i\phi}(Ae^{-i\beta b}-Be^{i\beta b}) &= \alpha(De^{i\alpha a}-Fe^{-i\alpha a}) \end{aligned}\right\} \tag{14.4}$$

Non-trivial solutions for the variables A, B, D, F are obtained only if the determinant of their coefficients vanishes, which gives the condition

$$\cos\phi = \begin{cases} \cos a\alpha.\cosh b\delta - \dfrac{\alpha^2-\delta^2}{2\alpha\delta}\sin a\alpha.\sinh b\delta, & \text{if} \quad 0 < E < V_0 \\ \cos a\alpha.\cos b\beta - \dfrac{\alpha^2+\beta^2}{2\alpha\beta}\sin a\alpha.\sin b\beta, & \text{if} \quad E > V_0 \end{cases} \tag{14.5}$$

where

$$\delta = \left[\frac{2m}{\hbar^2}(V_0-E)\right]^{1/2}.$$

Now the energy E, which appears in (14.5) through α, β and δ, has a value such that

$$-1 \leqslant \cos\phi \leqslant +1. \tag{14.6}$$

We have now to distinguish two cases:

(1) $0 < E < V_0$.

To get a picture of the structure of the energy spectrum we look at the particular case for whrch $b \to 0$, $V_0 \to \infty$ in such a manner that $b\delta^2$ remains finite. With the notation

$$P = \lim_{\substack{b \to 0 \\ \delta \to \infty}} \frac{ab\delta^2}{2},$$

(14.6) becomes

$$-1 \leqslant P \frac{\sin a\alpha}{a\alpha} + \cos a\alpha \leqslant +1.$$

From Fig. II.21, where $P = \frac{3\pi}{2}$, it can be seen that the energy spectrum consists of a series of separate regions, inside each of which the energy of the particle can vary continuously.

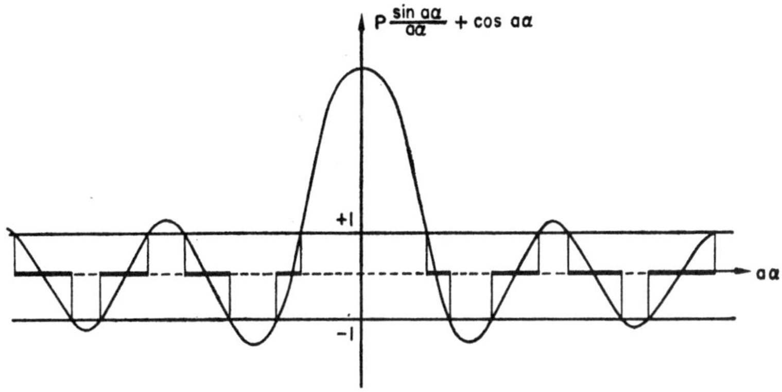

FIG. II.21.

These regions are called the "allowed bands", and the ones between them the "forbidden bands". It can be seen that the width of the allowed bands increases as the energy increases.

(2) $E > V_0$.

The possible values of the energy are determined by the condition

$$-1 \leqslant f(E) \leqslant +1, \tag{14.7}$$

where

$$f(E) = \cos a\alpha . \cos b\beta - \frac{\alpha^2+\beta^2}{2\alpha\beta} \sin a\alpha . \sin b\beta$$

$$= \cos (a\alpha + b\beta) - \frac{(\alpha-\beta)^2}{2\alpha\beta} \sin a\alpha . \sin b\beta. \tag{14.8}$$

The energy spectrum will also, in this case, have a band structure. The relation $\alpha a+\beta b = n\pi$, $n = 0, 1, 2, \ldots$, which would lead to a discrete spectrum does not satisfy the inequality (14.7). Indeed, by substituting $a\alpha = n\frac{\pi}{2}-\Phi$ and $b\beta = n\frac{\pi}{2}+\Phi$ into (14.8), we find that

$$f(E) = \begin{cases} 1+\dfrac{(\alpha-\beta)^2}{2\alpha\beta}\sin^2\Phi > 1 & \text{for } n \text{ even,} \\ -1-\dfrac{(\alpha-\beta)^2}{2\alpha\beta}\cos^2\Phi < -1 & \text{for } n \text{ odd,} \end{cases}$$

which is contrary to (14.7).

The results obtained above for rectangular potentials can be generalized for any periodic bounded potentials, such as occur, for instance, in ideal crystalline lattices.

15. By writing the positive constant V_0 in the form

$$V_0 = \frac{\hbar^2}{2m}\alpha^2\lambda(\lambda-1), \quad \lambda > 1, \tag{15.1}$$

the Schrödinger equation becomes

$$\frac{d^2\psi}{dx^2}+\left[k^2-\alpha^2\frac{\lambda(\lambda-1)}{\cos^2\alpha x}\right]\psi = 0 \tag{15.2}$$

where $k^2 = 2mE/\hbar^2$.

Changing the independent variable x in (15.2) to y given by

$$y = \sin^2\alpha x \tag{15.3}$$

we obtain the equation

$$y(1-y)\frac{d^2\psi}{dy^2}+\left(\frac{1}{2}-y\right)\frac{d\psi}{dy}+\frac{1}{4}\left[\frac{k^2}{\alpha^2}-\frac{\lambda(\lambda-1)}{1-y}\right]\psi = 0. \tag{15.4}$$

The possibility of introducing the new variable y requires discussion, since (15.3) does not establish a one-to-one correspondence between the variables x and y. Indeed, it transforms each point of the complex plane of x into points on a Riemann surface having an infinity of sheets in the plane of y. For $V_0 > 0$ (the upper half of Fig. II.22) the periodic potential (15a) consists of a series of valleys separated by walls of finite width and infinite height. Since these walls present impenetrable obstacles to the particle, we can restrict ourselves, in the study of the eigenvalue problem, to the range $-\frac{\pi}{2} < \alpha x < +\frac{\pi}{2}$, and use the boundary conditions $\psi\left(\pm\frac{\pi}{2}\right) = 0$. Since the range $\left(-\frac{\pi}{2}, +\frac{\pi}{2}\right)$ of αx corresponds to a single sheet of the Riemann surface of y, the transformation (15.3) will, with this restriction (warranted by the above physical considerations), be one-to-one. This argument is not trivial, since for $E > 0$ and $0 < \lambda < 1$, i.e. $V_0 < 0$ (the lower half of Fig. II.22), the particle can move over the whole of the plane $-\infty < \alpha x < +\infty$, and the above argument is then invalid.

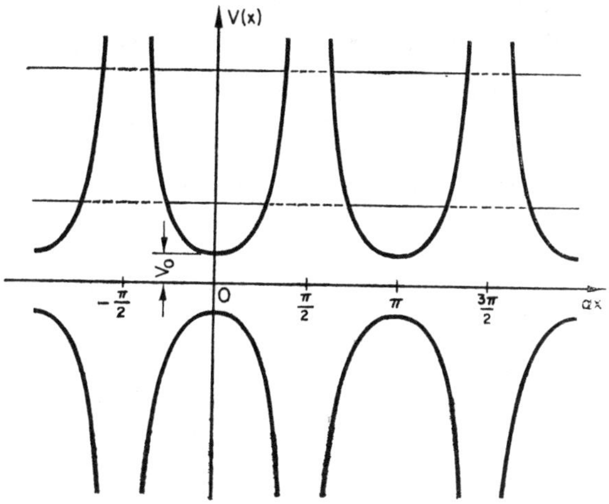

FIG. II.22.

Introducing a new function $f(y)$ through the relation

$$\psi = (1-y)^{\lambda/2} f(y), \tag{15.5}$$

we obtain from (15.4) the equation

$$y(1-y)\frac{d^2 f}{dy^2} + \left[\frac{1}{2} - (\lambda+1)y\right]\frac{df}{dy} + \frac{1}{4}\left(\frac{k^2}{\alpha^2} - \lambda^2\right) f = 0. \tag{15.6}$$

The general solution of this equation can be written in the form (A.53),

$$f = C_1 F(a, b, c; y) + C_2 y^{1-c} F(a+1-c, b+1-c, 2-c; y), \tag{15.7}$$

where

$$a = \frac{1}{2}\left(\lambda + \frac{k}{\alpha}\right), \quad b = \frac{1}{2}\left(\lambda - \frac{k}{\alpha}\right), \quad c = \frac{1}{2}. \tag{15.8}$$

According to (A.54), both F functions have a singularity at $y = 1$ of the form $(1-y)^{c-a-b} = (1-y)^{1/2-\lambda}$. Thus the function ψ, which differs from $f(y)$ by a factor $(1-y)^{\lambda/2}$, diverges like $(1-y)^{(1-\lambda)/2}$ near $y = 1$, i.e. $\alpha x = \pm\pi/2$. At these points, however, we must have $\psi = 0$. But from (A.54) and (A.51) it follows that near $y = 1$ the function ψ behaves like

$$(1-y)^{(1-\lambda)/2}\left[\frac{C_1 \Gamma(c)}{\Gamma(a)\Gamma(b)} + \frac{C_2 \Gamma(2-c)}{\Gamma(a+1-c)\Gamma(b+1-c)}\right]\Gamma(a+b-c). \tag{15.9}$$

Hence the coefficient of $(1-y)^{(1-\lambda)/2}$ must vanish. Note that all the above expressions are unchanged if $\alpha \to -\alpha$, with the exception of (15.8), in which $a \rightleftharpoons b$. The results which will be obtained below for a chosen value of b can thus be transformed, using the transformation $\alpha \to -\alpha$, which does not change the physical problem, into the results which would be obtained for a chosen value of a. Now, since the function $\Gamma(z)$ has simple poles at $z = 0, -1, -2, -3, \ldots$, the quantity (15.9) will vanish identically either if $b = -n\ (n = 0, 1, 2, 3, \ldots)$, and $C_2 = 0$, or if $b+1-c = -n$ and $C_1 = 0$. In the first case we obtain

$$\psi = (1-y)^{\lambda/2} F\left[\frac{1}{2}\left(\lambda+\frac{k}{\alpha}\right), \frac{1}{2}\left(\lambda-\frac{k}{\alpha}\right), \frac{1}{2}; y\right] \tag{15.10}$$

where

$$\frac{1}{2}\left(\lambda-\frac{k}{\alpha}\right) = -n, \quad \text{i.e.} \quad k^2 = \alpha^2(\lambda+2n)^2 \tag{15.11}$$

and in the second case

$$\psi = (1-y)^{\lambda/2} y^{1/2} F\left[\frac{1}{2}\left(\lambda+1+\frac{k}{\alpha}\right), \frac{1}{2}\left(\lambda+1-\frac{k}{\alpha}\right), \frac{3}{2}; y\right], \tag{15.12}$$

where

$$\frac{1}{2}\left(\lambda+1-\frac{k}{\alpha}\right) = -n, \quad \text{i.e.} \quad k^2 = \alpha^2(\lambda+2n+1)^2. \tag{15.13}$$

Thus the corresponding eigenvalues and eigenfunctions are

$$E_{2n} = \frac{V_0}{\lambda(\lambda-1)}(\lambda+2n)^2, \quad E_{2n+1} = \frac{V_0}{\lambda(\lambda-1)}(\lambda+2n+1)^2, \tag{15.14}$$

$$\begin{aligned} \psi_{2n} &= \cos^\lambda \alpha x . F(\lambda+n, -n, \tfrac{1}{2}; \sin^2 \alpha x), \\ \psi_{2n+1} &= \cos^\lambda \alpha x . \sin \alpha x . F(\lambda+1+n, -n, \tfrac{3}{2}; \sin^2 \alpha x). \end{aligned} \tag{15.15}$$

16. Arguments similar to those used in the preceding problem give $\psi = 0$ for $x = 0$ and for $\alpha x = \pi/2$.

Using the notation

$$V_1 = \frac{\hbar^2\alpha^2}{2m}\eta(\eta-1), \quad V_2 = \frac{\hbar^2\alpha^2}{2m}\lambda(\lambda-1), \quad k^2 = \frac{2m}{\hbar^2}E, \tag{16.1}$$

where $\eta, \lambda > 1$, the corresponding Schrödinger equation with the new variable

$$y = \sin^2 \alpha x \tag{16.2}$$

becomes

$$y(1-y)\frac{d^2\psi}{dy^2} + \left(\frac{1}{2}-y\right)\frac{d\psi}{dy} + \frac{1}{4}\left[\frac{k^2}{\alpha^2} - \frac{\eta(\eta-1)}{y} - \frac{\lambda(\lambda-1)}{1-y}\right]\psi = 0. \tag{16.3}$$

Let us change now to a new function $f(y)$ defined by the relation

$$\psi = y^{\eta/2}(1-y)^{\lambda/2} f(y). \tag{16.4}$$

For $f(y)$ the following equation is obtained

$$y(1-y)\frac{d^2f}{dy^2}+\left[\left(\eta+\frac{1}{2}\right)-y(\eta+\lambda+1)\right]\frac{df}{dy}+\frac{1}{4}\left[\frac{k^2}{\alpha^2}-(\eta+\lambda)^2\right]f=0, \tag{16.5}$$

whose general solution can be written in the form (15.7) with

$$a=\frac{\eta+\lambda}{2}+\frac{k}{2\alpha}, \quad b=\frac{\eta+\lambda}{2}-\frac{k}{2\alpha}, \quad c=\eta+\frac{1}{2}. \tag{16.6}$$

Near $y=1$, i.e. $\alpha x=\pi/2$, both F functions behave like $(1-y)^{1/2-\lambda}$, so that $\psi\sim(1-y)^{(1-\lambda)/2}$. Near $y=0$, i.e. $\alpha x=0$, we have $f=C_1+C_2y^{1-c}=C_1+C_2y^{1/2-\eta}$, and hence $\psi=C_1y^{\eta/2}+C_2y^{(1-\eta)/2}$. Since $\eta>1$, the boundary condition $\psi(0)=0$ requires that $C_2=0$. Thus both boundary conditions will be satisfied if we choose $C_2=0$, and $b=(\eta+\lambda)/2-k/2\alpha=-n$ ($n=0, 1, 2, \ldots$). The energy eigenvalues will thus be given by

$$E_n=\frac{\hbar^2\alpha^2}{2m}(\eta+\lambda+2n)^2,$$

and the corresponding eigenfunctions by

$$\begin{aligned}\psi_n&=y^{\eta/2}(1-y)^{\lambda/2}F(\eta+\lambda+n,\ -n,\ \eta+\tfrac{1}{2};\ y)\\&=\sin^\eta\alpha x\cos^\lambda\alpha x\,F(\eta+\lambda+n,\ -n,\ \eta+\tfrac{1}{2};\ \sin^2\alpha x).\end{aligned}$$

17. Introducing the dimensionless quantities

$$\xi=\frac{x}{x_0}, \quad \lambda=\frac{2E}{\hbar\omega},$$

where $x_0=(\hbar/m\omega)^{1/2}$, the Schrödinger equation for the oscillator

$$\frac{d^2\psi}{dx^2}+\frac{2m}{\hbar^2}\left(E-\frac{m\omega^2}{2}x^2\right)\psi=0$$

becomes

$$\frac{d^2\psi}{d\xi^2}+(\lambda-\xi)^2\psi=0. \tag{17.1}$$

Energy levels are obtained only for those values of the parameter λ for which the solutions of equation (17.1) vanish at infinity. We seek these solutions in the form

$$\psi=\psi_\infty(\xi)\,u(\xi) \tag{17.2}$$

where $\psi_\infty(\xi)=\exp(-\frac{1}{2}\xi^2)$ satisfies (17.1) in the asymptotic regions $\xi\to\pm\infty$, in which (17.1) reduces to

$$\frac{d^2\psi_\infty}{d\xi^2}-\xi^2\psi_\infty=0,$$

and tends toward zero as $\xi\to\pm\infty$.

By substituting (17.2) into (17.1), the following differential equation for $u(\xi)$ is obtained

$$\frac{d^2u}{d\xi^2} - 2\xi\frac{du}{d\xi} + (\lambda - 1)u = 0. \tag{17.3}$$

Looking for a solution of the form

$$u = \sum_{k=0}^{\infty} C_k \xi^k, \tag{17.4}$$

the following recurrence relation for the coefficients C_k is found:

$$C_{k+2} = \frac{2k+1-\lambda}{(k+1)(k+2)} C_k. \tag{17.5}$$

Now, for large values of k, $C_{k+2} \approx (2/k)C_k$. The latter recurrence relation holds also for the coefficients of the series

$$\exp(\xi^2) = \sum_{k=0,\,2,\,4,\,\dots} \frac{1}{\left(\frac{k}{2}\right)!}\xi^k. \tag{17.6}$$

Hence, for the same range of values of k, it follows that as $\xi \to \pm\infty$, $u(\xi)$ behaves like $\exp(\xi^2)$ and the wavefunction ψ diverges like $\exp(\frac{1}{2}\xi^2)$. To avoid this divergence the series (17.4) has to reduce to a polynomial. If $C_n \neq 0$, we can have $C_{k+2} = 0$ for all $k \geqslant n$ only if $\lambda = 2n+1$, from which follows the well-known quantization of the energy of the linear oscillator:

$$E_n = \hbar\omega(n+\tfrac{1}{2}), \qquad n = 0, 1, 2, \dots \tag{17.7}$$

To calculate the corresponding eigenfunctions, the coefficients C_0 and C_1 must be determined. Since, for the harmonic oscillator, $V(x) = V(-x)$, the eigenfunctions corresponding to bound states of energy E_n will be either even or odd, which corresponds either to $C_1 = 0$, $C_0 \neq 0$, or to $C_1 \neq 0$, $C_0 = 0$, respectively. Taking $C_n = 2^n$ (which means in effect specifying C_0 or C_1), the recurrence relation (17.5) with $\lambda = 2n+1$ gives for the solution u the Hermite polynomial

$$H_n(\xi) = (2\xi)^n - \frac{n(n-1)}{1!}(2\xi)^{n-2} + \frac{n(n-1)(n-2)(n-3)}{2!}(2\xi)^{n-4} - \dots . \tag{17.8}$$

The polynomial $H_n(\xi)$ satisfies equation (17.3) with $\lambda = 2n+1$, i.e.

$$\frac{d^2H_n}{d\xi^2} - 2\xi\frac{dH_n}{d\xi} + 2nH_n = 0. \tag{17.9}$$

$H_n(\xi)$ can also be written in the form

$$H_n(\xi) = (-1)^n e^{\xi^2}\frac{d^n e^{-\xi^2}}{d\xi^n}. \tag{17.10}$$

Consequently the time-independent wavefunctions of the oscillator are

$$\psi_n(x) = N_n H_n\left(\frac{x}{x_0}\right) \exp\left[-\frac{1}{2}\left(\frac{x}{x_0}\right)^2\right], \tag{17.11}$$

where the constant N_n is determined from the normalization condition

$$1 = \int_{-\infty}^{+\infty} \psi_n^* \psi_n \, dx = x_0 |N_n|^2 \int_{-\infty}^{+\infty} e^{-\xi^2} H_n^2(\xi) \, d\xi = (-1)^n x_0 |N_n|^2 \int_{-\infty}^{+\infty} H_n(\xi) \frac{d^n e^{-\xi^2}}{d\xi^n} \, d\xi.$$

Integrating by parts n times, we obtain

$$x_0 |N_n|^2 \int_{-\infty}^{+\infty} e^{-\xi^2} \frac{d^n H_n(\xi)}{d\xi^n} \, d\xi = 1.$$

But

$$\frac{d^n H_n(\xi)}{d\xi^n} = 2^n n! \quad \text{and} \quad \int_{-\infty}^{+\infty} e^{-\xi^2} \, d\xi = \sqrt{\pi},$$

so that, with the exception of a possible phase factor,

$$N_n = (2^n n! \sqrt{\pi}\, x_0)^{-1/2},$$

and consequently

$$\psi_n(x) = (2^n n!\, x_0 \sqrt{\pi})^{-1/2} H_n\left(\frac{x}{x_0}\right) \exp\left[-\frac{1}{2}\left(\frac{x}{x_0}\right)^2\right]. \tag{17.12}$$

The orthonormalization condition then holds for these wavefunctions, i.e.

$$\int_{-\infty}^{+\infty} \psi_m^* \psi_n \, dx = \delta_{mn}. \tag{17.13}$$

18. Since the general solution of the equation of motion of a classical oscillator, $\ddot{x} + \omega^2 x = 0$, is of the form $x = C \sin(\omega t + \phi)$, the total energy

$$E_1 = T + V = \frac{m\dot{x}^2}{2} + \frac{m\omega^2}{2} x^2$$

of such an oscillator is given by $E_1 = m\omega^2 c^2/2$.

Since $T \geqslant 0$, we have $E_1 \geqslant V$, which means that, classically, the particle can be found only in the range $-a \leqslant x \leqslant +a$. At the ends of this interval, where $E_1 = V$, its kinetic energy vanishes; the points $x = \pm a$ are called "turning points". Accordingly, $C^2 = a^2 = 2E_1/m\omega^2 = 3\hbar/m\omega$. The classical probability of finding the particle in the interval $(x, x+dx)$

is proportional to the time dt which it takes to pass through this interval. If the period of oscillation is $T = 2\pi/\omega$, then

$$W_{cl}(x)\,dx = 2\frac{dt}{T} = \frac{\omega}{\pi}\frac{dx}{\dot{x}} = \frac{\omega}{\pi}\frac{dx}{a\omega\cos(\omega t+\phi)} = \frac{1}{\pi a}\left(1-\frac{x^2}{a^2}\right)^{-1/2} dx,$$

which is the required expression.

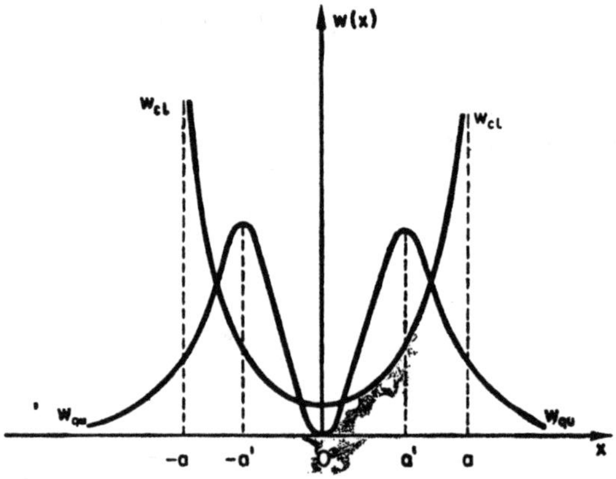

FIG. II.23.

It can be seen that this probability is greatest at the turning points $x = \pm a$ (Fig. II.23). According to quantum mechanics the probability of finding the particle in the interval $(x, x+dx)$ is

$$W_{qu}(x)\,dx = 2\pi^{-1/2}x_0^{-3}x^2\exp\left(-\frac{x^2}{x_0^2}\right)dx.$$

It should be noted that $W_{qu}(x)$ has maxima near the classical turning points $\left(a = \sqrt{3\hbar/m\omega},\ a' = \sqrt{\hbar/m\omega}\right)$, but, in contrast with the classical case, it does not vanish beyond these points. This phenomenon, of the penetration of a particle into regions with "negative kinetic energy" ($|x| > a$), does not lead to any contradiction because the equality $E = T+V$ in quantum mechanics is not a simple relation between numbers, but between operators; the kinetic and the potential energies cannot in fact be determined simultaneously.

For higher levels, it is found that the curve $2W_{cl}(x)$ becomes the envelope of the peaks of $W_{qu}(x)$ in the classical limit $n\to\infty$ (cf. Fig. II.24, which represents $W_{qu}(x) = |\psi_{10}(x)|^2$, $a = \sqrt{21\hbar/m\omega}$).

19. Since the particle cannot penetrate into the range $x < 0$, the eigenfunctions of the corresponding Schrödinger equation have to vanish for $x = 0$. On the other hand, in the range $x > 0$, these eigenfunctions are the same as those of the harmonic oscillator. Hence

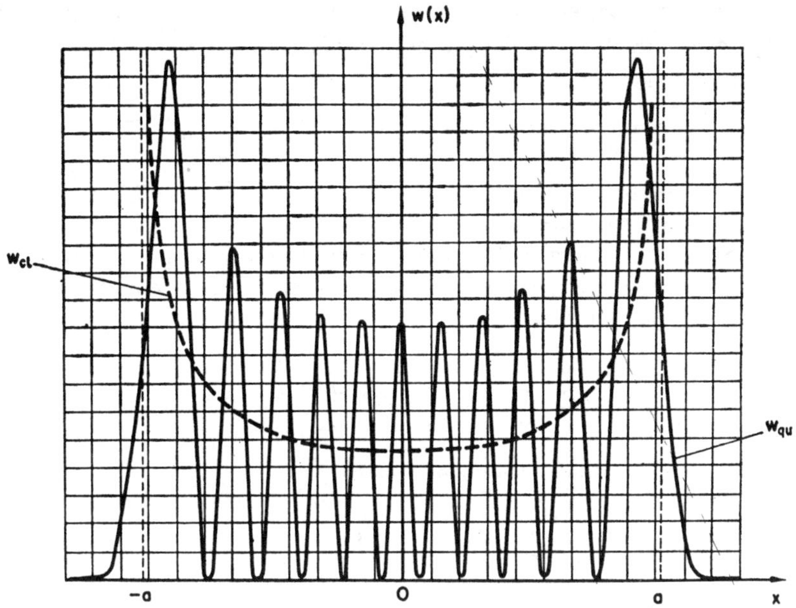

FIG. II.24.

the odd wavefunctions of the oscillator, with $n = 2k+1$, which vanish at $x = 0$, are the solutions of this problem. Therefore

$$E_k = \hbar\omega(2k+\tfrac{3}{2}), \quad k = 0, 1, 2, \ldots$$

20. Since $V(x, y, z) = V_1(x)+V_2(y)+V_3(z)$, this problem reduces to the problem of three independent harmonic oscillators of frequencies $\omega_1, \omega_2, \omega_3$, along the axes x, y, z respectively (see problem 1). Therefore

$$E_{n_1n_2n_3} = \hbar\omega_1(n_1+\tfrac{1}{2})+\hbar\omega_2(n_2+\tfrac{1}{2})+\hbar\omega_3(n_3+\tfrac{1}{2}), \tag{20.1}$$

$$\psi_{n_1n_2n_3}(x, y, z) = \left(\frac{m^3\omega_1\omega_2\omega_3}{\hbar^3\pi^3}\right)^{1/4}\left(\frac{2^{-(n_1+n_2+n_3)}}{n_1!n_2!n_3!}\right)^{1/2}\times$$

$$\times H_{n_1}(\xi_1)\, H_{n_2}(\xi_2)\, H_{n_3}(\xi_3) \exp\left[-\frac{1}{2}(\xi_1^2+\xi_2^2+\xi_3^2)\right], \tag{20.2}$$

where

$$\xi_1 = \left(\frac{m\omega_1}{\hbar}\right)^{1/2} x, \quad \xi_2 = \left(\frac{m\omega_2}{\hbar}\right)^{1/2} y, \quad \xi_3 = \left(\frac{m\omega_3}{\hbar}\right)^{1/2} z$$

and $n_1, n_2, n_3 = 0, 1, 2, \ldots$.

If the ratios of the eigenfrequencies are irrational, the energy levels are non-degenerate, otherwise they may be degenerate. The ground state E_{000} is always non-degenerate.

For the isotropic harmonic oscillator

$$E_n = \hbar\omega(n+\tfrac{3}{2}), \quad \text{where} \quad n = n_1+n_2+n_3 . \tag{20.3}$$

In this case all the energy levels with the exception of E_0 are degenerate. To calculate the degeneracy of the level of energy E_n, consider for the moment a particular value of the quantum number n_1. n_2 can then have any of the values 0, 1, ..., $n-n_1$, and the sum $n = n_1+n_2+n_3$ for given n and n_1 can be obtained in $n-n_1+1$ ways. Since $n_1 = 0, 1, 2, \ldots, n$, the degeneracy of E_n will be

$$\sum_{n_1=0}^{n} (n-n_1+1) = \tfrac{1}{2}(n+1)(n+2). \tag{20.4}$$

21. Since, as $x \to -\infty$, the Schrödinger equation

$$\frac{d^2\psi}{dx^2}+\frac{2m}{\hbar^2}(E+kx)\psi = 0 \tag{21.1}$$

has, for any value of E, a single bounded solution (the one which decreases exponentially as $x \to -\infty$), which, as $x \to +\infty$, oscillates endlessly, it follows that the energy spectrum of the particle in a homogeneous field is continuous and non-degenerate. In other words, to each energy value, in the range $-\infty < E < +\infty$, there corresponds a single solution, and this describes a motion of the particle which is limited in the negative direction of x and unlimited in the positive direction.

Introducing the dimensionless variable

$$y = \left(x+\frac{E}{k}\right)\left(\frac{2mk}{\hbar^2}\right)^{1/3}, \tag{21.2}$$

equation (21.1) becomes

$$\frac{d^2\psi}{dy^2}+y\psi = 0. \tag{21.3}$$

This equation does not contain the energy as a parameter. Therefore, after obtaining its correctly bounded solution, we can readily find the eigenfunction corresponding to any arbitrary value of the energy. The solution of (21.3), finite for any x, has the form

$$\psi = NA(-y), \tag{21.4}$$

where

$$A(y) = \frac{1}{\sqrt{\pi}}\int_0^\infty \cos\left(\frac{u^3}{y}+uy\right) du \tag{21.5}$$

is Airy's function, and $N = (2m\pi^{-3/2}k^{-1/2}\hbar^{-2})^{1/3}$ is a normalization factor (see problem 18, Chapter V). Thus, the wavefunction of a particle with energy E will be

$$\psi_E(x, t) = NA(-y)\, e^{-\frac{i}{\hbar}Et}, \tag{21.6}$$

where y is given by (21.2) above.

22. By making the substitution

$$\psi(x, t) = \psi(x)\, e^{-\frac{i}{\hbar}Et} \tag{22.1}$$

in the Schrödinger equation

$$i\hbar \frac{\partial \psi(x, t)}{\partial t} = \left[-\frac{\hbar^2}{2m}\frac{\partial^2}{\partial x^2} + V(x)\right]\psi(x, t),$$

we obtain, for $x < 0$,

$$\frac{d^2\psi_1}{dx^2} + \frac{2m}{\hbar^2}(E+V_0)\,\psi_1 = 0,$$

and, for $x > 0$,

$$\frac{d^2\psi_2}{dx^2} + \frac{2m}{\hbar^2}E\psi_2 = 0.$$

The general solutions of these equations are

$$\psi_1 = Ae^{\frac{i}{\hbar}qx} + Be^{-\frac{i}{\hbar}qx}, \quad q = [2m(E+V_0)]^{1/2}, \quad x < 0,$$

$$\psi_2 = Ce^{\frac{i}{\hbar}px} + De^{-\frac{i}{\hbar}px}, \quad p = \sqrt{2mE}, \quad x > 0.$$

According to classical mechanics, if $E > 0$, the electron has sufficient energy to overcome the potential barrier at the surface of separation and hence it will leave the metal. In the quantum-mechanical treatment the answer is not so simple.

The electron wavefunction is

$$\psi(x, t) = \begin{cases} Ae^{\frac{i}{\hbar}(qx-Et)} + Be^{-\frac{i}{\hbar}(qx+Et)}, & \text{if} \quad x < 0 \\ Ce^{\frac{i}{\hbar}(px-Et)} + De^{-\frac{i}{\hbar}(px+Et)}, & \text{if} \quad x > 0. \end{cases}$$

The term with the coefficient A represents a plane wave which arrives at the surface from the left (incident particle), the term with B represents the reflected wave, the term with C represents the transmitted wave and that with D represents a wave arriving at the surface from the right. Since such a wave does not exist under the conditions of this problem we put $D = 0$. The continuity conditions at $x = 0$ then yield the equations

$$A+B = C, \quad q(A-B) = pC, \tag{22.2}$$

whence

$$B = \frac{q-p}{q+p}A, \quad C = \frac{2q}{q+p}A.$$

It can be seen that the certainty of transmission (which would correspond to a total lack of reflection, i.e. to $B = 0$) prescribed by classical mechanics, occurs only in the trivial case $q = p$, i.e. $V_0 = 0$. Since

$$|\mathbf{j}_A| = |A|^2\frac{q}{m}, \quad |\mathbf{j}_B| = |B|^2\frac{q}{m}, \quad |\mathbf{j}_C| = |C|^2\frac{p}{m},$$

we obtain in fact

$$T = \frac{|\mathbf{j}_C|}{|\mathbf{j}_A|} = \left|\frac{C}{A}\right|^2\frac{p}{q} = \frac{4qp}{(q+p)^2} = \frac{4\sqrt{E(E+V_0)}}{(\sqrt{E+V_0}+\sqrt{E})^2}, \tag{22.3}$$

$$R = \frac{|\mathbf{j}_B|}{|\mathbf{j}_A|} = \left|\frac{B}{A}\right|^2 = \left(\frac{q-p}{q+p}\right)^2 = \frac{V_0^2}{(\sqrt{E+V_0}+\sqrt{E})^4}, \tag{22.4}$$

$$T+R = 1.$$

Note that, according to quantum mechanics, reflection occurs with a probability different from zero even if $E > 0$. However, if $E \gg V_0$ the reflection probability decreases rapidly with increasing energy:

$$R \approx \frac{V_0^2}{16E^2}. \tag{22.5}$$

On the other hand, if $0 < E \ll V_0$, we have the approximate formula

$$R \approx 1-4\sqrt{\frac{E}{V_0}}. \tag{22.6}$$

For the commonest metals, $V_0 \sim 10$ eV, and the reflection probability for an electron with $E = 0{\cdot}1$ eV is approximatively 60%.

If $-V_0 < E < 0$ the total energy of the electron is not sufficient for it to leave the metal, according to classical mechanics, so that we should have $T = 0$ and $R = 1$.

In fact, in this case, $q = [2m(V_0-|E|)]^{1/2}$, $p = i(2m|E|)^{1/2}$, and the solution bounded in the region $x > 0$ is

$$\psi_2 = Ce^{-(x/2d)} \quad \text{where} \quad d = \hbar(8m|E|)^{-1/2}.$$

The continuity conditions at $x = 0$ give

$$A+B = C, \quad \frac{i}{\hbar}q(A-B) = -\frac{1}{2d}C,$$

whence

$$\frac{B}{A} = -\frac{1+\dfrac{2i}{\hbar}qd}{1-\dfrac{2i}{\hbar}qd}, \quad \frac{C}{A} = -2\frac{\dfrac{2i}{\hbar}qd}{1-\dfrac{2i}{\hbar}qd}.$$

It follows from the expression for ψ_2 that $\mathbf{j}_C = 0$, and hence $T = 0$ and $R = |B/A|^2 = 1$. Thus, as in classical theory, an electron having a total energy smaller than the potential barrier height will be reflected with certainty. A new result, however, is that the probability of finding the electron outside the metal ($x > 0$) is different from zero, since

$$|\psi_2(x)|^2 = 4|A|^2\left(1+\frac{\hbar^2}{4q^2d^2}\right)^{-1} e^{-x/d} = 4|A|^2\frac{V_0-|E|}{V_0}e^{-x/d}.$$

This phenomenon is similar to that of the "total internal reflection" which occurs when the passage of light from a denser medium to a less dense one is impossible because the angle of incidence exceeds a certain critical angle. The theory of wave optics shows that, in the less dense medium, there is then a wave whose amplitude decreases exponentially, in analogy with the exponentially decreasing electron wave considered above.

23. Since the electrons, with momentum $p = \sqrt{2mE}$, encounter a potential drop $-V_0$ at the metal surface, they will all enter the metal, according to classical mechanics, and, because of the law of conservation of energy, they will acquire a final momentum $q = [2m(E+V_0)]^{1/2}$ after doing so.

According to quantum mechanics, on the other hand, some of the electrons may be reflected by the metal surface. Using the notation of the preceding problem, $D\exp(-i/\hbar px)$ now represents the incident wave, $C\exp(i/\hbar px)$ the reflected wave, and $B\exp(-i/\hbar qx)$ the transmitted wave inside the metal. In this case, $A = 0$. The continuity conditions at the point $x = 0$ give $C+D = B$, $p(C-D) = -qB$, whence

$$R = \frac{|\mathbf{j}_C|}{|\mathbf{j}_D|} = \left|\frac{C}{D}\right|^2 = \left(\frac{p-q}{p+q}\right)^2$$

and

$$T = \frac{|\mathbf{j}_B|}{|\mathbf{j}_D|} = \left|\frac{B}{D}\right|^2\frac{q}{p} = \frac{4pq}{(p+q)^2}.$$

Thus, if $E = 0{\cdot}1$ eV and $V_0 = 8$ eV, the reflection probability is

$$R = \left(\frac{\sqrt{E}-\sqrt{E+V_0}}{\sqrt{E}+\sqrt{E+V_0}}\right)^2 = \left(\frac{1-\sqrt{1+\dfrac{V_0}{E}}}{1+\sqrt{1+\dfrac{V_0}{E}}}\right)^2 = \left(\frac{8}{10}\right)^2 = 0{\cdot}64,$$

which is greater than the probability (0·36) of entering the metal. The higher the energy of the incident electrons, however, the less probable is their reflection. For example, the probability of reflection of an electron bombarding an anticathode in the usual Röntgen X-ray tube ($V_0 \approx 10$ eV, $E = 10^5$ eV) is approximately $6{\cdot}2 \times 10^{-10}$.

For greater familiarity with the phenomena of electron reflection at, and penetration into, metals, we suggest that the reader make a plot of the quantities R and T as functions of E.

24. By introducing the new variable

$$y = (1+e^{x/a})^{-1}, \tag{24.1}$$

the Schrödinger equation

$$\frac{d^2\psi}{dx^2}+\frac{2m}{\hbar^2}\left(E+\frac{V_0}{1+e^{x/a}}\right)\psi = 0 \tag{24.2}$$

becomes

$$y(y-1)\frac{d^2\psi}{dy^2}+(1-2y)\frac{d\psi}{dy}+\left(\frac{\eta^2}{y(1-y)}-\frac{\lambda^2}{y}\right)\psi = 0, \tag{24.3}$$

with the following notation:

$$\frac{2m(E+V_0)}{\hbar^2}a^2 = k^2a^2 = \eta^2; \qquad \frac{2mV_0}{\hbar^2}a^2 = \lambda^2. \tag{24.4}$$

Equation (24.3) has non-essential singularities at $y = 0$, 1 and ∞, and we accordingly introduce a new function $f(y)$ through the relation

$$\psi = y^\nu(1-y)^\mu f(y). \tag{24.5}$$

If we impose on ν and μ the conditions

$$\nu^2 = \lambda^2-\eta^2 \quad \text{and} \quad \mu^2 = -\eta^2, \tag{24.6}$$

the following differential equation for f is obtained

$$y(1-y)f''+[(2\nu+1)-(2\nu+2\mu+2)y]f'-(\mu+\nu)(\mu+\nu+1)f = 0. \tag{24.7}$$

A particular solution of this equation is the hypergeometric function

$$f = F(\mu+\nu,\ \mu+\nu+1,\ 2\nu+1;\ y). \tag{24.8}$$

Now, for

$$\psi = y^\nu(1-y)^\mu F(\mu+\nu,\ \mu+\nu+1,\ 2\nu+1;\ y) \tag{24.9}$$

to be a physically acceptable solution of equation (24.2), it has to satisfy the appropriate conditions as $x \to \pm\infty$. This can be achieved by establishing correctly the signs of ν and μ, which are not specified by (24.6).

If $-V_0 < E < 0$, we have $\lambda^2 > \eta^2$ and so ν is real. As $x \to +\infty$, $y \sim e^{-x/a} \to 0$, $F \to 1$ and $\psi \sim y \sim e^{-x/a}$. Thus the solution (24.9) vanishes as $x \to +\infty$, if we choose $\nu > 0$. If $E > 0$, we have $\lambda^2 < \eta^2$, and hence ν is pure imaginary, say $\nu = -i\sigma$. Then, as $x \to +\infty$, $\psi \sim y^{-i\sigma} \sim e^{i(x/a)\sigma} = e^{iKx}$, say, and represents a plane wave travelling in the positive direction of x, with a wave number K given by

$$K = \frac{\sigma}{a} = \frac{1}{a}\left(\frac{2m}{\hbar^2}E\right)^{1/2}. \tag{24.10}$$

As $x \to -\infty$, $1-y \sim e^{x/a} \to 0$. Using the transformation (A.54),

$$F(\mu+\nu, \mu+\nu+1, 2\nu+1; y) = \frac{\Gamma(2\nu+1)\Gamma(-2\mu)}{\Gamma(\nu-\mu)\Gamma(\nu-\mu+1)} F(\mu+\nu, \mu+\nu+1, 2\mu+1; 1-y)$$

$$+(1-y)^{-2\mu}\frac{\Gamma(2\nu+1)\Gamma(2\mu)}{\Gamma(\nu+\mu)\Gamma(\nu+\mu+1)} F(\nu-\mu, \nu-\mu+1, -2\mu+1; 1-y), \tag{24.11}$$

we have, as $x \to -\infty$,

$$\psi \sim \frac{\Gamma(2\nu+1)\Gamma(-2\mu)}{\Gamma(\nu-\mu)\Gamma(\nu-\mu+1)}(1-y)^{\mu} + \frac{\Gamma(2\nu+1)\Gamma(2\mu)}{\Gamma(\nu+\mu)\Gamma(\mu+\nu+1)}(1-y) \quad . \tag{24.12}$$

Choosing $\mu = i\eta$, in the asymptotic region $x \to -\infty$ we have $(1-y)^{\mu} \sim e^{i(x/a)\eta} = e^{ikx}$ say, and thus the first term of (24.12) represents a plane wave incident from the left, and the second term represents a plane wave reflected at the surface of the metal, thus

$$\psi \sim A_I e^{ikx} + A_R e^{-ikx} \qquad (x \to -\infty).$$

The reflection coefficient is then

$$R = \left|\frac{A_R}{A_I}\right|^2 = \left|\frac{\Gamma(2\mu)\Gamma(\nu-\mu)\Gamma(\nu-\mu+1)}{\Gamma(-2\mu)\Gamma(\nu+\mu)\Gamma(\mu+\nu+1)}\right|^2. \tag{24.13}$$

Since, if $-V_0 < E < 0$, ν is real and μ is pure imaginary, we have

$$[\Gamma(2\mu)\Gamma(\nu+\mu)\Gamma(\mu+\nu+1)]^* = \Gamma(2\mu)\Gamma(\nu-\mu)\Gamma(\nu-\mu+1)$$

and thus $R = 1$, i.e. there is total reflection of the electrons.

If $E > 0$ both μ and ν are pure imaginary, $\mu = i\eta$ and $\nu = -i\sigma$, say. In this case $\Gamma^*(2\mu) = \Gamma(-2\mu)$ and, using the well-known formula $\Gamma(z+1) = z\Gamma(z)$, we find that

$$R = \left|\frac{(\nu+\mu)\,|\Gamma(\nu-\mu+1)|^2}{(\nu-\mu)\,|\Gamma(\nu+\mu+1)|^2}\right|^2 = \left(\frac{\eta-\sigma}{\eta+\sigma}\right)^2\left|\frac{|\Gamma(1-i(\eta+\sigma))|^2}{|\Gamma(1+i(\eta-\sigma))|^2}\right|^2. \tag{24.14}$$

Bearing (A.40) in mind, we obtain finally the simple formula

$$R = \left[\frac{\sinh \pi(\eta-\sigma)}{\sinh \pi(\eta+\sigma)}\right]^2 = \left[\frac{\sinh \pi a(k-K)}{\sinh \pi a(k+K)}\right]^2. \tag{24.15}$$

In the limiting case $a \to 0$ we obtain once again the result of problem 22, viz.,

$$R = \left(\frac{k-K}{k+K}\right)^2 = \left(\frac{\sqrt{E+V_0}-\sqrt{E}}{\sqrt{E+V_0}+\sqrt{E}}\right)^2.$$

25. (1) $E > V_0$.
The general solution of the Schrödinger equation is

$$\psi(x) = \begin{cases} Ae^{\frac{i}{\hbar}px}+Be^{-\frac{i}{\hbar}px}; & x<0 \\ Ge^{\frac{i}{\hbar}qx}+Fe^{-\frac{i}{\hbar}qx}; & 0<x<a \\ Ce^{\frac{i}{\hbar}px}+De^{-\frac{i}{\hbar}px}; & x>a, \end{cases}$$

where $p = (2mE)^{1/2}$, $q = [2m(E-V_0)]^{1/2}$.

If a particle arrives at the barrier from the left, the terms with the coefficients A, B and C represent respectively the incident, the reflected, and the transmitted waves. The continuity conditions at $x = 0$ and at $x = a$ give

$$A+B = G+F; \qquad p(A-B) = q(G-F);$$

$$Ge^{\frac{i}{\hbar}qa}+Fe^{-\frac{i}{\hbar}qa} = Ce^{\frac{i}{\hbar}pa}; \qquad q\left(Ge^{\frac{i}{\hbar}qa}-Fe^{-\frac{i}{\hbar}qa}\right) = pCe^{\frac{i}{\hbar}pa}.$$

If we eliminate G and F from these equations, we obtain

$$\frac{B}{A} = \frac{(p^2-q^2)\left(1-e^{2\frac{i}{\hbar}qa}\right)}{(p+q)^2-(p-q)^2e^{2\frac{i}{\hbar}qa}}, \tag{25.1}$$

$$\frac{C}{A} = \frac{4pqe^{\frac{i}{\hbar}(q-p)a}}{(p+q)^2-(p-q)^2e^{2\frac{i}{\hbar}qa}} \tag{25.2}$$

whence

$$T = \left|\frac{C}{A}\right|^2 = \frac{4E(E-V_0)}{V_0^2\sin^2\frac{qa}{\hbar}+4E(E-V_0)} \tag{25.3}$$

and

$$R = \left|\frac{B}{A}\right|^2 = \frac{V_0^2\sin^2\frac{qa}{\hbar}}{V_0^2\sin^2\frac{qa}{\hbar}+4E(E-V_0)}. \tag{25.4}$$

Note that reflection occurs with a non-vanishing probability. If the energy of the particle equals the height of the potential barrier, i.e. if $E = V_0$, we have

$$T_0 = \left(1+\frac{ma^2V_0}{2\hbar^2}\right)^{-1} \quad \text{and} \quad R_0 = \left(1+\frac{2\hbar^2}{ma^2V_0}\right)^{-1}.$$

The barrier becomes completely transparent ($R = 0$, $T = 1$) if $\sin qa/\hbar = 0$, i.e. if $qa/\hbar = n\pi$, $n = 1, 2, 3, \ldots$.

This happens when stationary waves are formed inside the barrier, i.e. when

$$a = n\frac{h}{2q} = n\frac{\lambda}{2}, \quad n = 1, 2, 3, \ldots .$$

Thus the passage of particles through rectangular barriers leads to resonance phenomena of a type unknown in classical physics (Fig. II.25).

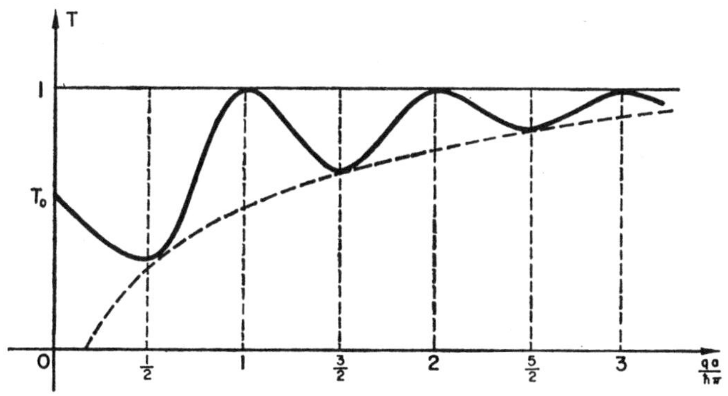

FIG. II.25.

(2) $0 < E < V_0$.

In this case, according to classical mechanics, the particle is reflected with certainty (to be inside the barrier, the particle would need to have a negative kinetic energy). Quantum mechanics puts in evidence here a new phenomenon. Because $E < V_0$, q is pure imaginary, since $q = i[2m(V_0-E)]^{1/2} = i(\hbar/2d)$ say, where $d = \hbar[8m(V_0-E)]^{-1/2}$.

Then

$$\psi(x) = \begin{cases} Ae^{\frac{i}{\hbar}px} + Be^{\frac{-i}{\hbar}px}, & x < 0 \\ Ge^{-x/2d} + Fe^{x/2d}, & 0 < x < a \\ Ce^{\frac{i}{\hbar}px}, & x > a. \end{cases}$$

Substituting $q = i(\hbar/2d)$ in (25.3) and (25.4), we obtain for R and T the following expressions

$$T = \frac{4E(V_0-E)}{V_0^2 \sinh^2 \frac{a}{2d} + 4E(V_0-E)}, \tag{25.5}$$

$$R = \frac{V_0^2 \sinh^2 \frac{a}{2d}}{V_0^2 \sinh^2 \frac{a}{2d} + 4E(V_0-E)}. \tag{25.6}$$

Since $T \neq 0$, the particle has a certain probability of passing through the barrier even if, classically, its energy is not sufficient for it to do so. This phenomenon is called the "tunnel effect". It is important only when $a/2d \sim 1$. If $d \ll a$, the transmission probability, and thus the tunnel effect, decreases rapidly according to the approximate formula

$$T \approx 16 \frac{E(V_0-E)}{V_0^2} e^{-a/d}. \tag{25.7}$$

In the classical limit $\hbar \to 0$, T tends towards zero. Thus the tunnel effect is a purely quantum-mechanical phenomenon.

26. Since the system is conservative, the particle passes through the barrier with constant total energy. The transmission probabilities T_i through the hatched regions will therefore be approximately independent, which means that the transmission coefficient through the whole barrier will be given by the product

$$T = \prod_i T_i = \exp\left[-\frac{1}{\hbar}\sum_i \sqrt{8m(V(x_i)-E)}\, \Delta x_i\right]. \tag{26.1}$$

Increasing the subdivision of the interval x_1, x_2 as far as is allowed by the condition $d_i \ll \Delta x_i$, the approximate formula

$$T = \exp\left[-\frac{1}{\hbar}\int_{x_1}^{x_2} \sqrt{8m(V(x)-E)}\, dx\right] \tag{26.2}$$

is obtained; this expression is known as "Gamow's penetrability factor".

27. The total force which acts on an electron outside the metal is the sum of the force $e\mathscr{E}$ due to the external applied field and of the "electrical image" force. The latter is due to the positive charges that the electron induces on the surface of the metal, through electrostatic forces. These charges attract the electron exactly as if there existed inside the metal, at a distance x from its surface, a charge $+|e|$, the "electrical image" of the electron. The total force on the electron is thus given by $F = e\mathcal{E} - e^2/4x^2$ and the potential energy of the electron in the field of this force is

$$V(x) = -e\mathcal{E}x - \frac{e^2}{4x}. \tag{27.1}$$

Note that $V(x)$ reaches its maximum value $V_m = -(e^3\mathcal{E})^{1/2}$ at $x = x_m = \frac{1}{2}(e/\mathcal{E})^{1/2}$ (Fig. II.26). Thus we see that under the influence of the external field the potential barrier height is decreased by $(e^3\mathcal{E})^{1/2}$. According to classical mechanics, electrons which have an energy E

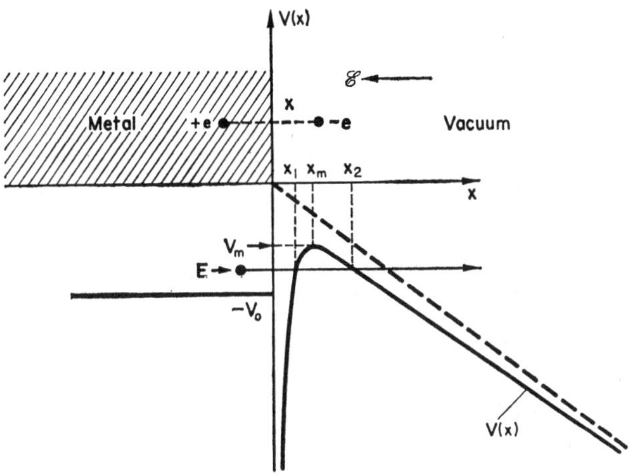

FIG. II.26.

in the range of $-V_0 < E < V_m$ cannot come out of the metal, while those which have an energy in the range $V_m < E < 0$ can (and do) come out. In the presence of the field, as we have seen, V_m decreases as the field increases. The current intensity therefore increases with the applied field, a deduction which is in qualitative agreement only with the experimental facts. According to quantum mechanics, and taking into account the tunnel effect, electrons having total energy E in the interval $-V_0 < E < V_m$ may pass through the potential barrier $V(x)$. Gamow's factor in this case is

$$T = \exp\left\{ -\frac{1}{\hbar} \int_{x_1}^{x_2} \sqrt{8m(|E| - e\mathcal{E}x - e^2/4x)}\, dx \right\} \tag{27.2}$$

where

$$x_1, x_2 = \frac{|E| \pm \sqrt{E^2 - e^3\mathcal{E}}}{2e\mathcal{E}}$$

By introducing a new variable $z = \frac{e\mathcal{E}}{|E|} x$, we obtain

$$T = \exp\left[-\frac{\sqrt{8m|E|^3}}{e\hbar\mathcal{E}}\phi(y)\right] \tag{27.3}$$

where

$$\phi(y) = \int_{z_1}^{z_2} \sqrt{1-\frac{y^2}{z}-z}\, dz, \quad y = \frac{\sqrt{e^2\mathcal{E}}}{2|E|}, \quad z_{1,2} = \frac{e\mathcal{E}}{|E|} x_{1,2} = \frac{1 \pm \sqrt{1-4y^2}}{2} \tag{27.4}$$

Denoting by J_0 the flux of electrons incident on the metal surface from inside, the "cold emission" current intensity is given by

$$J(\mathcal{E}) = J_0\bar{T} = J_0 \overline{\exp\left[-\frac{\sqrt{8m|E|}}{e\hbar\mathcal{E}}\phi(y)\right]}, \tag{27.5}$$

where the bar denotes a statistical average over the energies E of the conduction electrons (at a given temperature). This formula is valid only if the average width of the barrier is much greater than the relative penetration depth of the electrons, i.e. if, for all significant values of E,

$$\frac{\sqrt{8m|E|^3}}{e\hbar\mathcal{E}}\phi(y) \gg 1, \tag{27.6}$$

which corresponds to $\bar{T} \ll 1$.

If we neglect the action of the electrical image force, i.e. if we take $y = 0$, the elliptic integral reduces to a well-known integral having the value $\phi(0) = \frac{2}{3}$, and (27.5) becomes

$$J(\mathcal{E}) = J_0 \overline{\exp\left[-\frac{4}{3}\frac{\sqrt{2m|E|^3}}{e\hbar\mathcal{E}}\right]}. \tag{27.7}$$

A dependence of the form (27.5) or (27.7) of the cold emission current on the applied field $\mathcal{E}$, in the range in which (27.6) holds, has been verified experimentally.

28. Using the procedure of problem 26, we obtain

$$T \equiv G = \exp\left[-\frac{1}{\hbar}\int_{R_0}^{R} \sqrt{8m(V(r)-E)}\, dr\right], \tag{28.1}$$

where

$$e_1e_2/R = E. \tag{28.2}$$

Changing the variable r to $R\cos^2 x$ we have

$$G = \exp\left[-\frac{\sqrt{8mE}}{\hbar}\int_{R_0}^{R}\sqrt{\frac{R}{r}-1}\, dr\right] = \exp\left[-\frac{\sqrt{32mE}}{\hbar} R \int_0^{\arccos(R_0/R)} \sin^2 x\, dx\right]$$

$$= \exp\left[-\frac{\sqrt{8me_1e_2R}}{\hbar}\left(\arccos\sqrt{\frac{R_0}{R}} - \sqrt{\frac{R_0}{R}}\sqrt{1-\frac{R_0}{R}}\right)\right].$$

Since $E \ll V(R_0)$, i.e. $R_0 \ll R$, we can neglect terms of higher order than the first in $(R_0/R)^{1/2}$, and obtain

$$G = \exp\left[-\frac{\sqrt{8me_1e_2}}{\hbar}\left(\frac{\pi}{2}\sqrt{R}-\sqrt{R_0}\right)\right] = \exp\left[\frac{\sqrt{8me_1e_2R_0}}{\hbar}-\frac{1}{v}\frac{2\pi e_1e_2}{\hbar}\right], \quad (28.3)$$

in which we have put $E = e_1e_2/R = mv^2/2$, v being the velocity of the α-particle at a large distance r from the nucleus for which $V(r) \approx 0$.

Using G, the mean life τ of an α-emitting nucleus can be estimated by the following intuitive argument. Picture the α-particle as a classical particle moving freely inside the nucleus with velocity v_i and kinetic energy

$$\tfrac{1}{2}mv_i^2 = E+V_0 \approx V_0. \quad (28.4)$$

Since the α-particle is held inside the nucleus of radius R_0 by a potential barrier it will "collide" with this barrier $v_i/2R_0$ times per unit time. Now the probability per collision of penetrating the barrier is G, and hence the probability per unit time of the α-particle leaving the nucleus (= the inverse of the mean life τ) is

$$\frac{v_i}{2R_0}G = \frac{1}{\tau} \quad (28.5)$$

Using (28.3)–(28.5) we obtain finally

$$\ln\tau = \left\{\frac{1}{2}\ln\frac{2R_0^2m}{V_0}-\sqrt{\frac{8me_1e_2R_0}{\hbar^2}}\right\}+\frac{2\pi e_1e_2}{\hbar}\frac{1}{v}. \quad (28.6)$$

This relation gives an estimate of the mean life τ of an α-emitting nucleus in terms of the α-particle velocity v at large distances and the nuclear radius R_0. It is in good agreement with the observed data.

29. For a free particle, $V(r) = 0$ in the whole range $(0, \infty)$, and the Hamiltonian reduces to the kinetic energy term $H = \mathbf{p}^2/2m$. The simultaneous eigenfunction of the observables H, $\mathbf{l}^2$ and l_z, with energy eigenvalue $E = \hbar^2k^2/2m$ and angular momentum quantum numbers (l, m), can be written as $f_l(kr)Y_l^m(\theta, \phi)$, where $f_l(kr) = \dfrac{R_l(r)}{kr}$ is the solution bounded in the interval $(0, \infty)$ of equation (A.46),

$$\left[\frac{d^2}{d\varrho^2}+\frac{2}{\varrho}\frac{d}{d\varrho}+1-\frac{l(l+1)}{\varrho^2}\right]f_l(\varrho) = 0, \quad \varrho = kr. \quad (29.1)$$

Now the general solution of (29.1) is a linear combination of spherical Bessel functions, (A.45), but the only solution which is non-singular at the origin is $j_l(\varrho)$, which behaves like ϱ^l for small ϱ; the other three solutions have a pole of order $l+1$ (A.49). Their asymptotic behaviour is given by (A.48).

For $E < 0$ there is only one solution bounded at infinity; it is $h_l^{(+)}(i\alpha r)$, $\alpha = \left(\frac{2m}{\hbar^2}|E|\right)^{1/2}$ and it has a pole of order $l+1$ at the origin. Therefore we get an expected result, that there are no negative energy eigenstates for a free particle.

If $E > 0$ there is only one function bounded everywhere, namely, $j_l(kr)$. Hence for any positive value of the energy an eigenfunction of the form

$$Y_l^m(\theta, \phi)\, j_l(kr) \tag{29.2}$$

exists and corresponds to the state of angular momentum (l, m).

The ensemble of these spherical waves, when l and m take all their possible discrete values, and $0 < k < +\infty$, and each wave is suitably normalized, forms a complete orthonormal set of functions.

The simultaneous eigenfunctions of the observable H, p_x, p_y, p_z are planes waves, $\exp(i\mathbf{k}.\mathbf{r})$, which form (when suitably normalized) another complete orthonormal set. They describe states of linear momentum $\mathbf{p} = \hbar\mathbf{k}$ and energy $E = \hbar^2k^2/2m$ of the free particle, but do not have well defined angular momenta; conversely, the spherical waves (29.2) do not correspond to states of well defined linear momentum. This result, according to which the linear momentum and the angular momentum of a free particle cannot be specified at the same time, is only to be expected, since the operators $p_x, p_y, p_z, \mathbf{l}^2, l_z$ do not all commute with one another.

It can be seen that the energy eigenvalues of a free particle are infinitely degenerate. Because the spherical waves (29.2) form a complete set, the denumerable set of spherical waves corresponding to a given value of k spans the space of all eigenfunctions of given energy $E = \hbar^2k^2/2m$. It follows that a plane wave $\exp(i\mathbf{k}.\mathbf{r})$ can be expanded in a series of these spherical waves (see (A.50) and (A.50′)).

30. Let E be the energy of the particle. Using the notation

$$K = \left[\frac{2m}{\hbar^2}(E+V_0)\right]^{1/2}, \quad \varrho = Kr, \quad R_l(r) = \varrho f_l(\varrho),$$

the radial equation (II.15) for $0 < r < a$ becomes the equation of the spherical Bessel functions (A.46),

$$\left[\frac{d^2}{d\varrho^2} + \frac{2}{\varrho}\frac{d}{d\varrho} + 1 - \frac{l(l+1)}{\varrho^2}\right] f_l(\varrho) = 0. \tag{30.1}$$

The only solution finite at the origin is $Aj_l(Kr)$, A being a normalization constant. Outside the well, $r > a$, equation (30.1) becomes (29.1) written for a free particle with $\varrho = kr = \left(\frac{2m}{\hbar^2}E\right)^{1/2} r$.

In order to study the energy spectrum, two cases must be distinguished:

(1) $E < 0$.

The only solution bounded at infinity is the decreasing exponential $Bh_l^{(+)}(i\alpha r)$, where $\alpha = \left(\frac{2m}{\hbar^2}|E|\right)^{1/2}$. The continuity condition of the radial function at $r = a$ gives the ratio B/A, and that of the logarithmic derivative

$$\frac{1}{h_l^{(+)}(i\alpha r)}\frac{d}{dr}[h_l^{(+)}(i\alpha r)]\bigg|_{r=a} = \frac{1}{j_l(Kr)}\frac{d}{dr}[j_l(Kr)]\bigg|_{r=a} \tag{30.2}$$

determines the particle's possible energies for a given l. Relation (30.2) holds for certain discrete values of $E(< 0)$ only, which are the energy levels of the bound states.

For s-states ($l = 0$), using (A.47), (30.2) reduces to

$$-\alpha a = Ka \cot Ka, \tag{30.3}$$

which is clearly the same as (10.1). To solve it we can use the graphical procedure described in problem 9. The condition for the existence of at least one bound state is (10.2).

For higher values of the orbital quantum number l, the determination of the energy levels requires the solution of more complicated transcendental equations. For example, the possible energy values of a particle in a p-state ($l = 1$) are obtained from the equation

$$\frac{1}{X}\cot X - \frac{1}{X^2} = \frac{1+Y}{Y^2}, \tag{30.4}$$

where $X = Ka$, $Y = \alpha a$.

In general, the energies of states of given angular momentum l can be determined from the intersections of the curves (30.2), writing $X = Ka$, $Y = \alpha a$ as before, with the circle

$$X^2+Y^2 = \frac{2m}{2}a^2V_0. \tag{30.5}$$

The number of these states is always finite, and depends only on the product a^2V_0. If the potential has no bound states for $l = 0$, there will be no bound states for $l = 1, 2, \ldots$, either.

(2) $E > 0$.

Outside the well the particle moves freely. The solution of the radial equation (29.1) is bounded in the whole range $r > a$ and can be expressed as a linear combination of $j_l(kr)$ and $n_l(kr)$.

Putting it in the form

$$B[\cos\delta_l . j_l(kr)+\sin\delta_l . n_l(kr)], \tag{30.6}$$

the ratio B/A of the coefficients can be determined from the continuity condition at $r = a$, and the real quantities δ_l from the continuity condition of the logarithmic derivative

$$K\frac{j_l'(Ka)}{j_l(Ka)} = k\frac{\cos\delta_l . j_l'(ka)+\sin\delta_l . n_l'(ka)}{\cos\delta_l . j_l(ka)+\sin\delta_l . n_l(ka)}. \tag{30.7}$$

Thus, to each energy value $E = \frac{\hbar^2}{2m} k^2$, there corresponds one wavefunction only (to a constant factor).

Using (A.48) we can see that, as $r \to \infty$, (30.6) behaves like

$$B \frac{\sin (kr - l\pi/2 + \delta_l)}{kr}. \tag{30.8}$$

δ_l is called the "phase shift" of the spherical waves with angular momentum l.

31. Introducing the dimensionless variable

$$\xi = \frac{1}{\hbar} \sqrt{2mB} r^2,$$

the radial equation (II.15) for the potential (31a) becomes

$$\xi \frac{d^2 R_l}{d\xi^2} + \frac{1}{2} \frac{dR_l}{d\xi} + \frac{1}{4} \left(\alpha - \frac{\beta}{\xi} - \xi \right) R_l = 0, \tag{31.1}$$

where

$$\alpha = \sqrt{\frac{2m}{B}} \frac{E}{\hbar}; \quad \beta = l(l+1) + \frac{2mA}{\hbar^2}. \tag{31.2}$$

Since, as $\xi \to \infty$, the bounded solution behaves like $\exp\left(-\frac{1}{2}\xi\right)$, and since $\xi = 0$ is a singularity of (31.1), we look for a solution in the form

$$R_l = \xi^{\mu} e^{-(\xi/2)} u_l(\xi), \tag{31.3}$$

in which, on account of condition (II.16), μ has to be positive. Substituting (31.3) in (31.1) and taking

$$\mu = \frac{1}{4} \left[1 + \sqrt{(2l+1)^2 + \frac{8mA}{\hbar^2}} \right] \tag{31.4}$$

we get for u_l' the equation

$$\xi u_l'' + \left(2\mu + \frac{1}{2} - \xi \right) u_l' - \left(\mu + \frac{1}{4} - \frac{\alpha}{4} \right) u_l = 0. \tag{31.5}$$

The non-singular solution (at the origin) of this equation is, apart from a constant factor, the confluent hypergeometric series

$$u_l(\xi) = F\left(\mu + \frac{1}{4} - \frac{\alpha}{4}, 2\mu + \frac{1}{2}, \xi \right). \tag{31.6}$$

Bearing in mind the asymptotic behaviour (A.64), R_l increases without bound as $\xi \to \infty$ unless the series F reduces to a polynomial. This occurs only if

$$\mu + \frac{1}{4} - \frac{\alpha}{4} = -n', \quad n' = 0, 1, 2, \ldots, \tag{31.7}$$

i.e. if

$$E_{n'l} = \hbar\sqrt{\frac{B}{2m}}\left[4n'+2+\sqrt{(2l+1)^2+\frac{8mA}{\hbar^2}}\right]; \qquad n', l = 0, 1, 2, \ldots. \tag{31.8}$$

The eigenfunctions corresponding to these discrete energy levels are given by

$$\psi_{n'lm} = N_{n'l}r^{-1}R_{n'l}(r)Y_l^m(\theta, \phi), \tag{31.9}$$

where

$$R_{n'l} = \xi^{\mu}e^{-(\xi/2)}F(-n', 2\mu+\tfrac{1}{2}, \xi), \tag{31.10}$$

and $N_{n'l}$ is a normalization constant.

In the particular case $A = 0$, $B = \dfrac{m\omega^2}{2}$, we have

$$\mu = \tfrac{1}{2}(l+1),\ \xi = \frac{m\omega}{\hbar}r^2, \quad E_{n'l} = \omega\hbar(2n'+l+\tfrac{3}{2}), \quad \text{and} \tag{31.11}$$

$$\psi_{n'lm} = N_{n'l}r^l \exp\left(-\frac{m\omega}{2\hbar}r^2\right)F\left(-n', l+\tfrac{3}{2}, \frac{m\omega}{\hbar}r^2\right)Y_l^m(\theta, \phi). \tag{31.12}$$

Note that the energy eigenvalues now depend only on the sum $2n'+l$ of the quantum numbers n' and l, which sum can take any of the values 0, 1, 2, ...

Introducing the "principal quantum number" $n = 2n'+l$, we obtain the energy levels

$$E_n = \omega\hbar(n+\tfrac{3}{2}), \qquad n = 0, 1, 2, \ldots$$

of the isotropic three-dimensional harmonic oscillator, as already calculated in problem 20.

The degeneracy of the level E_n is given by (20.4). For example, the following six linearly independent eigenfunctions correspond to E_2:

$$\psi_{1,0,0},\quad \psi_{0,2,2},\quad \psi_{0,2,1},\quad \psi_{0,2,0},\quad \psi_{0,2,-1}\quad \text{and}\quad \psi_{0,2,-2}.$$

32. (1) In classical mechanics:

Let m_1, m_2 be the masses, $\mathbf{p}_1, \mathbf{p}_2$ the momenta, and $\mathbf{r}_1, \mathbf{r}_2$ the position vectors (relative to the origin) of the two particles in some frame of reference (S). The Hamiltonian of the two-particle system is then

$$H = \frac{\mathbf{p}_1^2}{2m_1}+\frac{\mathbf{p}_2^2}{2m_2}+V(\mathbf{r}_1-\mathbf{r}_2). \tag{32.1}$$

By introducing the new variables

$$\mathbf{R} = \frac{m_1\mathbf{r}_1+m_2\mathbf{r}_2}{m_1+m_2}, \quad \mathbf{P} = \mathbf{p}_1+\mathbf{p}_2, \quad M = m_1+m_2$$

$$\mathbf{r} = \mathbf{r}_1-\mathbf{r}_2, \quad \mathbf{p} = \frac{m_2\mathbf{p}_1-m_1\mathbf{p}_2}{m_1+m_2}, \quad m = \frac{m_1m_2}{m_1+m_2} \tag{32.2}$$

the motion of the two particles can be expressed in terms of two uncoupled motions. One of these motions is that of the centre of mass, which moves like a particle of mass, M, momentum $\mathbf{P}$, and position vector $\mathbf{R}$ in the system S. The other motion is that of particle 1 relative to particle 2, with relative position vector $\mathbf{r}$ and relative velocity $\mathbf{p}/m$ equal to the relative velocity $(\mathbf{p}_1/m_1)-(\mathbf{p}_2/m_2)$ of the two particles. The mass m defined above is called the "reduced mass" of the system. It can easily be verified that the total kinetic energy is

$$\frac{\mathbf{p}_1^2}{2m_1}+\frac{\mathbf{p}_2^2}{2m_2}=\frac{\mathbf{p}^2}{2m}+\frac{\mathbf{P}^2}{2M}$$

and that the total orbital angular momentum of the system about the origin of S is

$$\mathbf{l}_1+\mathbf{l}_2=\mathbf{l}+\mathbf{L} \tag{32.3}$$

where

$$\mathbf{l}_1=\mathbf{r}_1\times\mathbf{p}_1,\quad \mathbf{l}_2=\mathbf{r}_2\times\mathbf{p}_2,\quad \mathbf{l}=\mathbf{r}\times\mathbf{p},\quad \mathbf{L}=\mathbf{R}\times\mathbf{P}.$$

In the new variables the Hamiltonian becomes

$$H=\frac{\mathbf{P}^2}{2M}+\frac{\mathbf{p}^2}{2m}+V(\mathbf{r}) \tag{32.4}$$

Now the change of variables (32.2) leaves the classical Poisson brackets unchanged, i.e. it is a canonical transformation, and the equations of motion in the new variables can therefore be found from Hamilton's canonical equations, which yield

$$\dot{\mathbf{R}}=\frac{\mathbf{P}}{M},\quad \dot{\mathbf{P}}=0;\quad \dot{\mathbf{r}}=\frac{\mathbf{p}}{m},\quad \dot{\mathbf{p}}=-\operatorname{grad} V(\mathbf{r}).$$

Thus the motion of the centre-of-mass is completely separated from the relative motion. The former is the uniform motion in a straight line of a free particle of mass M, while the latter is such that a (non-rotating) observer sitting on particle 2 would conclude that particle 1 had a mass m and was moving in a potential $V(\mathbf{r})$ relative to himself as (fixed) origin. The above facts can of course also be established by a straightforward use of Newton's laws of motion.

Once the relative motion (i.e. the change in time of the vector $\mathbf{r}$) has been calculated, the motion of the particles relative to the centre of mass of the system ("motion in the centre of mass system", or "CMS") can be found immediately since the position vectors of particles 1 and 2 relative to the centre of mass are given by $\frac{m_2}{M}\mathbf{r}$ and $-\frac{m_1}{M}\mathbf{r}$ respectively.

(2) In quantum mechanics:

Let us introduce the new variables $\mathbf{r}$, $\mathbf{R}$, $\mathbf{p}$, $\mathbf{P}$ given by (32.2). The corresponding operators $\mathbf{r}\to\mathbf{r}$, $\mathbf{R}\to\mathbf{R}$, $\mathbf{p}\to -i\hbar\nabla_r$, $\mathbf{P}\to -i\hbar\nabla_R$ satisfy the usual commutation rules

$$[r_j, p_k]=i\hbar\delta_{jk},\quad [R_j, P_k]=i\hbar\delta_{jk},$$

and the Hamiltonian of the system becomes

$$H = H_R + H_r \equiv \left(-\frac{\hbar^2}{2M}\nabla_R^2\right) + \left(-\frac{\hbar^2}{2M}\nabla_r^2 + V(\mathbf{r})\right). \tag{32.5}$$

The Schrödinger equation in the $\{\mathbf{R}, \mathbf{r}\}$ representation can thus be written as

$$\left[\left(-\frac{\hbar^2}{2M}\nabla_R^2\right) + \left(-\frac{\hbar^2}{2m}\nabla_r^2 + V(r)\right)\right]\psi(\mathbf{R}, \mathbf{r}) = E\psi(\mathbf{R}, \mathbf{r}). \tag{32.6}$$

This equation is separable and has a complete set of eigenfunctions of the form $\psi(\mathbf{R}, \mathbf{r}) = \Phi(\mathbf{R})\,\phi(\mathbf{r})$, where Φ and ϕ satisfy the following equations:

$$H_R\Phi(\mathbf{R}) \equiv \left(-\frac{\hbar^2}{2M}\nabla_R^2\right)\Phi(\mathbf{R}) = E_R\Phi(\mathbf{R}), \tag{32.7}$$

$$H_r\,\phi(\mathbf{r}) \equiv \left(-\frac{\hbar^2}{2m}\nabla_r^2 + V(\mathbf{r})\right)\phi(\mathbf{r}) = E_r\phi(\mathbf{r}). \tag{32.8}$$

The total energy E of the two-particle system is then given by $E = E_R + E_r$ = the energy of motion of the centre of mass + the internal energy of the system. Since $\Phi(\mathbf{R})$ describes the motion of a free particle of mass M, the two-body problem in quantum mechanics reduces to the problem of the relative motion of the two particles under their mutual interaction (32.8). If the potential energy $V(\mathbf{r})$ of the interaction has spherical symmetry, and $\mathbf{r}$ is expressed in spherical coordinates, the radial variable can be separated from the angular ones and we are led to solve equation (II.15) with the condition (II.16) at the origin.

33. The hydrogen atom is a system of two particles moving under their mutual Coulomb interaction, the electron and the proton being subject to mutually attractive forces whose potential energy is $V(r) = -e^2/r$. If E_r is the internal energy of the atom (= total energy E of the atom in axes relative to which the centre of mass is at rest), then the wavefunction of the relative motion $\phi(\mathbf{r})$ (see the previous problem) satisfies the Schrödinger equation:

$$\left(-\frac{\hbar^2}{2m}\nabla^2 - \frac{e^2}{r}\right)\phi(\mathbf{r}) = E\,\phi(\mathbf{r}). \tag{33.1}$$

Using spherical coordinates, the eigenfunctions of the problem will have the form (II.14)

$$\phi_l^m(r, \theta, \phi) = \frac{R_l(r)}{r}Y_l^m(\theta, \phi), \tag{33.2}$$

where the R_l are those solutions of the radial equation

$$\frac{d^2R_l}{dr^2} + \frac{2m}{\hbar^2}\left[E_r - \left(\frac{l(l+1)\hbar^2}{2mr^2} - \frac{e^2}{r}\right)\right]R_l = 0, \tag{33.3}$$

which obey the condition (II.16).

If $E_r > 0$, R_l has an oscillatory behaviour at infinity. The corresponding functions ϕ_l^m can be accepted as eigenfunctions for any $E_r > 0$ and in fact they describe the relative motion of an electron and a proton in collision with given total energy E_r in the CMS. For $E_r < 0$, the solutions which satisfy the condition (II.16) at the origin behave at infinity like $C_1 \exp(+\lambda r) + C_2 \exp(-\lambda r)$, where $\lambda = \left(-\frac{2m}{\hbar^2}E\right)^{1/2}$. These solutions can be accepted as eigenfunctions only if $C_1 = 0$. This happens only for certain discrete values of E_r, which form the discrete spectrum of the hydrogen atom, and the eigenfunctions then represent the corresponding bound states. Since in this problem the reduced mass is very close to the electron mass [in fact $(m_e - m)/m_e \approx 5 \times 10^{-4}$], the true energy spectrum will coincide to a good approximation with that of an "atom" consisting of an electron in the field of a fixed (infinitely heavy) proton.

Introducing the new variable

$$x = 2\lambda r, \tag{33.4}$$

(33.3) becomes

$$\left[\frac{d^2}{dx^2} - \frac{l(l+1)}{x^2} + \frac{\nu}{x} - \frac{1}{4}\right] R_l = 0, \tag{33.5}$$

where ν is a dimensionless parameter

$$\nu = \frac{e^2}{\hbar c}\left(-\frac{mc^2}{2E_r}\right)^{1/2}. \tag{33.6}$$

The solution of (33.5) which satisfies the condition $R_l(0) = 0$ behaves like x^{l+1} near the origin. Since, as $x \to \infty$, the acceptable solution has the form $\exp(-x/2)$, we look for R_l in the form

$$R_l = x^{l+1} e^{-x/2} u_l(x). \tag{33.7}$$

Substituting (33.7) into (33.5), we find that $u_l(x)$ satisfies a Laplace type equation

$$\left[x\frac{d^2}{dx^2} + (2l+2-x)\frac{d}{dx} - (l+1-\nu)\right] u_l = 0.$$

The only solution of this equation regular at the origin is, apart from a constant factor, the confluent hypergeometric series

$$u_l(x) = F(l+1-\nu, 2l+2; x). \tag{33.8}$$

Now, as $x \to \infty$, F behaves in general as $x^{-l-1-\nu} e^x$, cf. (A.56); hence $R_l \sim x^{-\nu} \exp(x/2) \to \infty$, and cannot be the radial part of an eigenfunction, unless (33.8) reduces to a polynomial. This happens only if

$$l+1-\nu = -n', \qquad n' = 0, 1, 2, \ldots, \tag{33.9}$$

and thus, according to (A.65),

$$F(-n', 2l+2; \quad x) = AL_{n'}^{2l+1}(x), \tag{33.10}$$

where $L_{n'}^{2l+1}(x)$ is a Laguerre polynomial of order n' and A is a constant.

In this case, as $x \to \infty$, $R_l \sim x^{n'}e^{-(x/2)} \to 0$, and the corresponding solutions (33.2) of the Schrödinger equation can be accepted as eigenfunctions of the problem.

The condition (33.9) gives the (internal) energy levels of the hydrogen atom for the value l of the orbital quantum number, and n' of the radial quantum number:

$$E_{n'l} = -\left(\frac{e^2}{\hbar c}\right)^2 \frac{mc^2}{2(l+1+n')^2}. \tag{33.11}$$

Since $n' = 0, 1, 2, \ldots$, the discrete spectrum contains an infinite number of levels, which occur closer and closer together as the value of the radial quantum number n' increases for a given l. They tend towards the limit $E_r = 0$, from which point the continuous spectrum ($E_r > 0$) begins. The set of values $E_{n'l}$ for $n' = 0, 1, 2, \ldots$ and $l = 0, 1, 2, \ldots$ form the complete discrete spectrum of the hydrogen atom. Because these values depend only on the sum $l+n'$, a "principal quantum number" n can be defined such that $n = l+n'+1$ and then we can write for the possible (internal) energies of the atom

$$E_n = -\left(\frac{e^2}{\hbar c}\right)^2 \frac{mc^2}{2n^2}, \qquad n = 1, 2, \ldots. \tag{33.12}$$

For each value of $n = 1, 2, 3, \ldots$, i.e. for each possible energy E_n, the orbital quantum number l can have the values $l = 0, 1, 2, \ldots, n-1$ and, for each value of $l, m = -l, -l+1, \ldots, +l$, $(2l+1$ values). The degeneracy of the energy level E_n is therefore

$$\sum_{l=0}^{n-1} (2l+1) = n(n-1)+n = n^2. \tag{33.13}$$

The eigenfunctions belonging to E_n thus form a subspace of n^2 dimensions. In conclusion, and bearing in mind the relations (33.2), (33.4), (33.7), (33.9), (33.10) and (33.12), the wave-function of the state of relative motion (nlm) can be written as

$$\phi_{nlm} = a^{-3/2}N_{nl}K_{nl}\left(\frac{2r}{na}\right)Y_l^m(\theta, \phi), \tag{33.14}$$

where

$$K_{nl}(x) = x^l e^{-x/2}\, L_{n-l-1}^{2l+1}(x), \qquad a = \frac{\hbar^2}{me^2}, \tag{33.15}$$

and N_{nl} is a normalization constant.

Using (A.33) it is easy to show that the functions ϕ_{nlm} are orthonormal if

$$N_{nl} = \frac{2}{n^2}\sqrt{\frac{(n-l-1)!}{[(n+1)!]^3}}. \tag{33.16}$$

All the above results can be generalized for hydrogen-like ions (He^+, Li^{++}, ...) by substituting Ze^2 for e^2, where Ze is the nuclear charge. In Table II.1 some of the wavefunctions ϕ_{nlm} are given explicitly (see (33.14)–(33.16) and A.29)).

TABLE II.1

n	l	m	
1	0	0	$\phi_{100} = \dfrac{1}{\sqrt{\pi a^3}} \exp\left(-\dfrac{r}{a}\right)$
2	0	0	$\phi_{200} = \dfrac{1}{4\sqrt{2\pi a^3}} \left(2-\dfrac{r}{a}\right) \exp\left(-\dfrac{r}{2a}\right)$
	1	1	$\phi_{211} = \dfrac{1}{4\sqrt{4\pi a^3}} \left(\dfrac{r}{a}\right) \exp\left(-\dfrac{r}{2a}\right) \sin\theta e^{i\phi}$
		0	$\phi_{210} = \dfrac{1}{4\sqrt{2\pi a^3}} \left(\dfrac{r}{a}\right) \exp\left(-\dfrac{r}{2a}\right) \cos\theta$
		-1	$\phi_{21-1} = \dfrac{1}{4\sqrt{4\pi a^3}} \left(\dfrac{r}{a}\right) \exp\left(-\dfrac{r}{2a}\right) \sin\theta e^{-i\phi}$

34. By comparing (34a) with (II.12), we see that the energy eigenvalues of the rigid rotator are

$$E_l = \frac{\hbar^2}{2I} l(l+1), \qquad l = 0, 1, 2, \ldots \tag{34.1}$$

with the corresponding eigenfunctions $Y_l^m(\theta, \phi)$. Since $m = 0, \pm 1, \ldots, \pm l$, it follows that to each energy value E_l there correspond $2l+1$ linearly-independent eigenfunctions, i.e. the degeneracy of the level E_l is $2l+1$. When the rotator makes a transition to a lower state it emits a quantum of (angular) frequency

$$\omega_{ll'} = \frac{E_l - E_{l'}}{\hbar} = \frac{\hbar}{2I} [l(l+1) - l'(l'+1)]. \tag{34.2}$$

The set of emitted frequencies for all allowed quantum transitions constitutes the optical spectrum of the rotator. If the vibrations of the constituent atoms are neglected, a diatomic molecule can be regarded as a rigid rotator; I is then the moment of inertia of the molecule about its centre of mass, or, equivalently, that of a hypothetical single particle of "reduced" mass $m = m_1 m_2/(m_1+m_2)$ rotating about a hypothetical fixed origin at a distance from it equal to the distance between the two atoms of the molecule.

CHAPTER III

Mean Values and Uncertainty Relations

1. The Mean Values of Dynamical Variables

In quantum mechanics, to each dynamical variable of a system

$$\mathcal{A} = A(q_1, \ldots, q_N; p_1, \ldots, p_N) \tag{III.1}$$

there corresponds a linear Hermitian operator (or "observable"):

$$A\left(q_1, \ldots, q_N; \frac{\hbar}{i}\frac{\partial}{\partial q_1}, \ldots, \frac{\hbar}{i}\frac{\partial}{\partial q_N}\right), \tag{III.2}$$

which operates on the wavefunction $\Psi(q_1, \ldots, q_N; t)$ of the system. By definition, the "mean value" of this dynamical variable, when the physical system is in the dynamical state described by the wavefunction $\Psi(q_1, \ldots, q_N; t)$, is

$$\langle A \rangle = \frac{\langle \Psi, A\Psi \rangle}{\langle \Psi, \Psi \rangle}, \tag{III.3}$$

where

$$\langle \Psi, A\Psi \rangle \equiv \int \Psi^*(A\Psi)\, dq_1 \ldots dq_N,$$
$$\langle \Psi, \Psi \rangle \equiv \int \Psi^*\Psi\, dq_1 \ldots dq_N.$$

The function

$$\Phi(p_1, \ldots, p_N; t) = (2\pi\hbar)^{-N/2} \int \Psi(q_1, \ldots, q_N; t) \exp\left(-\frac{i}{\hbar}\sum_{i=1}^{N} p_i q_i\right) dq_1 \ldots dq_N, \tag{III.4}$$

which is the Fourier transform of Ψ, is called the "wavefunction in momentum space" and plays a part in the theory similar to that played by Ψ itself (see Chapter V, Section 2).

Thus, the dynamical variable (III.1) can be represented alternatively by the linear Hermitian operator

$$A\left(i\hbar\frac{\partial}{\partial p_1}, \ldots, i\hbar\frac{\partial}{\partial p_N}; p_1, \ldots, p_N\right), \tag{III.5}$$

which operates on the momentum-space wavefunction $\Phi(p_1, \ldots, p_N; t)$. By definition, its mean value, when the system is in the dynamical state described by the wavefunction

$\Phi(p_1, \ldots, p_N; t)$, is

$$\langle A \rangle = \frac{\langle \Phi, A\Phi \rangle}{\langle \Phi, \Phi \rangle}. \tag{III.6}$$

On account of the properties of Fourier transforms, the equivalence of the definitions (III.3) and (III.6) becomes obvious. Just as the functions Φ and Ψ are equivalent representations of the same dynamical state, so the observables (III.5) and (III.2) are equivalent representations of the same dynamical variable (III.1).

2. The Uncertainty Relations

Let u and v be two dynamical variables, and U and V their associated operators. By definition, the "uncertainties" ΔU and ΔV of these dynamical variables, when the system is in the state described by the wavefunction $\Psi(q_1, \ldots, q_N; t)$, are the root mean square deviations

$$\Delta U \equiv \sqrt{\langle (U - \langle U \rangle)^2 \rangle} = \sqrt{\langle U^2 \rangle - \langle U \rangle^2},$$
$$\Delta V \equiv \sqrt{\langle (V - \langle V \rangle)^2 \rangle} = \sqrt{\langle V^2 \rangle - \langle V \rangle^2}.$$

The following "uncertainty relation" then holds

$$\Delta U . \Delta V \geqslant \tfrac{1}{2}\hbar |\langle W \rangle| \tag{III.7}$$

where the operator W is defined to be $-\frac{i}{\hbar}(UV - VU)$.

Two dynamical variables are said to be *compatible* if they can be specified simultaneously with complete accuracy. Otherwise they are called *complementary* variables.

Compatible (complementary) dynamical variables are represented by commuting (non-commuting) operators.

The position coordinates q_k and their conjugate momenta p_k are complementary variables, since $[q_k, p_k] = i\hbar$. *Heisenberg's uncertainty relation* follows as a special case of (III.7):

$$\Delta q_k . \Delta p_k \geqslant \tfrac{1}{2}\hbar. \tag{III.8}$$

Heisenberg's uncertainty principle states that if, at a given moment, a dynamical variable has a well-defined value, then all the complementary dynamical variables of the same system are completely undetermined. Although it has a different meaning, a relation similar in appearance to (III.8), namely

$$\Delta t . \Delta E \geqslant \tfrac{1}{2}\hbar \tag{III.9}$$

is also valid. It is often called *the time-energy uncertainty relation*. The essential difference between (III.9) and (III.8) is the fact that while q_k and p_k cannot both be specified at the same time with complete accuracy, the energy of the system may have a well-defined value at every moment of time t. In (III.9), ΔE is the difference between two values E_1 and E_2 of the energy E measured at two different moments of time t_1 and t_2 ($\Delta t = t_2 - t_1$), and is not the uncertainty in the energy at a given moment of time.

Problems

1. Show that in the stationary states corresponding to the discrete spectrum of the Hamiltonian $H = \mathbf{p}^2/2m + V(\mathbf{r})$, the mean value of the momentum vanishes, i.e. $\langle \mathbf{p} \rangle = 0$.

2. Show that the mean value of x in a state described by the wavefunction $\psi(x)$, viz.,

$$\langle x \rangle = \int_{-\infty}^{+\infty} \psi^*(x) x \, \psi(x) \, dx, \tag{2a}$$

is equal to the value of a for which the expression

$$V(a) \equiv \int_{-\infty}^{+\infty} \psi^*(x+a) x^2 \psi(x+a) \, dx, \tag{2b}$$

is a minimum, and that this minimum has the value $V_{\min} = (\Delta x)^2 = \langle x^2 \rangle - \langle x \rangle^2$.

3. Study the possibility of generalizing the definitions given in the preceding problem (see the solution) to the case of an angular variable $-\pi \leqslant \phi \leqslant +\pi$, $|\psi(\phi)|^2$ being the probability density on a circle of unit radius.

4. Show that the energy of a harmonic oscillator in the state with quantum number n can be expressed in the form

$$E_n = m\omega^2 \langle x^2 \rangle_n . \tag{4a}$$

5. Following the procedure for solving the preceding problem, show that

$$\langle x^4 \rangle_n = \tfrac{3}{4} x_0^4 (2n^2 + 2n + 1). \tag{5a}$$

6. Calculate the mean value of r^{-2} in the quantum state (nlm) of the hydrogen atom.

7. Using the radial equation of the hydrogen atom (6.2), derive Kramers' relation:

$$\frac{s+1}{n^2} \langle r^s \rangle - (2s+1) \, a \langle r^{s-1} \rangle + \frac{s}{4} [(2l+1)^2 - s^2] \, a^2 \langle r^{s-2} \rangle = 0, \tag{7a}$$

where $\langle r^s \rangle$ is the mean value of r^s in the state (nlm), and $s > -2l-3$. As an application, calculate $\langle r^{-1} \rangle$, $\langle r \rangle$, $\langle r^2 \rangle$, $\langle r^{-3} \rangle$ and $\langle r^{-4} \rangle$.

8. Show that in any stationary state of the hydrogen atom, the mean value of the kinetic energy is equal to minus the total energy of that state, i.e. that

$$\left\langle \frac{\mathbf{p}^2}{2m} \right\rangle_{nlm} = -E_n .$$

9. Show that the time derivative of the mean value of any observable is given by

$$i\hbar \frac{d\langle A \rangle}{dt} = \langle [A, H] \rangle + i\hbar \left\langle \frac{\partial A}{\partial t} \right\rangle. \tag{9a}$$

10. When the motion of a particle is restricted to a finite region of space, the mean value of the scalar product $\mathbf{r}.\mathbf{p}$ has to be time-independent, i.e. $\frac{d}{dt}\langle \mathbf{r}.\mathbf{p} \rangle = 0$. Using this fact, derive the "quantum virial theorem":

$$2\langle T \rangle = \langle \mathbf{r}.\nabla V \rangle, \tag{10a}$$

where V is the potential energy and T is the kinetic energy operator of the particle.

11. Derive the uncertainty relation (III.7).

12. Explain the fact that the ground state energy of a particle in the potential well shown in Fig. II.1 is different from zero.

13. Using the uncertainty relation, evaluate the ground state energy of a hydrogen atom, and of a two-electrons atom whose nucleus has a charge Ze.

14. A beam of silver atoms in a Stern-Gerlach experiment emerges from an oven, which contains silver vapour at a temperature of 1200°K. The beam is collimated by being passed through a small circular aperture. Using Heisenberg's uncertainty relation, show that it is not possible, by narrowing the aperture, to decrease indefinitely the diameter of the spot on the screen. If the screen is at a distance of 1 m from the aperture, estimate the smallest diameter of the spot that can be obtained by varying the diameter of the aperture. Assume for simplicity that all atoms have the same momentum along the direction of the beam.

$$(m_{\mathrm{Ag}} = 1{\cdot}8 \times 10^{-22}\ \mathrm{g},\ k = 1{\cdot}38 \times 10^{-16}\ \mathrm{erg/^\circ K}).$$

15. Show that in the quantum state n of the harmonic oscillator, the product of the position and the momentum uncertainties is given by

$$\Delta x\, \Delta p = \hbar(n+\tfrac{1}{2}). \tag{15a}$$

16. Using a method similar to that used in problem 11, determine the normalized wavefunction $\psi(x)$ for which the product of the coordinate and the momentum uncertainties has the smallest value $\Delta x\, \Delta p = \hbar/2$.

17. In a Frank and Hertz experiment, hydrogen atoms are raised to their first excited state by colliding with a beam of electrons. Experimentally, it is observed that the energies (after collision) of the electrons which produced these transitions are not all the same, even if the incident beam of electrons is mono-energetic. Explain this phenomenon, bearing in mind that the lifetimes of excited atomic states are (normally) very short.

18. Experimental observations have shown that the fluctuations in the observed final energy of the electrons described in problem 17 above are of the order of 10^{-6} eV. Use this fact to evaluate the mean life of the excited atoms.

19. Using the time-energy uncertainty relation estimate the mass of a π meson, knowing that the effective distance over which nuclear forces act is about 1·4 fermi.

20. In spherical polar coordinates, the l_z component of the orbital angular momentum of a particle is represented by the operator $-i\hbar\partial/\partial\phi$. By analogy with the commutation rules for x and p_x, one would expect the following commutation rule and uncertainty relation to be valid

$$[\phi, l_z] = i\hbar, \tag{20a}$$

$$\Delta l_z \, \Delta\phi \geqslant \frac{\hbar}{2}. \tag{20b}$$

Show that (20a) is not in general valid, and that (20b) would contradict the Heisenberg uncertainty principle.

Solutions

1. Using the commutation rules $[x_j, p_k] = i\hbar\delta_{jk}$, it can easily be shown that

$$\mathbf{p} = \frac{im}{\hbar}(H\mathbf{r} - \mathbf{r}H). \tag{1.1}$$

The mean value of $\mathbf{p}$ in the state ψ corresponding to a discrete energy eigenvalue is

$$\langle \mathbf{p} \rangle = \frac{im}{\hbar}\langle \psi, (H\mathbf{r} - \mathbf{r}H)\psi \rangle. \tag{1.2}$$

Since H is Hermitian we have

$$\langle \mathbf{p} \rangle = \frac{im}{\hbar}[\langle H\psi, \mathbf{r}\psi \rangle - \langle \psi, \mathbf{r}H\psi \rangle]. \tag{1.3}$$

For stationary states $H\psi = E\psi$, and from (1.3) it follows then that $\langle \mathbf{p} \rangle = 0$.

2. Substituting $x - a$ for x in (2b), we find after some simple algebra that

$$V(a) = \langle x^2 \rangle - 2a\langle x \rangle + a^2. \tag{2.1}$$

From the equation $dV(a)/da = 0$ it follows that the function $V(a)$ reaches its minimum value for $a = \langle x \rangle$ and hence $V_{\min} = V(\langle x \rangle) = \langle x^2 \rangle - \langle x \rangle^2 = (\Delta x)^2$.

Consequently the definition (2a) is equivalent to the definition $\langle x \rangle = a$, where a is the value which makes $V(a)$ a minimum.

Similarly, the definition

$$(\Delta x)^2 = \int_{-\infty}^{+\infty} \psi^*(x)\,(x - \langle x \rangle)^2\,\psi(x)\,dx \tag{2.2}$$

is equivalent to $(\Delta x)^2 = V_{\min}$.

3. A formal change of variables in the results of the preceding problem would require the equivalence of the definition

$$\langle\phi\rangle = \int_{-\pi}^{\pi} \psi^*(\phi)\, \phi\psi(\phi)\, d\phi \equiv \langle\phi\rangle_1, \tag{3.1}$$

with $\langle\phi\rangle = \gamma \equiv \langle\phi\rangle_2$, where $-\pi \leqslant \gamma \leqslant +\pi$ and γ is the value for which the function

$$V(\gamma) = \int_{-\pi}^{\pi} \psi^*(\phi+\gamma)\, \phi^2\psi(\phi+\gamma)\, d\phi \tag{3.2}$$

reaches its minimum, and also the equivalence of the definition

$$(\Delta\phi)^2 = \int_{-\pi}^{\pi} \psi^*(\phi)\,(\phi-\langle\phi\rangle)^2\,\psi(\phi)\, d\phi \equiv (\Delta^{(1)}\phi)^2, \tag{3.3}$$

with $(\Delta\phi)^2 = V_{\min} \equiv (\Delta^{(2)}\phi)^2$.

But we can easily see that, in general, $\langle\phi\rangle_1 \neq \langle\phi_2\rangle$. Indeed $\langle\phi_1\rangle$ cannot be a measure of the mean value, since it depends on the choice of the origin from which the angle is measured: thus, if we put $\phi = \phi'+\alpha$, we find that $\langle\phi\rangle_1 = \langle\phi'\rangle_1+\alpha$. Thus, for any distribution $|\psi|^2$, by a suitable choice of origin, the "mean value" $\langle\phi\rangle_1$ may be allocated to any point of the circle. Hence, if the notion of the mean value of ϕ is to have a physical meaning, we are obliged to use $\langle\phi\rangle_2$, although, as opposed to the case of the variable $x\,(-\infty < x < +\infty)$, it might not be unique.

The definitions of mean square deviations, $\Delta^{(1)}\phi$ and $\Delta^{(2)}\phi$, are not equivalent, even if we take $\langle\phi\rangle$ to mean $\langle\phi\rangle_2$. The quantity $\Delta^{(1)}\phi$ cannot in fact be accepted as a measure of the uncertainty $\Delta\phi$ for the same reasons as apply in the case of $\langle\phi\rangle_1$. On the other hand, $\Delta^{(2)}\phi$ is independent of the choice of origin, and attains its maximum value $\Delta^{(2)}\phi = \pi/\sqrt{3}$ when the probability distribution is uniform, i.e. when $|\psi|^2 = \text{const.} = \frac{1}{2\pi}$. For these reasons, and from the formal analogy between (3.4) and the definition of Δx, it seems reasonable to take $\Delta^{(2)}\phi$ as a suitable measure of the uncertainty $\Delta\phi$ of the angular variable. (D. Judge, 1964).

4. We shall deduce (4a) by evaluating the integral

$$\langle x^2\rangle_n = \int_{-\infty}^{+\infty} \psi_n^*(x)\, x^2\psi_n(x)\, dx \tag{4.1}$$

and comparing the result obtained with (17.7) of Chapter II.

With the notation used in problem 17 of Chapter II, we have

$$\langle x^2\rangle_n = x_0^3 \int_{-\infty}^{+\infty} \psi_n^*\xi^2\psi_n\, d\xi. \tag{4.2}$$

Using (A.6) and (17.12) of Chapter II we can write

$$\xi\psi_n = \sqrt{\frac{n}{2}}\,\psi_{n-1}+\sqrt{\frac{n+1}{2}}\,\psi_{n+1}, \tag{4.3}$$

and hence

$$\xi^2\psi_n = \xi(\xi\psi_n) = \tfrac{1}{2}\sqrt{n(n-1)}\,\psi_{n-2}+(n+\tfrac{1}{2})\psi_n+\tfrac{1}{2}\sqrt{(n+1)(n+2)}\,\psi_{n+2}. \tag{4.4}$$

Substituting (4.4) into (4.2), and taking into account (17.13) of Chapter 2, we find that

$$\langle x^2\rangle_n = x_0^2(n+\tfrac{1}{2}). \tag{4.5}$$

But $x_0^2 = \hbar/m\omega$ and hence $E_n = m\omega^2\langle x^2\rangle_n$.

Since, for the harmonic oscillator,

$$H = \frac{p^2}{2m}+\frac{m\omega^2}{2}x^2 = T+V,$$

it follows from (4a) that in the quantum state n the mean value of the kinetic energy equals the mean value of the potential energy:

$$\langle T\rangle_n = \langle V\rangle_n = \frac{E_n}{2}. \tag{4.6}$$

5.

$$\langle x^4\rangle_n = \int_{-\infty}^{+\infty} \psi_n^* x^4\psi_n\,dx = x_0^5\int_{-\infty}^{+\infty} \psi_n^*\xi^4\psi_n\,d\xi. \tag{5.1}$$

Using the recurrence relation (4.4),

$$\begin{aligned}\xi^4\psi_n = \xi^2(\xi^2\psi_n) &= \tfrac{1}{4}\sqrt{n(n-1)(n-2)(n-3)}\,\psi_{n-4}+\tfrac{1}{2}\sqrt{n(n-1)^3}\,\psi_{n-2}+\tfrac{3}{4}(2n^2+2n+1)\psi_n\\ &+\sqrt{(n+1)(n+2)(n+3)^2}\,\psi_{n+2}+\tfrac{1}{4}\sqrt{(n+1)(n+2)(n+3)(n+4)}\,\psi_{n+4}.\end{aligned} \tag{5.2}$$

Using (5.1), (5.2) and (17.12) of Chapter II, (5a) follows immediately.

6. The mean value of r^s in the quantum state (nlm) can be calculated from

$$\begin{aligned}\langle r^s\rangle &= \int \psi_{nlm}^*(\mathbf{r})\,r^s\psi_{nlm}(\mathbf{r})\,d\mathbf{r}\\ &= \int_0^\infty\int_0^\pi\int_0^{2\pi} r^sR_{nl}^2(r)\,Y_l^{m*}(\theta,\phi)\,Y_l^m(\theta,\phi)\sin\theta\,dr\,d\theta\,d\phi = \int_0^\infty r^sR_{nl}^2(r)\,dr,\end{aligned} \tag{6.1}$$

where the radial function R_{nl} is obtained by comparing (33.2) with (33.14) of Chapter II.

For $s = -2$ we can also use a more elegant method. Introducing the variable $\varrho = r/a$, where $a = \hbar^2/me^2$, the radial equation (33.3) of Chapter II for $E = E_n$ becomes

$$\frac{d^2R_{nl}}{d\varrho^2}+\left\{\frac{2}{\varrho}-\frac{1}{n^2}-\frac{l(l+1)}{\varrho^2}\right\}R_{nl} = 0. \tag{6.2}$$

With this notation

$$\langle\varrho^s\rangle = a\int_0^\infty \varrho^s R_{nl}^2\, d\varrho. \tag{6.3}$$

Suppose for the moment that, in (6.2), l were a continuous variable, and that n depended on l in such a manner that $n-l$ remained an integer (equal to $n'+1$, see (33.9) Chapter II). Under these conditions the radial eigenfunctions R_{nl} would be continuous and differentiable with respect to l. From (6.2) we would then obtain

$$\frac{\partial}{\partial l}\left(\frac{R''_{nl}}{R_{nl}}\right) = \frac{\partial}{\partial l}\left\{\frac{1}{n^2}-\frac{2}{\varrho}+\frac{l(l+1)}{\varrho^2}\right\} = -\frac{2}{n^3}+\frac{2l+1}{\varrho^2}. \tag{6.4}$$

Multiplying this equation by R_{nl}^2 and integrating it over ϱ, we have

$$\int_0^\infty R_{nl}^2\frac{\partial}{\partial l}\left(\frac{R''_{nl}}{R_{nl}}\right) d\varrho = \int_0^\infty R_{nl}\frac{\partial R''_{nl}}{\partial l}\, d\varrho - \int_0^\infty R''_{nl}\frac{\partial R_{nl}}{\partial l}\, d\varrho$$

$$= \int_0^\infty R_{nl}\frac{\partial R''_{nl}}{\partial l}\, d\varrho - \int_0^\infty R_{nl}\frac{\partial R''_{nl}}{\partial l}\, d\varrho = 0 = -\frac{2}{n^3}\int_0^\infty R_{nl}^2\, d\varrho$$

$$+(2l+1)\int_0^\infty \varrho^{-2}R_{nl}^2\, d\varrho = -\frac{2}{an^3}+(2l+1)\,a\langle r^{-2}\rangle.$$

The desired result follows immediately:

$$\langle r^{-2}\rangle = \frac{1}{a^2n^3(l+\frac{1}{2})}. \tag{6.5}$$

7. Multiplying (6.2) by

$$\varrho^{s+1}\frac{dR_{nl}}{d\varrho}-\frac{1}{2}(s+1)\varrho^s R_{nl}, \tag{7.1}$$

and integrating over ϱ, we obtain

$$\int_0^\infty \left\{\varrho^{s+1}R'_{nl}R''_{nl}-\frac{s+1}{2}\varrho^s R_{nl}R''_{nl}+\left[-\frac{1}{n^2}\varrho^{s+1}+2\varrho^s-l(l+1)\varrho^{s-1}\right]R_{nl}R'_{nl}\right\} d\varrho$$

$$+\frac{s+1}{2a}\left[\frac{1}{n^2}\langle\varrho^s\rangle-2\langle\varrho^{s-1}\rangle+l(l+1)\langle\varrho^{s-2}\rangle\right] = 0. \tag{7.2}$$

Knowing that, as $\varrho\to 0$, $R_{nl}\sim\varrho^{l+1}$, and, as $\varrho\to\infty$, $R_{nl}\to 0$, we have

$$\int_0^\infty \varrho^{s+1}R'_{nl}R''_{nl}\, d\varrho = \frac{1}{2}\int_0^\infty \varrho^{s+1}(R'^2_{nl})'\, d\varrho = -\frac{s+1}{2}\int_0^\infty \varrho^s R'^2_{nl}\, d\varrho$$

if $s+2l+1 > 0$ and

$$-\frac{s+1}{2}\int_0^\infty \varrho^s R_{nl}R''_{nl}\, d\varrho = \frac{s+1}{2}\int_0^\infty R'_{nl}(\varrho^s R'_{nl}+s\varrho^{s-1}R_{nl})\, d\varrho$$

$$= \frac{s+1}{2}\int_0^\infty \varrho^s R'^2_{nl}\, d\varrho + \frac{s(s+1)}{2}\int_0^\infty \varrho^{s-1}R_{nl}R'_{nl}\, d\varrho, \quad \text{if} \quad s+2l+1 > 0.$$

Thus equation (7.2) becomes

$$\frac{s+1}{2a}\left[\frac{1}{n^2}\langle\varrho^s\rangle - 2\langle\varrho^{s-1}\rangle + l(l+1)\langle\varrho^{s-2}\rangle\right]$$

$$+\frac{1}{2}\int_0^\infty \left\{-\frac{1}{n^2}\varrho^{s+1}+2\varrho^s+\left[\frac{s(s+1)}{2}-l(l+1)\right]\varrho^{s-1}\right\}(R^2_{nl})'\, d\varrho = 0.$$

Integrating by parts we obtain

$$\frac{s+1}{2a}\left[\frac{1}{n^2}\langle\varrho^s\rangle - 2\langle\varrho^{s-1}\rangle + l(l+1)\langle\varrho^{s-2}\rangle\right]$$

$$+\frac{1}{2a}\left\{\frac{s+1}{n^2}\langle\varrho^s\rangle - 2s\langle\varrho^{s-1}\rangle - (s-1)\left[\frac{s(s+1)}{2}-l(l+1)\right]\langle\varrho^{s-2}\rangle\right\} = 0, \tag{7.3}$$

if $s+2l+3 > 0$.

Re-ordering the terms in (7.3), Kramers' relation is obtained:

$$\frac{s+1}{n^2}\langle r^s\rangle - (2s+1)\, a\langle r^{s-1}\rangle + \frac{s}{4}\left[(2l+1)^2 - s^2\right] a^2\langle r^{s-2}\rangle = 0, \tag{7.4}$$

with the condition $s > -2l-3$.

For s equal to 0, 1, 2, ..., we find after some simple calculation that

$$\langle r^{-1}\rangle = \frac{1}{an^2}, \quad \langle r\rangle = \frac{a}{2}\left[3n^2 - l(l+1)\right]$$

$$\langle r^2\rangle = \frac{a^2}{2}\left[5n^2+1-3l(l+1)\right]n^2. \tag{7.5}$$

For $s = -1, -2, \ldots$, the mean values of r^{-3}, r^{-4}, ... (in the states with $s > -2l-3$) can be obtained as functions of $\langle r^{-2}\rangle$, which was calculated in the preceding problem.

The final results are

$$\langle r^{-3}\rangle = \left[a^3 n^3 l(l+\tfrac{1}{2})(l+1)\right]^{-1},$$

$$\langle r^{-4}\rangle = \left[3n^2 - l(l+1)\right]\left\{a^4 n^5 l(l+\tfrac{1}{2})(l+1)\left[2l(l+1)-\frac{3}{2}\right]\right\}^{-1}. \tag{7.6}$$

8. For the hydrogen atom $H = \frac{\mathbf{p}^2}{2m} - \frac{e^2}{r} = T+V$, and hence $E_n = \langle T\rangle_{nlm} - e^2\langle r^{-1}\rangle$. Since $E_n = -e^2/2an^2$, we find by using (7.5) that

$$E_n = \langle T\rangle_{nlm} - \frac{e^2}{an^2} = \langle T\rangle_{nlm} + 2E_n,$$

whence

$$E_n = -\langle T\rangle_{nlm}.$$

The relation between the mean value of the potential energy and that of the kinetic energy is thus

$$\langle V\rangle_{nlm} = -2\langle T\rangle_{nlm}. \tag{8.1}$$

9. Consider Ψ to be normalized to unity. By differentiating the defining relation (III.3) we obtain

$$\frac{d\langle A\rangle}{dt} = \left\langle \frac{\partial \Psi}{\partial t}\Psi, A\right\rangle + \left\langle \Psi, A\frac{\partial \Psi}{\partial t}\right\rangle + \left\langle \Psi, \frac{\partial A}{\partial t}\Psi\right\rangle. \tag{9.1}$$

But, from the Schrödinger equation, we have that

$$\frac{\partial \Psi}{\partial t} = \frac{1}{i\hbar}H\Psi, \quad \frac{\partial \Psi^*}{\partial t} = -\frac{1}{i\hbar}(H\Psi)^*,$$

and thus, from (9.1), equation (9a) can be obtained.

If the observable A is not explicitly time-dependent, then $\frac{\partial A}{\partial t} = 0$ and equation (9a) has the form

$$i\hbar\frac{d\langle A\rangle}{dt} = \langle [A, H]\rangle. \tag{9.2}$$

This equation expresses a very important fact, namely, that the mean value of an observable which commutes with the Hamiltonian is a constant of the motion. In particular, if H is not explicitly time-dependent, the total energy of the system is conserved and the system is said to be conservative.

10. Since the Hamiltonian of the particle is

$$H = \frac{\mathbf{p}^2}{2m} + V(\mathbf{r}) = T + V(\mathbf{r}),$$

it follows from (9.2) that

$$\frac{d}{dt}\langle \mathbf{r}.\mathbf{p}\rangle = \frac{1}{i\hbar}\langle [\mathbf{r}.\mathbf{p}, H]\rangle = 2\langle T\rangle - \langle \mathbf{r}.\nabla V\rangle.$$

Taking into account the fact that $\frac{d}{dt}\langle \mathbf{r}.\mathbf{p}\rangle = 0$, the virial theorem follows immediately.

If the potential energy is proportional to r^n, we have

$$\langle \mathbf{r}.\nabla V\rangle = n\langle V\rangle$$

and (10a) becomes

$$2\langle T\rangle = n\langle V\rangle. \tag{10.1}$$

Note that for the harmonic oscillator potential ($n = 2$) and for the Coulomb potential ($n = -1$) the relations (4.6) and (8.1) follow directly from (10.1).

11. Consider an arbitrary operator B. The mean value in a state ψ of a physical quantity associated with the Hermitian operator BB^+ will be

$$\langle BB^+\rangle = \langle \psi, BB^+\psi\rangle = \langle B^+\psi, B^+\psi\rangle \geqslant 0. \tag{11.1}$$

Now let C and D be two Hermitian operators and λ a real number. Taking $B = C+i\lambda D$, we have $B^+ = C-i\lambda D$ and

$$0 \leqslant \langle BB^+\rangle = \langle C^2\rangle+\lambda^2\langle D^2\rangle-i\lambda\langle CD-DC\rangle = f(\lambda), \tag{11.2}$$

say, when function $f(\lambda)$ has no maximum, and its minimum is given by the condition $(df/d\lambda) = 0$. After some simple algebra we find that

$$f_{\min} = \langle C^2\rangle+\frac{\langle CD-DC\rangle^2}{4\langle D^2\rangle} \geqslant 0, \tag{11.3}$$

whence

$$\langle C^2\rangle\langle D^2\rangle \geqslant -\tfrac{1}{4}\langle CD-DC\rangle^2. \tag{11.4}$$

Let u and v be the dynamical variables associated with the Hermitian operators U and V of the uncertainty relation (III.7). The deviations

$$\delta U \equiv U-\langle U\rangle, \quad \delta V \equiv V-\langle V\rangle, \tag{11.5}$$

of U and V from their mean values $\langle U\rangle$ and $\langle V\rangle$ will also be Hermitian operators satisfying the commutation rule

$$[\delta U, \delta V] = [U, V]. \tag{11.6}$$

If we now take $C = \delta U$ and $D = \delta V$, (11.4) becomes

$$\langle(\delta U)^2\rangle\langle(\delta V)^2\rangle \geqslant -\tfrac{1}{4}\langle UV-VU\rangle^2. \tag{11.7}$$

Defining the root mean square deviations

$$\Delta U = \sqrt{\langle(\delta U)^2\rangle} \quad \text{and} \quad \Delta V = \sqrt{\langle(\delta V)^2\rangle},$$

we obtain from (11.7) the uncertainty relation

$$\Delta U.\Delta V \geqslant \sqrt{-\tfrac{1}{4}\langle[U, V]\rangle^2}\,. \tag{11.8}$$

If $[U, V] = i\hbar W$, say, then W is Hermitian and

$$\Delta U . \Delta V \geqslant \tfrac{1}{2}\hbar|\langle W\rangle|. \tag{11.9}$$

Thus the uncertainty relation expresses the impossibility of the exact simultaneous specification of two physical quantities represented by two non-commuting operators.

12. The infinite potential well localizes the particle in the range $-a < x < a$, a fact which, in accordance with the Heisenberg relation, leads to an uncertainty in its momentum.

Therefore, although inside the well the energy of the particle is purely kinetic, in the ground state it will have a non-vanishing value

$$E_1 \geqslant \hbar^2/8m(\Delta x)^2.$$

To satisfy ourselves that the right-hand side of this inequality is less than, or at most equal to, $\pi^2\hbar^2/8ma^2$ (see problem 7, Chapter II), we have to show that, in the ground state given by

$$\psi_1(x) = \frac{1}{\sqrt{a}} \cos \frac{\pi}{2a} x,$$

we have $(\Delta x)^2 \geqslant (a/\pi)^2$.

That this is so follows from a little simple algebra:

$$(\Delta x)^2 = \int_{-a}^{a} x^2\psi_1^2 dx - \left[\int_{-a}^{a} x\psi_1^2 dx^2\right]^2 = 2\int_{0}^{a} x^2\psi_1^2 dx = \left(\frac{a}{\pi}\right)^2\left(\frac{\pi^2}{3}-2\right) > \left(\frac{a}{\pi}\right)^2.$$

13. Let us denote by r_0 the "radius" of the hydrogen atom in its ground state, understanding by this that the wavefunction is mostly concentrated inside a sphere of radius r_0. In other words, the probability of finding the electron at a distance r from the proton is appreciable only if $r < r_0$, and is very small if $r > r_0$. This localization of the electron leads to an uncertainty in its momentum of the order of $\hbar/r_0$. The energy of the ground state can then be estimated by finding the value of r_0 for which the function

$$E(r_0) = \frac{\hbar^2}{2mr_0^2} - \frac{e^2}{r_0} \tag{13.1}$$

is a minimum. After some simple calculation we find for the "radius" of the hydrogen atom in its ground state the value

$$r_0 = \frac{\hbar^2}{ml^2} \approx 0{\cdot}529 \times 10^{-8} \text{ cm}, \tag{13.2}$$

and for the corresponding energy

$$E_1 = -\tfrac{1}{2}\frac{me^4}{\hbar^2} \approx -13{\cdot}5 \text{ eV}. \tag{13.3}$$

Since the above argument is only approximate, it can be expected to give only an order of magnitude for the ground-state energy of the hydrogen atom. The fact that E_1 has exactly the value obtained from (33.12) of Chapter II must be regarded as a coincidence.

In the case of two-electron atoms, let r_1 and r_2 be the radii of the "localization" regions for the two electrons. In accordance with Heisenberg's relation, the uncertainties in the respective momenta will be $\Delta p_1 \sim \hbar/r_1$, $\Delta p_2 \sim \hbar/r_2$, and the sum of the corresponding kinetic energies will be of order

$$\frac{\hbar^2}{2m}\left(\frac{1}{r_1^2}+\frac{1}{r_2^2}\right). \tag{13.4}$$

The potential energy of the interaction of the two electrons with the nucleus of charge Ze is

$$-Ze^2\left(\frac{1}{r_1}+\frac{1}{r_2}\right), \tag{13.5}$$

and their mutual interaction energy is of order $e^2/(r_1+r_2)$.

To find the ground-state energy we have to calculate the minimum of the estimated total energy

$$E(r_1, r_2) = \frac{\hbar^2}{2m}\left(\frac{1}{r_1^2}+\frac{1}{r_2^2}\right)-Ze^2\left(\frac{1}{r_1}+\frac{1}{r_2}\right)+\frac{e^2}{r_1+r_2}. \tag{13.6}$$

Since r_1 and r_2 play symmetrical roles, the minimum of $E(r_1, r_2)$ will occur when $r_1 = r_2 = r_m$, say.

After some simple calculation we find that

$$r_m = \frac{\hbar^2}{me^2}\left(Z-\frac{1}{4}\right)^{-1} \tag{13.7}$$

and

$$E_{\min} = -\left(Z-\frac{1}{4}\right)^2\frac{me^4}{\hbar^2} = 2\left(Z-\frac{1}{4}\right)^2 E_1, \tag{13.8}$$

where $E_1 = -13{\cdot}5$ eV is the ground-state energy of the hydrogen atom (13.3). The values o $E_{\min}$ calculated from (13.8) for different values of Z give good approximations to the experimental values (see Table III.1). The deviations from these values are due to the approximate nature of the arguments used.

TABLE III.1

	H^-	He	Li^+	Be^{++}	B^{+++}	C^{++++}
E_{calc}/E_1	1·125	6.125	15·12	28·12	45·12	66·12
E_{exp}/E_1	1·05	5·807	14·56	27·31	44·06	64·8

14. Let the atoms have a component of momentum p_x along the beam. Then, by the theorem of equipartition of energy, $p_x^2/2m = \frac{1}{2}kT$, and hence

$$p_x = \sqrt{mkT}. \tag{14.1}$$

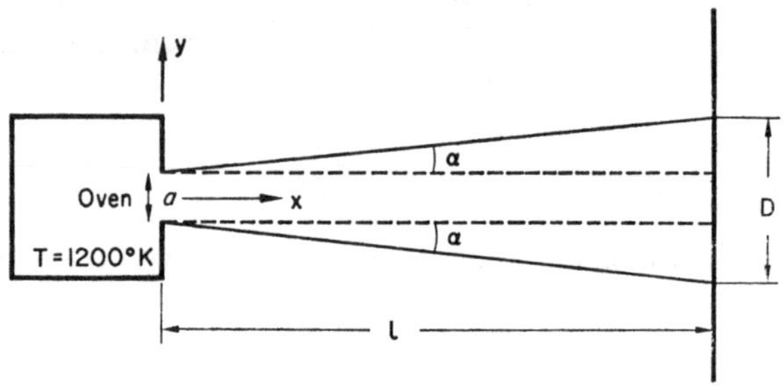

FIG. III.1.

The y-coordinate of an atom, at the moment when it passes through the aperture, is determined to a precision $\Delta y = a$, where a is the diameter of the aperture. Accordingly, there is an uncertainty Δp_y in the corresponding component p_y of the momentum, such that

$$\Delta p_y \geqslant \frac{\hbar}{2a}. \tag{14.2}$$

This leads to a spread of the beam with angle α, where

$$\tan\alpha = \frac{\frac{1}{2}\Delta p_y}{p_x} \geqslant \frac{\hbar}{4a\sqrt{mkT}} \tag{14.3}$$

Finally, for the diameter D of the spot obtained on the screen at a distance l from the aperture, we have that

$$D = a+2l\tan\alpha \geqslant a+\frac{l\hbar}{2a\sqrt{mkT}}. \tag{14.4}$$

From this expression it can be seen that the diameter of the spot cannot be made indefinitely small by narrowing the aperture. The right-hand side of the inequality (14.4) is a minimum for $a^2 = \hbar l(4mkT)^{-1/2}$ and thus

$$D_{\min} = \frac{\sqrt{2\hbar l}}{\sqrt[4]{mkT}}. \tag{14.5}$$

Using the numerical values given, $D_{\min}$ and the best diameter for the opening are given by $D_{\min} = 2a \approx 2\times10^{-3}$ cm.

15. Since the oscillator wavefunctions have well-defined parities, it follows that, in any steady state, $\langle x\rangle_n = \langle p\rangle_n = 0$, and hence $\Delta x = \sqrt{\langle x^2\rangle_n}$ and $\Delta p = \sqrt{\langle p^2\rangle_n}$. But, on account of (4a) and (4.6), we have $\langle p^2\rangle_n = mE_n$, $\langle x^2\rangle_n = E_n/m\omega^2$ and therefore

$$\Delta x\,\Delta p = \frac{E_n}{\omega} = \left(n+\frac{1}{2}\right)\hbar.$$

From this it can be seen that the product of coordinate and momentum uncertainty is a minimum for the oscillator ground state, for which $\Delta x\,\Delta p = \hbar/2$. In other words a simultaneous specification of coordinate and of momentum can be made with greater accuracy for this state than for any higher state. Thus the oscillator ground state is the nearest quantum analogue to the classical state of rest of the particle.

16. In the relation (11.8), the equality can hold only if, in (11.1),

$$B^{+}\psi(x) = (C-i\lambda D)\,\psi(x) = 0, \tag{16.1}$$

where λ has the value for which $f(\lambda) = f_{\min} = 0$, i.e.

$$\lambda = \frac{i}{2}\,\frac{\langle CD-DC\rangle}{\langle D^2\rangle}. \tag{16.2}$$

Using the notation of problem 11, it follows from (16.1) and (16.2) that

$$\left[U-\langle U\rangle+\frac{1}{2}\,\frac{\langle UV-VU\rangle}{(\Delta V)^2}\,(V-\langle V\rangle)\right]\psi(x) = 0. \tag{16.3}$$

For the special case $U = p$ and $V = x$, the following differential equation is obtained,

$$\left[-i\hbar\,\frac{d}{dx}-\langle p\rangle-\frac{i\hbar}{2}\,\frac{x-\langle x\rangle}{(\Delta x)^2}\right]\psi(x) = 0, \tag{16.4}$$

whose solution is

$$\psi(x) = N\exp\left[-\frac{(x-\langle x\rangle)^2}{4(\Delta x)^2}+\frac{i}{\hbar}\langle p\rangle\,x\right]. \tag{16.5}$$

From the normalization condition $\int_{-\infty}^{+\infty}|\psi(x)|^2 = 1$, we find for N the value $[2\pi(\Delta x)^2]^{-1/4}$ and hence

$$\psi(x) = [2\pi(\Delta x)^2]^{-1/4}\exp\left[-\frac{(x-\langle x\rangle)^2}{4(\Delta x)^2}+\frac{i}{\hbar}\langle p\rangle\,x\right]. \tag{16.6}$$

Wavefunctions of the form $\psi(x)$ are called "minimal wave-packets". $\psi(x)$ leads to a Gaussian probability distribution

$$|\psi(x)|^2 = [2\pi(\Delta x)^2]^{-1/2}\exp\left[-\frac{(x-\langle x\rangle)^2}{2(\Delta x)^2}\right], \tag{16.7}$$

of the measured values of x about their mean $\langle x\rangle$.

Since the state $\psi(x)$ allows the greatest permissible precision in the simultaneous determination of the coordinate and of the momentum, it can be said that a particle in such a state gives the best quantum mechanical approximation (consistent with the uncertainty relation) to a point-particle of classical mechanics.

In momentum space the minimal wave-packet may be obtained from (16.3) by substituting $U = x$ and $V = p$,

$$\phi(p) = [2\pi(\Delta p)^2]^{-1/4} \exp\left[-\frac{(p-\langle p\rangle)^2}{4(\Delta p)^2} - \frac{i}{\hbar}\langle x\rangle p\right]. \tag{16.8}$$

We see that in momentum space the minimal wave-packet also has a Gaussian probability distribution

$$|\phi(p)|^2 = [2\pi(\Delta p)^2]^{-1/2} \exp\left[-\frac{(p-\langle p\rangle)^2}{2(\Delta p)^2}\right] \tag{16.9}$$

of the measured momenta about their mean $\langle p\rangle$.

Let us define the width δ of a Gaussian distribution as half the distance between the two points of inflexion of the corresponding curve, i.e. as half the distance between the two zeros of the second derivative. After some simple calculation we then find that

$$\delta = \Delta x = h/2\Delta p. \tag{16.10}$$

Thus, if a Gaussian distribution "localizes" accurately a particle in configuration space (i.e. if Δx is very small), then, in agreement with Heisenberg's uncertainty relation, the "localization" in momentum space Δp becomes bad, and conversely (see Figs. III.2 and III.3).

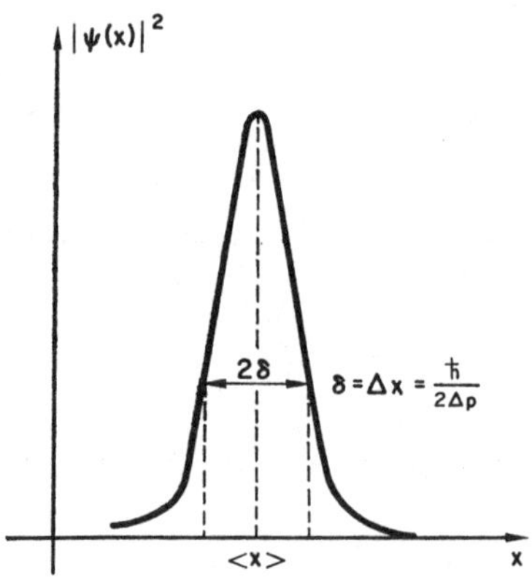

Fig. III.2.

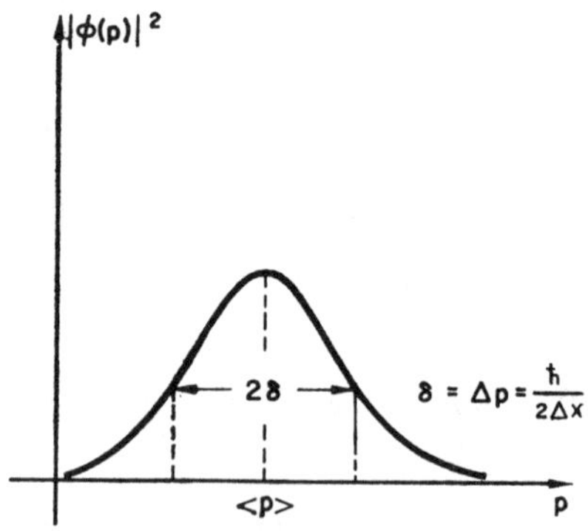

Fig. III.3.

17. Let τ be the mean life of the excited hydrogen atoms (this means that on average these atoms take a time τ to radiate their energy and return to the ground state). The excited state is thus located in time with a precision $\Delta t \approx \tau$, and therefore its energy, in accordance with (III.9), will have an uncertainty

$$\Delta E \geqslant \frac{\hbar}{2\tau}. \tag{17.1}$$

The resulting fluctuations in the energy of the excited atoms, because of their very short life, are appreciable, and corresponding fluctuations appear in the energy of the electrons which produced the excitations. It should be noted, incidentally, that it is the energy fluctuations ΔE due to the short lives of excited states which determine the natural width of spectral lines.

18. Using (17.1), the mean life τ of hydrogen atoms in the first excited state is found to be such that

$$\tau \geqslant \frac{\hbar}{2\,\Delta E} \approx 3{\cdot}1 \times 10^{-10} \text{ s}.$$

19. Since $\Delta t \sim a/c$, where a is the effective range of nuclear forces, and c is the velocity of light, we have from (III.9) that $\Delta E \sim \hbar c/a$. Thus

$$m_\pi \sim \frac{\Delta E}{c^2} \sim \frac{\hbar}{ac} \sim 270\, \text{m}_{\text{electron}}.$$

20. If we define the scalar product of the functions $f(\phi)$ and $g(\phi)$, where $-\pi \leqslant \phi \leqslant \pi$, by the expression

$$\langle f, g \rangle = \int_{-\pi}^{\pi} f^*(\phi)\, g(\phi)\, d\phi, \tag{20.1}$$

it can be seen that the operator $-i\hbar\partial/\partial\phi$ will be Hermitian only for the class of functions $\psi(\phi)$ which have equal values (with the exception of an arbitrary phase factor) at the upper and at the lower limits of integration, i.e. $\psi(\pi) = e^{i\alpha}\psi(-\pi)$. In general, if $\psi(\phi)$ is such a function, then $\phi\psi(\phi)$ is not, unless $\psi(\pi) = \psi(-\pi) = 0$. Therefore the operator $-i\hbar\partial/\partial\phi$, acting in (20a) on the function $\phi\psi(\phi)$, cannot represent the observable l_z, a fact which restricts the validity of the commutation relation to the special class of functions defined above.

It can be seen immediately that (20b) is not consistent with Heisenberg's uncertainty principle, as follows. Since ϕ belongs to a finite interval, the uncertainty $\Delta\phi$ has to be finite also. As was shown in problem 3, the maximum value of $\Delta\phi$ is $\pi/\sqrt{3}$ and it is obtained for a uniform probability distribution. Thus, if $\Delta l_z \to 0$, the left-hand side of (20b) tends towards zero, and we arrive at the contradiction $0 \geqslant \hbar/2$. From what has been said, it can be seen that an uncertainty relation between Δl_z and $\Delta\phi$, to be consistent with Heisenberg's principle, would have to have the following properties:

(1) When $\Delta\phi \to 0$, then $\Delta l_z \to \infty$.
(2) When $\Delta l_z \to 0$, then $\Delta\phi \to \pi/3$.

Such a relation was recently proposed by Judge (1963) and it has the form

$$\Delta l_z \frac{\Delta\phi}{1-\dfrac{3(\Delta\phi)^2}{\pi^2}} \geqslant 0{\cdot}16\hbar. \tag{20.2}$$

CHAPTER IV

The Semi-classical Approximation

1. The Wavefunction in the WKB Approximation

With the substitution

$$\psi(\mathbf{r}) = \exp\left[\frac{i}{\hbar} S(\mathbf{r})\right], \tag{IV.1}$$

the time-independent Schrödinger equation becomes

$$(\nabla S)^2 - i\hbar \nabla^2 S - 2m(E-V) = 0. \tag{IV.2}$$

The "semi-classical approximation" consists in writing the function $S(\mathbf{r})$ as a power series in $\hbar$:

$$S = S_0 + \frac{\hbar}{i} S_1 + \left(\frac{\hbar}{i}\right)^2 S_2 + \ldots. \tag{IV.3}$$

For the one-dimensional case, $V = V(x)$, $S = S(x)$, the functions S_i $(i = 0, 1, 2, \ldots)$ are then determined in succession from the system of differential equations obtained by substituting (IV.3) into (IV.2) and equating powers of $\hbar$:

$$\begin{aligned} S_0'^2 &= 2m(E-V) \\ 2S_1' &= -S_0''/S_0' \\ 2S_2' &= -(S_1''+S_1'^2)/S_0' \\ &\ldots\ldots\ldots\ldots\ldots\ldots \end{aligned} \tag{IV.4}$$

The first equation of the set (IV.4) shows that the wavefunction in zero-order approximation is a linear combination of the exponentials $\exp\left[\pm i/\hbar \int p(x)\,dx\right]$, where $p(x) = \sqrt{2m[(E-V(x)]}$ is the (classical) momentum of the particle. This approximation (i.e. the replacement of S by S_0 and the omission in (IV.2) of $-i\hbar S_0''$ in comparison with $S_0'^2$) is a good one if

$$\hbar\left|\frac{S_0''}{S_0'^2}\right| \ll 1, \tag{IV.5}$$

i.e. if

$$\left|\frac{d\lambda\!\!\!^{-}}{dx}\right| \ll 1,$$

where $\lambda\!\!\!{}^{-} = \lambda/2\pi$, and $\lambda = 2\pi\hbar/p(x)$ is the de Broglie wavelength corresponding to the classical momentum of the particle at each point x.

The approximate expressions obtained for the wavefunction $\psi(x)$ by using (IV.1), (IV.3) and (IV.4) to any given order are valid only for sufficiently large values of p. They cannot be used near the classical *turning points*, at which $p = 0$, i.e. $E = V(x)$.

The so-called "WKB approximation" (G. Wentzel, H. A. Kramers, L. Brillouin, 1926) is the approximation obtained by taking the first two terms only of the series (IV.3), i.e. by omitting terms in $\hbar^2$ and higher powers of $\hbar$. Using the second equation of the set (IV.4), one obtains $S_1 = -\frac{1}{2}\ln p + \text{const.}$, and hence the wavefunction $\psi(x)$ in the WKB approximation becomes

$$\psi = \frac{A}{\sqrt{p}} e^{\frac{i}{\hbar}\int p\,dx} + \frac{B}{\sqrt{p}} e^{-\frac{i}{\hbar}\int p\,dx} \tag{IV.6}$$

The condition for the validity of the WKB approximation is that $\hbar|S_2| \ll 1$. Since $S_0' = \pm\hbar/\lambda\!\!\!{}^{-}$, it follows from the equation

$$-2S_0'S_2' = \frac{3}{4}\left(\frac{S_0''}{S_0'}\right)^2 - \frac{1}{2}\frac{S_0'''}{S_0'},$$

obtained from the system (IV.4), that

$$-\hbar S_2' = \pm\left(\frac{1}{4}\lambda\!\!\!{}^{-}{}'' - \frac{1}{8}\frac{\lambda\!\!\!{}^{-}{}'^2}{\lambda\!\!\!{}^{-}}\right)$$

whence

$$\hbar S_2 = \mp\left(\frac{1}{4}\lambda\!\!\!{}^{-}{}' - \frac{1}{8}\int\frac{\lambda\!\!\!{}^{-}{}'^2}{\lambda\!\!\!{}^{-}}\,dx\right) \tag{IV.7}$$

The condition $\hbar|S_2| \ll 1$ is thus satisfied if $|\lambda\!\!\!{}^{-}{}'| \ll 1$, cf. (IV.5). The effect of higher approximations shows in the appearance, in the factors multiplying the exponentials $\exp\left[\pm\frac{i}{\hbar}\int p\,dx\right]$, of terms of higher order in $\hbar$. In regions of space which are classically "inaccessible", i.e. in which $E < V(x)$, the function $p(x)$ is imaginary. The wavefunction (IV.6) in these regions can then be written in the form

$$\psi = \frac{A'}{\sqrt{|p|}} e^{-\frac{1}{\hbar}\int|p|\,dx} + \frac{B'}{\sqrt{|p|}} e^{\frac{1}{\hbar}\int|p|\,dx}. \tag{IV.8}$$

2. Formulae for connecting WKB Wavefunctions on Opposite Sides of Turning Points

The WKB wavefunctions (IV.6) and (IV.8) give approximate solutions of the Schrödinger equation for regions sufficiently far to the left and/or to the right of any classical turning points, but they contain arbitrary constants in each such region which are unrelated to those in other regions. To relate these constants, the WKB solutions have to be connected across the turning points. If, near the turning point $x = a$, say, the potential $V(x)$ behaves as shown in Fig. IV.1, then the properly connected WKB solutions are found to have the form

$$\psi_1 = \begin{cases} \dfrac{1}{\sqrt{p}} \sin\left(\dfrac{1}{\hbar}\displaystyle\int_x^a p\,dx + \dfrac{\pi}{4}\right) & \text{for} \quad x < a \qquad \text{(IV.9)} \\[2ex] \dfrac{1}{2\sqrt{|p|}} \exp\left(-\dfrac{1}{\hbar}\displaystyle\int_a^x |p|\,dx\right) & \text{for} \quad x > a. \qquad \text{(IV.9')} \end{cases}$$

If, on the other hand, $V(x)$ behaves as shown in Fig. IV.2, then the properly connected solutions are found to have the form

$$\psi_2 = \begin{cases} \dfrac{1}{\sqrt{p}} \sin\left(\dfrac{1}{\hbar}\displaystyle\int_a^x p\,dx + \dfrac{\pi}{4}\right) & \text{for} \quad x > a, \qquad \text{(IV.10)} \\[2ex] \dfrac{1}{2\sqrt{|p|}} \exp\left(-\dfrac{1}{\hbar}\displaystyle\int_x^a |p|\,dx\right) & \text{for} \quad x < a. \qquad \text{(IV.10')} \end{cases}$$

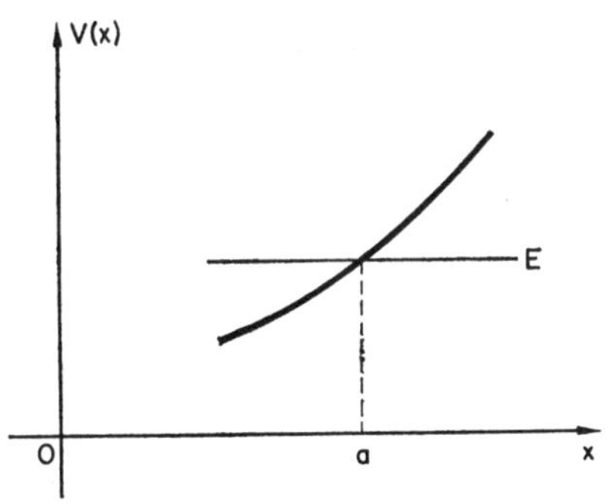

FIG. IV.1.

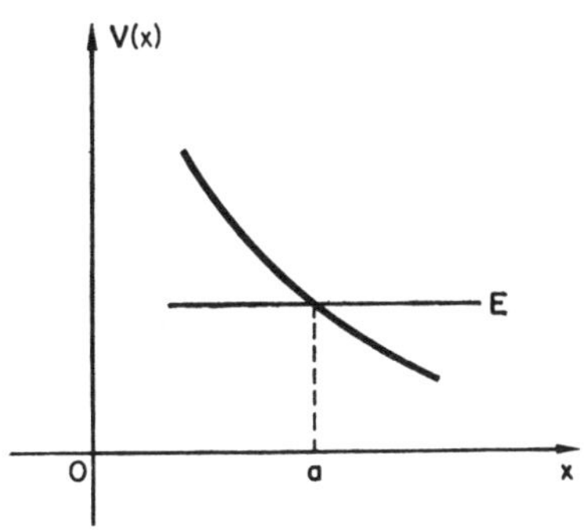

FIG. IV.2.

Problems

1. Determine the wavefunction in the semi-classical approximation up to terms of order $\hbar$ in the factor which multiplies the exponential $\exp\left(\frac{i}{\hbar}\int p\,dx\right)$ in (IV.6).

2. Find the solution of the one-dimensional Schrödinger equation

$$\frac{d^2\psi}{dx^2} + \frac{2m}{\hbar^2}[E - V(x)]\psi = 0, \qquad \text{(2a)}$$

which in the WKB approximation has the form

$$\frac{1}{\sqrt{p}} \exp\left(\pm\frac{i}{\hbar}\int_x^a p\,dx\right) \tag{2b}$$

to the left of the turning point shown in Fig. IV.1.

3. Find the solution of equation (2a) which, in the WKB approximation, has the form

$$\frac{1}{\sqrt{|p|}} \exp\left(\pm\frac{1}{\hbar}\int_x^a |p|\,dx\right)$$

to the left of the turning point shown in Fig. IV.2.

4. A particle of mass m moves in the potential well $V = V(x)$ (Fig. IV.3). Suppose that $V(x)$ is such that, for any energy $E > V_{\min}$, there are two and only two turning points. Show that, in the WKB approximation, the discrete energy levels are determined by the condition

$$\int_a^b p(x)\,dx = \pi\hbar(n+\tfrac{1}{2}), \quad n = 0, 1, 2, \ldots. \tag{4a}$$

where a and b are the turning points given by $V(a) = V(b) = E$, and $b > a$.

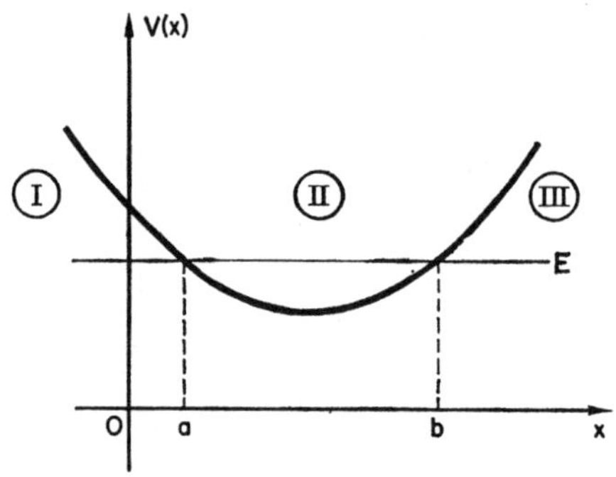

FIG. IV.3.

5. Determine, in the WKB approximation, the discrete energy levels for a particle in the following potentials

(a) $V(x) = m\omega^2 x^2/2$ (harmonic oscillator),

(b) $V(x) = -V_0/\cosh^2\left(\frac{x}{c}\right)$ (see Fig. IV.4).

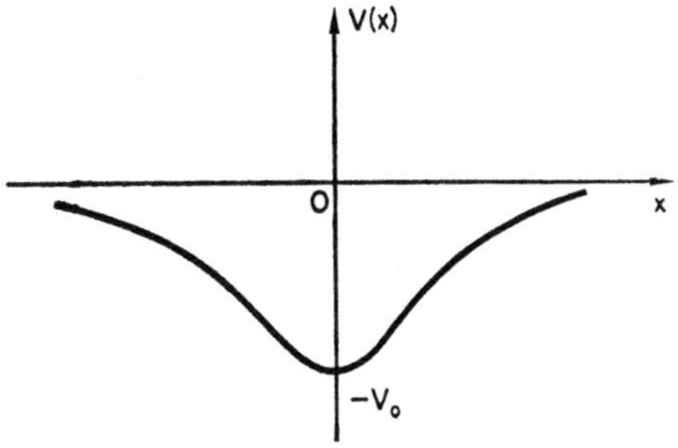

FIG. IV.4.

6. Determine, in the WKB approximation, the potential energy function $V = V(x)$ which yields a given discrete energy spectrum E_n, on the assumption that $V(x)$ is an even function of x, i.e. that $V(x) = V(-x)$, and that, for $x > 0$, $V(x)$ increases monotonically.

7. Apply the results of the preceding problem to the energy spectrum of a one-dimensional harmonic oscillator.

8. Show that, in the WKB approximation, the mean value of the kinetic energy in the bound state $\psi_n(x)$ of the potential well $V = V(x)$ (Fig. IV.3), is given by

$$\langle T\rangle_n = \frac{1}{2}\left(n+\frac{1}{2}\right)\frac{dE_n}{dn}\,. \tag{8a}$$

9. Using the quantum virial theorem (see problem 10, Chapter III), which states that

$$2\langle T\rangle_n = \langle \mathbf{r}\,.\,\nabla V\rangle_n\,, \tag{9a}$$

determine, in the WKB approximation, the discrete energy levels of a particle moving in the potential $V(x) = V_0 x^s$.

10. Show that, in the WKB approximation, the transmission coefficient for a particle of mass m and energy E through the potential barrier $V = V(x)$ (Fig. IV.5) is given by the expression

$$T = e^{-2L}(1+\tfrac{1}{4}e^{-2L})^{-2}, \tag{10a}$$

where

$$L = \frac{1}{\hbar}\int_a^b |p|\,dx. \tag{10b}$$

The condition (IV.5) can be assumed to hold.

11. Calculate the transmission coefficient (10a) for the parabolic potential barrier shown in Fig. IV.6:

$$V(x) = \begin{cases} V_0(1-x^2/x_0^2), & \text{if } \quad |x| < x_0\,, \\ 0, & \text{if } \quad |x| > x_0\,. \end{cases} \tag{11a}$$

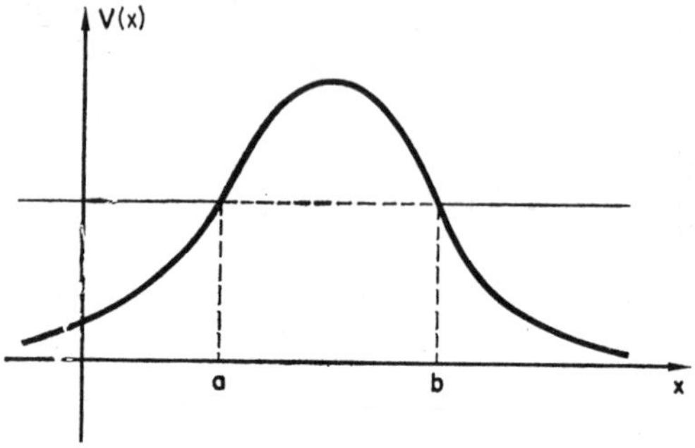

FIG. IV.5.

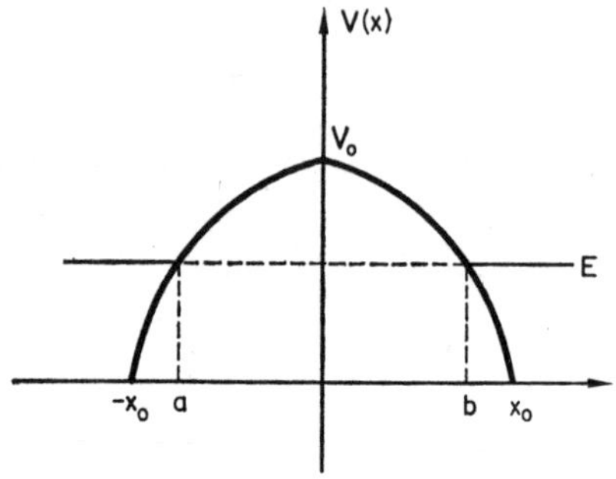

FIG. IV.6.

What restrictions must be imposed on the parameters x_0 and V_0 for the result obtained to be valid?

12. Find the general solution of the radial equation (II.15) in the WKB approximation, supposing that as $r \to \infty$ the potential energy $V(r)$ tends to zero faster than r^{-1}, and that, at the point $r = 0$, it has a singularity weaker than r^{-2}.

Solutions

1. Remembering that $\lambda\!\!\!^{-} = \hbar/p$, we find from (IV.7) that

$$S_2 = \frac{p'}{4p^2} + \frac{1}{8}\int \frac{p'^2}{p^3}\,dx. \tag{1.1}$$

Thus the wavefunction to the required approximation is

$$\psi = \exp\left(\frac{i}{\hbar}S\right) \approx \exp\left(\frac{i}{\hbar}S_0 + S_1 - i\hbar S_2\right) = \frac{1}{\sqrt{p}}\exp\left(\frac{i}{\hbar}\int p\,dx - \frac{i\hbar}{4}\frac{p'}{p^2} - \frac{i\hbar}{8}\int \frac{p'^2}{p^3}\,dx\right).$$

But, on account of (IV.5), we have $\hbar|p'/p^2| \ll 1$, $\hbar(p'^2/p^3) \ll p/\hbar$ and thus, to the required approximation, we find

$$\psi \approx \frac{1}{\sqrt{p}}\left(1 - \frac{i\hbar}{4}\frac{p'}{p^2} - \frac{i\hbar}{8}\int \frac{p'^2}{p^3}\,dx\right)\exp\left(\frac{i}{\hbar}\int p\,dx\right). \tag{1.2}$$

2. In order to solve this problem we try to find another WKB solution such that it and (IV.9, 9′) together form a system of two linearly independent solutions. We shall seek it in the form

$$\tilde{\psi}_1 = \begin{cases} \dfrac{1}{\sqrt{p}}\cos\left(\dfrac{1}{\hbar}\displaystyle\int_x^a p\,dx + \dfrac{\pi}{4}\right) & \text{for} \quad x < a, \qquad (2.1) \\[2ex] \dfrac{C}{\sqrt{|p|}}\exp\left(\dfrac{1}{\hbar}\displaystyle\int_a^x |p|\,dx\right) & \text{for} \quad x > a. \qquad (2.1') \end{cases}$$

To determine the constant C we use the fact that the Wronskian of two linearly independent solutions of (2a), with the same value of E, does not depend on the variable x. It follows that, in the present case, we require C to be such that

$$W(\psi_1,\ \tilde{\psi}) = \begin{vmatrix} \psi_1, \psi_1' \\ \tilde{\psi}_1, \tilde{\psi}_1' \end{vmatrix} \tag{2.2}$$

be a constant independent of x.

Now, in the range $x < a$, i.e., for the solutions (IV.9) and (2.1) we have

$$W(x) = \begin{vmatrix} \dfrac{1}{\sqrt{p}} \sin\left(\dfrac{1}{\hbar}\displaystyle\int_x^a p\,dx+\dfrac{\pi}{4}\right); & -\dfrac{1}{2}\dfrac{p'}{\sqrt{p^3}}\sin\left(\dfrac{1}{\hbar}\displaystyle\int_x^a p\,dx+\dfrac{\pi}{4}\right)-\dfrac{\sqrt{p}}{\hbar}\cos\left(\dfrac{1}{\hbar}\displaystyle\int_x^a p\,dx+\dfrac{\pi}{4}\right) \\ \dfrac{1}{\sqrt{p}} \cos\left(\dfrac{1}{\hbar}\displaystyle\int_x^a p\,dx+\dfrac{\pi}{4}\right); & -\dfrac{1}{2}\dfrac{p'}{\sqrt{p^3}}\cos\left(\dfrac{1}{\hbar}\displaystyle\int_x^a p\,dx+\dfrac{\pi}{4}\right)+\dfrac{\sqrt{p}}{\hbar}\sin\left(\dfrac{1}{\hbar}\displaystyle\int_x^a p\,dx+\dfrac{\pi}{4}\right) \end{vmatrix}.$$

Since, according to (IV.5), $|p'/\sqrt{p^3}| \ll \sqrt{p}/\hbar$, the terms of order $p'/\sqrt{p^3}$ can be neglected in comparison with those of order $\sqrt{p}/\hbar$, and we find then that $W(x) = 1/\hbar$ for $x < a$.

In the range $x > a$, i.e. for the solutions (IV.9′) and (2.1′), we find after a similar calculation that $W(x) = C/\hbar$.

In accordance with (2.2), it follows that $C = 1$. Solution (2b) is then obtained in the form of a linear combination

$$\psi = (\tilde{\psi}_1 \pm i\psi_1)e^{\mp i\frac{\pi}{4}},$$

i.e. finally,

$$\psi = \begin{cases} \dfrac{1}{\sqrt{p}} \exp\left(\pm\dfrac{i}{\hbar}\displaystyle\int_x^a p\,dx\right), & \text{for} \quad x < a \\ \dfrac{1}{\sqrt{|p|}} \exp\left(\dfrac{1}{\hbar}\displaystyle\int_a^x |p|\,dx \mp i\dfrac{\pi}{4}\right) + \dfrac{1}{2\sqrt{|p|}} \exp\left(-\dfrac{1}{\hbar}\displaystyle\int_a^x |p|\,dx \mp i\dfrac{\pi}{4}\right), & \text{for} \quad x > a. \end{cases}$$

3. Starting from the function (IV.10, 10′) and following step by step the procedure used in the preceding problem we obtain

$$\tilde{\psi}_2 = \begin{cases} \dfrac{1}{\sqrt{p}} \cos\left(\dfrac{1}{\hbar}\displaystyle\int_a^x p\,dx+\dfrac{\pi}{4}\right) & \text{for} \quad x > a, \quad (3.1) \\ \dfrac{1}{\sqrt{|p|}} \exp\left(\dfrac{1}{\hbar}\displaystyle\int_x^a |p|\,dx\right) & \text{for} \quad x < a. \quad (3.1') \end{cases}$$

Let us now consider the two linear combinations

$$\psi^{(+)} = (\tilde{\psi}_2 + i\psi_2)\exp\left(-i\frac{\pi}{4}\right); \qquad \psi^{(-)} = (\tilde{\psi}_2 - i\psi_2)\exp\left(i\frac{\pi}{4}\right).$$

After some simple calculation, we find that

$$\psi^{(+)} = \begin{cases} \dfrac{1}{\sqrt{p}}\exp\left(\dfrac{i}{\hbar}\displaystyle\int_a^x p\,dx\right), & \text{for} \quad x > a, \\[2ex] \dfrac{1}{\sqrt{|p|}}\exp\left(\dfrac{1}{\hbar}\displaystyle\int_x^a |p|\,dx - i\frac{\pi}{4}\right) + \dfrac{1}{2\sqrt{|p|}}\exp\left(-\dfrac{1}{\hbar}\displaystyle\int_x^a |p|\,dx + i\frac{\pi}{4}\right), & \text{for} \quad x < a; \end{cases}$$

$$\psi^{(-)} = \begin{cases} \dfrac{1}{\sqrt{p}}\exp\left(-\dfrac{i}{\hbar}\displaystyle\int_a^x p\,dx\right), & \text{for} \quad x > a, \\[2ex] \dfrac{1}{\sqrt{|p|}}\exp\left(\dfrac{1}{\hbar}\displaystyle\int_x^a |p|\,dx + i\frac{\pi}{4}\right) + \dfrac{1}{2\sqrt{|p|}}\exp\left(-\dfrac{1}{\hbar}\displaystyle\int_x^a |p|\,dx - i\frac{\pi}{4}\right), & \text{for} \quad x < a. \end{cases}$$

The solutions which, under the conditions of this problem, have the forms

$$\frac{1}{\sqrt{|p|}}\exp\left(\frac{1}{\hbar}\int_x^a |p|\,dx\right) \quad \text{and} \quad \frac{1}{\sqrt{|p|}}\exp\left(-\frac{1}{\hbar}\int_x^a |p|\,dx\right)$$

can be constructed from the linear combinations

$$\psi = \tfrac{1}{2}(\psi^{(+)} - i\psi^{(-)})\exp\left(i\frac{\pi}{4}\right) \quad \text{and} \quad \psi = (\psi^{(+)} + i\psi^{(-)})\exp\left(-i\frac{\pi}{4}\right)$$

respectively.

Thus, in the first case we have

$$\psi = \begin{cases} \dfrac{1}{\sqrt{|p|}}\exp\left(\dfrac{1}{\hbar}\displaystyle\int_x^a |p|\,dx\right), & \text{for} \quad x < a, \\[2ex] \dfrac{1}{2\sqrt{p}}\exp\left(\dfrac{i}{\hbar}\displaystyle\int_a^x p\,dx + i\frac{\pi}{4}\right) + \dfrac{1}{2\sqrt{p}}\exp\left(-\dfrac{i}{\hbar}\displaystyle\int_a^x p\,dx - i\frac{\pi}{4}\right), & \text{for} \quad x > a, \end{cases}$$

and in the second case

$$\psi = \begin{cases} \dfrac{1}{\sqrt{|p|}} \exp\left(-\dfrac{1}{\hbar}\displaystyle\int_x^a |p|\, dx\right), & \text{for} \quad x < a, \\ \dfrac{1}{\sqrt{p}} \exp\left(\dfrac{i}{\hbar}\displaystyle\int_a^x p\, dx - i\dfrac{\pi}{4}\right) + \dfrac{1}{\sqrt{p}} \exp\left(-\dfrac{i}{\hbar}\displaystyle\int_a^x p\, dx + i\dfrac{\pi}{4}\right), & \text{for} \quad x > a. \end{cases}$$

4. In the WKB approximation the determination of discrete energy levels in the potential well $V = V(x)$ reduces to finding the conditions under which real exponential WKB solutions vanish asymptotically in regions I and III. Thus in region I we require the solution (IV.10′), and in region III the solution (IV.9′), so that at any point x $(a < x < b)$ situated sufficiently far from the turning points a and b, the solution (IV.9),

$$\psi_1 = \frac{A}{\sqrt{p}} \sin\left(\frac{1}{\hbar}\int_x^b p\, dx + \frac{\pi}{4}\right), \quad a < x < b,$$

has to coincide with (IV.10′),

$$\psi_2 = \frac{B}{\sqrt{p}} \sin\left(\frac{1}{\hbar}\int_a^x p\, dx + \frac{\pi}{4}\right), \quad a < x < b.$$

Equating these solutions and their derivatives at a point x in the interval (a, b), we obtain

$$A \sin\left(\frac{1}{\hbar}\int_x^b p\, dx + \frac{\pi}{4}\right) - B \sin\left(\frac{1}{\hbar}\int_a^x p\, dx + \frac{\pi}{4}\right) = 0,$$

$$A\left[\frac{\sqrt{p}}{\hbar} \cos\left(\frac{1}{\hbar}\int_x^b p\, dx + \frac{\pi}{4}\right) + \frac{1}{2}\frac{p'}{\sqrt{p^3}} \sin\left(\frac{1}{\hbar}\int_x^b p\, dx + \frac{\pi}{4}\right)\right]$$
$$+ B\left[\frac{\sqrt{p}}{\hbar} \cos\left(\frac{1}{\hbar}\int_a^x p\, dx + \frac{\pi}{4}\right) - \frac{1}{2}\frac{p'}{\sqrt{p^3}} \sin\left(\frac{1}{\hbar}\int_a^x p\, dx + \frac{\pi}{4}\right)\right] = 0.$$

As was shown in problem 2, we have, on the strength of condition (IV.5), that $\left|p'/\sqrt{p^3}\right| \ll \sqrt{p}/\hbar$, and hence the terms containing $p'/\sqrt{p^3}$ can be neglected in comparison with those containing $\sqrt{p}/\hbar$. For the homogeneous system so obtained to have non-trivial solutions it is necessary that its determinant should vanish, i.e. that

$$\sin\left(\frac{1}{\hbar}\int_a^b p\, dx + \frac{\pi}{2}\right) = 0. \tag{4.1}$$

Since $p = \sqrt{2m(E-V)} \geqslant 0$, the integral $\int_a^b p\,dx$ is non-negative, so that the condition becomes

$$\frac{1}{\hbar}\int_a^b p\,dx+\frac{\pi}{2} = (n+1)\pi, \quad n = 0, 1, 2, \ldots, \tag{4.2}$$

which is what we had to prove.

Note that, according to classical mechanics, the particle can move only in the range $a \leqslant x \leqslant b$, and performs a periodic motion in this range with a period T given by

$$T = 2\int_a^b \frac{dx}{v} = 2m\int_a^b \frac{dx}{p}.$$

Thus condition (4a) can be written in the form

$$\oint p\,dx = 2\pi\hbar(n+\tfrac{1}{2}), \tag{4.3}$$

where the integral $\oint = 2\int_a^b$ is extended over a complete period of the classical motion. The relation (4.3) is just *the Bohr–Sommerfeld quantization rule* of the old Quantum Theory.

5. The energy levels of an harmonic oscillator in the WKB approximation are obtained from condition (4a), in which the limits of the integration are the solutions of the equation $p(x) = \left[2m\left(E-\frac{m\omega^2}{2}x^2\right)\right]^{1/2} = 0$, i.e.

$$a = -(2E/m\omega^2)^{1/2}, \quad b = +(2E/m\omega^2)^{1/2}.$$

Making the substitution $x = (2E/m\omega^2)^{1/2}t$, (4a) becomes

$$\int_a^b p\,dx = \frac{2E}{\omega}\int_{-1}^{1}\sqrt{1-t^2}\,dt = \frac{\pi E}{\omega} = \pi\hbar\left(n+\frac{1}{2}\right),$$

whence

$$E \equiv E_n = \hbar\omega(n+\tfrac{1}{2}), \quad n = 0, 1, 2, \ldots.$$

Note that the energy levels of an oscillator calculated in the WKB approximation coincide with those obtained by solving the Schrödinger equation exactly (see problem 17, Chapter II).

For the potential $V(x) = -V_0/\cosh^2\left(\frac{x}{c}\right)$, the energy levels in the WKB approximation are also obtainable from (4a), a and b being now the solutions of the equation

$$p(x) = \left[2m\left(E+\frac{V_0}{\cosh^2\left(\frac{x}{c}\right)}\right)\right]^{1/2} = 0. \tag{5.1}$$

To evaluate the integral

$$J(E) = \int_a^b \left[2m\left(E+\frac{V_0}{\cosh^2\left(\frac{x}{c}\right)}\right)\right]^{-1/2} dx \tag{5.2}$$

let us differentiate both sides with respect to E. Then, because of equation (5.1), the derivatives of the integral with respect to its upper and lower limits (which depend on E!) vanish, and thus

$$\frac{dJ}{dE} = m \int_a^b \left[2m\left(E+\frac{V_0}{\cosh^2\left(\frac{x}{c}\right)}\right)\right]^{-1/2} dx. \tag{5.3}$$

Making the substitution $\sinh\left(\frac{x}{c}\right) = z$, (5.3) becomes

$$\frac{dJ}{dE} = mc \int_{z_1}^{z_2} \{2m[E(1+z^2)+V_0]\}^{-1/2}\,dz = \frac{mc\pi}{\sqrt{-2mE}},$$

whence we find

$$J(E) = -\pi c(-2mE)^{1/2}+K.$$

The constant K can be determined by observing that for $E = -V_0$ the range of integration in (5.2) reduces to a point (see Fig. IV.4), so that $J(-V_0) = 0$, and so

$$J(E) = \pi\sqrt{2mc^2}\,(\sqrt{V_0}-\sqrt{-E}) = \pi\hbar(n+\tfrac{1}{2}).$$

Finally, the energy levels in the WKB approximation are given by

$$E_n = -\frac{\hbar^2}{2mc^2}\left[\sqrt{\frac{2mc^2V_0}{\hbar^2}}-\left(n+\frac{1}{2}\right)\right]^2, \qquad n = 0, 1, 2, \ldots.$$

6. The energy spectrum for the potential $V = V(x)$ is given by relation (4a):

$$\pi\hbar(n+\tfrac{1}{2}) = \int_a^b \sqrt{2m[E-V(x)]}\,dx. \tag{4a}$$

Now, since the potential is even, $a = -b$, and (4a) becomes

$$\pi\hbar(n+\tfrac{1}{2}) = 2\int_0^b \sqrt{2m[E-V(x)]}\,dx. \tag{6.1}$$

Our problem reduces to that of solving this integral equation for $V(x)$. For this purpose let us differentiate (6.1) with respect to E, and, instead of x, let us take V as the independent variable. Then, since $V(b) = E$, we have

$$\frac{\pi\hbar}{\sqrt{2m}}\,\frac{dn}{dE} = \int_0^b \frac{dx}{\sqrt{E-V}} = \int_0^E \frac{dx}{dV}\,\frac{dV}{\sqrt{E-V}}. \tag{6.2}$$

Multiplying (6.2) by $(\alpha-E)^{1/2}$ and integrating over E from 0 to α, where α is a parameter to be determined later, we obtain

$$\frac{\pi\hbar}{\sqrt{2m}}\int_0^\alpha \frac{dn}{dE}\frac{dE}{\sqrt{\alpha-E}} = \int_0^\alpha dE\int_0^E \frac{dx}{dV}\frac{dV}{\sqrt{(\alpha-E)(E-V)}}$$

$$= \int_0^\alpha \frac{dx}{dV}\,dV\int_V^\alpha \frac{dE}{\sqrt{(\alpha-E)(E-V)}} = \pi x(\alpha).$$

Putting $\alpha = V$, we find the following expression for the inverse function $x = x(V)$ of $V = V(x)$:

$$x(V) = \frac{\hbar}{\sqrt{2m}}\int_0^V \frac{dE}{\dfrac{dE}{dn}\sqrt{V-E}}. \tag{6.3}$$

Knowing the dependence $E = E(n)$, equation (6.3) allows $x(V)$, and hence the potential energy $V(x)$, to be determined (see problem 7).

7. Since the energies E_n of an harmonic oscillator as calculated in the WKB approximation coincide with those obtained by solving the Schrödinger equation exactly, i.e. $E_n = \omega\hbar\left(n+\frac{1}{2}\right)$, we may expect that if E in (6.3) is taken to have this dependence on n, then $V(x)$ will be found to be the harmonic oscillator potential $V(x) = m\omega^2x^2/2$. That this is in fact the case can be seen readily, since (6.3) becomes

$$x = \frac{\hbar}{\sqrt{2m}}\int_0^V \frac{dE}{\omega\hbar\sqrt{V-E}} = \sqrt{\frac{2V}{m\omega^2}},$$

whence $V = m\omega^2x^2/2$.

8. The mean value of the kinetic energy in the bound state ψ_n (assuming the wavefunction to be real) is

$$\langle T\rangle_n = -\frac{\hbar^2}{2m}\int_{-\infty}^{+\infty}\psi_n\frac{d^2\psi_n}{dx^2}\,dx = \frac{\hbar^2}{2m}\int_{-\infty}^{+\infty}\left(\frac{d\psi_n}{dx}\right)^2 dx. \tag{8.1}$$

In the WKB approximation, the wavefunction in the range in which classical motion is possible, i.e. $a \leqslant x \leqslant b$, has the form

$$\psi_n = \frac{N_n}{\sqrt{p}}\sin\left(\frac{1}{\hbar}\int_a^x p\,dx+\frac{\pi}{4}\right), \quad p = \sqrt{2m[E_n-V(x)]}.$$

Hence

$$\frac{d\psi_n}{dx} = \frac{\sqrt{p}}{\hbar} N_n \cos\left(\frac{1}{\hbar}\int_a^x p\,dx + \frac{\pi}{4}\right) - \frac{N_n}{2}\frac{p'}{\sqrt{p^3}} \sin\left(\frac{1}{\hbar}\int_a^x p\,dx + \frac{\pi}{4}\right). \tag{8.2}$$

Substituting this expression into (8.1), the range of integration can be reduced to the interval $[a, b]$, since outside this interval the function ψ_n tends exponentially to zero. Replacing the squares of the trigonometric functions by their mean value $\frac{1}{2}$, and neglecting the integral of the oscillating factor

$$\sin\left(\frac{1}{\hbar}\int_a^x p\,dx + \frac{\pi}{4}\right)\cos\left(\frac{1}{\hbar}\int_a^x p\,dx + \frac{\pi}{4}\right) = \tfrac{1}{2}\sin\left(\frac{2}{\hbar}\int_a^x p\,dx + \frac{\pi}{2}\right),$$

we obtain finally the expression

$$\langle T\rangle_n = \frac{N_n^2}{4m}\int_a^b \left(p + \frac{\hbar^2}{4}\frac{p'^2}{p^3}\right) dx.$$

Because of (IV.5), $\hbar^2 p'^2/p^3 \ll p$, and thus

$$\langle T\rangle_n \approx \frac{N_n^2}{4m}\int_a^b p\,dx = \frac{N_n^2}{4m}\pi\hbar(n+\tfrac{1}{2}). \tag{8.3}$$

The constant N_n is determined from the normalization condition

$$\int_{-\infty}^{+\infty} \psi_n^2\,dx \approx N_n^2 \int_a^b \frac{1}{p}\sin^2\left(\frac{1}{p}\int_a^x p\,dx + \frac{\pi}{4}\right) dx \approx \frac{N_n^2}{2}\int_a^b \frac{dx}{p} = 1.$$

On the other hand, by differentiating (4a) with respect to n, we obtain

$$m\frac{dE_n}{dn}\int_a^b \frac{dx}{\sqrt{2m(E_n - V)}} = m\frac{dE_n}{dn}\int_a^b \frac{dx}{d} = \pi\hbar,$$

hence $N_n^2 = (2m/\pi\hbar)(dE_n/dn)$, and the expression (8a) then follows from (8.3). Note that for an harmonic oscillator the accurate result $\langle T\rangle_n = \dfrac{\omega\hbar}{2}(n+\tfrac{1}{2})$ is found again (see III.4.6).

9. From the quantum virial theorem it follows that $2\langle T\rangle_n = s\langle V\rangle_n$, and hence that

$$E_n = \langle T\rangle_n + \langle V\rangle_n = \frac{2+s}{s}\langle T\rangle_n. \tag{9.1}$$

Replacing $\langle T \rangle_n$ by its value obtained from (8a), the following differential equation is obtained

$$E_n = \frac{2+s}{2s}\left(n+\tfrac{1}{2}\right)\frac{dE_n}{dn},$$

whence

$$E_n = C\left(n+\tfrac{1}{2}\right)^{2s/(2+s)}. \tag{9.2}$$

If s is an even integer, the constant C can be determined by using (6.1),

$$\pi\hbar\left(n+\tfrac{1}{2}\right) = 2\int_0^b \sqrt{2m(E_n - V_0 x^s)}\, dx, \tag{9.3}$$

where b is the solution of $b^s = E_n/V_0$.

Let us differentiate with respect to E_n, and introduce a new variable $y = V_0 x^s/E_n$. Using (9.2) we obtain

$$C = \left[\sqrt{\frac{\pi}{2m}}\, s\hbar V_0^{1/s} \frac{\Gamma\left(\frac{3}{2}+\frac{1}{s}\right)}{\Gamma\left(\frac{1}{s}\right)}\right]^{2s/(2+s)} \tag{9.4}$$

If $s = 2$ and $V_0 = m\omega^2/2$ (the one-dimensional harmonic oscillator potential) then $C = \hbar\omega$ and $E_n = \omega\hbar\left(n+\frac{1}{2}\right)$.

10. Suppose that the particle arrives at the potential barrier from the negative direction of the x-axis (Fig. IV.5). In the WKB approximation, the wave transmitted through the barrier has the form

$$\frac{C}{\sqrt{p}} \exp\left(\frac{i}{\hbar}\int_b^x p\, dx\right)$$

in the range $x > b$. For later convenience let us write $C = 2D \exp\left(-i\frac{\pi}{4}\right)$, so that

$$\psi = \frac{2D}{\sqrt{p}} \exp\left[i\left(\frac{1}{\hbar}\int_b^x p\, dx - \frac{\pi}{4}\right)\right], \qquad x > b. \tag{10.1}$$

In the same way that we solved problem 2, we now use the two linearly independent WKB solutions [see (IV.10.10′)]:

$$\psi_2 = \begin{cases} \dfrac{1}{\sqrt{p}} \sin\left(\dfrac{1}{\hbar}\displaystyle\int_b^x p\, dx + \dfrac{\pi}{4}\right), & x > b, \\[2ex] \dfrac{1}{2\sqrt{|p|}} \exp\left(-\dfrac{1}{\hbar}\displaystyle\int_x^b |p|\, dx\right), & a < x < b, \end{cases}$$

and [see (3.1, 1′)]:

$$\tilde{\psi}_2 = \begin{cases} \dfrac{1}{\sqrt{p}} \cos\left(\dfrac{1}{\hbar}\displaystyle\int_x^b p\,dx + \dfrac{\pi}{4}\right), & x > b, \\[2ex] \dfrac{1}{\sqrt{|p|}} \exp\left(\dfrac{1}{\hbar}\displaystyle\int_x^b |p|\,dx\right), & a < x < b, \end{cases}$$

to show that the solution which has the form (10.1) for $x > b$, has, for $a < x < b$, the form

$$\psi = \frac{D}{\sqrt{|p|}}\left[\exp\left(-\frac{1}{\hbar}\int_x^b |p|\,dx\right) - 2i\exp\left(\frac{1}{\hbar}\int_x^b |p|\,dx\right)\right], \quad a < x < b,$$

or, using the notation of (10b),

$$\psi = \frac{D}{\sqrt{|p|}}\left[e^{-L}\exp\left(\frac{1}{\hbar}\int_a^x |p|\,dx\right) - 2i\,e^{L}\exp\left(-\frac{1}{\hbar}\int_a^x |p|\,dx\right)\right], \quad a < x < b. \quad (10.2)$$

We have now only to determine the solution in the range $x < a$, knowing that for $a < x < b$ it has the form (10.2). For this purpose we multiply (IV.9, 9′) by $-4iDe^{L}$:

$$-4iDe^{L}\psi_1 = \begin{cases} -\dfrac{4iD}{\sqrt{p}} e^{L} \sin\left(\dfrac{1}{\hbar}\displaystyle\int_x^a p\,dx + \dfrac{\pi}{4}\right), & x < a, \\[2ex] -\dfrac{2iD}{\sqrt{|p|}} e^{L} \exp\left(-\dfrac{1}{\hbar}\displaystyle\int_a^x |p|\,dx\right), & a < x < b, \end{cases} \quad (10.3)$$

and the function (2.1, 1′), with $C = 1$, by De^{-L}:

$$De^{-L}\tilde{\psi}_1 = \begin{cases} \dfrac{D}{\sqrt{p}} e^{-L} \cos\left(\dfrac{1}{\hbar}\displaystyle\int_x^a p\,dx + \dfrac{\pi}{4}\right), & x < a, \\[2ex] \dfrac{D}{\sqrt{|p|}} e^{-L} \exp\left(\dfrac{1}{\hbar}\displaystyle\int_a^x |p|\,dx\right), & a < x < b. \end{cases} \quad (10.4)$$

By adding (10.3) and (10.4) we obtain, for $a < x < b$, the required solution (10.2), and, in the range $x < a$, the solution

$$\psi = \frac{4D}{i\sqrt{p}} e^{L} \sin\left(\frac{1}{\hbar}\int_x^a p\,dx + \frac{\pi}{4}\right) + \frac{D}{\sqrt{p}} e^{-L} \cos\left(\frac{1}{\hbar}\int_x^a p\,dx + \frac{\pi}{4}\right), \quad x < a. \quad (10.5)$$

Using (10b), (10.5) can be written, after some elementary manipulation, as

$$\psi = \frac{2D}{i\sqrt{p}}\left\{\left(e^{L}-\frac{1}{4}e^{-L}\right)\exp\left[i\left(\frac{1}{\hbar}\int_x^a p\,dx-\frac{\pi}{4}\right)\right]\right.$$
$$\left.+\left(e^{L}+\frac{1}{4}e^{-L}\right)\exp\left[-i\left(\frac{1}{\hbar}\int_x^a p\,dx-\frac{\pi}{4}\right)\right]\right\}, \quad x < a. \tag{10.6}$$

The solution (10.6) represents a superposition of two waves travelling in opposite directions along the x-axis; one of them (given by the first term) is the incident wave and the other (given by the second term) is the wave reflected at the barrier. Using (II.7), (10.1) and (10.6), we obtain, for the fluxes of the transmitted and of the incident waves, and for the transmission coefficient, the following expressions

$$|\mathbf{j}_T| = 4|D|^2\frac{1}{m}, \quad |\mathbf{j}_I| = 4|D|^2\frac{1}{m}\left(e^{L}+\frac{1}{4}e^{-L}\right)^2, \tag{10.7}$$

$$T = \frac{|\mathbf{j}_T|}{|\mathbf{j}_I|} = e^{-2L}\left(1+\frac{1}{4}e^{-2L}\right)^{-2}. \tag{10a}$$

If the barrier is broad enough, then $e^{-2L} \ll 1$, and from (10a) we find that

$$T = e^{-2L} = \exp\left[-\frac{1}{\hbar}\int_a^b \sqrt{8m(V(x)-E)}\,dx\right] \ll 1. \tag{10.8}$$

11. The upper and lower limits of the integral are the two solutions of the equation $V_0(1-x^2/x_0^2)-E = 0$, i.e. $b = -a = x_0(1-E/V_0)^{1/2}$.

Making the substitution $x = b\sin z$, the expression (10.8), as applied to the parabolic potential (11a), becomes

$$T = \exp\left[-x_0\sqrt{8mV_0}\,\frac{1}{\hbar}\left(1-\frac{E}{V_0}\right)\int_{-\pi/2}^{+\pi/2}\cos^2 z\,dz\right]$$
$$= \exp\left[-\pi x_0\sqrt{2mV_0}\,\frac{1}{\hbar}\left(1-\frac{E}{V_0}\right)\right]. \tag{11.1}$$

The condition for the validity of this result is that

$$\pi x_0\sqrt{2mV_0}\,\frac{1}{\hbar}\left(1-\frac{E}{V_0}\right) \gg 1. \tag{11.2}$$

It is thus valid for large enough x_0 and/or V_0 ($T \ll 1$).

12. Using the customary notation $2mE/\hbar^2 = k^2$, and $2mV(r)/\hbar = U(r)$, the radial equation (II.15) becomes

$$\frac{d^2R_l}{dr^2}+\left[k^2-U(r)-\frac{l(l+1)}{r^2}\right]R_l = 0. \tag{12.1}$$

Instead of r, let us introduce a variable x defined by

$$r = k^{-1}e^x. \tag{12.2}$$

Since r lies in the range $(0, +\infty)$, it follows that x lies in the range $(-\infty, +\infty)$.

Making the change of independent variable (12.2) and the change of function,

$$R_l(r) = e^{x/2}\,u(x), \tag{12.3}$$

equation (12.1) becomes

$$\frac{d^2u}{dx^2}+Q^2(x)u = 0, \tag{12.4}$$

where

$$Q^2(x) = e^{2x}(1-k^{-2}U)-(l+\tfrac{1}{2})^2. \tag{12.5}$$

Note that for large enough positive x we have $Q^2 > 0$, and for large enough negative x we have $Q^2 < 0$. For simplicity let us suppose that $Q^2(x)$ vanishes at a single point $x = x_0$ only. Then

$$\begin{aligned} Q^2(x) &> 0 \quad \text{for} \quad x > x_0, \\ Q^2(x) &< 0 \quad \text{for} \quad x < x_0. \end{aligned} \tag{12.6}$$

An approximate solution of (12.4) can be sought in the form

$$u(x) = \frac{1}{\sqrt{q(x)}}\exp\left[\pm\int_{x_0}^{x} q(s)\,ds\right], \tag{12.7}$$

where the function $q(x)$ is to be determined. By substituting (12.7) into (12.4) we obtain the equation

$$q^2(x)+f(x) = Q^2(x) \tag{12.8}$$

where

$$f(x) = \frac{1}{2}\frac{q''}{q}-\frac{3}{4}\frac{q'^2}{q^2}. \tag{12.9}$$

In the WKB approximation, $f(x)$ is neglected, and it then follows that

$$q^2(x) = Q^2(x). \tag{12.10}$$

The condition of validity of this approximation is that

$$|f(x)| \ll |Q^2(x)|. \tag{12.11}$$

The general solution of (12.4) in the WKB approximation is, in the range $x > x_0$, of the form

$$u = \frac{1}{\sqrt{Q}}\left[C_+ \exp\left(+i\int_{x_0}^{x} Q(s)\,ds\right) + C_- \exp\left(-i\int_{x_0}^{x} Q(s)\,ds\right)\right], \tag{12.12}$$

where C_+ and C_- are constants, and $Q > 0$. In the range $x < x_0$ the corresponding general solution has the form

$$u = \frac{1}{\sqrt{|Q|}}\left[D_+ \exp\left(+\int_{x}^{x_0} |Q(s)|\,ds\right) + D_- \exp\left(-\int_{x}^{x_0} |Q(s)|\,ds\right)\right], \tag{12.12'}$$

where D_+ and D_- are also constants.

The condition (II.16) requires that $u(x) \to 0$ as $x \to -\infty$, and hence we must put $D_+ = 0$ in (12.12'). Thus, in the range $x < x_0$, we have finally

$$u(x) = \frac{1}{\sqrt{|Q(x)|}} D_- \exp\left(-\int_{x}^{x_0} |Q(s)|\,ds\right). \tag{12.13}$$

Remarks: The point $x = x_0$ is not a singular point of the differential equation (12.4), so that solving this equation presents no difficulty (in principle). From the continuity conditions at $x = x_0$ it should be possible to determine the constants C_+, C_- as functions of D_-. However, it follows from (12.11) that the WKB approximation fails at the turning point $x = x_0$. Hence (12.12) and (12.13) cannot easily be joined, since we have first to solve (12.4) in the neighbourhood of the turning point. As a result of some calculations (see Goldberger-Watson, 1964) it is found that

$$C_+ = \frac{D_-}{i} e^{i(\pi/4)}, \quad C_- = -\frac{D_-}{i} e^{-i(\pi/4)},$$

and thus that

$$u(x) = \begin{cases} \dfrac{2D}{\sqrt{Q(x)}} \sin\left(\displaystyle\int_{x_0}^{x} Q(s)\,ds + \frac{\pi}{4}\right) & \text{for } x > x_0, \\[2ex] \dfrac{D_-}{\sqrt{|Q(x)|}} \exp\left(-\displaystyle\int_{x}^{x_0} |Q(s)|\,ds\right) & \text{for } x < x_0. \end{cases} \tag{12.14}$$

From (12.2) and (12.5) we have $Q^2(x) = r^2 p^2(r)$, where

$$p^2(r) = k^2 - U(r) - (l+\tfrac{1}{2})^2 r^{-2}. \tag{12.15}$$

Thus for $r > r_0 = e^{x_0}/k$, the radial function in the WKB approximation becomes

$$R_l(r) = \sqrt{kr}\,u = 2D_-(k/p)^{1/2} \sin\left(\int_{r_0}^{r} p(r')\,dr' + \frac{\pi}{4}\right). \tag{12.16}$$

As $r \to \infty$, $p \to k$, and (12.16) has the asymptotic form

$$R_l(r) \sim 2D_- \sin\left[\frac{\pi}{4} + \int_{r_0}^{\infty} (p(r') - k)\,dr' + k(r - r_0)\right]. \tag{12.17}$$

CHAPTER V

Pictures and Representations

1. The Schrödinger, the Heisenberg and the Interaction Pictures

In quantum mechanics, the state of a system at any given time is described by a unit vector in a Hilbert space, in which sets of axes can be defined by the eigenvectors of complete sets of observables of the system. Any change with time in the state of the system can be investigated by keeping the axes fixed and allowing the state vector to rotate, or by keeping the state vector fixed and allowing the axes to rotate, or by permitting simultaneous rotation of the state vector and of the axes, using in each case the appropriate equations of motion of the vectors concerned. The three possibilities described above are called the Schrödinger, the Heisenberg and the interaction "pictures" respectively.

Table V.1 gives the equations of motion of the state vector $|\psi\rangle$ and of any observable A of

TABLE V. 1.

Schrödinger picture	$i\hbar \frac{\partial \lvert\psi(t)\rangle}{\partial t} = H\lvert\psi(t)\rangle$ (V.1) $\lvert\psi(t)\rangle = U(t, t_0)\lvert\psi(t_0)\rangle$ (V.2)
Heisenberg picture	$\frac{\partial \lvert\psi_H\rangle}{\partial t} = 0, \quad i\hbar \frac{dA_H}{dt} = [A_H, H_H] + i\hbar U^+ \frac{\partial A_H}{\partial t} U$ (V.3) $\lvert\psi_H\rangle = U^+(t, t_0)\lvert\psi(t)\rangle,$ $A_H(t) = U^+(t, t_0)\, A U(t, t_0)$ (V.4)
Interaction picture	$i\hbar \frac{\partial \lvert\psi_I(t)\rangle}{\partial t} = H_I'\lvert\psi_I(t)\rangle$ $i\hbar \frac{dA_I}{dt} = [A_I, H_{0I}] + i\hbar U^{(0)+} \frac{\partial A_I}{\partial t} U^{(0)}$ (V.5) $\lvert\psi_I(t)\rangle = U^{(0)+}(t, t_0)\lvert\psi(t)\rangle,$ $A_I(t) = U^{(0)+}(t, t_0)\, A U^{(0)}(t_1, t_0).$ (V.6)

the system in each of the three pictures (using the subscripts H and I to denote "Heisenberg" and "interaction" respectively. Quantities without subscripts refer to the Schrödinger picture.) Relations between corresponding entities in the different pictures are also given.

In the interaction picture $H_I = H_{0I}+H_I'$, i.e., the Hamiltonian is split into two parts: H_{0I}, the "free" Hamiltonian, and H_I', the "interaction" Hamiltonian. $U(t, t_0)$ and $U^{(0)}(t, t_0)$ are unitary operators satisfying the differential equations

$$i\hbar\frac{\partial U(t, t_0)}{\partial t} = HU(t, t_0); \qquad i\hbar\frac{\partial U^{(0)}(t, t_0)}{\partial t} = H_0U^{(0)}(t, t_0), \tag{V.7}$$

with the initial conditions $U(t_0, t_0) = 1$, $U^{(0)}(t_0, t_0) = 1$. If H and H_0 are time-independent, we obtain as solutions

$$U(t, t_0) = e^{-\frac{i}{\hbar}H(t-t_0)}, \qquad U^{(0)}(t, t_0) = {}^{-\frac{i}{\hbar}H_0(t-t_0)}. \tag{V.8}$$

2. Representations

A "representation" is a method of specifying any arbitrary vector by means of the coefficients which occur when it is expressed as a linear combination of a given complete set of orthogonal unit vectors ("base vectors"), the latter being the set of all eigenvectors of some specified complete set of commuting observables of the system. If the vector thus specified is a state vector, its representation is called a "wavefunction". The observables of the system are represented by matrices (continuous or discrete), which operate on the sets of coefficients treated as one-column matrices. To pass from one representation to another, a unitary matrix is used. In any representation, the observables of the complete commuting set which defines that representation are represented by diagonal matrices, for obvious reasons.

2.1. The Coordinate Representation $\{q\}$

The position coordinates q_1, q_2, q_3 of a particle form a complete set of commuting observables for that particle.†

Denoting the base vectors of the representation (i.e. the simultaneous eigenvectors of the observables q_1, q_2, q_3) by $|q'\rangle = |q_1'q_2'q_3'\rangle$, the following orthonormality and closure relations hold:

$$\langle q'|q''\rangle = \prod_{i=1}^{3}\langle q_i'|q_i''\rangle = \prod_{i=1}^{3}\delta(q_i'-q_i'') \equiv \delta(q'-q''), \qquad \int |q'\rangle\, dq'\langle q'| = 1. \tag{V.9}$$

The state vector $|\psi\rangle$ is then represented by a one-column continuous matrix which is in this case the usual *wavefunction in co-ordinate representation:*

$$\langle q'|\psi\rangle \equiv \psi(q'), \tag{V.10}$$

† For simplicity we restrict our considerations to the case of a single spinless particle.

where the implied time-dependence of $\psi(q')$ is not shown explicitly. In the Schrödinger picture in $\{q\}$ representation the rate of change with time of $\psi(q')$ is given by the usual *Schrödinger equation.*

It should be noted that although the symbol q, suggestive of the "generalized coordinates" of classical Lagrangian and Hamiltonian mechanics, is traditionally used in abstract argument to denote position variables, extreme care is necessary in practice if the q's are used to denote anything other than the Cartesian coordinates of individual particles of a system.

2.2. The Momentum Representation $\{p\}$

The base vectors of this representation, $|p'\rangle \equiv |p'_1p'_2p'_3\rangle$, are the simultaneous eigenvectors of the complete set of observables p_1, p_2, p_3 (the momenta conjugate to the coordinates q_1, q_2, q_3). The orthonormality and closure relations are similar to those given in (V.9), (V.10). The state vector $|\psi\rangle$ is represented by the one-column continuous matrix $\langle p'|\psi\rangle \equiv \Phi(p')$, called the "wavefunction in the $\{p\}$ representation" (the implied time-dependence of $\Phi(p')$ is not shown explicitly).

The following relations hold between these two representations

$$\left.\begin{aligned}\langle q'|p_n|\psi\rangle &= \frac{\hbar}{i}\frac{\partial}{\partial q'_n}\psi(q') \\ \langle p'|q_n|\Phi\rangle &= i\hbar\frac{\partial}{\partial p'_n}\Phi(p')\end{aligned}\right\} \quad n = 1, 2, 3 \tag{V.11}$$

$$\langle q'|p'\rangle = \langle p'|q'\rangle^* = (2\pi\hbar)^{-3/2}e^{\frac{i}{\hbar}(p'_1q'_1+p'_2q'_2+p'_3q'_3)}$$

$$\Phi(p') = (2\pi\hbar)^{-3/2}\int \psi(q')e^{-\frac{i}{\hbar}(p'_1q'_1+p'_2q'_2+p'_3q'_3)}\,dq'.$$

2.3. The Energy Representation

For conservative systems, representations in which the Hamiltonian is diagonal are often useful. The base vectors are denoted by $|E\alpha\rangle$, where E is an eigenvalue of the energy operator H, and α denotes the ensemble of the eigenvalues of the other observables, which, together with H, form a complete set.

If $|\psi(t)\rangle$ is a state vector in the Schrödinger picture, the wavefunction in the energy representation is

$$\psi(E, \alpha; t) = \langle E\alpha|\psi(t)\rangle. \tag{V.12}$$

The time-dependence of this wavefunction is of the form

$$\psi(E, \alpha; t) = \psi(E, \alpha; t_0)e^{-\frac{i}{\hbar}E(t-t_0)} \tag{V.13}$$

In the case of a discrete spectrum, any observable A is represented by an ordinary matrix with the help of the base vectors $|E_n\alpha\rangle \equiv |n\alpha\rangle$. The elements of this matrix, $\langle n\alpha|A|m\alpha'\rangle$, are readily obtained using the coordinate or the momentum representations of the vectors $|n\alpha\rangle$ and of the observable A (see problem 21).

3. The Density Operator

In quantum mechanics, the dynamical state of a system is completely determined if the state vector is completely determined. Such states are called *pure states*. When the information about the system is incomplete, one usually merely knows that the system has certain probabilities $p_1, p_2, \ldots, p_m$, of being at a given time t_0 in the states represented by the unit vectors $|1\rangle, |2\rangle, \ldots, |m\rangle$, respectively. In this case the dynamical state of the system is represented by a statistical mixture of vectors (*mixed state*).

The average value of a measurement of a physical quantity A at time t_0 is then given by

$$\langle A\rangle = \sum_m p_m\langle m|A|m\rangle \equiv \mathrm{Tr}\,(\varrho A), \tag{V.14}$$

where

$$\varrho = \sum_m |m\rangle p_m\langle m|, \quad p_m \geqslant 0, \quad \sum_m p_m = 1, \tag{V.15}$$

ϱ is called the *density operator*. The properties of the system at any other time t can be investigated by using the time-dependent density operator

$$\varrho_t = \sum_m U(t, t_0)|m\rangle p_m\langle m|U^+(t, t_0), \tag{V.15'}$$

which satisfies the equation of motion

$$i\hbar\frac{\partial\varrho_t}{\partial t} = [H, \varrho_t]. \tag{V.16}$$

A representation of the density operator using the eigenvectors of an observable is called a *density matrix*.

Problems

1. Show that, if the observables L, M, N satisfy the commutation relation $[L, M] = iN$ in the Schrödinger picture, this relation is valid also in the other pictures.

2. Treating the coordinate x as an operator in the Schrödinger picture, determine the corresponding operators x_H in the Heisenberg picture (i) for the free particle and (ii) for the harmonic oscillator.

3. Determine the coordinate and momentum operators in the Heisenberg picture for a one-dimensional harmonic oscillator whose centre of force moves uniformly with velocity a along the direction of oscillation. The relation (V.3) may be used.

4. Evaluate the commutators

$$[p_H(t_1), x_H(t_2)], \quad [p_H(t_1), p_H(t_2)], \quad [x_H(t_1), x_H(t_2)]$$

for a harmonic oscillator.

5. For conservative systems, show that if, at $t = 0$, the state vector $|\psi(t)\rangle$ is an eigenvector of the observable A with eigenvalue a, then, for $t > 0$, $|\psi(t)\rangle$ will be an eigenvector of the operator $A_H(-t)$ with the same eigenvalue a.

6. Let us denote by $S(t, t_0) = U^{(0)+}(t, t_0)\, U(t, t_0)$ the transformation operator between the Heisenberg and the interaction pictures. Show that this operator is the solution of the differential equation

$$i\hbar \frac{\partial S(t, t_0)}{\partial t} = H_I' S(t, t_0) \tag{6a}$$

with the initial condition $S(t_0, t_0) = 1$.

7. Determine explicitly the operator $S(t, 0)$ of the preceding problem, for a one-dimensional harmonic oscillator of mass m and charge e placed in a constant uniform electric field $\mathcal{E}$, with

$$H = \frac{p^2}{2m} + \frac{m\omega^2 x^2}{2} - e\mathcal{E}x; \qquad H_0 = \frac{p^2}{2m} + \frac{m\omega^2 x^2}{2}; \qquad H' = -e\mathcal{E}x, \tag{7a}$$

8. Assume that, at time $t = 0$, the wavefunction $\psi(x, t)$ of a particle is of the form (cf. problem 16, Chapter III):

$$\psi(x, 0) = \frac{1}{(2\pi\delta^2)^{1/4}} \exp\left(-\frac{x^2}{4\delta^2}\right), \qquad \delta^2 = (\Delta x)^2. \tag{8a}$$

Investigate the change in time of this wave-packet if, for $t > 0$, no forces act on the particle.

9. Solve the preceding problem assuming that for $t > 0$ the particle moves under the action of a constant force.

10. Determine the Green's function $G(\xi, x, t)$ of the Schrödinger equation for a harmonic oscillator.

11. Investigate the spread of the minimal wave-packet of a harmonic oscillator, for which, at $t = 0$,

$$\psi(x, 0) = \sqrt{\frac{\alpha}{\sqrt{\pi}}} \exp\left[-\frac{\alpha^2(x-a)^2}{2}\right], \quad \alpha = \sqrt{\frac{m\omega}{\hbar}},$$

and whose centre is thus initially at a distance a from the origin.

12. Show that the Schrödinger equation is invariant under any Galilean transformation of the coordinate system.

13. Show that the solution of the Schrödinger equation $i\hbar \frac{\partial \psi}{\partial t} = H\psi$ can, whether the Hamiltonian H is time-dependent or not, be written in the form of an integral:†

$$\psi(\mathbf{r}_2, t_2) = \int G^+(\mathbf{r}_2, t_2; \mathbf{r}_1, t_1)\, \psi(\mathbf{r}_1, t_1)\, dV_1; \qquad t_2 > t_1 \tag{13a}$$

which satisfies the condition at time t_1:

$$\psi(\mathbf{r}, t_2)\Big|_{t_2=t_1} = \psi(\mathbf{r}, t_1) \tag{13b}$$

provided the Green's function $G^+(\mathbf{r}_2, t_2; \mathbf{r}_1, t_1)$ is taken to be the solution of the equation

$$\left(i\hbar \frac{\partial}{\partial t_2} - H(2)\right) G^+(\mathbf{r}_2, t_2; \mathbf{r}_1, t_1) = i\hbar\delta(t_2 - t_1)\, \delta(\mathbf{r}_2 - \mathbf{r}_1) \tag{13c}$$

which vanishes for $t_2 < t_1$. Here the notation $H(2)$ means that the Hamiltonian operator H is to act only on the variables $\mathbf{r}_2$, t_2 of G.

14. Show that the differential equation (13c) for the Green's function of the Schrödinger equation, $G^+(\mathbf{r}_2, t_2; r_1, t_1) \equiv G^+(2, 1)$ say, can be written in the form of an integral equation:

$$G^+(2, 1) = G_0^+(2, 1) - \frac{i}{\hbar} \int_{-\infty}^{+\infty} G_0^+(2, 3)\, H^{(1)}(3)\, G^+(3, 1)\, d^4x_3 \tag{14a}$$

in which $G_0^+(2, 1)$ satisfies the equations

$$\left(i\hbar \frac{\partial}{\partial t_2} - H^{(0)}(2)\right) G_0^+(2, 1) = i\hbar\delta^4(2, 1),$$

$$G_0^+(t_2, t_1) = 0 \quad \text{for} \quad t_2 < t_1, \tag{14b}$$

and in which

$$\delta^4(2, 1) = \delta(x_2 - x_1)\, \delta(y_2 - y_1)\, \delta(z_2 - z_1)\, \delta(t_2 - t_1), \qquad d^4x_3 = dx_3\, dy_3\, dz_3\, dt_3,$$

and $H = H^{(0)} + H^{(1)}$ is any decomposition of the Hamiltonian into the sum of two operators.

† R. P. Feynman, *Phys. Rev.* **76**, 749 (1949).

15. Show that, for a time-independent Hamiltonian H, the Green's function $G^+(\mathbf{r}_2, t_2; \mathbf{r}_1, t_1)$ of problem 13 can be written in the form

$$G^+(\mathbf{r}_2, t_2; \mathbf{r}_1, t_1) = \sum_n \psi_n(\mathbf{r}_2)\, \psi_n^*(\mathbf{r}_1)\, e^{-\frac{i}{\hbar} E_n(t_2-t_1)} \tag{15a}$$

where E_n and $\psi_n(\mathbf{r})$ are the eigenvalues and the eigenfunctions respectively of the Hamiltonian H, i.e., $H\psi_n(\mathbf{r}) = E_n\psi_n(\mathbf{r})$.

16. Show that, in the $\{p\}$ representation, the Schrödinger equation

$$i\hbar \frac{\partial |\psi\rangle}{\partial t} = H|\psi\rangle, \quad H = \frac{\mathbf{p}^2}{2m} + V(\mathbf{r}),$$

can be transformed into the integro-differential equation

$$i\hbar \frac{\partial}{\partial t} \Phi(\mathbf{p}, t) = \frac{\mathbf{p}^2}{2m} \Phi(\mathbf{p}, t) + \int U(\mathbf{p}-\mathbf{p}')\, \Phi(\mathbf{p}', t)\, d\mathbf{p}', \tag{16a}$$

$\Phi(\mathbf{p}, t)$ being the wavefunction in momentum representation, and

$$U(\mathbf{p}) = (2\pi\hbar)^{-3} \int V(\mathbf{r}) e^{-\frac{i}{\hbar}\mathbf{p}.\mathbf{r}}\, d\mathbf{r}. \tag{16b}$$

17. Solve the problem of the harmonic oscillator in momentum representation.

18. Solve the problem of the motion of a particle moving under a constant force in momentum representation.

19. Express the operator $1/p$ in coordinate representation, and the operator $1/x$ in momentum representation, for motion in one-dimension.

20. Find the law of transformation of wavefunctions in momentum representation for Galilean transformations of the coordinate system.

21. Find the representations of the coordinate and of the momentum operators of a harmonic oscillator in the energy representation.

22. Find the coordinate and the momentum matrices in the energy representation of a particle in the potential well shown in Fig. II.1.

23. Let $H = (\mathbf{p}^2/2\mu) + V(\mathbf{r})$ be the Hamiltonian of a particle of mass μ moving in a potential $V(\mathbf{r})$.

Show that

$$\sum_n (E_n - E_m)|x_{nm}|^2 = \frac{\hbar^2}{2\mu}, \tag{23a}$$

where the sum is taken over all states n of the particle, and x is the x-component of the position vector $\mathbf{r}$ in a Cartesian coordinate system (the Thomas–Reiche–Kuhn sum rule).

24. Show, by using the double commutator

$$[[H, e^{i\mathbf{k}.\mathbf{r}}], e^{-i\mathbf{k}.\mathbf{r}}],$$

that the relation

$$\sum_n (E_n - E_s) |\langle n | e^{i\mathbf{k}.\mathbf{r}} | s\rangle|^2 = 3\frac{\hbar^2 k^2}{2\mu} \tag{24a}$$

can be obtained as a generalization of the TRK sum rule discussed in the previous problem.

25. Denoting by ϱ the density operator of a system, show

(a) that ϱ and $1-\varrho$ are positive definite operators,

(b) that $\mathrm{Tr}\,(\varrho^2) \leqslant 1$, and that $\mathrm{Tr}\,(\varrho^2) = 1$ is the necessary and sufficient condition for ϱ to represent a pure state.

26. Verify that, in the Heisenberg picture, the density operator is time-independent.

27.† Suppose that the density operator of some given system can be represented by an $N \times N$ matrix. It can then be expressed as a linear combination of any "complete" set of $N \times N$ matrices which form a "basis" in the space of square matrices of order N, a basis which necessarily contains N^2 linearly independent matrices. Suppose further that we have succeeded in determining such a basis in the form of N^2 Hermitian matrices $\Omega^{(r)}$, $r = 1, 2, \ldots, N^2$, which satisfy the "orthonormality relations"

$$\mathrm{Tr}\,(\Omega^{(r)}\Omega^{(s)}) = \delta_{rs}. \tag{27a}$$

Show that, if

$$\varrho_t = \sum_r \varrho_t^{(r)}\Omega^{(r)}, \tag{27b}$$

then the coefficients $\varrho_t^{(r)} \equiv \langle\Omega^{(r)}\rangle$ satisfy the following equations of motion

$$\frac{d\varrho_t^{(s)}}{dt} = \frac{i}{\hbar}\sum_r \mathrm{Tr}\,\{H[\Omega^{(s)}, \Omega^{(r)}]\}\,\varrho_t^{(r)}; \qquad s = 1, 2, \ldots, N^2. \tag{27c}$$

Solutions

1. In virtue of the relations (V.4), (V.6), and also of the unitarity of the operators $U(t, t_0)$, $U^{(0)}(t, t_0)$, we obtain $[L_H, M_H] = iN_H$ and $[L_I, M_I] = iN_I$.

2. Since the Hamiltonian is in both cases time-independent, we shall use the relations (V.4) and (V.8), and take $t_0 = 0$.

For a free particle

$$x_H = e^{\frac{i}{\hbar}Ht}\,x\,e^{-\frac{i}{\hbar}Ht}, \quad H = -\frac{\hbar^2}{2m}\frac{\partial^2}{\partial x^2}, \tag{2.1}$$

$$[H, x]\,\psi(x) = -\frac{\hbar^2}{2m}\left[\frac{\partial^2}{\partial x^2}(x\psi(x)) - x\frac{\partial^2}{\partial x^2}(\psi(x))\right] = -\frac{\hbar^2}{m}\frac{\partial\psi(x)}{\partial x}$$

† Other problems concerning the density operator will be found in Chapter VI (problems 38, 39).

and consequently

$$[H, x] = -\frac{\hbar^2}{m}\frac{\partial}{\partial x}, \quad [H, [H, x]] = 0, \quad \big[H, \big[H, [H, x]\big]\big] = 0, \ldots \tag{2.2}$$

Using relation (I.27a) we find from (2.1)

$$x_H(t) = x - \frac{i\hbar}{m} t \frac{\partial}{\partial x}. \tag{2.3}$$

For the oscillator $H = -\frac{\hbar^2}{2m}\frac{\partial^2}{\partial x^2} + \frac{m\omega^2 x^2}{2}$. Working out the commutators (2.2) and regrouping the terms, one obtains

$$x_H(t) = x \cos \omega t - \frac{i\hbar}{m\omega} \sin \omega t \frac{\partial}{\partial x}. \tag{2.4}$$

With the help of the momentum operator $p = -i\hbar\frac{\partial}{\partial x}$, one can write the relations (2.3) and (2.4) in the form

$$x_H(t) = x + \frac{p}{m} t \tag{2.5}$$

$$x_H(t) = x \cos \omega t + \frac{p}{m\omega} \sin \omega t. \tag{2.6}$$

3. Since the Schrödinger coordinate and momentum operators do not depend explicitly on time, we have

$$\frac{dx_H}{dt} = \frac{i}{\hbar}[H_H, x_H] = \frac{i}{\hbar} U^+(t)[H, x]U(t)$$

$$\frac{dp_H}{dt} = \frac{i}{\hbar}[H_H, p_H] = \frac{i}{\hbar} U^+(t)[H, p]U(t), \tag{3.1}$$

where

$$H = \frac{p^2}{2m} + \frac{m\omega^2}{2}(x - at)^2 \qquad p = -i\hbar\frac{\partial}{\partial x}. \tag{3.2}$$

From (3.1), some simple algebra leads to the equations

$$\frac{dx_H}{dt} = \frac{1}{m} p_H; \quad \frac{dp_H}{dt} = -m\omega^2(x_H - at) \tag{3.3}$$

which are analoguous to the corresponding equations of motion of classical Hamiltonian mechanics. The equations (3.3) have as solutions:

$$x_H(t) = C_1 \sin \omega t + C_2 \cos \omega t + at$$

$$p_H(t) = m\omega C_1 \cos \omega t - m\omega C_2 \sin \omega t + am, \tag{3.4}$$

where the constants C_1 and C_2 are determined from the conditions $x_H(0) = x$, $p_H(0) = p$.

Finally we obtain

$$x_H(t) = x\cos\omega t + \frac{p}{m\omega}\sin\omega t + at$$

$$p_H(t) = p\cos\omega t - m\omega x\sin\omega t + am. \tag{3.5}$$

If $a = 0$ (i.e. if the centre of elastic force remains fixed) we have

$$x_H(t) = x\cos\omega t + \frac{p}{m\omega}\sin\omega t$$

$$p_H(t) = p\cos\omega t - m\omega x\sin\omega t. \tag{3.6}$$

As expected, the first of the equations (3.6) is the same as (2.6).

4. Using the equations (3.6), we obtain

$$[p_H(t_1), x_H(t_2)] = [p, x]\cos\omega t_1\cos\omega t_2 - [x, p]\sin\omega t_1\sin\omega t_2$$
$$= -i\hbar\cos(\omega t_1 - \omega t_2), \tag{4.1}$$

$$[p_H(t_1), p_H(t_2)] = -m\omega[p, x]\cos\omega t_1\sin\omega t_2 - m\omega[x, p]\sin\omega t_1\cos\omega t_2$$
$$= -im\omega\hbar\sin(\omega t_1 - \omega t_2) \tag{4.2}$$

$$[x_H(t_1), x_H(t_2)] = \frac{1}{m\omega}[x, p]\cos\omega t_1\cos\omega t_2 + \frac{1}{m\omega}[p, x]\sin\omega t_1\cos\omega t_2$$
$$= -\frac{i\hbar}{m\omega}\sin(\omega t_1 - \omega t_2). \tag{4.3}$$

5. Taking $t_0 = 0$, and writing for simplicity $U(t, 0) = U(t)$, we have from (V.4) that

$$|\psi(t)\rangle = U(t)|\psi(0)\rangle; \qquad A_H(t) = U^+(t)AU(t). \tag{5.1}$$

Also, in this problem, $A|\psi(0)\rangle = a|\psi(0)\rangle$. By making the substitution $|\psi(0)\rangle = U^+(t)|\psi(t)\rangle$ and multiplying both sides of the resulting equation from the left by $U(t)$ we have

$$U(t)AU^+(t)|\psi(t)\rangle = a|\psi(t)\rangle. \tag{5.2}$$

By comparison with (5.1) and (V.8), it follows that

$$A_H(-t)|\psi(t)\rangle = a|\psi(t)\rangle. \tag{5.3}$$

6. The operator $S(t, t_0)$ performs the transformation from the Heisenberg to the interaction picture. Using (V.4) and (V.7) we obtain

$$i\hbar\frac{\partial S}{\partial t} = i\hbar\left(\frac{\partial U^{(0)+}}{\partial t}U + U^{(0)+}\frac{\partial U}{\partial t}\right) = U^{(0)+}H'U = U^{(0)+}U(U^+H'U)$$
$$= U^{(0)+}UH'_H = U^{0+}UH'_HU^+U^{(0)+}U^{(0)}U = SH'_HS^+S = H'_IS.$$

The condition $S(t_0, t_0) = 1$ follows directly.

With these facts established, we see that the operator $S(t, t_0)$ describes the change in time of the wavefunction in the interaction picture,

$$|\psi_I(t)\rangle = S(t, t_0)|\psi(t_0)\rangle. \tag{6.1}$$

7. The equations of motion (V.5) for the coordinate and the momentum operators can be written in the form

$$\frac{dx_I}{dt} = \frac{i}{\hbar}[H_{0I}, x_I], \quad \frac{dp_I}{dt} = \frac{i}{\hbar}[H_{0I}, p_I]. \tag{7.1}$$

Knowing H_{0I}, we obtain from (7.1) the equations

$$\frac{dx_I}{dt} = \frac{i}{\hbar}p_I, \quad \frac{dp_I}{dt} = -m\omega^2 x_I \tag{7.2}$$

which are of the same form as equations (3.3), with $a = 0$.

Integrating these equations we obtain

$$\begin{aligned} x_I(t) &= x \cos \omega t + \frac{p}{m\omega}\sin \omega t \\ p_I(t) &= p \cos \omega t - m\omega x \sin \omega t. \end{aligned} \tag{7.3}$$

Taking into account (6a) and (7a), it follows that

$$i\hbar \frac{\partial S(t, 0)}{\partial t} = -e\mathscr{E}\left(x \cos \omega t + \frac{p}{m\omega} \sin \omega t\right) S(t, 0). \tag{7.4}$$

The solution of this differential operator equation, which satisfies the initial condition $S(0, 0) = 1$, is

$$S(t, 0) = \exp\left[\frac{ie\mathscr{E}}{\hbar}\left(\frac{x}{\omega}\sin \omega t - \frac{p}{m\omega^2}\cos \omega t + \frac{p}{m\omega^2}\right)\right], \tag{7.5}$$

the exponential being here defined by means of its power series. In the case of a weak external field, we need retain only the first few terms of the expansion.

8. It is necessary to determine the wavefunction $\psi(x, t)$ which satisfies the Schrödinger equation

$$i\hbar \frac{\partial \psi(x, t)}{\partial t} = H\psi(x, t),$$

and which, at time $t = 0$, is the given function $\psi(x, 0)$. With that end in view we expand $\psi(x, 0)$ in terms of the set of orthonormal time-independent eigenfunctions $\psi_n(x)$, $(H\psi_n(x) = E_n\psi_n(x))$ (see p. 204, footnote), thus:

$$\psi(x, 0) = \sum_n a_n\psi_n(x), \qquad a_n = \int \psi_n^*(x)\,\psi(x, 0)\,dx. \tag{8.1}$$

The function $\sum_n a_n \psi_n(x) \exp\left(-\frac{i}{\hbar} E_n t\right)$ then satisfies the Schrödinger equation, and, at time $t = 0$, coincides with $\psi(x, 0)$. Hence

$$\psi(x, t) = \sum_n a_n \psi_n(x) e^{-\frac{i}{\hbar} E_n t}, \tag{8.2}$$

i.e.

$$\psi(x, t) = \int G(\xi, x, t) \psi(\xi, 0)\, d\xi \tag{8.3}$$

where

$$G(\xi, x, t) = \sum_n \psi_n^*(\xi)\, \psi_n(x)\, e^{-\frac{i}{\hbar} E_n t}. \tag{8.4}$$

Since, in the case of free motion, the eigenfunctions are

$$\psi_p(x) = \frac{1}{(2\pi\hbar)^{1/2}} \exp\left(\frac{i}{\hbar} px\right), \tag{8.5}$$

the Green's function (8.4) becomes (with p continuous)

$$\begin{aligned} G(\xi, x, t) &= \int \frac{1}{2\pi\hbar} \exp\left\{\frac{i}{\hbar}\left[p(x-\xi) - \frac{p^2 t}{2m}\right]\right\} dp \\ &= \left(\frac{m}{2\pi i \hbar t}\right)^{1/2} e^{\frac{im}{2\hbar t}(x-\xi)^2}. \end{aligned} \tag{8.6}$$

From (8.3) and (8a) it follows that

$$\psi(x, t) = \int \left(\frac{m}{2\pi i \hbar t}\right)^{1/2} \frac{1}{(2\pi\delta^2)^{1/4}} \exp\left\{-\frac{\xi^2}{4\delta^2} + \frac{im}{2\hbar t}(x-\xi)^2\right\} d\xi,$$

whence we obtain finally, for the wave function,

$$\psi(x, t) = \frac{1}{(2\pi\delta^2)^{1/4}\left(1 + \frac{\hbar^2 t^2}{4m^2\delta^4}\right)^{1/4}} \exp\left\{-\frac{x^2}{4\delta^2\left(1 + \frac{\hbar^2 t^2}{4m^2\delta^4}\right)}\left(1 - \frac{i\hbar t}{2m\delta^2}\right)\right\}, \tag{8.7}$$

and for the probability density

$$|\psi(x, t)|^2 = \left[2\pi\delta^2\left(1 + \frac{\hbar^2 t^2}{4m^2\delta^4}\right)\right]^{-1/2} \exp\left\{-\frac{x^2}{2\delta^2\left(1 + \frac{\hbar^2 t^2}{4m^2\delta^4}\right)}\right\}. \tag{8.8}$$

This expression has the same form as the initial probability density

$$|\psi(x, 0)|^2 = \frac{1}{(2\pi\delta^2)^{1/2}} \exp\left\{-\frac{x^2}{2\delta^2}\right\}, \tag{8.9}$$

with the difference that, instead of $\delta^2 = (\Delta x)^2$, there appears

$$\delta_t^2 \equiv \delta^2 + \frac{\hbar^2 t^2}{4m^2\delta^2} = (\Delta x)^2 + \frac{(\Delta p)^2 t^2}{m^2}. \tag{8.10}$$

Thus the centre of the wave-packet remains at the point $x = 0$, ($\langle x \rangle = 0$), but its width increases with time in both directions of the x-axis. The smaller the initial coordinate uncertainty, the larger is the momentum uncertainty and the faster the packet spreads. It is interesting to observe that the term which determines the spreading of the wave-packet, $\frac{(\Delta p)t}{m}$, is exactly equal to the distance which would be covered in time t by a classical particle in uniform rectilinear motion with momentum Δp. We can make an estimate of the time required for the wave-packet to spread appreciably. From (8.10) it can be seen that this time is of the order $\tau \sim \delta^2 m/\hbar$. Thus, for an electron initially localized with a precision of $\delta \sim 10^{-8}$ cm, τ is of the order of 10^{-16} sec, while for a particle of mass $m = 1$ g, and $\delta \sim 10^{-5}$ cm, τ is of the order of 10^{17} sec, i.e. about 3000 million years.

9. The potential energy in this problem is $V = -kx$, and the eigenfunctions of the Hamiltonian $H = -\frac{\hbar^2}{2m}\frac{d^2}{dx^2} - kx$ then have the form (see problem 21, Chapter II)

$$\psi_E(x) = A \int_{-\infty}^{+\infty} e^{i\left(\frac{u^3}{3} - uy\right)} du, \tag{9.1}$$

where

$$y = \left(x + \frac{E}{k}\right)\alpha, \quad \alpha = \frac{2mk}{\hbar^2}^{1/3}, \quad A = \frac{\alpha}{2\pi\sqrt{k}}.$$

The Green's function (8.4) is then given (with E continuous) by

$$G(\xi, x, t) = \int_{-\infty}^{+\infty} \psi_E^*(\xi)\psi_E(x) e^{-\frac{i}{\hbar}Et} dE$$

$$= A^2 \int_{-\infty}^{+\infty} e^{-\frac{i}{\hbar}Et} dE \iint_{-\infty}^{+\infty} e^{-i\left(\frac{v^3}{3} - vz\right) + i\left(\frac{u^3}{3} - uy\right)} du\, dV, \tag{9.2}$$

where $z = (\xi + E/k)\alpha$. We can also write

$$G(\xi, x, t) = A^2 \iint_{-\infty}^{+\infty} e^{-i\frac{v^3}{3} + i\frac{u^3}{3} + iv\xi\alpha - iux\alpha} du\, dv \int_{-\infty}^{+\infty} e^{-\frac{i}{\hbar}E\alpha\left(u + \frac{kt}{\alpha\hbar} - v\right)} dE$$

$$= A^2 \iint_{-\infty}^{+\infty} e^{-i\frac{v^3}{3} + i\frac{u^3}{3} - iv\xi\alpha - iux\alpha} \frac{2\pi k}{\alpha} \delta\left(u + \frac{kt}{\alpha\hbar} - v\right) du\, dv$$

We then integrate over v, using the properties of the δ-function, and group the terms in the following manner:

$$G(\xi, x, t) = \frac{2\pi k}{\alpha} A^2 \int_{-\infty}^{+\infty} \exp\left\{-i\frac{kt}{\alpha\hbar}\left[u+\frac{kt}{2\alpha\hbar}+\frac{\alpha^2\hbar}{2kt}(x-\xi)^2\right]\right.$$

$$\left.-\frac{i}{12}\left(\frac{kt}{\alpha\hbar}\right)^3+\frac{i}{2}\frac{kt}{\hbar}(x+\xi)+\frac{i\alpha^3\hbar}{4kt}(x-\xi)^2\right\} du.$$

Finally, integrating over u, we obtain

$$G(\xi, x, t) = \left(\frac{m}{2\pi i\hbar t}\right)^{1/2} \exp\left\{\frac{i}{12}\left(\frac{kt}{\hbar\alpha}\right)^3+\frac{ikt}{2\hbar}(x+\xi)+\frac{im}{2\hbar t}(x-\xi)^2\right\}. \tag{9.3}$$

As is to be expected, the expression (9.3) goes over into (8.6) as $k \to 0$. From (8.3) we have

$$\psi(x, t) = \left[2\pi\delta^2\left(1+\frac{\hbar^2t^2}{4m^2\delta^4}\right)\right]^{-1/4} \exp\left[-\frac{\left(x-\frac{kt^2}{2m}\right)^2}{4\delta^2\left(1+\frac{\hbar^2t^2}{4m^2\delta^4}\right)}\left(1-\frac{i\hbar t}{2\delta^2 m}\right)+\frac{ikx}{\hbar}t-\frac{ik^2}{6\hbar m}t^3\right], \tag{9.4}$$

whence

$$|\psi(x, t)|^2 = \left[2\pi\delta^2\left(1+\frac{\hbar^2t^2}{4m^2\delta^4}\right)\right]^{-1/2} \exp\left[-\frac{\left(x-\frac{kt^2}{2m}\right)^2}{2\delta^2\left(1+\frac{\hbar^2t^2}{4m^2\delta^4}\right)}\right]. \tag{9.5}$$

Thus the Gaussian form of the probability density does not change, but the centre of the wave-packet travels with constant acceleration in accordance with the classical law of motion of a particle moving under a constant force. The width of the packet changes with time according to the equation

$$\delta_t^2 = \delta^2\left(1+\frac{\hbar^2t^2}{4m^2\delta^4}\right), \tag{9.6}$$

which is the same as (8.10).

10. One can solve this problem by using (8.4), and taking the $\psi_n(x)$ to be the eigenfunctions of the harmonic oscillator.

Another possibility is the following. From (8.4) we have immediately

$$G(\xi, x, 0) = \delta(\xi-x). \tag{10.1}$$

Regarding $G(\xi, x, t)$ as a function of x, we observe that it is a solution of the appropriate Schrödinger equation with (10.1) as initial condition. From (10.1) it can be seen that

$$xG(\xi, x, 0) = x\delta(\xi-x) = \xi\delta(\xi-x) = \xi G(\xi, x, 0) \tag{10.2}$$

since the δ function has the property $x\delta(x) = 0$.

If, in (10.2), we treat x as a coordinate operator and ξ as a constant, it can be seen that the Green's function $G(\xi, x, 0)$ is an eigenfunction of the coordinate operator with the eigenvalue ξ.

From the result of problem 5, it then follows that $G(\xi, x, t)$ is an eigenfunction of the coordinate operator $x_H(-t)$, with the same eigenvalue ξ.

Using (2.4) we have thus that

$$x_H(-t)G = \left(x \cos \omega t + \frac{i\hbar}{m\omega} \frac{\partial}{\partial x}\right) G = \xi G. \tag{10.3}$$

The solution of this equation is of the form

$$G(\xi, x, t) = G_0(t) \exp \left[\frac{im\omega(x^2 \cos \omega t - 2x\xi)}{2\hbar \sin \omega t}\right], \tag{10.4}$$

where $G_0(t)$ is a time-dependent integration constant.

In order to determine $G_0(t)$, we impose on (10.4) the requirement that it be a solution of the Schrödinger equation

$$\left(-\frac{\hbar^2}{2m} \frac{\partial^2}{\partial x^2} + \frac{m\omega^2 x^2}{2}\right) G = i\hbar \frac{\partial G}{\partial t}. \tag{10.5}$$

After some simple calculation we obtain the equation

$$\frac{dG_0}{dt} = \left(\frac{\omega}{2} \cot \omega t + \frac{im\omega^2 \xi^2}{2\hbar \sin^2 \omega t}\right) G_0 = 0 \tag{10.6}$$

whence

$$G_0(t) = \frac{C}{\sqrt{\sin \omega t}} \exp \left(\frac{im\omega\xi^2}{2\hbar} \cot \omega t\right), \tag{10.7}$$

C being an integration constant which can be found from the initial condition (10.1), or, more easily, by comparing (10.7) with the free-particle Green's function (8.6), since, for $t \ll (1/\omega)$, the Green's function of the oscillator must go over to the Green's function of the free particle. We find thus that

$$C = \sqrt{-\frac{im\omega}{2\pi\hbar}}.$$

Finally we have that

$$G(\xi, x, t) = \sqrt{-\frac{im\omega}{2\pi\hbar \sin \omega t}} \exp \left[i \frac{m\omega}{2\hbar} \cot \omega t (x^2 - 2\xi x \sin \omega t + \xi^2)\right]. \tag{10.8}$$

11. It is possible to solve this problem by taking the function $G(\xi, x, t)$ found in the preceding one and substituting it into (8.3) There is, however, a more straightforward way, as follows. Let $\psi(x, t)$ be expanded in terms of the set of oscillator eigenfunctions, thus

$$\psi(x, t) = \sum_{n=0}^{\infty} A_n \psi_n(x)\, e^{-\frac{iE_n t}{\hbar}} = e^{-\frac{i\omega t}{2}} \sum_{n=0}^{\infty} A_n \psi_n(x)\, e^{-in\omega t}. \tag{11.1}$$

The constants A_n are given by the integrals

$$A_n = \int_{-\infty}^{+\infty} \psi_n^*(x)\psi(x, 0)\, dx = \frac{N_n}{\pi^{1/4}\alpha^{1/2}} \int_{-\infty}^{+\infty} H_n(\xi)\, e^{-\xi^2/2-(\xi-\xi_0)^2/2}\, d\xi \tag{11.2}$$

where $N_n = (\pi^{1/2}2^n n!\alpha^{-1})^{-1/2}$, $\xi_0 = \alpha a$ (see (17.12) Chapter II).

The integral in (11.2) can be evaluated with the help of the generating function of the Hermite polynomials (A.3), by equating coefficients in the following expansions

$$\int_{-\infty}^{+\infty} e^{-s^2+2s\xi} e^{-(\xi^2-\xi\xi_0+\xi_0^2/2)}\, d\xi = \sum_{n=0}^{\infty} \frac{s^n}{n!} \int_{-\infty}^{+\infty} H_n(\xi)\, e^{-(\xi^2-\xi\xi_0+\xi_0^2/2)}\, d\xi$$

$$\pi^{1/2}\, e^{-\xi_0^2/4+s\xi_0} = \pi^{1/2}\, e^{-\xi_0^2/4} \sum_{n=0}^{\infty} \frac{(s\xi_0)^n}{n!}.$$

We find in this way that

$$A_n = \frac{\xi_0^n\, e^{-\xi_0^2/4}}{(2^n n!)^{1/2}}. \tag{11.3}$$

Substituting into (11.1) the coefficients A_n given by (11.3) and the oscillator wavefunctions (17.12) of Chapter II, we obtain

$$\begin{aligned}
\psi(x, t) &= \sqrt{\frac{\alpha}{\sqrt{\pi}}}\, e^{-(\xi^2/2)-(\xi_0^2/4)-i\omega t/2} \sum_{n=0}^{\infty} \frac{H_n(\xi)}{n!} \left(\frac{1}{2}\,\xi_0\, e^{-i\omega t}\right)^n \\
&= \sqrt{\frac{\alpha}{\sqrt{\pi}}} \exp\left(-\frac{1}{2}\,\xi^2 - \frac{1}{4}\,\xi_0^2 - \frac{1}{2}\,i\omega t - \frac{1}{4}\,\xi_0^2\, e^{-2i\omega t} + \xi\xi_0\, e^{-i\omega t}\right) \\
&= \sqrt{\frac{\alpha}{\sqrt{\pi}}} \exp\left[-\frac{1}{2}\,(\xi - \xi_0 \cos \omega t)^2 - i\left(\frac{1}{2}\,\omega t + \xi\xi_0 \sin \omega t - \frac{1}{4}\,\xi_0^2 \sin 2\omega t\right)\right]
\end{aligned} \tag{11.4}$$

The probability density is then

$$|\psi(x, t)|^2 = \frac{\alpha}{\sqrt{\pi}}\, e^{-\alpha^2(x-a\cos\omega t)^2}. \tag{11.5}$$

The expression (11.5) shows that the wavefunction ψ describes a wave-packet whose centre oscillates about the point $x = 0$ with amplitude a and classical frequency ω, while the wave-packet keeps its shape and size. As $a \to 0$, $\psi(x, t)$ tends to the ground state wavefunction

$\psi_0(x) \exp(-\frac{1}{2}\omega t)$. As a increases, more and more stationary states contribute to form the wavepacket, and the number n_0, for which the coefficient A_{n_0} is the greatest coefficient, increases too. In order to establish this for the case $n \gg 1$, we derive an approximate expression for A_n by the use of Stirling's formula:

$$\ln A_n \approx n(\ln \xi_0 - \tfrac{1}{2}\ln 2) - \tfrac{1}{2}n(\ln n - 1)$$

whence

$$n_0 = \tfrac{1}{2}\xi_0^2 = \omega m a^2/2\hbar.$$

It follows that the eigenfunction which gives the greatest contribution to $\psi(x, t)$ belongs to the energy level

$$E_{n_0} = (n_0 + \tfrac{1}{2})\hbar\omega \approx \frac{m\omega^2 a^2}{2}. \tag{11.6}$$

It is interesting to note that E_{n_0} is the energy of a classical oscillator having the same amplitude and frequency as the centre of the wave-packet. Hence it is possible to say that the packet is "grafted" onto the classical oscillator of energy E_{n_0}.

12. For the sake of simplicity we shall solve this problem for the one-dimensional Schrödinger equation. Let us consider two frames of reference, $K(x, t)$ and $K'(x', t')$, which are in uniform motion relative to each other with constant velocity v; $x = x' + vt$, $t = t'$. Denoting by $V(x, t)$ and $V'(x', t')$ the potential energies in the two frames of reference, we have that

$$V'(x - vt, t) = V(x, t). \tag{12.1}$$

In the frame K' the Schrödinger equation has the form

$$-\frac{\hbar^2}{2m}\frac{\partial^2\psi'}{\partial x'^2} + V'\psi' = i\hbar\frac{\partial\psi'}{\partial t'}, \tag{12.2}$$

where $\psi'(x', t')$ is the wavefunction in K'. Since the probability density does not depend on the frames of reference, we have that

$$|\psi(x, t)|^2 = |\psi'(x', t')|^2 \tag{12.3}$$

and it follows that

$$\psi(x, t) = e^{iS}\psi'(x', t'), \tag{12.4}$$

where S is a real function of (x, t), or, equivalently, of (x', t').

By substituting ψ' from (12.4) into (12.2) and transforming to the new variables x and t, we find that

$$-\frac{\hbar^2}{2m}\frac{\partial^2\psi}{\partial x^2} + i\hbar\left(\frac{\hbar}{m}\frac{\partial S}{\partial x} - v\right)\frac{\partial\psi}{\partial x}$$
$$+\left[V(x, t) + i\frac{\hbar^2}{2m}\frac{\partial^2 S}{\partial x^2} + \frac{\hbar^2}{2m}\left(\frac{\partial S}{\partial x}\right)^2 - \hbar v\frac{\partial S}{\partial x} - \hbar\frac{\partial S}{\partial t}\right]\psi = i\hbar\frac{\partial\psi}{\partial t}. \tag{12.5}$$

The Schrödinger equation will keep the same form, i.e. will be invariant under the Galilean transformation, if there exists a function $S(x, t)$ such that equation (12.5) reduces to

$$-\frac{\hbar^2}{2m}\frac{\partial^2\psi}{\partial x^2}+V(x, t)\psi = i\hbar\frac{\partial\psi}{\partial t}. \tag{12.6}$$

For this to be so, S must satisfy the equations

$$\frac{\hbar}{m}\frac{\partial S}{\partial x}-v = 0; \qquad i\frac{\hbar^2}{2m}\frac{\partial^2 S}{\partial x^2}+\frac{\hbar^2}{2m}\left(\frac{\partial S}{\partial x}\right)^2-\hbar v\frac{\partial S}{\partial x}-\hbar\frac{\partial S}{\partial t} = 0. \tag{12.7}$$

It can easily be verified that this is the case if we take

$$S(x, t) = \frac{mv}{\hbar}x-\frac{mv^2}{2\hbar}t.$$

Thus, finally, the relation between the wavefunction $\psi(x, t)$ and its Galilean transform, $\psi'(x', t')$, is, except for a possible constant phase factor,

$$\psi(x, t) = \exp\left(i\frac{mv}{\hbar}x-i\frac{mv^2}{2\hbar}t\right)\psi'(x-vt, t). \tag{12.8}$$

13. Let us show that the function $\psi(\mathbf{r}_2, t_2)$ defined by the integral (13a) satisfies the Schrödinger equation and the given initial condition. To do this, we apply the operator $i\hbar(\partial/\partial t_2)-H(2)$ to both sides of (13a).

Putting $t_2 > t_1$ and using (13c), we conclude that $\psi(\mathbf{r}_2, t_2)$ satisfies the Schrödinger equation. It will also satisfy the initial condition (13b) if

$$G^+(\mathbf{r}_2, t; \mathbf{r}_1, t) = \delta(\mathbf{r}_2-\mathbf{r}_1). \tag{13.1}$$

To show that this is the case, let us integrate equation (13c) with respect to t_2 over a small time interval $t_1-\Delta t$ to $t_1+\Delta t$, $(\Delta t > 0)$; then

$$\int_{t_1-\Delta t}^{t_1+\Delta t} dt_2\left(i\hbar\frac{\partial}{\partial t_2}-H_2\right) G^+(\mathbf{r}_2, t_2; \mathbf{r}_1, t_1) = i\hbar\,\delta(\mathbf{r}_2-\mathbf{r}_1) \tag{13.2}$$

Now the integral of the second term on the left-hand side vanishes as $\Delta t \to 0$. To evaluate the integral of the first term we take into account the fact that $G^+(\mathbf{r}_2, t_1-\Delta t; \mathbf{r}_1, t_1) = 0$, and we then obtain the value $i\hbar G^+(\mathbf{r}_2, t_1; \mathbf{r}_1 t_1)$. Equation (13.1) is therefore satisfied.

Remarks: The Green's function used above is defined in coordinate representation. It is possible also to define a Green's function in momentum representation, and indeed to define one which dispenses with any particular representation. This is the so-called "Green's function operator" $G(t)$, defined as the solution of the operator equation

$$\left(i\hbar\frac{\partial}{\partial t}-H\right)G(t) = i\hbar\,\delta(t) \tag{13.3}$$

which obeys the condition $G(t) = 0$ for $t < 0$. If we write $G(t_2-t_1) \equiv G(t_2, t_1)$ and express all operators in coordinate representation, equation (13.3) becomes (13c) on account of the closure relation. The operator $G(t)$ is such that

$$|\psi(t_2)\rangle = G(t_2-t_1)|\psi(t_1)\rangle, \tag{13.4}$$

which leads to equation (13a) for the coordinate representations (wavefunctions) of states. The above considerations are of importance in the formal theory of scattering (see Chapter IX, Section 2).

14. Applying the operator $i\hbar(\partial/\partial t_2)-H^{(0)}(2)$ to both sides of equation (14a), and taking into account (14b), we obtain

$$\left(i\hbar\frac{\partial}{\partial t_2}-H^{(0)}(2)\right)G^+(2, 1) = i\hbar\,\delta^4(2, 1)+H^{(1)}(2)\,G^+(2, 1)$$

and hence

$$\left(i\hbar\frac{\partial}{\partial t_2}-H(2)\right)G^+(2, 1) = i\hbar\,\delta^4(2, 1). \tag{14.1}$$

This equation is the same as (13c). The condition $G^+(2, 1) = 0$ for $t_2 < t_1$ follows from equation (14.1) since $G_0^+(2, 1) = 0$ for $t_2 < t_1$.

Remarks: Equation (14a) can be used to develop time-dependent perturbation theory. Treating $H^{(1)}$ as a perturbation, we have, in zero order approximation, ($H^{(1)} = 0$),

$$G^+(2, 1) = G_0^+(2, 1). \tag{14.2}$$

The first order correction is obtained from (14a) by introducing $G^+(3, 1)$ (in zero order proximation) in the right-hand side:

$$G_1^+(2, 1) = -\frac{i}{\hbar}\int G_0^+(2, 3)\,H^{(1)}(3)\,G_0^+(3, 1)\,d^4x_3\,. \tag{14.3}$$

Similarly, the second order correction can be found:

$$G_2^+(2, 1) = \left(-\frac{i}{\hbar}\right)^2\int G_0^+(2, 3)\,H^{(1)}(3)\,G_0^+(3, 4)\,H^{(1)}(4)\,G_0^+(4, 1)\,d^4x_3\,d^4x_4\,. \tag{14.4}$$

Finally the Green's function has the form of a series in powers of the perturbation:

$$G^{+}(2,1) = G_0^{+}(2,1) - \frac{i}{\hbar}\int G_0^{+}(2,3)\,H^{(1)}(3)\,G_0^{+}(3,1)\,d^4x_3$$
$$+\left(-\frac{i}{\hbar}\right)^2\int G_0^{+}(2,3)\,H^{(1)}(3)\,G_0^{+}(3,4)\,H^{(1)}(4)\,G_0^{+}(4,1)\,d^4x_3\,d^4x_4 + \ldots \qquad (14.5)$$

Equation (14.5) may be understood in the following manner. Suppose for simplicity that $H^{(0)}$ is the Hamiltonian of a free particle; the zero order term then describes free-particle motion from point 1 to point 2. The first order term describes free-particle motion from point 1 to point 3, at which point the perturbation $H^{(1)}(3)$ acts, after which the particle moves freely from 3 to 2. The second order term gives the effect of the perturbation acting twice, viz., at points 3 and 4, and so on.

It is possible to give a simple diagrammatic representation of equation (14.5). For this let us suppose that the time-axis is vertical, directed upwards towards increasing time. Let the Green's function ("propagator") of the actual problem be represented by a thick line, and that of a free particle by a thin line. Let each interaction with the perturbing Hamiltonian $H^{(1)}$ be represented by a kink in the thin line. With these conventions, the series (14.5) can be represented as shown in Fig. V.1. Note that, using the same conventions, the diagram for the relation (14a) is as shown in Fig. V.2.

15. Let us write the function $\psi(\mathbf{r}, t)$, which is a solution of the Schrödinger equation, in the form

$$\psi(\mathbf{r}, t) = \sum_n c_n(t)\,\psi_n(\mathbf{r}). \qquad (15.1)$$

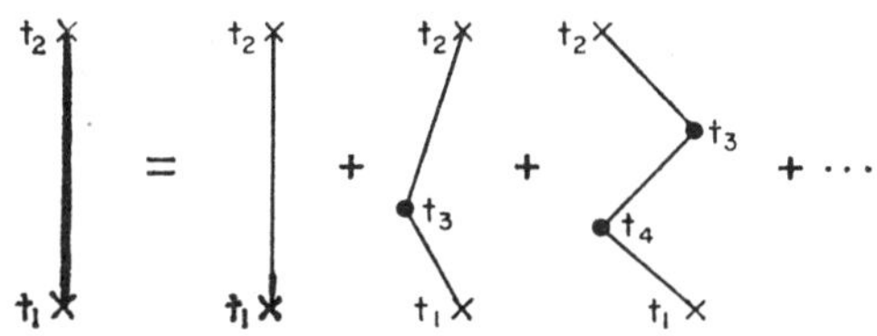

FIG. V.1.

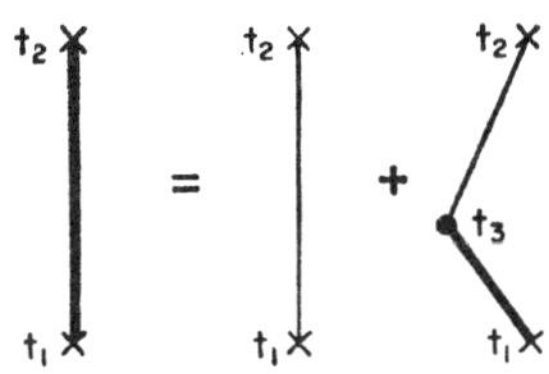

FIG. V.2.

In order to determine the time-dependence of the amplitudes $c_n(t)$ we substitute (15.1) for $\psi(\mathbf{r}, t)$ into the Schrödinger equation $i\hbar(\partial\psi/\partial t) = H\psi$. We then obtain

$$i\hbar \frac{dc_n(t)}{dt} = E_n c_n(t), \tag{15.2}$$

whence

$$c_n(t) = c_n(t_1) e^{-\frac{i}{\hbar} E_n (t-t_1)} \tag{15.3}$$

We can determine the values of the amplitudes $c_n(t_1)$ at the initial time t_1, using (15.1), thus

$$c_n(t_1) = \int \psi(\mathbf{r}_1, t_1)\, \psi^*(\mathbf{r}_1)\, dV_1 . \tag{15.4}$$

Substituting (15.3) into (15.1) and using (15.4), we find an expression for the wavefunction $\psi(\mathbf{r}_2, t_2)$ at time $t_2 \geqslant t_1$:

$$\psi(\mathbf{r}_2, t_2) = \sum_n \psi_n(\mathbf{r}_2) e^{-\frac{i}{\hbar} E_n (t_2 - t_1)} \int \psi(\mathbf{r}_1, t_1)\, \psi_n^*(\mathbf{r}_1)\, dV_1 . \tag{15.5}$$

By bringing the integral sign to the front, and comparing the result with (13a), we find that

$$G^+(\mathbf{r}_2, t_2; \mathbf{r}_1, t_1) = \sum_n \psi_n(\mathbf{r}_2)\, \psi_n^*(\mathbf{r}_1) e^{-\frac{i}{\hbar} E_n (t_2 - t_1)} . \tag{15.6}$$

According to the closure relation, the Green's function (15.6) becomes $\delta(\mathbf{r}_2 - \mathbf{r}_1)$ when $t_2 = t_1$, in agreement with (13.1). Note that this form of Green's function (for the one-dimensional case) has been used already, viz., in problems 8–10, to study the spread of wave-packets.

16. From relation (V.11) the wavefunction in momentum (or "$\{p\}$") representation can be written as

$$\Phi(\mathbf{p}') = \langle \mathbf{p}' \,|\, \psi \rangle = (2\pi\hbar)^{-3/4} \int e^{-\frac{i}{\hbar}(\mathbf{p}' . \mathbf{r}')} \psi(\mathbf{r}')\, d\mathbf{r}' , \tag{16.1}$$

where, for simplicity, the time-dependence of the wavefunctions is not shown explicitly. The Schrödinger equation in the $\{p\}$ representation becomes

$$i\hbar \frac{\partial}{\partial t} \langle \mathbf{p}' \,|\, \psi \rangle = \int \langle \mathbf{p}' \,|\, H \,|\, \mathbf{p}'' \rangle \langle \mathbf{p}'' \,|\, \psi \rangle\, d\mathbf{p}'' . \tag{16.2}$$

It remains for us to calculate $\langle \mathbf{p}' \,|\, H \,|\, \mathbf{p}'' \rangle$. We have that

$$\begin{aligned}
\langle \mathbf{p}' \,|\, V(\mathbf{r}) \,|\, \mathbf{p}'' \rangle &= \iint \langle \mathbf{p}' \,|\, \mathbf{r}' \rangle\, d\mathbf{r}' \langle \mathbf{r}' \,|\, V(\mathbf{r}) \,|\, \mathbf{r}'' \rangle\, d\mathbf{r}'' \langle \mathbf{r}'' \,|\, \mathbf{p}'' \rangle \\
&= (2\pi\hbar)^{-3} \iint e^{-\frac{i}{\hbar}(\mathbf{p}' . \mathbf{r}')}\, d\mathbf{r}' V(\mathbf{r}')\, \delta(\mathbf{r}' - \mathbf{r}'')\, d\mathbf{r}'' e^{\frac{i}{\hbar}(\mathbf{p}'' . \mathbf{r}'')} \\
&= (2\pi\hbar)^{-3} \int V(\mathbf{r}') e^{-\frac{i}{\hbar}(\mathbf{p}' - \mathbf{p}'') . \mathbf{r}'}\, d\mathbf{r}' = U(\mathbf{p}' - \mathbf{p}'').
\end{aligned} \tag{16.3}$$

It follows that the operator H will have, in the $\{p\}$ representation, the following matrix representation:

$$\langle \mathbf{p}' | H | \mathbf{p}'' \rangle = \frac{\mathbf{p}'^2}{2m} \delta(\mathbf{p}' - \mathbf{p}'') + U(\mathbf{p}' - \mathbf{p}''). \tag{16.4}$$

Introducing explicitly the time-dependence of the wavefunction, we obtain, from (16.1), (16.2) and (16.4), the equation

$$i\hbar \frac{\partial}{\partial t} \Phi(\mathbf{p}', t) = \frac{\mathbf{p}'^2}{2m} \Phi(\mathbf{p}', t) + \int U(\mathbf{p}' - \mathbf{p}'') \Phi(\mathbf{p}'', t)\, d\mathbf{p}'',$$

which is, *mutatis mutandis*, equation (16a).

17. The oscillator Hamiltonian, in momentum representation, is

$$H = \frac{\mathbf{p}^2}{2m} - \frac{m\omega^2\hbar^2}{2} \frac{\partial^2}{\partial p^2}, \tag{17.1}$$

and hence the Schrödinger equation in this representation is

$$i\hbar \frac{\partial}{\partial t} \Phi(p, t) = \left(\frac{\mathbf{p}^2}{2m} - \frac{m\omega^2\hbar^2}{2} \frac{\partial^2}{\partial p^2} \right) \Phi(p, t). \tag{17.2}$$

Note that equation (17.2) may also be obtained as a special case of equation (16a).

Looking for solutions of the form

$$\Phi(p, t) = \phi(p) e^{-\frac{i}{\hbar} Et} \tag{17.3}$$

we obtain for $\phi(p)$ the eigenvalue equation

$$\frac{d^2\phi}{dp^2} + \frac{2}{m\omega^2\hbar^2} \left(E - \frac{p^2}{2m} \right) \phi = 0, \tag{17.4}$$

i.e.

$$\frac{d^2\phi}{d\eta^2} + (\lambda - \eta^2)\phi = 0, \tag{17.4'}$$

where

$$\eta = \frac{p}{p_0}; \quad p_0 = \sqrt{m\omega\hbar}; \quad \lambda = 2E/\hbar\omega. \tag{17.5}$$

By comparing (17.4') and (17.5) with (17.1) of Chapter II, and using (17.12) of Chapter II it follows that the eigenfunctions corresponding to the energy eigenvalues $E_n = \omega\hbar(n + \frac{1}{2})$, $n = 0, 1, 2, \ldots$, are given by

$$\phi_n(p) = (2^n n!\, p_0 \sqrt{\pi})^{-1/2} H_n \left(\frac{p}{p_0} \right) e^{-1/2(p/p_0)^2}. \tag{17.6}$$

18. For a homogeneous field, $V(x) = -kx$, and equation (17.4) is replaced by

$$i\hbar k \frac{d\phi(p)}{dp} + \left(-\frac{p^2}{2m} + E\right)\phi(p) = 0. \tag{18.1}$$

The solution of this equation with eigenvalue E is

$$\phi_E(p) = Ce^{-\frac{i}{\hbar k}\left(\frac{p^3}{6m} - E_p\right)}. \tag{18.2}$$

Since the energy spectrum is continuous, the appropriate normalization condition is

$$\int \phi_E(p)^* \phi_{E'}(p)\, dp = \delta(E-E')$$

and it follows that

$$C = (2\pi\hbar k)^{-1/2}.$$

To relate these results to those of problem 21 of Chapter II, we shall change to coordinate representation. Omitting the time-dependent factor in (V.11) we have that

$$\psi_E(x) = (2\pi\hbar)^{-1/2} \int_{-\infty}^{+\infty} \phi_E(p) e^{\frac{i}{\hbar}px}\, dp = \frac{1}{2\pi\hbar\sqrt{k}} \int_{-\infty}^{+\infty} e^{-\frac{i}{\hbar k}\left(\frac{p^3}{6m} - E_p\right)} e^{\frac{i}{\hbar}px}\, dp. \tag{18.3}$$

Changing the variable of integration to $u = -(2m\hbar k)^{-1/3}p$, we obtain

$$\psi_E(x) = \frac{\alpha}{2\pi\sqrt{k}} \int_{-\infty}^{+\infty} e^{i(u^3/3) - iuy}\, du = \frac{\alpha}{\pi\sqrt{k}} \int_0^{\infty} \cos\left(\frac{u^3}{3} - uy\right) dy \tag{18.4}$$

where $y = \left(x + \frac{E}{k}\right)\alpha$ and $\alpha = (2mk\hbar^{-2})^{1/3}$.

19. Let us denote by $\psi_t(x)$ the result of the operation p^{-1} on the function $\psi(x)$, i.e.

$$\psi_t(x) = p^{-1}\psi(x). \tag{19.1}$$

Our problem consists in finding $\psi_t(x)$ when $\psi(x)$ is given. It is convenient to change to momentum representation. Let $f(p)$ and $f_t(p)$ be the Fourier transforms of the functions $\psi(x)$ and $\psi_t(x)$ respectively. Equation (19.1) then becomes

$$f_t(p) = \frac{1}{p} f(p). \tag{19.2}$$

This equation shows that the function $f_t(p)$ has, in general, a pole at $p = 0$, and thus does not in general have the properties of continuity, etc., which are required of wavefunctions. In order not to violate these properties it is necessary that $f(0) = 0$, a fact which, in coordinate representation, becomes

$$\int_{-\infty}^{+\infty} \psi(x)\, dx = 0. \tag{19.3}$$

From (19.1) and (19.2) we have

$$\psi_t(x) = (2\pi\hbar)^{-1/2} \int_{-\infty}^{+\infty} \frac{f(p)}{p} e^{\frac{i}{\hbar}px}\, dp. \tag{19.4}$$

Let us suppose the integration along the real axis to be carried out in the complex p plane, and the contour of integration to be changed into one which avoids the origin by following a small semicircle in the lower half-plane (Fig. V.3).

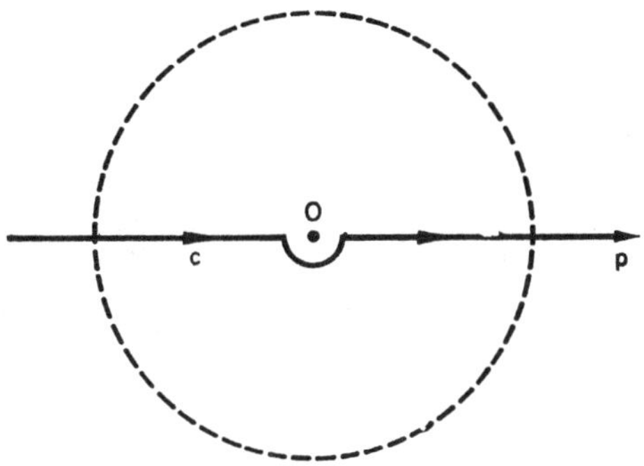

FIG. V.3.

Since the function under the integral sign is holomorphic in the whole plane (including the origin), the value of the integral is unchanged, i.e.

$$\psi_t(x) = (2\pi\hbar)^{-1/2} \int_C \frac{f(p)}{p} e^{\frac{i}{\hbar}px}\, dp. \tag{19.5}$$

Now, by definition,

$$f(p) = (2\pi\hbar)^{-1/2} \int_{-\infty}^{+\infty} \psi(x') e^{-\frac{i}{\hbar}px'}\, dx'. \tag{19.6}$$

Substituting for $f(p)$ in (19.5) and changing the order of integration, we obtain

$$\psi_t(x) = \frac{1}{2\pi\hbar} \int_{-\infty}^{+\infty} \psi(x')\, dx' \int_C \frac{e^{-\frac{i}{\hbar}p(x-x')}}{p}\, dp. \tag{19.7}$$

The second integral in (19.7) can be calculated immediately if we complete the contour C by a semicircle having a sufficiently large radius and situated in the upper, or the lower, half-plane, according to whether $x > x'$, or $x < x'$, respectively. From the theory of residues, we find then that

$$\int_C \frac{e^{\frac{i}{\hbar}p(x-x')}}{p}\, dp = \begin{cases} 2\pi i \text{ if } x > x', \\ 0 \text{ if } x < x'. \end{cases}$$

Substituting in (19.7) we have

$$\psi_t(x) = \frac{i}{\hbar}\int_{-\infty}^{x} dx'\, \psi(x') \tag{19.8}$$

and hence the operator relation

$$p^{-1} = \frac{i}{k}\int_{-\infty}^{x} dx'. \tag{19.9}$$

This result is not unexpected, since p^{-1} is the inverse of the differential operator $p = -i\hbar\frac{\partial}{\partial x}$. For the operator x^{-1} in the $\{p\}$ representation we obtain similarly the operator relation

$$x^{-1} = \frac{1}{i\hbar}\int_{-\infty}^{p'} dp'. \tag{19.10}$$

It must be remembered that, for the above results to hold, the condition $f(0)=0$ is essential. If this condition were not satisfied, the operators would not be well defined. Thus, if we take the semicircle around the origin to be in the upper half-plane and perform the same calculations as above, we obtain

$$\psi_t(x) = -\frac{i}{\hbar}\int_{x}^{+\infty} dx'\, \psi(x').$$

This relation is equivalent to (19.8) if, and only if, $\int_{-\infty}^{+\infty} \psi(x')\, dx' = 0$.

From (19.6), we see that this condition is equivalent to the condition $f(0) = 0$.

20. In problem 12 it was shown that the transformation of the wavefunction $\psi(x, t)$ in coordinate representation, which corresponds to a Galilean transformation of the coordinates $x = x'+vt$, $t = t'$, is given by the expression

$$\psi'(x', t') = \exp\left(-i\frac{mv}{\hbar}x+i\frac{mv^2}{2\hbar}t\right)\psi(x, t). \tag{20.1}$$

Now the wavefunction in momentum representation is given by

$$\Phi(p, t) = (2\pi\hbar)^{-1} \int_{-\infty}^{+\infty} \psi(x, t) e^{-\frac{i}{\hbar}px}\, dx.$$

Substituting for $\psi(x, t)$ from (20.1) we have

$$\Phi(p, t) = \exp\left[\left(i\frac{mv^2}{2\hbar} - i\frac{pv}{\hbar}\right)t\right] (2\pi\hbar)^{-1} \int_{-\infty}^{+\infty} \psi'(x', t') e^{-\frac{i}{\hbar}(p-mv)x'}\, dx'$$

and hence

$$\Phi(p, t) = \exp\left[\left(i\frac{mv^2}{2\hbar} - i\frac{pv}{\hbar}\right)t\right] \Phi(p-mv, t). \tag{20.2}$$

21. Since, for the one-dimensional harmonic oscillator, the Hamiltonian by itself forms a complete set of observables, there is only one energy representation. Let us denote by $|E_n\rangle = |n\rangle$, say, the base vectors of this representation, in the Schrödinger picture. Taking into account (V. 9) and (V.10), the matrix elements of the coordinate operator x in this representation (remembering that $x|x'\rangle = x'|x'\rangle$) are given by

$$\begin{aligned}\langle n|x|m\rangle &= \int \langle n|x|x'\rangle\langle x'|m\rangle\, dx' = \int \psi_n^*(x)\, x\psi_m(x)\, dx \\ &= x_0^2 N_n N_m \int_{-\infty}^{+\infty} \xi H_n(\xi) H_m(\xi) e^{-\xi^2}\, d\xi.\end{aligned} \tag{21.1}$$

In the last expression the notation of problem 17 of Chapter II is used. The integral on the right-hand side can be evaluated by using the generating function (A.3), from which it follows that

$$\int_{-\infty}^{+\infty} e^{-s^2+2s\xi} e^{-t^2+2t\xi} e^{-\xi^2}\, d\xi = \sum_{n=0}^{\infty} \sum_{m=0}^{\infty} \frac{s^n t^m}{n!m!} \int_{-\infty}^{+\infty} \xi H_n(\xi) H_m(\xi) e^{-\xi^2}\, d\xi. \tag{21.2}$$

Now

$$\int_{-\infty}^{+\infty} \xi e^{-s^2+2s\xi-t^2+2t\xi-\xi^2}\, d\xi = \sqrt{\pi}(s+t)e^{2st},$$

and

$$\sqrt{\pi}(s+t)e^{2st} = \sqrt{\pi} \sum_{n=0}^{\infty} \frac{2^n(s^{n+1}t^n + s^n t^{n+1})}{n!}, \tag{21.3}$$

so that, on equating corresponding coefficients of powers of s and t in the two expansions (21.2) and (21.3), we obtain

$$\langle n|x|m\rangle = \begin{cases} x_0\left(\dfrac{n+1}{2}\right)^{1/2}, & m = n+1, \\ x_0\left(\dfrac{n}{2}\right)^{1/2}, & m = n-1, \end{cases} \tag{21.4}$$

or, in explicitly tabulated matrix form,

$$(x) = x_0 \begin{bmatrix} 0 & \sqrt{\frac{1}{2}} & 0 & 0 & \dots \\ \sqrt{\frac{1}{2}} & 0 & \sqrt{\frac{2}{2}} & 0 & \dots \\ 0 & \sqrt{\frac{2}{2}} & 0 & \sqrt{\frac{3}{2}} & \dots \\ \cdot & \cdot & \cdot & \cdot & \cdot \end{bmatrix}. \tag{21.5}$$

In a similar manner the matrix representation of the momentum operator can be obtained. We have to calculate the integrals

$$\langle n|p|m\rangle = -i\hbar \int_{-\infty}^{+\infty} \psi_n(x) \frac{d\psi_m(x)}{dx} dx. \tag{21.6}$$

After calculations similar to those made above we obtain finally

$$\langle n|p|m\rangle = i\hbar x_0^{-2}(n-m)\langle n|x|m\rangle, \tag{21.7}$$

and hence, in tabulated matrix form,

$$(p) = \frac{i\hbar}{x_0} \begin{bmatrix} 0 & -\sqrt{\frac{1}{2}} & 0 & 0 & \dots \\ \sqrt{\frac{1}{2}} & 0 & -\sqrt{\frac{2}{2}} & 0 & \dots \\ 0 & \sqrt{\frac{2}{2}} & 0 & -\sqrt{\frac{3}{2}} & \dots \\ \cdot & \cdot & \cdot & \cdot & \cdot \end{bmatrix}. \tag{21.8}$$

Remarks: The matrix representation of p can be obtained also in the following way. Since $H_H = \frac{1}{2\mu} p_H^2 + V(x_H)$ and $[x, p] = [x_H, p_H] = i\hbar$ (see problem 1), we have from (V.3) that

$$\frac{dx_H}{dt} = \frac{1}{i\hbar}[x_H, H_H] = \frac{1}{2i\mu\hbar}[x_H, p_H^2] = \frac{p_H}{\mu}, \tag{21.9}$$

and hence, in particular,

$$\frac{d}{dt}\langle n|x_H|m\rangle = \frac{1}{\mu}\langle n|p_H|m\rangle. \tag{21.10}$$

Taking into account the relations $\langle n|x_H|m\rangle = \langle n|x|m\rangle e^{i\omega_{nm}t}$ and $\langle n|p_H|m\rangle = \langle n|p|m\rangle e^{i\omega_{nm}t}$, where $\omega_{nm} = (E_n - E_m)/\hbar = \omega(n-m)$, we obtain once again the relation (21.7).

22. The wavefunctions found in problem 7 of Chapter II, viz.,

$$\psi_n^{(p)}(x) = \sqrt{\frac{1}{a}} \cos \frac{n\pi}{2a} x, \quad \text{for } n \text{ even},$$

$$\psi_n^{(i)}(x) = \sqrt{\frac{1}{a}} \sin \frac{n\pi}{2a} x, \quad \text{for } n \text{ odd},$$

can be used to find the matrix elements of the coordinate operator x. We have to distinguish carefully the results obtained for states of different parity. Thus, for n and m even, we find that

$$\langle n|x|m\rangle = x_{nm} = \frac{1}{a}\int_{-a}^{a} x\cos\frac{n\pi}{2a}x\cos\frac{m\pi}{2a}x\,dx = 0. \tag{22.1}$$

For n and m odd, we find that

$$x_{nm} = \frac{1}{a}\int_{-a}^{a} x\sin\frac{n\pi}{2a}x\sin\frac{m\pi}{2a}x\,dx = 0. \tag{22.2}$$

For even n and odd m,

$$\begin{aligned} x_{nm} &= \frac{1}{a}\int_{-a}^{a} x\cos\frac{n\pi}{2a}x\sin\frac{m\pi}{2a}x\,dx \\ &= \frac{1}{2a}\int_{-a}^{a} x\left[\sin\frac{(n+m)\pi}{2a}x-\sin\frac{(n-m)\pi}{2a}x\right]dx \\ &= \frac{4a}{\pi^2}\left[\frac{1}{(n+m)^2}\sin\frac{(n+m)\pi}{2}+\frac{1}{(n-m)^2}\sin\frac{(n-m)\pi}{2}\right]. \end{aligned} \tag{22.3}$$

For odd n and even m,

$$x_{mn} = \frac{4a}{\pi^2}\left[\frac{1}{(n+m)^2}\sin\frac{(n+m)\pi}{2}-\frac{1}{(n-m)^2}\sin\frac{(n-m)\pi}{2}\right]. \tag{22.4}$$

From (22.1)–(22.4) it follows that the diagonal elements of the (x) matrix vanish. The non-vanishing elements correspond to an odd difference $(n-m)$, namely to a transition $n \to m$ associated with a change in the parity of the wavefunction.

In order to find the momentum matrix we can use the method described in problem 21; thus, from the relation

$$\langle n|p_H|m\rangle = \mu\frac{d}{dt}\langle n|x_H|m\rangle = \mu\frac{i}{\hbar}(E_n-E_m)\langle n|x_H|m\rangle$$

it follows that

$$p_{nm} = \langle n|p|m\rangle = \mu\frac{i}{\hbar}(E_n-E_m)\langle n|x|m\rangle = \mu\frac{i}{\hbar}(E_n-E_m)x_{nm}. \tag{22.5}$$

23. As was noted in problem 21, we have

$$\langle n|x_H|m\rangle = \langle n|x|m\rangle e^{\frac{i}{\hbar}(E_n-E_m)t}, \tag{23.1}$$

and hence

$$\frac{d}{dt}\langle n|x_H|m\rangle = \frac{i}{\hbar}(E_n-E_m)\langle n|x_H|m\rangle. \tag{23.2}$$

Using these relations we can write

$$\sum_n (E_n-E_m)|\langle n|x|m\rangle|^2 = \sum_n (E_n-E_m)|\langle n|x_H|m\rangle|^2$$
$$= \sum_n (E_n-E_m)\langle n|x_H|m\rangle\langle m|x_H|n\rangle$$
$$= \frac{1}{2}\frac{\hbar}{i}\sum_n \langle m|x_H|n\rangle\frac{d}{dt}\langle n|x_H|m\rangle - \frac{1}{2}\frac{\hbar}{i}\sum_n \langle n|x_H|m\rangle\frac{d}{dt}\langle m|x_H|n\rangle$$
$$= \frac{1}{2}\frac{\hbar}{i}\langle m|(x_H\dot{x}_H-\dot{x}_H x_H)|m\rangle = \frac{1}{2}\frac{\hbar}{i}\langle m|(x\dot{x}-\dot{x}x)|m\rangle = \frac{\hbar^2}{2\mu}.$$

24. The double commutator (24a) can be written in the form

$$[[H, e^{i\mathbf{k}.\mathbf{r}}]] = He^{i\mathbf{k}.\mathbf{r}}e^{-i\mathbf{k}.\mathbf{r}} - e^{i\mathbf{k}.\mathbf{r}}He^{-i\mathbf{k}.\mathbf{r}} - e^{-i\mathbf{k}.\mathbf{r}}He^{i\mathbf{k}.\mathbf{r}} + e^{-i\mathbf{k}.\mathbf{r}}e^{i\mathbf{k}.\mathbf{r}}H. \tag{24.1}$$

On the other hand, taking into account (I.27a), we have that

$$[[H, e^{i\mathbf{k}.\mathbf{r}}], e^{-i\mathbf{k}.\mathbf{r}}] = 2H - e^{i\mathbf{k}.\mathbf{r}}He^{-i\mathbf{k}.\mathbf{r}} - e^{-i\mathbf{k}.\mathbf{r}}He^{i\mathbf{k}.\mathbf{r}}$$
$$= -2\frac{(ik)^2}{2!}[\mathbf{r}, [\mathbf{r}, H]] - 2\frac{(ik)^4}{4!}\big[\mathbf{r}, [\mathbf{r}, [\mathbf{r}, [\mathbf{r}, H]]]\big] - \dots \tag{24.2}$$

Since

$$[\mathbf{r}, H] = \frac{i\hbar}{\mu}\mathbf{p}, \quad [\mathbf{r}, \mathbf{p}] = 3i\hbar,$$

it follows from (24.2) that

$$[[H, e^{i\mathbf{k}.\mathbf{r}}], e^{-i\mathbf{k}.\mathbf{r}}] = -\frac{3\hbar^2k^2}{\mu}. \tag{24.3}$$

Let us now take the average value of both sides of (24.3) for the state $|m\rangle$. We obtain the result that

$$\sum_n \{E_m\langle m|e^{i\mathbf{k}.\mathbf{r}}|n\rangle\langle n|e^{-i\mathbf{k}.\mathbf{r}}|m\rangle - E_n\langle m|e^{i\mathbf{k}.\mathbf{r}}|n\rangle\langle n|e^{i\mathbf{k}.\mathbf{r}}|m\rangle$$
$$-E_n\langle m|e^{-i\mathbf{k}.\mathbf{r}}|n\rangle\langle n|e^{i\mathbf{k}.\mathbf{r}}|m\rangle + E_m\langle m|e^{-i\mathbf{k}.\mathbf{r}}|n\rangle\langle n|e^{i\mathbf{k}.\mathbf{r}}|m\rangle\} = -3\hbar^2k^2/\mu. \tag{24.4}$$

From this result, the relation (24a) is easily obtained, if we note that

$$\langle m|e^{i\mathbf{k}.\mathbf{r}}|n\rangle\langle n|e^{-i\mathbf{k}.\mathbf{r}}|m\rangle = \langle n|e^{-i\mathbf{k}.\mathbf{r}}|m\rangle^*\langle n|e^{-i\mathbf{k}.\mathbf{r}}|m\rangle$$
$$= |\langle n|e^{-i\mathbf{k}.\mathbf{r}}|m\rangle|^2 = |\langle n|e^{i\mathbf{k}.\mathbf{r}}|m\rangle|^2. \tag{24.5}$$
$$\langle m|e^{-i\mathbf{k}.\mathbf{r}}|n\rangle\langle n|e^{i\mathbf{k}.\mathbf{r}}|m\rangle = |\langle n|e^{i\mathbf{k}.\mathbf{r}}|m\rangle|^2. \tag{24.6}$$

The last equality in (24.5) follows from the relations $e^{\pm i\mathbf{k}.\mathbf{r}} = \cos\mathbf{k}.\mathbf{r} \pm i\sin\mathbf{k}.\mathbf{r}$.

If we retain only the first term in the expansion of the exponential $e^{i\mathbf{k}.\mathbf{r}}$ (the "dipole approximation") in (24a), we obtain (23a) again.

25. (a) In accordance with the definition (V.15), for any $|u\rangle$ we have that

$$\langle u\,|\,\varrho\,|\,u\rangle = \sum_m \langle u\,|\,m\rangle p_m\langle m\,|\,u\rangle = \sum_m p_m\,|\langle u\,|\,m\rangle|^2 \geqslant 0. \tag{25.1}$$

Further, if we use the Schwartz inequality (see problem 1 of Chapter I):

$$|\langle u\,|\,m\rangle|^2 \leqslant \langle u\,|\,u\rangle\langle m\,|\,m\rangle = \langle u\,|\,u\rangle,$$

it then follows that

$$\langle u|\,(1-\varrho)\,|u\rangle = \langle u\,|\,u\rangle = \sum_m p_m\,|\langle u\,|\,m\rangle|^2 \geqslant \langle u\,|\,u\rangle - \langle u\,|\,u\rangle \sum_m p_m = 0. \tag{25.2}$$

(b) Let us suppose that the system of base vectors $|1\rangle$, $|2\rangle$, .., $|m\rangle$, is such that ϱ is represented by a diagonal matrix. In this representation, then,

$$\text{Tr}\,(\varrho^2) = \sum_l \langle l\,|\,\varrho^2\,|\,l\rangle = \sum_l \langle l\,|\,\varrho\,|\,l\rangle^2 \leqslant \left[\sum_l \langle l\,|\,\varrho\,|\,l\rangle\right]^2 = (\text{Tr}\,\varrho)^2 = 1. \tag{25.3}$$

Since the trace of an operator does not depend on the representation, (25.3) is valid in any representation. If the system is in a pure state, $|X\rangle$ say, the density operator is of the form $\varrho = |X\rangle\langle X|$, provided the norm of $|X\rangle$ is equal to unity. We see immediately that in this case $\varrho^2 = \varrho$, whence $\text{Tr}\,(\varrho^2) = 1$. Conversely, if $\text{Tr}\,(\varrho^2) = 1$, then, since this is true in any representation, let us take one in which ϱ is diagonal, and in which, therefore, $\sum_l \langle l\,|\,\varrho\,|\,l\rangle^2 = 1 = \sum_l \langle l\,|\,\varrho\,|\,l\rangle$.

Since the quantities $\langle l\,|\,\varrho\,|\,l\rangle$ are all smaller than, or at most equal to, unity, the above relation implies that there is one and only one non-vanishing diagonal element of ϱ, and that this is equal to unity. But this is true only for a system in a pure state.

26. From (V.4) we have $\varrho_H = U^+(t, t_0)\varrho U(t, t_0)$. By calculating the time-derivative of ϱ_H and using relation (V.16) we find that

$$\frac{\partial \varrho_H}{\partial t} = \frac{\partial U^+}{\partial t}\varrho U + U^+\frac{\partial \varrho}{\partial t}U + U^+\varrho\frac{\partial U}{\partial t} = U^+\left\{\frac{\partial \varrho}{\partial t} + \frac{i}{\hbar}[H, \varrho]\right\}U = 0.$$

27. Any operator Ω can be written as a linear combination of the operators $\Omega^{(r)}$, $r = 1, 2, \ldots, N^2$, in the form

$$\Omega = \sum_r \omega_r \Omega^{(r)} \text{ where } \omega_r = \text{Tr}\,(\Omega\Omega^{(r)}).$$

In particular, the density matrix in the Schrödinger picture can be written as

$$\varrho_t = \sum_r \varrho_t^{(r)}\Omega^{(r)} \text{ where } \varrho_t^{(r)} = \text{Tr}\,(\varrho_t\Omega^{(r)}) = \langle\Omega^{(r)}\rangle.$$

Thus, if the operators $\Omega^{(r)}$ are given, the dynamical state corresponding to the matrix ϱ_t is described by the set of average values $\varrho_t^{(r)} = \langle \Omega^{(r)} \rangle$. Taking now the time-derivative of relation (27b), using (V.16) and (27a), multiplying on the left by $\Omega^{(s)}$, and taking the trace, we obtain finally

$$\frac{i}{\hbar} \operatorname{Tr} \{\Omega^{(s)} [\varrho_t, H]\} = \frac{d\varrho_t^{(s)}}{dt}. \tag{27.1}$$

The equations of motion (27c) are obtained from (27.1) by substituting for ϱ_t from (27b) and permuting cyclically the operators under the trace sign. The operators $\Omega^{(r)}$ and the equations of motion (27c) are given explicitly for a particular case in problem 38 of Chapter VI.

CHAPTER VI

Orbital Angular Momentum and Spin

1. Properties of Angular Momentum Operators

A vector **J** is called an *angular momentum operator* if its components are observables which satisfy the commutation rules:†

$$[J_x, J_y] = iJ_z, \quad [J_y, J_z] = iJ_x, \quad [J_z, J_x] = iJ_y. \tag{VI.1}$$

Simultaneous eigenvectors $|\tau jm\rangle$ of $\mathbf{J}^2$ and of J_z can be found such that

$$\mathbf{J}^2|\tau jm\rangle = j(j+1)|\tau jm\rangle, \quad J_z|\tau jm\rangle = m|\tau jm\rangle \tag{VI.2}$$

$$J_\pm|\tau jm\rangle = \sqrt{j(j+1)-m(m\pm1)}\,|\tau, j, m\pm1\rangle \tag{VI.3}$$

where $J_\pm = J_x \pm iJ_y$, and τ stands for those indices which, together with j and m, specify completely the states of the system under consideration. The number j can in general have integral or half-integral values and, for any given j, m can have the following $2j+1$ values: $-j, -j+1, \ldots, j$. For a single particle, the vector operator $\mathbf{l} = \mathbf{r}\times\mathbf{p}$ satisfies the commutation rules (VI.1) and represents the orbital angular momentum of the particle about the origin of **r** (see Chapter II, Section 3). When the location of the origin is obvious (e.g., at the centre of a central field of force), **l** is referred to simply as *the* orbital angular momentum operator of the particle.

2. Spin

The electron has an intrinsic angular momentum **s** called the "spin", of magnitude $\frac{1}{2}$ ("spin $\frac{1}{2}$"), with which a magnetic moment $\mu_s = g_s \dfrac{e}{2mc}\,\mathbf{s}$ is associated. The value of the constant g_s can be determined empirically or from the relativistic theory of the electron, and in fact $g_s \approx 2$. Experimental evidence shows that various other particles also have an

† In this chapter it is assumed that one is working with units such that $\hbar = 1$ (except in the problems 32–39).

intrinsic angular momentum (spin) whose magnitude may have an integral or half-integral value. Since **s** obeys the rule (VI.1), a representation of the electronic spin operator **s** by "Pauli matrices" can be obtained by putting $j = \frac{1}{2}$ in equations (VI.2) and (VI.3), and writing

$$\mathbf{s} = \frac{1}{2}\sigma; \qquad \sigma_x = \begin{bmatrix} 0 & 1 \\ 1 & 0 \end{bmatrix}, \quad \sigma_y = \begin{bmatrix} 0 & -i \\ i & 0 \end{bmatrix}, \quad \sigma_z = \begin{bmatrix} 1 & 0 \\ 0 & -1 \end{bmatrix}. \tag{VI.4}$$

The corresponding wavefunction of a spin-$\frac{1}{2}$ particle in coordinate representation then has two components $\begin{pmatrix} \langle \mathbf{r}, j = \frac{1}{2}, m = +\frac{1}{2} \rangle \\ \langle \mathbf{r}, j = \frac{1}{2}, m = -\frac{1}{2} \rangle \end{pmatrix} \equiv \begin{bmatrix} \psi_+(\mathbf{r}) \\ \psi_-(\mathbf{r}) \end{bmatrix}$ say, and is called a "spinor".

3. Angular Momentum and Rotations of Coordinate Axes. The Addition of Angular Momenta

3.1. The Rotation Operator of a Physical System

Let D be the (unitary) operator which represents a rotation of a given physical system about a given axis, i.e. the operator which is such that the state vectors and the observables of the rotated system are related to those of the original system by the relations

$$|a'\rangle = D|a\rangle, \quad Q' = DQD^+. \tag{VI.5}$$

If **J** is the operator which represents the total angular momentum of the system about a given origin O, then the operators $D(\mathbf{u}, d\phi)$ and $D(\mathbf{u}, \phi)$ which represent an infinitesimal rotation $d\phi$ of the system, and a finite rotation ϕ, respectively, about an axis through O in the direction of the unit vector **u**, can be written in the form

$$D(\mathbf{u}, d\phi) = 1 - i\, d\phi(\mathbf{J} \cdot \mathbf{u}); \qquad D(\mathbf{u}, \phi) = e^{-i\phi(\mathbf{J}\cdot\mathbf{u})}. \tag{VI.6}$$

For a rotation $R(\alpha, \beta, \gamma)$ about the origin, given by the Euler angles α, β, γ (Fig. VI.1), we have, in an obvious notation, that

$$D(\alpha, \beta, \gamma) = e^{-i\alpha J_z} e^{-\beta J_y} e^{-i\gamma J_z}. \tag{VI.7}$$

The operator $D(\alpha, \beta, \gamma)$ has the following matrix representation‡

$$D^{(j)}_{mm'}(\alpha, \beta, \gamma) \equiv \langle jm | D(\alpha, \beta, \gamma) | jm' \rangle = e^{i\alpha m}\, d^{(j)}_{mm'}(\beta) e^{-i\gamma m'}, \tag{VI.8}$$

where the coefficients $d^{(j)}_{mm'}(\beta)$ are given by *Wigner's formula* (see problem 14).

‡ The matrix $D^{(j)} \equiv (D^{(j)}_{mm'})$ is of $2j+1$ dimensions and is called the irreducible representation in $2j+1$ dimensions of the rotation group. This name arises from the fact that applying the operator $D(\alpha, \beta, \gamma)$ to the state $|jm\rangle$ yields a linear superposition of all the $2j+1$ states $|jm'\rangle$ with $m' = -j, -j+1, \ldots, j$, the value of j itself being unchanged.

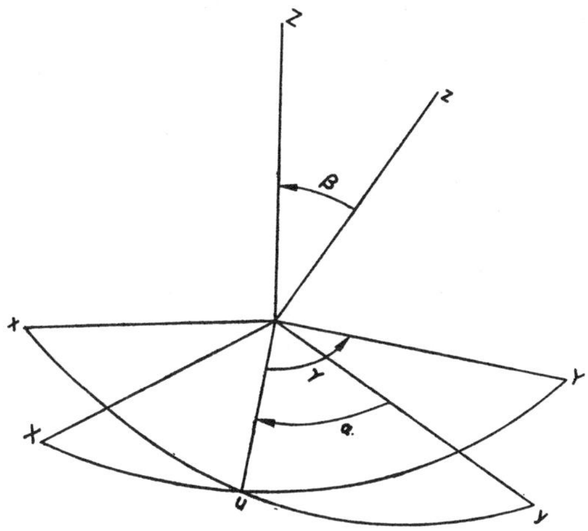

FIG. VI.1.

3.2. THE ADDITION OF ANGULAR MOMENTA

Let us denote by $\mathbf{J} = \mathbf{j}_1+\mathbf{j}_2$ the sum of the angular momentum operators of two subsystems which together form the system under study, by $|\tau j_1 j_2 m_1 m_2\rangle$ the simultaneous eigenvectors of the set of operators $\mathbf{j}_1^2, \mathbf{j}_2^2, j_{1z}, j_{2z}$, and by $|\tau j_1 j_2 JM\rangle$ the simultaneous eigenvectors of the set of operators $\mathbf{j}_1^2, \mathbf{j}_2^2, \mathbf{J}^2, J_z$. Here τ represents the indices which, together with j_1, j_2, m_1, m_2, or j_1, j_2, J, M, respectively, specify completely the states of the system. We have then that

$$|\tau j_1 j_2 JM\rangle = \sum_{m_1, m_2} |\tau j_1 j_2 m_1 m_2\rangle \langle j_1 j_2 m_1 m_2 | JM\rangle, \tag{VI.9}$$

where $|j_1-j_2| \leqslant J \leqslant j_1+j_2$, $m_1+m_2 = M$, and the coefficients $\langle j_1 j_2 m_1 m_2 | JM\rangle$ are called the *Wigner* or *Clebsch–Gordan coefficients*. The coefficients most frequently used in applications are given in the following tables.

TABLE VI.1

Clebsch–Gordan Coefficients for $j_2 = 1$

	$m_2 = 1$	$m_2 = 0$	$m_2 = -1$
$J = j_1+1$	$\left[\frac{(j_1+M)(j_1+M+1)}{(2j_1+1)(2j_1+2)}\right]^{1/2}$	$\left[\frac{(j_1-M+1)(j_1+M+1)}{(2j_1+1)(j_1+1)}\right]^{1/2}$	$\left[\frac{(j_1-M)(j_1-M+1)}{(2j_1+1)(2j_1+2)}\right]^{1/2}$
$J = j_1$	$-\left[\frac{(j_1+M)(j_1-M+1)}{2j_1(j_1+1)}\right]^{1/2}$	$\left[\frac{M^2}{j_1(j_1+1)}\right]^{1/2}$	$\left[\frac{(j_1-M)(j_1+M+1)}{2j_1(j_1+1)}\right]^{1/2}$
$J = j_1-1$	$\left[\frac{(j_1-M)(j_1-M+1)}{2j_1(2j_1+1)}\right]^{1/2}$	$-\left[\frac{(j_1-M)(j_1+M)}{j_1(2j_1+1)}\right]^{1/2}$	$\left[\frac{(j_1+M+1)(j_1+M)}{2j_1(2j_1+1)}\right]^{1/2}$

TABLE VI.2

Clebsch–Gordan Coefficients for $j_2 = \frac{1}{2}$

	$m_2 = \frac{1}{2}$	$m_2 = -\frac{1}{2}$
$J = j_1+\frac{1}{2}$	$\left[\frac{j_1+M+\frac{1}{2}}{2j_1+1}\right]^{1/2}$	$\left[\frac{j_1-M+\frac{1}{2}}{2j_1+1}\right]^{1/2}$
$J = j_1-\frac{1}{2}$	$-\left[\frac{j_1-M+\frac{1}{2}}{2j_1+1}\right]^{1/2}$	$\left[\frac{j_1+M+\frac{1}{2}}{2j_1+1}\right]^{1/2}$

3.3. IRREDUCIBLE TENSOR OPERATORS

By definition, the $2k+1$ operators $T_q^{(k)}$ ($q = -k, -k+1, \ldots, k$) form the components of an *irreducible tensor operator* of order k, $\mathbf{T}^{(k)}$, say, if they transform under a rotation of axes according to the relation

$$RT_q^{(k)}R^{-1} = \sum_{q'} T_{q'}^{(k)}R_{q'q}^{(k)}. \tag{VI.10}$$

The relation (VI.10) is equivalent to the following commutation rules between the $T_q^{(k)}$ and the components of the total angular momentum operator $\mathbf{J}$:

$$\begin{aligned}[][J_\pm, T_q^{(k)}] &= \sqrt{k(k+1)-q(q\pm 1)}\,T_{q\pm 1}^{(k)},\\ [J_z, T_q^{(k)}] &= qT_q^{(k)}.\end{aligned} \tag{VI.11}$$

3.4. THE WIGNER–ECKART THEOREM

In a representation $\{\mathbf{J}^2, J_z\}$ in which the base vectors are $|\tau JM\rangle$, the matrix element $\langle \tau JM|T_q^{(k)}|\tau' J'M'\rangle$ is equal to the product of the Clebsch–Gordan coefficient $\langle J'kM'q|JM\rangle$ and a quantity independent of M, M' and q, i.e.

$$\langle \tau JM|T_q^{(k)}|\tau' J'M'\rangle = \frac{1}{\sqrt{2j+1}}\langle \tau J||\mathcal{T}^{(k)}||\tau' J'\rangle\langle J'kM'q|JM\rangle. \tag{VI.12}$$

Problems

1. Show that, between the simultaneous eigenvectors of the operators $\mathbf{J}^2$ and J_z, there exist the following relations

$$|\tau, j, \pm m\rangle = \sqrt{\frac{(j+m)!}{(2j)!\,(j-m)!}}\,(J_\mp)^{j-m}|\tau, j, \pm j\rangle,$$

$$|\tau, j, \pm j\rangle = \sqrt{\frac{(j+m)!}{(2j)!\,(j-m)!}}\,(J_\pm)^{j-m}|\tau, j, \pm m\rangle.$$

2. Show that:

(a) In any representation in which J_x and J_z are real matrices, J_y is of the form $i \times$ a real antisymmetrical matrix.

(b) If an operator commutes with two of the components of the angular momentum operator, then it commutes also with the third component.

3. Prove the validity of the following commutation rules:

$$\text{(a)}\quad [l_i, x_k] = i\varepsilon_{ikl}x_l, \qquad \text{(b)}\quad [l_i, p_k] = i\varepsilon_{ikl}p_l, \qquad \text{(c)}\quad [l_i, l_k] = i\varepsilon_{ikl}l_l,$$
$$\text{(d)}\quad [\mathbf{l}, p^2] = 0, \qquad \text{(e)}\quad [\mathbf{l}, r^2] = 0$$

where ε_{ikl} stands for the following totally anti-symmetrical tensor: $\varepsilon_{ikl} = 0$ if any two indices are equal, and $\varepsilon_{ikl} = +1$ (or -1) if the indices i, k, l can be obtained from 1, 2, 3 by an even (or an odd) number of permutations. (The indices 1, 2 and 3 stand for the x-, y- and z-components of vectors, and repeated suffixes are to be summed over.)

4. Show that if a system is in an eigenstate of J_z, then the mean values of the operators J_x and J_y vanish.

5. Show that for a system in the eigenstate $|\tau jm\rangle$ of the operator J_z, the mean value of the component of angular momentum along a direction z', which makes an angle θ with the z-axis, is equal to $m \cos \theta$.

6. Since the components of the angular momentum operator do not commute, their simultaneous measurement is not possible. Show that in a state $|\tau jm\rangle$ the greatest accuracy of measurement of the components J_x and J_y is obtained when $|m| = j$.

7. Obtain expressions for the operators l_x, l_y, l_z and $\mathbf{l}^2$ in spherical coordinates (i) by using a direct method, and (ii) by starting from the fact that l_x, l_y, l_z are related to infinitesimal rotation operators.

8. Let $\mathbf{l}$ be the orbital angular momentum operator, r, θ and ϕ a set of polar coordinates, and P the parity operator of a particle, all referred to the same origin. P performs a reflection about the origin, so that its action on any function of position of the form $F(r, \theta, \phi)$ is given by the relation

$$PF(r, \theta, \phi) = F(r, \pi-\theta, \phi+\pi).$$

Show that $[P, \mathbf{l}] = 0$ and, starting from this fact, prove that each of the spherical harmonics has a well-defined parity, which depends only on the quantum number l. Find this parity as a function of l.

9. Establish the identity $(\boldsymbol{\sigma}.\mathbf{A})(\boldsymbol{\sigma}.\mathbf{B}) = \mathbf{A}.\mathbf{B}+i\boldsymbol{\sigma}(\mathbf{A}\times\mathbf{B})$, where $\boldsymbol{\sigma} = (\sigma_x, \sigma_y, \sigma_z)$ are the Pauli matrices, and $\mathbf{A}$ and $\mathbf{B}$ are vector operators which commute with $\boldsymbol{\sigma}$, but do not necessarily commute with each other.

10. Find the eigenvalues and the eigenvectors of the operators s_x, s_y, s_z, in the representation by Pauli matrices in which s_z is diagonal.

11. Find the matrices corresponding to the spin operator components s_x, s_y, s_z, for a particle of spin $s = 1$, in the representation in which the operators $\mathbf{s}^2$ and s_z are diagonal.

12. Using the transformation matrix associated with a rotation about the origin given by the Euler angles α, β, γ (Fig. VI.1), viz.,

$$R(\alpha, \beta, \gamma) = \begin{bmatrix} \cos\gamma\cos\beta\cos\alpha - \sin\gamma\sin\alpha & -\sin\gamma\cos\beta\cos\alpha - \cos\gamma\sin\alpha & \sin\beta\cos\alpha \\ \cos\gamma\cos\beta\sin\alpha + \sin\gamma\cos\alpha & -\sin\gamma\cos\beta\sin\alpha + \cos\gamma\cos\alpha & \sin\beta\sin\alpha \\ -\cos\gamma\sin\beta & \sin\gamma\sin\beta & \cos\beta \end{bmatrix}$$

show that the components of the angular momentum operator transform according to the relations

$$J'_z = \tfrac{1}{2}\sin\beta(e^{-i\alpha}J_+ + e^{-i\alpha}J_-) + \cos\beta J_z, \tag{12a}$$

$$J'_\pm = e^{\mp i\gamma}\left[\frac{1+\cos\beta}{2}\, e^{\mp i\alpha}J_\pm - \frac{1-\cos\beta}{2}\, e^{\pm i\alpha}J_\mp - \sin\beta J_z\right]. \tag{12b}$$

13. Find the matrix elements of the rotation operator for states having $j = \frac{1}{2}$ directly from the definition

$$D^{(j)}_{mm'}(\alpha, \beta, \gamma) \equiv \langle jm|\, D(\alpha, \beta, \gamma)\,|jm'\rangle.$$

14. Show that the rotation matrix elements $d^{(j)}_{mm'}$ are given by the Wigner formula

$$d^{(j)}_{mm'}(\beta) = \left[\frac{(j+m)!\,(j-m)!}{(j+m')!\,(j-m')!}\right]^{1/2} \sum_t \binom{j+m'}{j-m-t}\binom{j-m'}{t}.$$

$$(-1)^{j-m'-t}(\cos\beta/2)^{2t+m+m'}(\sin\beta/2)^{2j-2t-m-m'}. \tag{14a}$$

15. Suppose an electron to be in a state in which the component of its spin along the z-axis is $+\frac{1}{2}$. What is the probability that the component of the spin along an axis z' (which makes an angle θ with the z-axis) will have the value $+\frac{1}{2}$ or $-\frac{1}{2}$? What is the average value of the component of the spin along this axis?

16. Find how the spherical harmonics Y_1^1, Y_1^0, Y_1^{-1} transform under a rotation given by the Euler angles α, β, γ.

17. In a Stern–Gerlach type of experiment, a beam of atoms (each with total angular momentum j) is passed through an inhomogeneous magnetic field. Each atom is deflected through an angle which depends on the component of angular momentum of the atom in the direction of the magnetic field. If the atoms all have initially a specified value of the component of angular momentum along an axis which is not in the direction of the magnetic field, the beam will split into $2j+1$ component beams. Determine the relative intensities of these component beams if $j = 1$.

18. A spin-$\frac{1}{2}$ particle moves in a central field of force. By solving the appropriate eigenfunction equations, find those steady state wavefunctions of the particle which are also eigenfunctions of the operators $\mathbf{l}^2$, $\mathbf{j}^2$ and j_z, where $\mathbf{j} = \mathbf{l}+\mathbf{s}$. Show that these wavefunctions can alternatively be obtained by using Clebsch–Gordan coefficients.

19. Using the vector model of angular momentum (cf. the answer to problem 4), determine the possible values of the angle between the vectors $\mathbf{l}$ and $\mathbf{s}$ of the preceding problem.

20. Let the energy eigenfunctions of an electron which moves in a central field of force be characterized also by the quantum numbers l, j, m_j, where j and m_j are the quantum numbers of the total angular momentum. Determine the possible values of the z-components of the orbital angular momentum and of the spin and also their probabilities and their average values.

21. Let $\mathbf{s}_1$ and $\mathbf{s}_2$ be the spin operators of two spin-$\frac{1}{2}$ particles. Find the simultaneous eigenfunctions of the operators $\mathbf{s}^2$ and s_z, where $\mathbf{s} = \mathbf{s}_1+\mathbf{s}_2$. Show that these are also eigenfunctions of the operator $\mathbf{s}_1.\mathbf{s}_2$.

22. Show that the operator $(\boldsymbol{\sigma}_1.\boldsymbol{\sigma}_2)^n$, where $\boldsymbol{\sigma}_1$ and $\boldsymbol{\sigma}_2$ are Pauli matrices, depends linearly on the product $(\boldsymbol{\sigma}_1.\boldsymbol{\sigma}_2)$. Find the explicit form of this dependence.

23. Consider a system of two nucleons (protons or neutrons). Let $\mathbf{r} = \mathbf{r}_1-\mathbf{r}_2$ be their relative position vector, and $\frac{1}{2}\boldsymbol{\sigma}_1$, $\frac{1}{2}\boldsymbol{\sigma}_2$ their spin operators. Now it has been found that part of the interaction energy between nucleons is similar in form to the classical interaction between two dipoles, i.e. it can be written in the form

$$V = V(r)\left[3\frac{(\boldsymbol{\sigma}_1.\mathbf{r})(\boldsymbol{\sigma}_2.\mathbf{r})}{r^2}-\boldsymbol{\sigma}_1.\boldsymbol{\sigma}_2\right]. \tag{23a}$$

This is the so-called "tensor force". It is not of course electromagnetic in origin. Show that the operator

$$S_{12} = \left[3\frac{(\boldsymbol{\sigma}_1.\mathbf{r})(\boldsymbol{\sigma}_2.\mathbf{r})}{r^2}-\boldsymbol{\sigma}_1.\boldsymbol{\sigma}_2\right], \tag{23b}$$

which expresses the spin dependence of the "tensor force", may be written in the form

$$S_{12} = 2\left[\frac{3(\mathbf{S}.\mathbf{r})}{r^2}-\mathbf{S}^2\right], \tag{23c}$$

where $\mathbf{S} = \frac{1}{2}(\boldsymbol{\sigma}_1+\boldsymbol{\sigma}_2)$ is the total spin operator.

24. The Hamiltonian of relative motion of the nucleons of the preceding problem is

$$H = \frac{\mathbf{p}^2}{2M_0}+V,$$

where M_0 is the "reduced mass" of the system ($\approx \frac{1}{2}\times$mass of one nucleon), $\mathbf{p}$ is the relative momentum, and V is the potential energy of the mutual interaction. Show that if we take

$V = V(r)S_{12}$, there will exist simultaneous eigenfunctions of H and of P, $\mathbf{S}^2$, $\mathbf{J}^2$ and J_z. where P is the parity operator, and $\mathbf{J} = \mathbf{l}+\mathbf{S}$, $\mathbf{l} = \mathbf{r}\times\mathbf{p}$.

25. Show that the Schrödinger equation for a system of two nucleons with an interaction potential energy $V = V_1(r)+V_2(r)(\boldsymbol{\sigma}_1.\boldsymbol{\sigma}_2)$ can be split into two Schrödinger-type equations, one corresponding to the potential V_1-3V_2, and the other to the potential V_1+V_2.

26. Show that the operator $S_{12} = 2[3[(\mathbf{S}.\mathbf{r})/r^2]-\mathbf{S}^2]$, regarded as a function of $\mathbf{r}$, depends only on the polar angles θ and ϕ, and that this dependence has the form of a spherical harmonic with $l = 2$.

27. Express the operators $\mathbf{r}$, $\mathbf{p}$ and $\mathbf{l} = \mathbf{r}\times\mathbf{p}$ as irreducible tensor operators.

28. Let $\mathbf{l}$ be the orbital angular momentum of a spinless particle, and $\mathbf{A}$ an operator whose components satisfy the commutation rules:

$$[l_i, A_k] = i\varepsilon_{ikl}A_l. \tag{28a}$$

Using the Wigner–Eckart theorem, find the matrix elements of the components of the operator $\mathbf{A}$ in the $\{\mathbf{l}^2, l_z\}$ representation.

Find the selection rules for "allowed" transitions (i.e. those with non-vanishing first order matrix elements) due to a perturbation $\alpha\mathbf{A}$.

29. An expression from which the components of a vector can be calculated is said to define a "polar" vector if these components change sign under the coordinate transformation $x \to -x$, $y \to -y$, $z \to -z$ (so that the geometrical vector so defined is unchanged), and an "axial" vector if the components are unchanged (so that the geometrical vector so defined has the opposite direction to the one defined in the original coordinate system). Show that, in the $\{\mathbf{l}^2, l_z\}$ representation, the following allowed transition selection rules for the orbital angular momentum are valid:

$\Delta l = \pm 1$ for a polar vector perturbation operator,
$\Delta l = 0$ for an axial vector perturbation operator,

provided these vector operators satisfy the commutation rules (28a).

30. Find the matrix elements of the z-component of the unit vector $\mathbf{n}$ in the direction of $\mathbf{r}$, for a spinless particle, in terms of the eigenfunctions of angular momentum of the particle.

31. Show that the operator (23c) may be written in the following form

$$S_{12} = \left(\frac{24\pi}{5}\right)^{1/2}(\mathbf{S}^{(2)}.\mathbf{J}^{(2)}), \tag{31a}$$

where $\mathbf{S}^{(2)}$ and $\mathbf{J}^{(2)}$ are second order irreducible tensor operators which are functions of the spin and of the orbital variables respectively, and the bracket denotes their scalar product defined as follows:

$$(\mathbf{S}^{(2)}.\mathbf{J}^{(2)}) = \sum_{q=-2}^{+2} (-1)^q S_q^{(2)} T_{-q}^{(2)}. \tag{31b}$$

32. Show that the Schrödinger equation of a system of electrons in a homogeneous magnetic field $\mathbf{H}$ can be written as

$$i\hbar \frac{\partial \psi}{\partial t} = \left\{ \frac{1}{2m} \sum_k \mathbf{p}_k^2 - \frac{e}{2mc} (\mathbf{L}+2\mathbf{S}).\mathbf{H} + \frac{e^2}{8mc^2} \sum_k (\mathbf{H}\times\mathbf{r}_k)^2 + U \right\} \psi,$$

where $\mathbf{L} = \sum_k \mathbf{l}_k$ is the total orbital angular momentum operator of the system, $\mathbf{S} = \sum_k \mathbf{s}_k$ is the total spin operator, and U is the potential energy of all interactions other than those with the field $\mathbf{H}$.

33. Show that, in a magnetic field which is homogeneous but not necessarily constant in time, the wavefunction of a particle with spin which satisfies the Pauli equation (32.3) can be written as the product of a coordinate function and a spin function.

34. Find the energy spectrum of a spin-$\frac{1}{2}$ particle in a constant homogeneous magnetic field.

35. Find the wavefunctions of a spinless particle in a homogeneous magnetic field for states in which (i) the component of linear momentum, and (ii) the component of angular momentum, both in the direction of the magnetic field, have well-defined values.

36. Find the wavefunction of a spin-$\frac{1}{2}$ particle which is uncharged, but has a magnetic moment, in a homogeneous time-dependent magnetic field $\mathbf{H}(t)$.

37. The plane $x = 0$ separates two regions of space; in the region in which $x > 0$, there exists a uniform magnetic field $H_x = H_y = 0$, $H_z = H$, and, in the region in which $x < 0$, there is no field. A beam of polarized neutrons of momentum $\mathbf{p}$ is incident on the plane $x = 0$ from the region $x < 0$. Find the reflection coefficient of the neutrons.

38. Taking into account the results of problem 27 of Chapter V, show that, to describe a system of independent spin-$\frac{1}{2}$ particles, we can choose operators Ω of the form

$$\Omega^{(0)} = \frac{1}{\sqrt{2}} I, \quad \Omega^{(\alpha)} = \frac{1}{\sqrt{2}} \sigma_\alpha; \qquad \alpha = 1, 2, 3,$$

where σ_1, σ_2, σ_3 are the Pauli matrices.

Writing $\langle \sigma_\alpha \rangle = P_\alpha$, say, show that, if the system is in a constant external magnetic field $\mathbf{H}$, the equation of motion of the vector $\mathbf{P} \equiv (P_1, P_2, P_3)$ is

$$\frac{d\mathbf{P}}{dt} = \gamma(\mathbf{P}\times\mathbf{H}), \tag{38a}$$

where γ is the gyromagnetic ratio.

39. The "magnetic resonance" phenomena observed in a "spin system" placed in a constant magnetic field $\mathbf{H}_0$ (see problem 7 of Chapter IX) are due to the absorption of energy by the system from a second, rotating, magnetic field $\mathbf{H}_1$ ($H_{1x} = H_1 \cos \omega t$, $H_{1y} = H_1 \sin \omega t$), and to the occurrence of "relaxation" processes which have the effect of continually tending to establish a thermal equilibrium (Boltzmann) distribution of the spin orientations of the system in the magnetic fields $\mathbf{H}_0$ and $\mathbf{H}_1$.

Show that if the relaxation processes are such that in a constant field they would lead exponentially to the equilibrium state with a "relaxation" time τ, then the real and the imaginary parts of the magnetic susceptibility of the system are given by Bloch's relations †

$$\chi' = \chi_0 - \chi_0 \omega\tau \frac{(\omega_0+\omega)\,\tau}{1+(\omega_0+\omega)^2\tau^2+\omega_1^2\tau^2}$$

$$\chi'' = \chi_0 \omega\tau \frac{1}{1+(\omega_0+\omega)^2\tau^2+\omega_1^2\tau^2}, \tag{39a}$$

where $\chi_0 = (N\gamma^2\hbar^2)/(4kT)$ is the static susceptibility (N = number of spins, γ = gyromagnetic ratio).

Solutions

1. We will consider only the relation

$$|\tau jm\rangle = \sqrt{\frac{(j+m)!}{(2j)!\,(j-m)!}}\,(J_-)^{j-m}|\tau jj\rangle, \tag{1.1}$$

since the other three relations can be derived by similar reasoning. According to (VI.3) we have that

$$J_-|\tau, j, m+1\rangle = \sqrt{j(j+1)-m(m+1)}\,|\tau jm\rangle = \sqrt{(j-m)(j+m+1)}\,|\tau jm\rangle$$

and hence

$$\begin{aligned}(J_-)^2|\tau, j, m+2\rangle &= J_-\sqrt{(j-m-1)(j+m+2)}\,|\tau, j, m+1\rangle \\ &= \sqrt{(j-m-1)(j-m)(j+m+1)(j+m+2)}\,|\tau jm\rangle\end{aligned}$$

and, in general,

$$(J_-)^{j-m}|\tau jj\rangle = \sqrt{1.2\ldots(j-m-1)(j-m)(j+m+1)(j+m+2)\ldots(2j-1)(2j)}\,|\tau jm\rangle. \tag{1.2}$$

In order to obtain (1.1) it is sufficient to observe that the product under the square root sign can be written in the form $[(2j)!\,(j-m)!]/[(j+m)!]$.

† F. Bloch, *Phys. Rev.* **70**, 460 (1946).

2. (a) In any representation in which the matrices which represent J_x and J_z are real as well as Hermitian, they are also symmetrical, i.e. $\tilde{J}_x = J_x$, $\tilde{J}_z = J_z$. That the matrix representing J_y is then of the form $i \times$ a real antisymmetrical matrix follows immediately from the commutation rule $[J_z, J_x] = iJ_y$, since

$$\widetilde{[J_z, J_x]} = \widetilde{J_zJ_x} - \widetilde{J_xJ_z} = \tilde{J}_x\tilde{J}_z - \tilde{J}_z\tilde{J}_x = [J_x, J_z] = -[J_z, J_x].$$

Hence $J_y = -i[J_z, J_x] = i \times$ a real antisymmetrical matrix. Note that the representation in which $\mathbf{J}^2$ and J_z are diagonal is of the above type, since from (VI.2) and (VI.3) one can obtain, by a little simple algebra, the matrix elements

$$\langle \tau, j, m\pm1 | J_x | \tau jm \rangle = \tfrac{1}{2}\sqrt{(j\mp m)(j\pm m+1)} \tag{2.1}$$

$$\langle \tau, jm\pm1 | J_y | \tau jm \rangle = \mp\frac{i}{2}\sqrt{(j\mp m)(j\pm m+1)} \tag{2.2}$$

$$\langle \tau j, m | J_z | \tau jm \rangle = m. \tag{2.3}$$

(b) From the commutation rules $AJ_x = J_xA$, $AJ_y = J_yA$, and $J_xJ_y - J_yJ_x = iJ_z$, we find that $i(-J_xAJ_y + J_yAJ_x) = AJ_z = J_zA$, i.e. that $[A, J_z] = 0$.

3. Using the commutation rules between the coordinates and the momenta, we have

$$l_xy - yl_x = (yp_z - zp_y)y - y(yp_z - zp_y) = -z[p_y, y] = iz.$$

In a similar manner we find that

$$\begin{array}{lll} [l_x, x] = 0 & [l_x, y] = iz & [l_x, z] = -iy \\ [l_y, y] = 0 & [l_y, z] = ix & [l_y, x] = -iz \\ [l_z, z] = 0 & [l_z, x] = iy & [l_z, y] = -ix. \end{array}$$

All these results can be combined in the equation

$$[l_i, x_k] = i\varepsilon_{ikl}x_l\,.$$

By similar reasoning, the results (b), (c), (d) and (e) can be obtained.

4. For any state $|\tau jm\rangle$, we have $\langle \tau jm | J_+ | \tau jm \rangle = 0$. Separating the real and the imaginary parts, the problem is solved. We can depict diagrammatically the vanishing of the mean values of J_x and of J_y for systems in eigenstates of J_z, by resorting to the "vector model" of angular momentum, much used in the old Quantum Mechanics (Fig. VI.2a). The eigenstate $|\tau jm\rangle$ is represented diagrammatically by an "angular momentum vector" of length $\sqrt{j(j+1)}$, which rotates continuously around a cone of height m. It follows that the (time) average values of the components J_x, J_y, vanish. The angle ϱ between the angular momentum vector and the z-axis is given by the relation $\cos\varrho = \dfrac{m}{\sqrt{j(j+1)}}$, and hence

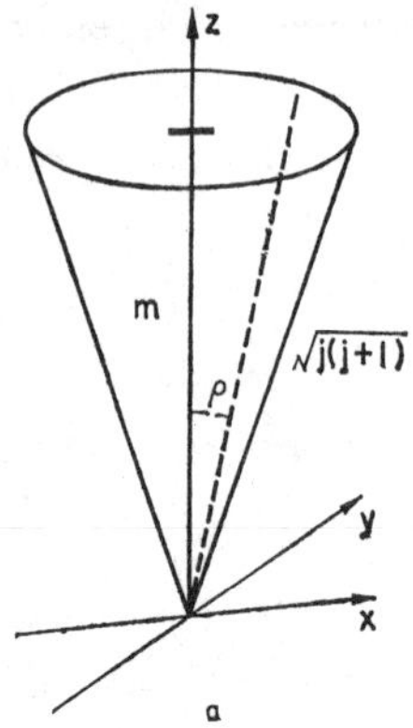

FIG. VI.2a.

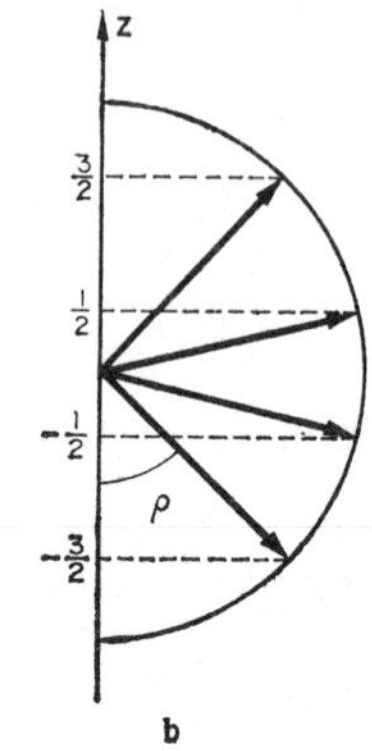

FIG. VI.2b.

the angle ϱ can have only a finite number of values. All possible orientations of the angular momentum vector relative to the z-axis are shown in Fig. VI.2b (in which for purposes of illustration j is taken to have the value $\frac{3}{2}$).

5. Using the vector model of angular momentum (see the preceding problem), we have that

$$J_{z'} = J_x \cos(x, z') + J_y \cos(y, z') + J_z \cos\theta. \tag{5.1}$$

Since $|\tau jm\rangle$ is an eigenvector of J_z, we obtain

$$\rangle \tau jm|J_x|\tau jm\rangle = \langle \tau jm|J_y|\tau jm\rangle = 0 \quad \text{and} \quad \langle \tau jm|J_{z'}|\tau jm\rangle = \langle \tau jm|J_z|\tau jm\rangle \cos\theta = m\cos\theta. \tag{5.2}$$

6. The uncertainty in the angular momentum J_x is (see problem 4)

$$\Delta J_x = \sqrt{\langle J_x^2\rangle - \langle J_x\rangle^2} = \sqrt{\langle J_x^2\rangle} = \sqrt{\langle \tau jm|J_x^2|\tau jm\rangle}, \tag{6.1}$$

and similarly

$$\Delta J_y = \sqrt{\langle J_y^2\rangle - \langle J_y\rangle^2} = \sqrt{\langle J_y^2\rangle} = \sqrt{\langle \tau jm|J_y^2|\tau jm\rangle}. \tag{6.2}$$

Now

$$\langle J^2\rangle = \langle \tau jm|J^2|\tau jm\rangle = j(j+1) = \langle J_x^2\rangle + \langle J_y^2\rangle + \langle J_z^2\rangle = (\Delta J_x)^2 + (\Delta J_y)^2 + m^2,$$

and hence

$$(\Delta J_x)^2 + (\Delta J_y)^2 = j^2 + j - m^2. \tag{6.3}$$

From (6.3) we see that the minimum value of the combined uncertainties of J_x and of J_y is obtained when $|m| = j$, namely, in terms of the vector model of angular momentum, when the angular momentum vector is nearest to the z-axis. The components J_x and J_y are well-defined only if the total angular momentum j vanishes.

7. We have $\mathbf{l} = \mathbf{r}\times\mathbf{p}$, whence

$$l_x = -i\hbar\left(y\frac{\partial}{\partial z}-z\frac{\partial}{\partial y}\right),\quad l_y = -i\hbar\left(z\frac{\partial}{\partial x}-x\frac{\partial}{\partial z}\right),\quad l_z = -i\hbar\left(x\frac{\partial}{\partial y}-y\frac{\partial}{\partial x}\right). \tag{7.1}$$

Let ψ be a function of position, which can therefore be expressed in terms of Cartesian coordinates x, y, z or of polar coordinates r, θ, ϕ. We have then that

$$\frac{\partial\psi}{\partial\phi} = \frac{\partial\psi}{\partial x}\frac{\partial x}{\partial\phi}+\frac{\partial\psi}{\partial y}\frac{\partial y}{\partial\phi}+\frac{\partial\psi}{\partial z}\frac{\partial z}{\partial\phi} = -\frac{\partial\psi}{\partial x}r\sin\theta\sin\phi+\frac{\partial\psi}{\partial y}r\sin\theta\cos\phi$$

$$= x\frac{\partial\psi}{\partial y}-y\frac{\partial\psi}{\partial x}. \tag{7.2}$$

By comparing (7.2) with (7.1) we find that

$$l_z = -i\hbar\frac{\partial}{\partial\phi}. \tag{7.3}$$

Similarly we can write

$$\frac{\partial\psi}{\partial\theta} = \cot\theta\left(x\frac{\partial\psi}{\partial x}+y\frac{\partial\psi}{\partial y}\right)-\tan\theta\left(z\frac{\partial\psi}{\partial z}\right). \tag{7.4}$$

On the other hand, from the relations

$$x+iy = r\sin\theta(\cos\phi+i\sin\phi) = re^{i\phi}\sin\theta,\qquad z = r\cos\theta,$$

we have that

$$l_x+il_y = \hbar e^{i\phi}\left[\cot\theta\left(x\frac{\partial\psi}{\partial x}+y\frac{\partial\psi}{\partial y}\right)-\tan\theta\left(z\frac{\partial\psi}{\partial z}\right)+i\cot\theta\left(x\frac{\partial\psi}{\partial y}-y\frac{\partial\psi}{\partial x}\right)\right]. \tag{7.5}$$

From (7.4) and (7.5) one obtains

$$l_x\pm il_y = \pm\hbar e^{\pm i\phi}\left(\frac{\partial}{\partial\theta}\pm i\cot\theta\frac{\partial}{\partial\phi}\right), \tag{7.6}$$

whence we find that

$$l_x = i\hbar\left(\sin\phi\frac{\partial}{\partial\theta}+\cot\theta\cos\phi\frac{\partial}{\partial\phi}\right) \tag{7.7}$$

$$l_y = -i\hbar\left(\cos\phi\frac{\partial}{\partial\theta}-\cot\theta\sin\phi\frac{\partial}{\partial\phi}\right) \tag{7.8}$$

$$\mathbf{l}^2 = l_x^2+l_y^2+l_z^2 = -\hbar^2\left[\frac{1}{\sin\theta}\frac{\partial}{\partial\theta}\left(\sin\theta\frac{\partial}{\partial\theta}\right)+\frac{1}{\sin^2\theta}\frac{\partial^2}{\partial\phi^2}\right]. \tag{7.9}$$

Let us now derive the same expressions starting from the fact that l_x, l_y and l_z are related to the infinitesimal rotation operators about the coordinate axes x, y and z respectively.

For a rotation through an angle $d\alpha$ about any axis $\mathbf{u}$, the wavefunction $\psi(\mathbf{r})$ of a spinless particle of orbital angular momentum $\mathbf{l}$ becomes, according to (VI.6),

$$\psi'(\mathbf{r}) = \left[1 - \frac{i\,d\alpha}{\hbar}(\mathbf{l}.\mathbf{u})\right]\psi(\mathbf{r}), \tag{7.10}$$

provided of course that the axis passes through the origin to which $\mathbf{l}$ is referred. For a start let us suppose that $\mathbf{u}$ is the unit vector $\mathbf{k}$ in the direction of the positive z-axis. Since any rotation of a physical system is equivalent to a rotation of the coordinate axes in the opposite direction, we have that

$$\psi'(r, \theta, \phi) = \psi(r, \theta, \phi - d\alpha) = \psi(r, \theta, \phi) - \frac{\partial\psi}{\partial\phi}\,d\alpha. \tag{7.11}$$

Comparing (7.10) with (7.11) we obtain (7.3). In order to obtain l_x we perform a rotation about the x-axis

$$\psi'(r, \theta, \phi) = \psi(r, \theta - d\theta, \phi - d\phi) = \left[1 - \left(\frac{d\theta}{d\alpha}\frac{\partial}{\partial\theta} + \frac{d\phi}{d\alpha}\frac{\partial}{\partial\phi}\right)d\alpha\right]\psi(r, \theta, \phi),$$

whence

$$l_x = -i\hbar\left(\frac{d\theta}{d\alpha}\frac{\partial}{\partial\theta} + \frac{d\phi}{d\alpha}\frac{\partial}{\partial\phi}\right). \tag{7.12}$$

Finally, we require expressions for $d\theta/d\alpha$ and $d\phi/d\alpha$. Using Fig. VI.3 one can see that

$$dz = z_1 - z = y\,d\alpha, \quad dy = y_1 - y = -z\,d\alpha. \tag{7.13}$$

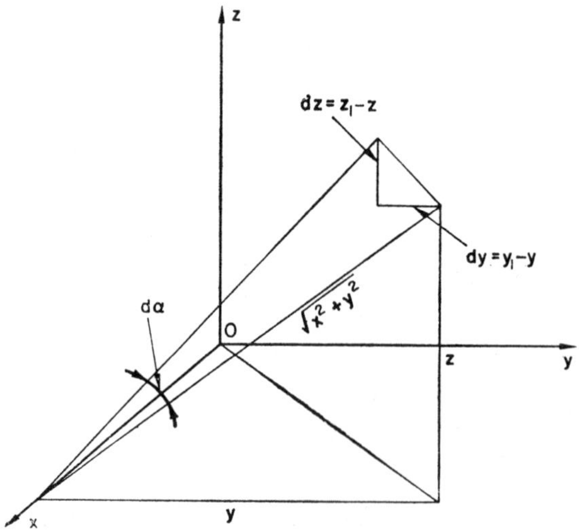

FIG. VI.3.

Then, by differentiating the relations $z = r\cos\theta$, $y = r\sin\theta\sin\phi$, and comparing the results with (7.13) we obtain

$$\frac{d\theta}{d\alpha} = -\sin\phi, \quad \frac{d\phi}{d\alpha} = -\cot\theta\cos\phi. \tag{7.14}$$

From (7.12) and (7.14) we find (7.7); (7.8) can be found in the same way.

8. The first part of the problem is solved immediately if we take into account the expressions for the angular momentum operator components l_x, l_y, l_z in polar coordinates, (7.3), (7.7) and (7.8). In order to prove the second part we write

$$[P, l_z]Y_l^m(\theta, \phi) = mY_l^m(\pi-\theta, \phi+\pi) - l_z Y_l^m(\pi-\theta, \phi+\pi) = 0,$$

whence it follows that $Y_l^m(\pi-\theta, \phi+\pi)$ is proportional to $Y_l^m(\theta, \phi)$, the proportionality constant, $C(l, m)$ say, having unit modulus, since the normalization condition of the spherical harmonics Y_l^m gives

$$\int_0^{2\pi}\int_0^{\pi} |Y_l^m(\pi-\theta, \phi+\pi)|^2 \sin\theta\, d\theta\, d\phi = |C(l, m)|^2 \int_0^{2\pi}\int_0^{\pi} |Y_l^m(\theta, \phi)|^2 \sin\theta\, d\theta\, d\phi = 1,$$

whence $|C(l, m)|^2 = 1$.

Let us show now that $C(l, m)$ is in fact independent of m. To do this, we write

$$[P, l_+]Y_l^m(\theta, \phi) = \sqrt{l(l+1)-m(m+1)}\,Y_l^{m+1}(\pi-\theta, \phi+\pi) - l_+ Y_l^m(\pi-\theta, \phi+\pi) = 0 \tag{8.1}$$

and similarly

$$[P, l_-]Y_l^m(\theta, \phi) = \sqrt{l(l+1)-m(m-1)}\,Y_l^{m-1}(\pi-\theta, \phi+\pi) - l_- Y_l^m(\pi-\theta, \phi+\pi) = 0. \tag{8.2}$$

From (8.1) and (8.2) we have that

$$C(l, m) = C(l, m+1) = C(l, m-1), \tag{8.3}$$

i.e. $C(l, m)$ does not depend on the integer m, and can thus be written as $C(l)$. For this reason, it is sufficient, in determining its value, to limit ourselves to the spherical harmonics with $m = 0$, (A.24), i.e. to

$$Y_1^0(\theta, \phi) = \sqrt{\frac{2l+1}{4\pi}}\,P_l(\cos\theta).$$

For the Legendre polynomials (A.8) we clearly have that $P_l(-u) = (-1)^l P_l(u)$, so that $Y_1^0(\pi-\theta, \phi+\pi) = (-1)^l Y_1^0(\theta, \phi)$. Since $Y_1^0(\pi-\theta, \phi+\pi) = C(l)Y_1^0(\theta, \phi)$, we find that $C(l) = (-1)^l$. Hence the parity of the spherical harmonics is $(-1)^l$. Since the angular part of the wavefunction of a particle in a central field is a spherical harmonic (see Chapter

II, section 3), it follows that the parity of the wavefunction in such a field depends only on the orbital quantum number l and is equal to $(-1)^l$.

9. The required identity is obtained immediately if we express the product $(\boldsymbol{\sigma}.\mathbf{A})(\boldsymbol{\sigma}.\mathbf{B})$ in Cartesian components and regroup the terms, taking into account known relations such as $\sigma_x^2 = \sigma_y^2 = \sigma_z^2 = 1$, $\sigma_x\sigma_y = -\sigma_y\sigma_x = i\sigma_z$, etc. An important particular case is

$$(\boldsymbol{\sigma}.\mathbf{r})(\boldsymbol{\sigma}.\mathbf{p}) = \mathbf{r}.\mathbf{p}+i\boldsymbol{\sigma}(\mathbf{r}\times\mathbf{p}). \tag{9.1}$$

10. In this representation we have (omitting the factor $\hbar$):

$$s_x = \frac{1}{2}\begin{bmatrix} 0 & 1 \\ 1 & 0 \end{bmatrix}, \quad s_y = \frac{1}{2}\begin{bmatrix} 0 & -i \\ i & 0 \end{bmatrix}, \quad s_z = \frac{1}{2}\begin{bmatrix} 1 & 0 \\ 0 & -1 \end{bmatrix}. \tag{10.1}$$

To obtain the eigenvectors $\chi_{s'_x} = \begin{bmatrix} A \\ B \end{bmatrix}$ say, and $\chi_{s'_y} = \begin{bmatrix} C \\ D \end{bmatrix}$ say, of s_x and of s_y respectively, with eigenvalues s'_x and s'_y, we have to solve the matrix eigenvalue equations:

$$\frac{1}{2}\begin{bmatrix} 0 & 1 \\ 1 & 0 \end{bmatrix}\begin{bmatrix} A \\ B \end{bmatrix} = s'_x\begin{bmatrix} A \\ B \end{bmatrix} \tag{10.2}$$

$$\frac{1}{2}\begin{bmatrix} 0 & -i \\ i & 0 \end{bmatrix}\begin{bmatrix} C \\ D \end{bmatrix} = s'_y\begin{bmatrix} C \\ D \end{bmatrix} \tag{10.3}$$

The solutions are easily found to be

$$\chi_{s'_x=+1/2} = \frac{1}{\sqrt{2}}e^{i\alpha_1}\begin{bmatrix} 1 \\ 1 \end{bmatrix}, \quad \chi_{s'_x=-1/2} = \frac{1}{\sqrt{2}}e^{i\beta_1}\begin{bmatrix} 1 \\ -1 \end{bmatrix} \tag{10.4}$$

$$\chi_{s'_y=+1/2} = \frac{1}{\sqrt{2}}e^{i\alpha_2}\begin{bmatrix} 1 \\ i \end{bmatrix}, \quad \chi_{s'_y=-1/2} = \frac{1}{\sqrt{2}}e^{i\beta_2}\begin{bmatrix} 1 \\ -i \end{bmatrix} \tag{10.5}$$

and similarly, for s_z,

$$\chi_{s'_z=+1/2} = e^{i\alpha_3}\begin{bmatrix} 1 \\ 0 \end{bmatrix}, \quad \chi_{s'_z=-1/2} = e^{i\beta_3}\begin{bmatrix} 0 \\ 1 \end{bmatrix}, \tag{10.6}$$

where the αs and the βs are arbitrary real phase angles.

It is easy to check that the two eigenvectors of each of the operators s_x, s_y and s_z are normal, mutually orthogonal, and form a complete set, i.e. any arbitrary eigenvector can be written as a linear combination of the members of any one of these pairs of eigenvectors. The square moduli of the coefficients of such a linear combination then give the probability of obtaining the particular value $\pm\frac{1}{2}$ of s'_x, of s'_y or of s'_z, according to the pair of eigenvectors used. Thus, for example, since the z- and the x-axes are perpendicular to each other, the probabilities of finding $s'_z = \pm\frac{1}{2}$ if a system is known to be in an eigenstate of s_x are equal to $\frac{1}{2}$.

11. We have to use in turn the relations (2.1), (2.2) and (2.3), for $j = 1$. We then find that

$$s_x = \frac{1}{\sqrt{2}}\begin{bmatrix} 0 & 1 & 0 \\ 1 & 0 & 1 \\ 0 & 1 & 0 \end{bmatrix}, \qquad s_y = \frac{1}{\sqrt{2}}\begin{bmatrix} 0 & -i & 0 \\ i & 0 & -i \\ 0 & i & 0 \end{bmatrix} \tag{11.1}$$

$$s_z = \begin{bmatrix} 1 & 0 & 0 \\ 0 & 0 & 0 \\ 0 & 0 & -1 \end{bmatrix}, \qquad s^2 = 2\begin{bmatrix} 1 & 0 & 0 \\ 0 & 1 & 0 \\ 0 & 0 & 1 \end{bmatrix}. \tag{11.2}$$

12. Under the rotation R, the cartesian components of any vector $\mathbf{V}$ transform according to the relations

$$V_i' = R_{ij}V_j, \tag{12.1}$$

in which the coefficients R_{ij} are the elements of the matrix associated with the rotation. Hence, if $\mathbf{A}_j$ is the transform of the unit vector $\mathbf{a}_j$ in the direction of the jth coordinate axis ($j = 1, 2, 3$), we have that

$$\mathbf{A}_j = R_{ij}\mathbf{a}_i; \qquad R_{ij} = \mathbf{a}_i \cdot \mathbf{A}_j. \tag{12.2}$$

The matrix R associated with the rotation has the following properties

$$R^* = R, \quad \tilde{R} = R^{-1}, \quad \det R = 1. \tag{12.3}$$

The rotation operator D of a physical system, as it is defined in paragraph 3, causes each state $|\rangle$ of the system to have a transform $D|\rangle$ under the rotation R. Similarly, any observable Q (cf. VI.5) has a transform DQD^{-1}. Let $\mathbf{J} \equiv (J_x, J_y, J_z)$ be the total angular momentum operator (about the origin) of a physical system, and let $J_i \equiv \mathbf{J} \cdot \mathbf{a}_i$. Then the transform

$$J_i' = DJ_iD^{-1} = \mathbf{J} \cdot \mathbf{A}_i = J_jR_{ji}. \tag{12.4}$$

Taking in turn $i = 3, 2, 1$, we find that

$$\begin{aligned} J_z' &= \sin\beta\cos\alpha\, J_x + \sin\beta\sin\alpha J_y + \cos\beta J_z \\ &= \tfrac{1}{2}\sin\beta(e^{-i\alpha}J_+ + e^{-i\alpha}J) + \cos\beta J_z, \end{aligned} \tag{12.5}$$

$$\begin{aligned} J_y' &= (-\sin\gamma\cos\beta\cos\alpha - \cos\gamma\sin\alpha)J_x \\ &\quad -(\sin\gamma\cos\beta\sin\alpha - \cos\gamma\sin\alpha)J_y + \sin\gamma\sin\beta J_z, \end{aligned} \tag{12.6}$$

$$\begin{aligned} J_x' &= (\cos\gamma\cos\beta\cos\alpha - \sin\gamma\sin\alpha)J_x \\ &\quad +(\cos\gamma\cos\beta\sin\alpha + \sin\gamma\cos\alpha)J_y - \cos\gamma\sin\beta J_z. \end{aligned} \tag{12.7}$$

From (12.6) and (12.7) we easily obtain

$$J_\pm' = e^{\mp i\gamma}\left[\frac{1+\cos\beta}{2}e^{\mp i\alpha}J_\pm - \frac{1-\cos\beta}{2}e^{\pm i\alpha}J_\mp - \sin\beta J_z\right]. \tag{12.8}$$

Relations (12.5) and (12.8) can be used to calculate the rotation matrix elements $D^{(j)}_{mm'}(\alpha, \beta, \gamma)$ (VI.8).

13. According to (VI.8) we have that

$$D^{(j)}_{mm'}(\alpha, \beta, \gamma) = e^{-i\alpha m}\, d^{(j)}_{mm'}(\beta) e^{-i\gamma m'}, \tag{13.1}$$

where

$$d^{(j)}_{mm'}(\beta) = \langle jm | e^{-i\beta J_y} | jm' \rangle. \tag{13.2}$$

Let us define a matrix $A = \beta/2\begin{bmatrix} 0 & 1 \\ -1 & 0 \end{bmatrix}$. It is then easily shown that

$$A^{2n} = (-1)^n (\beta/2)^{2n}\begin{bmatrix} 1 & 0 \\ 0 & 1 \end{bmatrix} \quad \text{and} \quad A^{2n+1} = (-1)^n (\beta/2)^{2n+1}\begin{bmatrix} 0 & 1 \\ -1 & 0 \end{bmatrix}$$

where n is an integer. Since, in the $\{\mathbf{J}^2, J_z\}$ representation, $i\beta J_y = A$, we have

$$\langle jm | e^{-i\beta J_y} | jm' \rangle = \begin{array}{c|cc} m & m' = \frac{1}{2} & -\frac{1}{2} \\ \hline \frac{1}{2} & \cos\beta/2 & -\sin\beta/2 \\ -\frac{1}{2} & \sin\beta/2 & \cos\beta/2 \end{array}, \tag{13.3}$$

and hence, omitting for typographical simplicity the suffixes m and m',

$$D^{1/2}(\alpha, \beta, \gamma) = \begin{bmatrix} e^{-(i/2)\alpha}\cos\beta/2\, e^{-(i/2)\gamma} & -e^{-(i/2)\alpha}\sin\beta/2\, e^{(i/2)\gamma} \\ e^{(i/2)\alpha}\sin\beta/2\, e^{-(i/2)\gamma} & e^{(i/2)\alpha}\cos\beta/2\, e^{(i/2)\gamma} \end{bmatrix}. \tag{13.4}$$

14. The essence of the derivation given below consists in establishing a one-to-one correspondence between angular momentum operators and certain linear differential operators. Thus, consider the following operators:

$$\frac{\partial}{\partial\chi_+} \equiv \partial_+, \quad \frac{\partial}{\partial\chi_-} \equiv \partial_-, \tag{14.1}$$

in the use of which χ_+ and χ_- are to be treated as if they were real number variables until all differentiations have been carried out, after which they are to be given the values $\chi_+ = \begin{bmatrix} 1 \\ 0 \end{bmatrix}$ and $\chi_- = \begin{bmatrix} 0 \\ 1 \end{bmatrix}$, i.e. they are to become the eigenvectors of the operator J_z (with the eigenvalues $+\frac{1}{2}$ and $-\frac{1}{2}$) for the particular case $j = \frac{1}{2}$. Now consider the one-to-one correspondence

$$\begin{aligned} J_x &\sim \tfrac{1}{2}(\chi_-\partial_+ + \chi_+\partial_-) \\ J_y &\sim \frac{i}{2}(\chi_-\partial_+ - \chi_+\partial_-) \\ J_z &\sim \tfrac{1}{2}(\chi_+\partial_+ - \chi_-\partial_-) \\ J_+ &\sim \chi_+\partial_-, \quad J_- \sim \chi_-\partial_+. \end{aligned} \tag{14.2}$$

It is easily verified that the operators on the right-hand side satisfy the same commutation rules as the corresponding angular momentum operators. For example, we have that

$$\left[\frac{1}{2}(\chi_- \partial_+ + \chi_+ \partial_-),\ \frac{i}{2}(\chi_- \partial_+ - \chi_+ \partial_-)\right] = \frac{i}{2}(\chi_+ \partial_+ - \chi_- \partial_-),$$

which relation corresponds to $[J_x, J_y] = iJ_z$. The square of the angular momentum operator will also correspond to a differential operator, thus

$$\begin{aligned}\mathbf{J}^2 = J_z(J_z-1)+J_+J_- \sim \tfrac{1}{4}(&\chi_+\chi_+ \partial_+ \partial_+ + \chi_-\chi_- \partial_- \partial_- + 2\chi_+\chi_- \partial_+ \partial_- \\ &+3\chi_+ \partial_+ + 3\chi_- \partial_-) = k(k+1),\end{aligned} \tag{14.3}$$

say, where

$$k = \tfrac{1}{2}(\chi_+ \partial_+ + \chi_- \partial_-). \tag{14.4}$$

These differential operators act on a two-dimensional spin space and lead to a useful method of representing angular momentum eigenvectors. Let us consider an arbitrary product of powers of χ_+ and of χ_-, for example $(\chi_+)^x (\chi_-)^y$. It can easily be verified that this product is a simultaneous eigenvector of J_z and of $\mathbf{J}^2$, represented in the above form, with the eigenvalues $\frac{1}{2}(x-y)$ and $\frac{x+y}{2}\left(\frac{x+y}{2}+1\right)$, respectively. As for the operators J_+ and J_-, their action on the product is to change the values of the exponents, in such a way that their sum remains unchanged.

It follows that the products $\chi_+^{j+m}\chi_-^{j-m}$, with $m = -j,\ -j+1, \ldots, j-1,\ j$, form a basis (in $2j+1$ dimensions) which is equivalent to the basis formed by the states $|jm\rangle$, with $m = -j,\ -j+1, \ldots, j-1,\ j$. In order to satisfy the relations (VI.3), we must normalize the products

$$|jm\rangle \sim u(jm) \equiv \frac{\chi_+^{j+m}\chi_-^{j-m}}{[(j+m)!\,(j-m)!]^{1/2}}. \tag{14.5}$$

Consider now a rotation of a physical system given by the Euler angles $\alpha = 0$, $\beta \neq 0$, $\gamma = 0$. Taking into account (13.3), we have that

$$\begin{aligned} D(0, \beta, 0)\, u(jm') &= \frac{\left(\chi_+ \cos\frac{\beta}{2} + \chi_- \sin\frac{\beta}{2}\right)^{j+m'} \left(-\chi_+ \sin\frac{\beta}{2} + \chi_- \cos\frac{\beta}{2}\right)^{j-m'}}{[(j+m')!\ (j-m')!]^{1/2}} \\ &= \sum_m D^{(j)}_{mm'}(0, \beta, 0)\, u(jm) = \sum_m d^{(j)}_{mm'}(\beta) \frac{\chi_+^{j+m}\chi_-^{j-m}}{[(j+m)!\,(j-m)!]^{1/2}}, \end{aligned} \tag{14.6}$$

whence, by equating the coefficients of the base vectors, one obtains the Wigner formula (14a).

15. Under a rotation given by the Euler angles α, β, γ, the state $|jm\rangle$ transforms in general into the state $|jm'\rangle = D|jm\rangle$, the inverse relation being $|jm\rangle = D^{-1}|jm\rangle'$. In the present case, using (13.4), one finds that

$$\left|\frac{1}{2},\ \frac{1}{2}\right\rangle = e^{(i/2)\alpha}\cos\frac{\beta}{2}\,e^{(i/2)\gamma}\left|\frac{1}{2},\ \frac{1}{2}\right\rangle' + e^{(i/2)\alpha}\sin\frac{\beta}{2}\,e^{-(i/2)\gamma}\left|\frac{1}{2},\ -\frac{1}{2}\right\rangle'.$$

The required probabilities are therefore

$$\omega\left(+\frac{1}{2}\right) = \cos^2\frac{\beta}{2},\quad \omega\left(-\frac{1}{2}\right) = \sin^2\frac{\beta}{2}. \tag{15.1}$$

The average value of the z'-spin projection, according to (5.2), is equal to $\frac{1}{2}\cos\beta$.
This average value may also be obtained by using the result (15.1),

$$\langle s_{z'}\rangle = \frac{1}{2}\omega\left(+\frac{1}{2}\right)+\left(-\frac{1}{2}\right)\omega\left(-\frac{1}{2}\right) = \frac{1}{2}\cos^2\frac{\beta}{2}-\frac{1}{2}\sin^2\frac{\beta}{2} = \frac{1}{2}\cos\beta. \tag{15.2}$$

16. The spherical harmonics Y_1^1, Y_1^0, Y_1^{-1} are the simultaneous eigenfunctions of the operators $\mathbf{l}^2$ and l_z for $l = 1$. Their transformation under a rotation is given by the rotation matrix $D^{(1)}$ whose elements are $D^{(1)}_{mm'}(\alpha, \beta, \gamma) = e^{-i\alpha m}d^{(1)}_{mm'}(\beta)e^{-i\gamma m'}$. Using the Wigner formula (14a), we find that

$$d^{(1)}(\beta) = \begin{bmatrix} \frac{1}{2}(1+\cos\beta) & -\frac{\sqrt{2}}{2}\sin\beta & \frac{1}{2}(1-\cos\beta) \\ \frac{\sqrt{2}}{2}\sin\beta & \cos\beta & -\frac{\sqrt{2}}{2}\sin\beta \\ \frac{1}{2}(1-\cos\beta) & \frac{\sqrt{2}}{2}\sin\beta & \frac{1}{2}(1+\cos\beta) \end{bmatrix} \tag{16.1}$$

and therefore

$$\begin{bmatrix} Y_1^{1'} \\ Y_1^{0'} \\ Y_1^{-1'} \end{bmatrix} = \begin{bmatrix} \frac{1}{2}e^{-i\alpha}(1+\cos\beta)e^{-i\gamma} & -\frac{\sqrt{2}}{2}e^{-i\alpha}\sin\beta & \frac{1}{2}e^{-i\alpha}(1-\cos\beta)e^{i\gamma} \\ \frac{\sqrt{2}}{2}\sin\beta e^{-i\gamma} & \cos\beta & -\frac{\sqrt{2}}{2}\sin\beta e^{i\gamma} \\ \frac{1}{2}e^{i\alpha}(1-\cos\beta)e^{-i\gamma} & \frac{\sqrt{2}}{2}e^{i\alpha}\sin\beta & \frac{1}{2}e^{i\alpha}(1+\cos\beta)e^{i\gamma} \end{bmatrix} \begin{bmatrix} Y_1^1 \\ Y_1^0 \\ Y_1^{-1} \end{bmatrix} \tag{16.2}$$

17. Choosing the z-axis along the direction of the magnetic field, let the z'-axis, along which the component of the total angular momentum has a well-defined value m, make an angle β with the z-axis. Now, as in problem 15, in order to determine the relative intensities of the $2j+1 = 3$ beams, we must use the inverse of the transformation (16.2). The required intensities are, for $m = 1$:

$$\omega(+1) = \cos^4\frac{\beta}{2}, \quad \omega(0) = \frac{1}{2}\sin^2\beta, \quad \omega(-1) = \sin^4\frac{\beta}{2},$$

for $m = 0$:

$$\omega(+1) = \tfrac{1}{2}\sin^2\beta, \quad \omega(0) = \cos^2\beta, \quad \omega(-1) = \tfrac{1}{2}\sin^2\beta,$$

and for $m = -1$:

$$\omega(+1) = \sin^4\frac{\beta}{2}, \quad \omega(0) = \frac{1}{2}\sin^2\beta, \quad \omega(-1) = \cos^4\frac{\beta}{2}.$$

18. First, let us find the form of the eigenfunctions of the operator j_z. In matrix form we have

$$j_z = \begin{bmatrix} l_z & 0 \\ 0 & l_z \end{bmatrix} + \begin{bmatrix} \frac{1}{2} & 0 \\ 0 & -\frac{1}{2} \end{bmatrix} = \begin{bmatrix} l_z+\frac{1}{2} & 0 \\ 0 & l_z-\frac{1}{2} \end{bmatrix}. \tag{18.1}$$

Since $l_z = -i\dfrac{\partial}{\partial\phi}$, the differential equation satisfied by the eigenfunctions of j_z can be written in the form

$$\begin{bmatrix} -i\dfrac{\partial}{\partial\phi}+\dfrac{1}{2} & 0 \\ 0 & -i\dfrac{\partial}{\partial\phi}-\dfrac{1}{2} \end{bmatrix} \begin{bmatrix} \psi_1 \\ \psi_2 \end{bmatrix} = m_j \begin{bmatrix} \psi_1 \\ \psi_2 \end{bmatrix} \tag{18.2}$$

i.e.

$$-i\frac{\partial\psi_1}{\partial\phi}+\frac{1}{2}\psi_1 = m_j\psi_1, \qquad -i\frac{\partial\psi_2}{\partial\phi}-\frac{1}{2}\psi_2 = m_j\psi_2. \tag{18.3}$$

It follows from (18.3) that the ϕ-dependences of the functions ψ_1 and ψ_2 take the form of the factors $e^{i(m_j-1/2)\phi}$, and $e^{i(m_j+1/2)\phi}$, respectively. Since $\begin{bmatrix} \psi_1 \\ \psi_2 \end{bmatrix}$ is also an eigenfunction of the operator $\mathbf{l}^2$, we can infer that

$$\psi \equiv \begin{bmatrix} \psi_1 \\ \psi_2 \end{bmatrix} = \begin{bmatrix} R_1(r)Y_l^{m_j-1/2}(\theta, \phi) \\ R_2(r)Y_l^{m_j+1/2}(\theta, \phi) \end{bmatrix}, \tag{18.4}$$

where the $Y_l^{m_j\pm1/2}(\theta, \phi)$ are the usual spherical harmonics (note that m_j is a half-integer), and $R_1(r)$, $R_2(r)$ are radial functions. We can find a relation between $R_1(r)$ and $R_2(r)$ from the condition that ψ is also an eigenfunction of the square of the total angular momentum, i.e. that $\mathbf{j}^2\psi = j(j+1)\psi$.

Since

$$\mathbf{j}^2 = (\mathbf{l}+\mathbf{s})(\mathbf{l}+\mathbf{s}) = \mathbf{l}^2+\mathbf{s}^2+2\mathbf{l}.\mathbf{s} = \mathbf{l}^2+\mathbf{s}^2+2(l_xs_x+l_ys_y+l_zs_z)$$

or, in matrix form

$$\mathbf{j}^2 = \begin{bmatrix} \mathbf{l}^2 & 0 \\ 0 & \mathbf{l}^2 \end{bmatrix} + \begin{bmatrix} \frac{3}{4} & 0 \\ 0 & \frac{3}{4} \end{bmatrix} + \begin{bmatrix} l_z & l_x - il_y \\ l_x + il_y & -l_z \end{bmatrix},$$

we require that

$$\begin{bmatrix} \mathbf{l}^2 + \frac{3}{4} + l_z & l_- \\ l_+ & \mathbf{l}^2 + \frac{3}{4} - l_z \end{bmatrix} \begin{bmatrix} R_1(r) Y_l^{m_j - 1/2}(\theta, \phi) \\ R_2(r) Y_l^{m_j + 1/2}(\theta, \phi) \end{bmatrix} = j(j+1) \begin{bmatrix} R_1(r) Y_l^{m_j - 1/2}(\theta, \phi) \\ R_2(r) Y_l^{m_j + 1/2}(\theta, \phi) \end{bmatrix}. \tag{18.5}$$

Since, further, the spherical harmonics are such that

$$l_+ Y_l^m = \sqrt{(l-m)(l+m+1)}\, Y_l^{m+1},$$
$$l_- Y_l^m = \sqrt{(l+m)(l-m+1)}\, Y_l^{m-1},$$

it follows that $R_1(r)$ and $R_2(r)$ must satisfy the following homogeneous equations

$$\begin{aligned} \left[l(l+1) - j(j+1) + m_j + \tfrac{1}{4}\right] R_1 + \sqrt{(l+\tfrac{1}{2})^2 - m_j^2}\, R_2 &= 0, \\ \sqrt{(l+\tfrac{1}{2})^2 - m_j^2}\, R_1 + \left[l(l+1) - j(j+1) - m_j + \tfrac{1}{4}\right] R_2 &= 0. \end{aligned} \tag{18.6}$$

From the condition that the determinant of this system must vanish, we obtain an equation which gives two possible values of j, viz., $j = l+\frac{1}{2}$ and $j = l-\frac{1}{2}$.

Choosing $j = l+\frac{1}{2}$, we find that

$$R_1(r) = \sqrt{l+m_j+\tfrac{1}{2}}\, R(r), \qquad R_2(r) = \sqrt{l-m_j+\tfrac{1}{2}}\, R(r), \tag{18.7}$$

while, for $j = l-\frac{1}{2}$,

$$R_1(r) = -\sqrt{l-m_j+\tfrac{1}{2}}\, R(r), \quad R_2(r) = \sqrt{l+m_j+\tfrac{1}{2}}\, R(r), \tag{18.8}$$

where $R(r)$ is a function which can be determined when the central field is specified.

Finally, then

$$\psi_{l,\, j=l+1/2,\, m_j} = R(r) \begin{bmatrix} \sqrt{\dfrac{l+m_j+\frac{1}{2}}{2l+1}}\, Y_l^{m_j-1/2} \\ \sqrt{\dfrac{l-m_j+\frac{1}{2}}{2l+1}}\, Y_l^{m_j+1/2} \end{bmatrix} \tag{18.9}$$

$$\psi_{l,\, j=l-1/2,\, m_j} = R(r) \begin{bmatrix} -\sqrt{\dfrac{l-m_j+\frac{1}{2}}{2l+1}}\, Y_l^{m_j-1/2} \\ \sqrt{\dfrac{l+m_j+\frac{1}{2}}{2l+1}}\, Y_l^{m_j+1/2} \end{bmatrix} \tag{18.10}$$

The factor $1/(\sqrt{2l+1})$ arises from the normalization of ψ, as can easily be verified, since

$$\frac{l+m_j+\frac{1}{2}}{2l+1} + \frac{l-m_j+\frac{1}{2}}{2l+1} = 1.$$

Denoting the spin wavefunctions by χ_+ and χ_-, the result may be written in the following form

$$\psi_{l,j=l+1/2,m_j} = \sqrt{\frac{l+m_j+\frac{1}{2}}{2l+1}}\, R(r) Y_l^{m_j-1/2}\chi_+ + \sqrt{\frac{l-m_j+\frac{1}{2}}{2l+1}}\, R(r) Y_l^{m_j+1/2}\chi_- \qquad (18.11)$$

$$\psi_{l,j=l-1/2,m_j} = -\sqrt{\frac{l-m_j+\frac{1}{2}}{2l+1}}\, R(r) Y_l^{m_j-1/2}\chi_+ + \sqrt{\frac{l+m_j+\frac{1}{2}}{2l+1}}\, R(r) Y_l^{m_j+1/2}\chi_-. \qquad (18.12)$$

The expressions (18.11) and (18.12) can also readily be obtained by using the theory of the addition of angular momenta (see section 3) and the appropriate Clebsch–Gordan coefficients (Table VI.2).

19. In the "vector model", eigenstates of angular momentum are represented diagrammatically by vectors whose magnitudes and z-components only are supposed to be known (Fig. VI.4). Some of the properties of the quantum mechanical addition of angular mo-

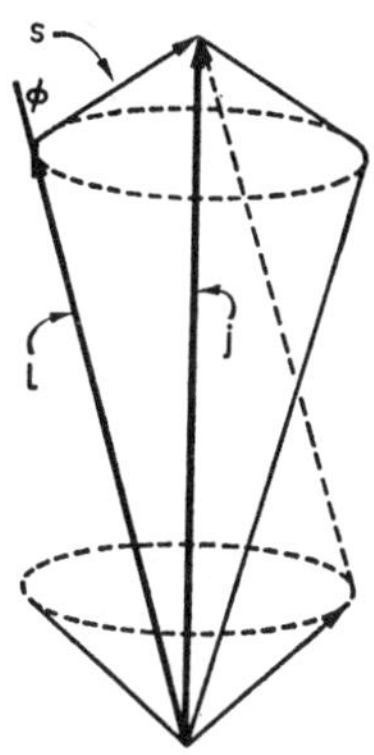

FIG. VI.4.

menta can then be represented by the addition of the corresponding vectors. Thus, in the present problem, we are asked only for the relative orientations of the vectors **l** and **s**, in an eigenstate of the total angular momentum $\mathbf{j} = \mathbf{l}+\mathbf{s}$. Consider a fixed direction of the vector representing the latter. **l** and **s** can then be regarded as "precessing" about this direction (Fig. VI.4), with constant magnitudes $\sqrt{l(l+1)}$ and $\sqrt{s(s+1)}$ respectively. Then, by simple trigonometry, if ϕ be the angle between the directions of **l** and of **s**,

$$\cos\phi = \frac{j(j+1)-l(l+1)-s(s+1)}{2\sqrt{l(l+1)\,s(s+1)}}, \qquad (19.1)$$

where $j = l\pm\frac{1}{2}$, $l = 0, 1, \ldots$, $s = \frac{1}{2}$.

The expression (19.1) can be used to determine the energy corrections ("fine structure") in the spectra of hydrogen-like ions due to spin–orbit interaction.

20. The possible values of the z-components of the orbital and of the spin angular momenta, and the corresponding probabilities, can be obtained immediately from the results of problem 18 (viz., the expressions (18.11) and (18.12)). These probabilities are tabulated as functions of m_j for given l, and $j = l+\frac{1}{2}$ or $j = l-\frac{1}{2}$, in Table VI.3.

TABLE VI.3

Orbital angular momentum (z-component)	Spin (z-component)	Value of j	
		$j=l+\frac{1}{2}$	$j=l-\frac{1}{2}$
$m_j-\frac{1}{2}$	$\frac{1}{2}$	$\dfrac{l+m_j+\frac{1}{2}}{2l+1}$	$\dfrac{l-m_j+\frac{1}{2}}{2l+1}$
$m_j+\frac{1}{2}$	$-\frac{1}{2}$	$\dfrac{l-m_j+\frac{1}{2}}{2l+1}$	$\dfrac{l+m_j+\frac{1}{2}}{2l+1}$

The tabulated probabilities can be used to calculate the average values of l_z and s_z, thus

$$\langle l_z\rangle_{j=l+1/2} = \left(m_j-\frac{1}{2}\right)\frac{l+m_j+\frac{1}{2}}{2l+1}+\left(m_j+\frac{1}{2}\right)\frac{l-m_j+\frac{1}{2}}{2l+1} = \frac{2lm_j}{2l+1}, \tag{20.1}$$

and, similarly,

$$\langle l_z\rangle_{j=l-1/2} = \frac{2(l+1)\,m_j}{2l+1}, \tag{20.2}$$

$$\langle s_z\rangle_{j=l+1/2} = \frac{m_j}{2l+1}, \quad \langle s_z\rangle_{j=l-1/2} = -\frac{m_j}{2l+1}. \tag{20.3}$$

21. This problem can be solved in an elementary way by using the Pauli matrices to represent the spin operators (see, for example, problem 10) or, more directly, by using the Clebsch–Gordan coefficients. With the values given in Table VI.2, the results can be written down systematically as follows:

TABLE VI.4

Simultaneous eigenfunctions of $\mathbf{s}^2$ and of s_z	Eigenvalues	
	s	s_z
$\chi_+^{(1)}\chi_+^{(2)}$	1	1
$\dfrac{1}{\sqrt{2}}[\chi_+^{(1)}\chi_-^{(2)}+\chi_-^{(1)}\chi_+^{(2)}]$	1	0
$\chi_-^{(1)}\chi_-^{(2)}$	1	-1
$\dfrac{1}{\sqrt{2}}[\chi_+^{(1)}\chi_-^{(2)}-\chi_-^{(1)}\chi_+^{(2)}]$	0	0

The possible states thus consist of a triplet of states with total spin $s = 1$, and a singlet with $s = 0$.

One can verify directly that the above spin functions are also eigenfunctions of the operator $\mathbf{s}_1 . \mathbf{s}_2$, with the eigenvalues $+\frac{1}{4}$ and $-\frac{3}{4}$ for the triplet and for the singlet states respectively. Thus, e.g., for $s = 0$

$$(\mathbf{s}_1 . \mathbf{s}_2)\,\chi_0 = \tfrac{1}{2}(\mathbf{s}^2 - \mathbf{s}_1^2 - \mathbf{s}_2^2)\,\chi_0 = -\tfrac{1}{2}(\mathbf{s}_1^2 + \mathbf{s}_2^2)\,\chi_0 = -\tfrac{1}{2}(\tfrac{3}{4} + \tfrac{3}{4})\,\chi_0 = -\tfrac{3}{4}\,\chi_0 .$$

22. The operator $\boldsymbol{\sigma}_1 . \boldsymbol{\sigma}_2$ has the eigenvalues $+1$ and -3 for the triplet and for the singlet states respectively (see problem 21), i.e.

$$(\boldsymbol{\sigma}_1 . \boldsymbol{\sigma}_2)\,\chi_{s=1} = +\chi_{s=1}, \quad (\boldsymbol{\sigma}_1 . \boldsymbol{\sigma}_2)\,\chi_{s=0} = -3\chi_{s=0} . \tag{22.1}$$

By iteration we obtain the results

$$(\boldsymbol{\sigma}_1 . \boldsymbol{\sigma}_2)^n\,\chi_{s=1} = \chi_{s=1}, \quad (\boldsymbol{\sigma}_1 . \boldsymbol{\sigma}_2)^n\,\chi_{s=0} = (-3)^n\,\chi_{s=0} . \tag{22.2}$$

Since the three triplet states and the singlet state together form a complete orthonormal basis in the space of the spin states of the two-particle system, it is sufficient to verify the linear dependence of $(\boldsymbol{\sigma}_1 . \boldsymbol{\sigma}_2)^n$ on $\boldsymbol{\sigma}_1 . \boldsymbol{\sigma}_2$ for these states. Let us suppose then that

$$(\boldsymbol{\sigma}_1 . \boldsymbol{\sigma}_2)^n = A + B(\boldsymbol{\sigma}_1 . \boldsymbol{\sigma}_2), \tag{22.3}$$

where the coefficients A and B are to be determined. Applying both sides of the operator relation (22.3) to the triplet states and to the singlet state, we obtain

$$\chi_{s=1} = (A+B)\,\chi_{s=1}, \quad (-3)^n\,\chi_{s=0} = (A-3B)\,\chi_{s=0}, \tag{22.4}$$

whence $A + B = 1$, $A - 3B = (-3)^n$, so that

$$A = \tfrac{1}{4}[3 + (-3)^n], \quad B = \tfrac{1}{4}[1 - (-3)^n]. \tag{22.5}$$

Since the operator relation (22.3) is then valid for the complete set of spin states considered, it will be valid for all spin states of the two-particle system.

Note that, for $n = 2$

$$(\boldsymbol{\sigma}_1 . \boldsymbol{\sigma}_2)^2 = 3 - 2(\boldsymbol{\sigma}_1 . \boldsymbol{\sigma}_2), \tag{22.6}$$

a result which is sometimes useful.

23. From the identity

$$(\mathbf{S} . \mathbf{r})^2 = \tfrac{1}{4}(\boldsymbol{\sigma}_1 . \mathbf{r} + \boldsymbol{\sigma}_2 . \mathbf{r})^2 = \tfrac{1}{4}[(\boldsymbol{\sigma}_1 . \mathbf{r})^2 + (\boldsymbol{\sigma}_2 . \mathbf{r})^2 + 2(\boldsymbol{\sigma}_1 . \mathbf{r})(\boldsymbol{\sigma}_2 . \mathbf{r})] = \tfrac{1}{2}[(\boldsymbol{\sigma}_1 . \mathbf{r})(\boldsymbol{\sigma}_2 . \mathbf{r}) + r^2],$$

it follows that

$$(\boldsymbol{\sigma}_1 . \mathbf{r})(\boldsymbol{\sigma}_2 . \mathbf{r}) = 2(\mathbf{S} . \mathbf{r})^2 - r^2,$$

and thus the "tensor force" operator S_{12} can be put into the required form

$$S_{12} = 2\left[3\frac{(\mathbf{S}\cdot\mathbf{r})^2}{r^2} - \mathbf{S}^2\right]. \tag{23.1}$$

24. The Hamiltonian H is invariant under rotation and reflection (under a reflection, $\mathbf{r}$ and $\mathbf{p}$ transform into $-\mathbf{r}$ and $-\mathbf{p}$, and the spin operators remain unchanged). It follows that $[H, \mathbf{J}] = 0$ and $[H, P] = 0$, respectively. There remains to be shown only that $[H, \mathbf{S}^2] = 0$. This clearly follows from the results of the preceding problem, with the observation that $[S_{12}, \mathbf{S}^2] = 0$.

Note that $\mathbf{l}^2$ and H do not commute, so that the eigenfunctions of the problem do not have a well-defined orbital angular momentum.

Remarks: The interaction between nucleons represented by the "tensor force" operator is only one of many needed for the construction of a "phenomenological" two-body potential adequate for the description of nuclear phenomena; such a potential is usually taken to be a linear combination of operators of the following forms

$$V_1(r) \tag{24.1}$$

$$V_2(r)(\boldsymbol{\sigma}_1 \cdot \boldsymbol{\sigma}_2) \tag{24.2}$$

$$V_3(r)(\mathbf{l}\cdot\mathbf{S}) \tag{24.3}$$

$$V(r)S_{12}. \tag{24.4}$$

Note that if V contains only terms of the form (24.1) and (24.3), the energy eigenfunctions can be taken to be simultaneous eigenfunctions of the operators P, $\mathbf{l}^2$, $\mathbf{S}^2$ and J_z, and the degeneracy is $(2J+1)$-fold. For terms of the form (24.1) and (24.2), the energy eigenfunctions can be taken to be simultaneous eigenfunctions of P, $\mathbf{l}^2$, $\mathbf{S}^2$, l_z and S_z, since (24.1) and (24.2) are both invariant under rotations performed separately in coordinate space and in spin space, and hence commute with $\mathbf{l}$ and with $\mathbf{S}$. The degeneracy is then $(2L+1)(2S+1)$. If, however, a term of the form (24.4) is present as well as any or all of the others, then only P, $\mathbf{S}^2$, $\mathbf{J}^2$ and J_z remain of the operators which commute with V.

25. As mentioned in the remarks made at the end of the preceding problem, the Hamiltonian in this case commutes with $\mathbf{l}$ and $\mathbf{S}$, so that the eigenfunctions can be taken to be simple products of spin and coordinate functions, the latter corresponding to well-defined angular momenta (l, m). If we take the spin functions to be those given in the results of problem 21, then the potential energy term in the equation for the coordinate functions has two different values, viz., $V_1 - 3V_2$ and $V_1 + V_2$, for the states with $S = 0$ and those with $S = 1$ respectively. Thus, instead of having to solve the full Schrödinger equation for V we need solve only two equations, for the coordinate parts of the eigenfunctions of the $S = 0$ states and of the $S = 1$ states respectively.

26. In spherical polar coordinates, $\mathbf{r} = (r \sin\theta \cos\phi, r\sin\theta\sin\phi, r\cos\theta)$, we have

$$S_x \cos\phi + S_y \sin\phi = \tfrac{1}{2}(S_+ e^{-i\phi} + S_- e^{i\phi}),$$

$$\frac{1}{r^2}(\mathbf{S}\cdot\mathbf{r})^2 = \left[\frac{1}{2}(S_+ e^{-i\phi} + S_- e^{i\phi})\sin\theta + S_z\cos\theta\right]^2$$

$$= \frac{1}{4}S_+^2 e^{-2i\phi}\sin^2\theta + \frac{1}{4}S_-^2 e^{2i\phi}\sin^2\theta + \frac{1}{2}(S_+S_z + S_zS_+) + e^{-i\phi}\sin\theta\cos\theta$$

$$+\frac{1}{2}(S_-S_z + S_zS_-)e^{i\phi}\sin\theta\cos\theta + S_z^2\cos^2\theta + \frac{1}{4}(S_+S_- + S_-S_+)\sin^2\theta.$$

Using the relations

$$[S_+, S_-] = 2S_z, \quad \mathbf{S}^2 = S_+S_- + S_z^2 - S_z, \quad S_+S_- + S_-S_+ = 2S_+S_- - 2S_z = 2(\mathbf{S}^2 - S_z^2),$$

we have that

$$S_{12} = \tfrac{3}{2}S_+^2 e^{-2i\phi}\sin^2\theta + 3(S_+S_z + S_zS_+)e^{-i\phi}\sin\theta\cos\theta$$
$$-(3S_z^2 - \mathbf{S}^2)(1-3\cos^2\theta) + 3(S_-S_z + S_zS_-)e^{i\phi}\sin\theta\cos\theta + \tfrac{3}{2}S_-^2 e^{2i\phi}\sin^2\theta.$$

In terms of the spherical harmonics with $l = 2$ (A.29), we find that

$$S_{12} = \sqrt{\frac{24\pi}{5}}\left[S_+^2 Y_2^{-2} + (S_+S_z + S_zS_+)Y_2^{-1} + \sqrt{\frac{2}{3}}(3S_z^2 - \mathbf{S}^2)Y_2^0 \right.$$
$$\left. -(S_-S_z + S_zS_-)Y_2^1 + S_-^2 Y_2^2\right]. \tag{26.1}$$

27. The components of the operators **r**, **p** and **l** satisfy the following commutation rules (see problem 3):

$$[l_i, x_k] = i\varepsilon_{ikl}x_l, \quad [l_i, p_k] = i\varepsilon_{ikl}p_l, \quad [l_i, l_k] = i\varepsilon_{ikl}l_l. \tag{27.1}$$

Let A_x, A_y, A_z be the components of any vector operator, which is such that

$$[l_i, A_k] = i\varepsilon_{ikl}A_l, \tag{27.2}$$

where the indices 1, 2, 3 stand for the x-, y- and z-components as usual. Let us define the quantities

$$A_1^{(1)} = -\frac{1}{\sqrt{2}}(A_x + iA_y), \quad A_0^{(1)} = A_z, \quad A_{-1}^{(1)} = \frac{1}{\sqrt{2}}(A_x - iA_y). \tag{27.3}$$

Then, after some elementary calculation, we find that

$$[l_\pm, A_q^{(1)}] = \sqrt{2 - q(q\pm 1)}A_{q\pm 1}^{(1)} \tag{27.4}$$

$$[l_z, A_q^{(1)}] = qA_q^{(1)}. \tag{27.5}$$

For example, one of the two relations (27.4) can be established as follows:

$$[l_+, A_0^{(1)}] = [l_1, A_3]+i[l_2, A_3] = -iA_2-A_1 = \sqrt{2}\,A_1^{(1)}.$$

Taking into account (27.4), (27.5) and the definition of the irreducible tensor operators (VI.11), the problem is solved.

28. Using the Wigner–Eckart theorem (VI.12), we obtain the required matrix elements in the form

$$\langle l, m \,|\, A_q^{(1)} \,|\, l', m' \rangle = \frac{1}{\sqrt{2l+1}} \langle l \,||\, \mathcal{A}^{(1)} \,||\, l' \rangle \langle l, 1, m', q \,|\, l, m \rangle,$$

where the $A_q^{(1)}$ are defined in (27.3) above, and the expressions $\langle l \,||\, \mathcal{A}^{(1)} \,||\, l' \rangle$ depend, for a given **A**, only on l and l'.

In the following we shall require the Clebsch–Gordan coefficients $\langle l', 1, m', q \,|\, l, m \rangle$ given in Table VI.1. Suppose first that $q = -1$. Then, since $m'+q = m$, we obtain $m' = m+1$. For l' we have three possibilities, viz., $l = l' \pm 1$ or l', so that, conversely, $l' = l \mp 1$ or l. Then, we can write directly, for the non-vanishing matrix elements,

$$\begin{aligned}\langle l, m \,|\, A_{-1}^{(1)} \,|\, l-1, m+1 \rangle &= \frac{a_{l,\,l-1}}{\sqrt{2l+1}} \langle l-1, 1, m+1, -1 \,|\, l, m \rangle \\ &= \frac{a_{l,\,l-1}}{\sqrt{2l+1}} \left[\frac{(l-m-1)(l-m)}{(2l-1)2l}\right]^{1/2} = a_{l,\,l-1} \left[\frac{(l-m-1)(l-m)}{(2l-1)\,2l(2l+1)}\right]^{1/2},\end{aligned} \tag{28.1}$$

where the coefficients $a_{l,\,l-1} = \langle l \,||\, \mathcal{A}^{(1)} \,||\, l-1 \rangle$ are independent of m. Similarly

$$\langle l, m \,|\, A_{-1}^{(1)} \,|\, l, m+1 \rangle = a_{l,\,l} \left[\frac{(l-m)(l+m+1)}{(2l+1)\,2l(l+1)}\right]^{1/2} \tag{28.2}$$

$$\langle l, m \,|\, A_{-1}^{(1)} \,|\, l+1, m+1 \rangle = a_{l,\,l+1} \left[\frac{(l+m+2)(l+m+1)}{(2l+1)(2l+2)(2l+3)}\right]^{1/2}. \tag{28.3}$$

For $q = 0$, $m' = m$

$$\langle l, m \,|\, A_0^{(1)} \,|\, l-1, m \rangle = a_{l,\,l-1} \left[\frac{(l-m)(l+m)}{(2l+1)\,l(2l+1)}\right]^{1/2} \tag{28.4}$$

$$\langle l, m \,|\, A_0^{(1)} \,|\, l, m \rangle = a_{l,\,l} \frac{m}{[(2l+1)\,l(l+1)]^{1/2}} \tag{28.5}$$

$$\langle l, m \,|\, A_0^{(1)} \,|\, l+1, m \rangle = -a_{l,\,l+1} \left[\frac{(l-m+1)(l+m+1)}{(2l+1)(l+1)(2l+3)}\right]^{1/2}. \tag{28.6}$$

Finally, the matrix elements for $q = +1$ can be obtained in the same way, using the relevant Clebsch–Gordan coefficients.

The selection rules for the orbital and for the "magnetic" quantum numbers are thus $l \to l$ or $l \to l \pm 1$, and $m \to m$ or $m \to m \pm 1$, respectively.

29. Let us denote the parity operator by P, i.e. $P\psi(\mathbf{r}), = \psi(-\mathbf{r})$. Then, if $\mathbf{A}$ is a polar vector operator, it anticommutes with P. But P is diagonal in the $\{\mathbf{l}^2, l_z\}$ representation (see problem 8) and thus, in this representation, $\mathbf{A}$ has non-vanishing matrix elements only off the main diagonal. Therefore, taking into account the result of the preceding problem, the selection rules giving changes in the orbital quantum number for transitions induced by polar vector operators are $\Delta l = \pm 1$. Clearly, for axial vector operators, the selection rule is $\Delta l = 0$.

30. Using the results of problem 27, one can write the components of the vector $\mathbf{n}$ in terms of irreducible tensor operators, and thus the required matrix elements can be determined using the Wigner–Eckart theorem (see problem 28). Since $\mathbf{n}$ is a polar vector, the dependence on m of the non-vanishing matrix elements is given by (28.4) and (28.6).

There remains for us only to determine $a_{l,\,l+1}$ and $a_{l,\,l-1}$.

Putting $m = 0$ in (28.6), we have that

$$\langle l, 0|\, n_z\,| l+1, 0\rangle = -a_{l,\,l+1}\left[\frac{l+1}{(2l+1)(2l+3)}\right]^{1/2}, \tag{30.1}$$

whereas, by direct calculation,

$$\langle l, 0\,|\, n_z\,|\, l+1, 0\rangle = \langle l, 0\,|\cos\theta\,|\, l+1, 0\rangle = \int Y_l^{0*}(\theta,\phi)\cos\theta Y_{l+1}^0(\theta,\phi)\sin\theta\, d\theta\, d\phi. \tag{30.2}$$

But, according to (A.28),

$$\cos\theta Y_{l+1}^0 = \frac{l+2}{\sqrt{(2l+3)(2l+5)}}Y_{l+2}^0 + \frac{l+1}{\sqrt{(2l+1)(2l+3)}}Y_l^0.$$

Using this result in (30.2), the integration is easily performed. By comparing the result with (30.1) we find that $a_{l,\,l+1} = -\sqrt{l+1}$. Similarly, $a_{l,\,l-1} = \sqrt{l}$.

31. If the expressions (31a) and (31b) are compared with (26.1), it becomes evident that we have to prove that the quantities

$$T_2^{(2)} = Y_2^2,\quad T_1^{(2)} = Y_2^1,\quad T_0^{(2)} = Y_2^0,\quad T_{-1}^{(2)} = Y_2^{-1},\quad T_{-2}^{(2)} = Y_2^{-2},$$

and

$$S_2^{(2)} = (S_+)^2,\quad S_1^{(2)} = S_+S_z+S_zS_+,\quad S_0^{(2)} = \sqrt{\tfrac{2}{3}}\,[3(S_z)^2-\mathbf{S}^2],$$
$$S_{-1}^{(2)} = S_-S_z+S_zS_-,\quad S_{-2}^{(2)} = (S_-)^2,$$

are in fact the components of two second order irreducible tensor operators $\mathbf{T}^{(2)}$ and $\mathbf{S}^{(2)}$. This statement is evidently valid for $\mathbf{T}^{(2)}$; for $\mathbf{S}^{(2)}$ we need to verify the validity of the following commutation rules:

$$[S_\pm, S_q^{(2)}] = \sqrt{6-q(q\pm 1)}S_{q\pm 1}^{(2)} \tag{31.1}$$

$$[S_z, S_q^{(2)}] = qS_q^{(2)}. \tag{31.2}$$

Consider, for example, the commutation rule

$$[S_+, S^{(2)}_{-2}] = [S_+, (S_-)^2] = \sqrt{6-(-2)(-2+1)}S^{(2)}_{-1} = 2(S_-S_z+S_zS_-), \tag{31.3}$$

which is one of the set (31.1). This can be verified directly by writing $S_+S_- - S_-S_+ = 2S_z$, multiplying on the left and then on the right by S_-, and adding together the relations thus obtained. The remaining rules can be verified in a similar way.

32. The Schrödinger equation for an electron in a potential V and a magnetic field $\mathbf{H} =$ curl $\mathbf{A}$ is

$$i\hbar\frac{\partial\psi}{\partial t} = H\psi, \quad H = \frac{1}{2m}\left(\mathbf{p}-\frac{e}{c}\mathbf{A}\right)^2 + V + V_{\text{spin}}, \tag{32.1}$$

where V_{spin} is the operator which represents the potential energy of the interaction between the intrinsic magnetic moment of the electron and the field $\mathbf{H}$. The wavefunction ψ is of course a function of the position and of the spin orientation of the electron.

By analogy with the classical expression for the potential energy of a magnetic dipole in a field $\mathbf{H}$, the term V_{spin} is written as

$$V_{\text{spin}} = -\left(\frac{e}{mc}\mathbf{s}\right).\mathbf{H} = -\frac{e}{mc}(\mathbf{s}.\mathbf{H}), \quad \mathbf{s} = \frac{1}{2}\hbar\boldsymbol{\sigma} \tag{32.2}$$

Equation (32.1) then becomes

$$i\hbar\frac{\partial\psi}{\partial t} = \frac{1}{2m}\left(\mathbf{p}-\frac{e}{c}\mathbf{A}\right)^2\psi - \frac{e}{mc}(\mathbf{s}.\mathbf{H})\psi + V\psi, \tag{32.3}$$

which is sometimes called the Pauli equation.

Consider now the particular case of a homogeneous magnetic field $\mathbf{H}$. The vector potential for such a field can be written in the form

$$\mathbf{A} = \tfrac{1}{2}(\mathbf{H}\times\mathbf{r}). \tag{32.4}$$

It can be seen from (32.4) that in this case, div $\mathbf{A} = 0$, and hence that $\mathbf{p}.\mathbf{A}-\mathbf{A}.\mathbf{p} = -i\hbar$ div $\mathbf{A} = 0$. Equation (32.3) then becomes

$$i\hbar\frac{\partial\psi}{\partial t} = \frac{\mathbf{p}^2}{2m}\psi - \frac{e}{2mc}(\mathbf{H}\times\mathbf{r}).\mathbf{p}\psi - \frac{e}{mc}(\mathbf{s}.\mathbf{H})\psi + \frac{e^2}{8mc^2}(\mathbf{H}\times\mathbf{r})^2\psi + V\psi. \tag{32.5}$$

But $(\mathbf{H}\times\mathbf{r}).\mathbf{p} = \mathbf{H}.(\mathbf{r}\times\mathbf{p}) = \mathbf{H}.\mathbf{l}$, where $\mathbf{l}$ is the orbital angular momentum operator of the electron. Hence (32.5) becomes

$$i\hbar\frac{\partial\psi}{\partial t} = \frac{\mathbf{p}^2}{2m}\psi - (\mu_l.\mathbf{H})\psi - (\mu_s.\mathbf{H})\psi + \frac{e^2}{8mc^2}(\mathbf{H}\times\mathbf{r})^2\psi + V\psi, \tag{32.6}$$

in which

$$\mu_l = \frac{e}{2mc}\mathbf{l} \tag{32.7}$$

arises as an operator for the effective magnetic moment due to orbital motion (in contrast to μ_s, which is the intrinsic magnetic moment operator). Note that the ratio of the orbital magnetic moment μ_l to the mechanical angular momentum $\mathbf{l}$ is (as in classical physics) equal to $\frac{e}{2mc}$. For the corresponding quantities attributed to the spin, this ratio is twice as large, a fact which was considered very puzzling until explained by Dirac's theory of the electron.

For a system of electrons in a homogeneous field $\mathbf{H}$, Pauli's equation takes the form

$$i\hbar\frac{\partial\psi}{\partial t} = \left\{\frac{1}{2m}\sum_k \mathbf{p}_k^2 - \frac{e}{2mc}(\mathbf{L}.\mathbf{H}) - \frac{e}{mc}(\mathbf{S}.\mathbf{H}) + \frac{e^2}{8mc^2}\sum_k(\mathbf{H}\times\mathbf{r}_k)^2 + U\right\}\psi, \tag{32.8}$$

where $\mathbf{L} = \sum_k \mathbf{l}_k$ is the total orbital angular momentum operator, $\mathbf{S} = \sum_k \mathbf{s}_k$ the total spin operator, and U is the potential energy of all interactions other than those with the field $\mathbf{H}$.

33. Let us seek a solution of the Pauli equation for one particle in the form

$$\psi = \begin{bmatrix}\psi_1\\ \psi_2\end{bmatrix} = \phi(x, y, z, t)\begin{bmatrix}\chi_+(t)\\ \chi_-(t)\end{bmatrix}, \tag{33.1}$$

where $\phi(x, y, z, t)$ depends only on the particle's position, and $\chi(t)$ only on its spin orientation. One finds immediately that $\phi(x, y, z, t)$ has to satisfy the equation

$$i\hbar\frac{\partial\phi}{\partial t} = \frac{1}{2m}\left(\mathbf{p} - \frac{e}{c}\mathbf{A}\right)^2\phi + V\phi, \tag{33.2}$$

while χ has to satisfy the equation

$$i\hbar\frac{\partial}{\partial t}\begin{bmatrix}\chi_+(t)\\ \chi_-(t)\end{bmatrix} = -\frac{e}{mc}(\mathbf{s}.\mathbf{H})\begin{bmatrix}\chi_+(t)\\ \chi_-(t)\end{bmatrix}. \tag{33.3}$$

34. Choosing the z-axis to be in the direction of the field, the vector potential $\mathbf{A}$ can be written in the form

$$A_x = -Hy, \quad A_y = A_z = 0, \tag{34.1}$$

and the Hamiltonian of the Pauli equation, except for the square term containing $\frac{e^2}{c^2}$, is

$$H = \frac{1}{2m}\left(p_x + \frac{eHy}{c}\right)^2 + \frac{p_y^2}{2m} + \frac{p_z^2}{2m} - \frac{e\hbar}{mc}s_zH. \tag{34.2}$$

We observe first that H is time-independent and commutes with s_z. It follows that the time-independent wavefunctions are given (cf. the case of the spinless Schrödinger equation) by a "time-independent Pauli equation"

$$H\psi = E\psi, \tag{34.3}$$

and also that the z-component of the spin in the state ψ has well-defined values $s = \pm\frac{1}{2}$. By separating the position-dependent and the spin-dependent parts of ψ ($= \phi\chi$ say), we find that, for a given value of s, ϕ obeys the following equation:

$$\frac{1}{2m}\left[\left(p_x+\frac{eHy}{c}\right)^2+p_y^2+p_z^2\right]\phi-\frac{e\hbar}{mc}sH\phi = E\phi. \tag{34.4}$$

Now the Hamiltonian of this problem is independent of x and of z, and it therefore come mutes with the momentum operator components p_x and p_z. These components are therefore constants of the motion, and we accordingly seek a solution ϕ of the form

$$\phi = e^{\frac{i}{\hbar}(p_x x+p_z z)}\Phi(y). \tag{34.5}$$

Substituting (34.5) into (34.4), we obtain the following differential equation for $\Phi(y)$:

$$\Phi''+\frac{2m}{\hbar^2}\left[E+\frac{e\hbar}{mc}sH-\frac{p_z^2}{2m}-\frac{m}{2}\left(\frac{eH}{mc}\right)^2(y-y_0)^2\right]\Phi = 0, \tag{34.6}$$

in which we have written $y_0 = -cp_x/eH$. Equation (34.6) is formally the same as the Schrödinger equation of a linear oscillator of frequency $\omega = |e|H/mc$, whose centre of oscillation is at the point $y = y_0$. We can therefore, from (17.7) of Chapter II, write down directly the possible values of the quantity $\left(E+\dfrac{e\hbar}{mc}sH-\dfrac{p_z^2}{2m}\right)$, since these values are just the energy eigenvalues of the oscillator; thus

$$E+\frac{e\hbar}{mc}sH-\frac{p_z^2}{2m} = \left(n+\frac{1}{2}\right)\hbar\omega,$$

whence

$$E = \left(n+\frac{1}{2}+s\right)\frac{|e|\,\hbar}{mc}H+\frac{p_z^2}{2m}, \qquad n = 1, 2, \ldots \tag{34.7}$$

The corresponding eigenfunction are (except for a normalization constant):

$$\phi = e^{\frac{i}{\hbar}(p_x x+p_z z)}\exp\left[-\frac{|e|\,H}{2c\hbar}(y-y_0)^2\right]H_n\left[\sqrt{\frac{|e|\,H}{c\hbar}}(y-y_0)\right], \tag{34.8}$$

where H_n is a Hermite polynomial.

We observe that there is a degeneracy in the energy levels (34.7) owing to the spin. Thus, the energies of the states for a given $n = n'$ and $s = \frac{1}{2}$, and for $n = n'+1$ and $s = -\frac{1}{2}$, respectively, are equal, although the corresponding wavefunctions are clearly different.

35. This problem is best solved in cylindrical coordinates ϱ, ϕ, z, in which the z-axis is taken to be in the direction of the magnetic field. Suitable vector potential components are then $A_z = A_\varrho = 0$ and $A_\phi = \frac{1}{2}H\varrho$, and the Schrödinger equation becomes (cf. problem 32)

$$-\frac{\hbar^2}{2\mu}\left[\frac{1}{\varrho}\frac{\partial}{\partial\varrho}\left(\varrho\frac{\partial\psi}{\partial\varrho}\right)+\frac{\partial^2\psi}{\partial z^2}+\frac{1}{\varrho^2}\frac{\partial^2\psi}{\partial\phi^2}\right]+\frac{ie\hbar}{2\mu c}H\frac{\partial\psi}{\partial\phi}+\frac{e^2H^2}{8\mu c^2}\varrho^2\psi = E\psi. \tag{35.1}$$

Since the variables ϕ and z do not appear explicitly in equation (35.1), the operators l_z and p_z commute with H, i.e. the z-components of the angular momentum and of the linear momentum are constants of the motion. For this reason we seek a solution of the form

$$\psi(\varrho, \phi, z) = NR(\varrho)e^{im\phi}e^{\frac{ip_z z}{\hbar}}. \tag{35.2}$$

Substituting (35.2) into (35.1), we obtain for the radial part of the wavefunction

$$\frac{d^2R}{\varrho^2}+\frac{1}{\varrho}\frac{dR}{d\phi}+\left[\frac{2\mu E-p_z^2}{\hbar^2}-\left(\frac{eH}{2c\hbar}\right)^2\varrho^2-\frac{eH}{c\hbar}m-\frac{m^2}{\varrho^2}\right]R = 0.$$

Let us introduce the new independent variable $\xi = (|e|H/2c\hbar)\varrho^2$. We then obtain

$$\xi R''+R'+\left(-\frac{\xi}{4}+\beta-\frac{m^2}{4\xi}\right)R = 0, \quad \beta = \frac{(2\mu E-p_z^2)c}{2\hbar|e|H}+\frac{m}{2}. \tag{35.3}$$

If can be seen that, as $\xi \to +\infty$, and as $\xi \to 0$, the solution of equation (35.3) behaves as $e^{-\xi/2}$ and as $\xi^{|m|/2}$, respectively. We therefore take a trial solution of the form

$$R = e^{-\xi/2}\xi^{|m|/2}\omega(\xi). \tag{35.4}$$

Substituting this into (35.3), we obtain for $\omega(\xi)$ the differential equation

$$\xi\omega''+(|m|+1-\xi)\omega'+\left(\beta-\frac{|m|+1}{2}\right)\omega = 0 \tag{35.5}$$

whence, by (A.26),

$$\omega = F\left[-\left(\beta-\frac{|m|+1}{2}\right), |m|+1, \xi\right]. \tag{35.6}$$

For the wavefunction to remain finite as $\xi \to +\infty$, the number $\beta-(|m|+1)/2$ must be a positive integer. Denoting this integer by n_ϱ, we find the energy levels, from (35.3), to be given by

$$E = \frac{|e|H}{\mu c}\hbar\left(n_\varrho+\frac{|m|-m+1}{2}\right)+\frac{p_z^2}{2\mu}. \tag{35.7}$$

This expression is the same as (34.7) if we take $s = 0$ and $n_\varrho+(|m|-m)/2 = n$.

36. The wavefunction is a two-component spinor $\chi = \begin{bmatrix}\chi_+\\ \chi_-\end{bmatrix}$ satisfying the equation

$$i\hbar \frac{\partial \chi}{\partial t} = V_{\text{spin}}\,\chi, \tag{36.1}$$

where $V_{\text{spin}} = -\boldsymbol{\mu}\,.\,\mathbf{H}$ ($\boldsymbol{\mu}$ being the magnetic moment of the particle).

Choosing the z-axis to be in the direction of the magnetic field, equation (36.1) becomes

$$i\hbar \frac{\partial \chi_+}{\partial t} = -\mu H \chi_+, \quad i\hbar \frac{\partial \chi_-}{\partial t} = \mu H \chi_-, \tag{36.2}$$

whence

$$\chi_+ = c_1 \exp\left(\frac{i\mu}{\hbar}\int H\,dt\right), \quad \chi_- = c_2 \exp\left(-\frac{i\mu}{\hbar}\int H\,dt\right). \tag{36.3}$$

The constants c_1 and c_2 remain to be determined from the initial conditions and the normalization $|\chi_+|^2+|\chi_-|^2 = 1$.

37. The polarization of the beam depends on the relative numbers of neutrons in different spin states. Let us take as a basis for these spin states the eigenstates of s_z, with the eigenvalues $\pm\frac{1}{2}$. To be specific, let us suppose that all the neutrons of the incident beam are in the $s_z = +\frac{1}{2}$ state, namely in the $\chi_+ = \begin{bmatrix}1\\0\end{bmatrix}$ state. Since there are no forces which could reverse the spins on reflection or on penetration through the surface, the incident, the reflected and the transmitted waves will be of the form

$$A\chi_+ e^{i\mathbf{k}\cdot\mathbf{r}}, \quad B\chi_+ e^{i\mathbf{k}'\cdot\mathbf{r}}, \quad C\chi_+ e^{i\mathbf{k}''\cdot\mathbf{r}}; \qquad \mathbf{k} = \frac{\mathbf{p}}{\hbar}. \tag{37.1}$$

The quantities $\mathbf{k}$, $\mathbf{k}'$ and $\mathbf{k}''$ are related to the total energy E and to the magnetic moment μ_0 of the neutrons by the relations

$$\frac{\hbar^2 k^2}{2m} = E, \quad \frac{\hbar^2 k'^2}{2m} = E, \quad \frac{\hbar^2 k''^2}{2m} = E+\mu_0 H. \tag{37.2}$$

From the usual continuity conditions on the wavefunction, and on its first derivative with respect to x, we find that

$$\begin{aligned} k_y = k_y = k_y'', \quad k_z = k_z' = k_z'' \\ A+B = C, \quad k_x A + k_x' B = k_x'' C. \end{aligned} \tag{37.3}$$

From these relations it follows that $k_x' = -k_x$, and hence that the angles of incidence and of reflection are equal; also that

$$\frac{B}{A} = \frac{k_x - k_x''}{k_x + k_x''}, \quad \frac{C}{A} = \frac{2k_x}{k_x + k_x''}, \tag{37.4}$$

and, from (37.2),

$$k_x'' = k_x \sqrt{1+\frac{2m}{\hbar^2 k_x^2}\mu_0 H}\,. \tag{37.5}$$

The reflection and the transmission coefficients at the surface are then given by

$$R = \left(\frac{k_x - k_x''}{k_x + k_x''}\right)^2, \qquad T = \frac{4k_x k_x''}{(k_x+k_x'')^2}\,. \tag{37.6}$$

If the spins of the incident neutrons are all in the $\chi_- = \begin{bmatrix}0\\1\end{bmatrix}$ state, the results of the calculation are the same, except for a change of sign in the expression for k_x'', which becomes

$$k_x'' = k_x \sqrt{1-\frac{2m}{\hbar^2 k_x^2}\mu_0 H}\,. \tag{37.7}$$

38. Restricting our considerations to spin orientations, the density operator can be represented by a 2×2 matrix $\langle s_z | \varrho | s_z' \rangle$; $s_z, s_z' = \pm\hbar/2$. Now, if we denote the Pauli matrices by σ_α ($\alpha = 1, 2, 3$) and the 2×2 unit matrix by I, it is easy to verify that

$$\sigma_\alpha \sigma_\beta = \delta_{\alpha\beta} I + i \sum_{\gamma=1}^{3} \varepsilon_{\alpha\beta\gamma}\sigma_\gamma\,, \tag{38.1}$$

where $\varepsilon_{\alpha\beta\gamma}$ is the completely anti-symmetrical tensor of order 3 (see problem 3). From (38.1) it follows that

$$\text{Tr}\,(\sigma_\alpha) = \text{Tr}\,(I\sigma_\alpha) = 0, \qquad \text{Tr}\,(\sigma_\alpha\sigma_\beta) = 2\delta_{\alpha\beta}\,. \tag{38.2}$$

These relations show that, in this case,

$$\Omega^{(0)} = \frac{1}{\sqrt{2}} I, \qquad \Omega^{(\alpha)} = \frac{1}{\sqrt{2}}\sigma_\alpha\,, \qquad \alpha = 1, 2, 3, \tag{38.3}$$

can be chosen as the required orthogonal operators.

The expansion of the density matrix in terms of these operators (in the Schrödinger representation) is then

$$\varrho_t = \varrho_t^{(0)}\Omega^{(0)} + \sum_{\alpha=1}^{3} \varrho_t^{(\alpha)}\Omega^{(\alpha)}; \qquad \varrho_t^{(0)} = \frac{1}{\sqrt{2}}\langle I\rangle = \frac{1}{\sqrt{2}}, \quad \varrho_t^{(\alpha)} = \frac{1}{\sqrt{2}}\langle\sigma_\alpha\rangle. \tag{38.4}$$

Writing $\langle\sigma_\alpha\rangle = P_\alpha$, so that P_α denotes the components of the polarization vector $\mathbf{P} = (P_1, P_2, P_3)$, the expressions (38.4) become†

$$\varrho_t = \frac{1}{2}\left(I + \sum_{r=1}^{3} P_\alpha\sigma_\alpha\right) = \frac{1}{2}(I + \mathbf{P}\cdot\boldsymbol{\sigma}) = \frac{1}{2}\begin{bmatrix} 1+P_3 & P_1 - iP_2 \\ P_1 + iP_2 & 1-P_3 \end{bmatrix}. \tag{38.5}$$

† The ϱ matrix occurs in the form (38.5) in the study of photon polarization; the quantities P_1, P_2, P_3 are then called Stokes' coefficients.

In order to determine the equation of motion of **P**, we observe that, using equation (38.1), it is possible to write

$$D_{\alpha\beta} = \frac{i}{\hbar}\frac{1}{2}\,\mathrm{Tr}\,\{H[\sigma_\alpha, \sigma_\beta]\} = \frac{1}{\hbar}\sum_{\gamma=1}^{3} \varepsilon_{\beta\alpha\gamma}\,\mathrm{Tr}\,(H\sigma_\gamma),$$
$$D_{\alpha 0} = D_{0\beta} = 0. \tag{38.6}$$

Then, from (V.16), it follows that

$$\frac{dP_\alpha}{dt} = \frac{1}{\hbar}\sum_{\beta,\gamma=1}^{3} \varepsilon_{\beta\alpha\gamma}\,\mathrm{Tr}\,(H\sigma_\gamma)P_\beta. \tag{38.7}$$

For a spin system in an external constant magnetic field **H**, we have

$$H = -\boldsymbol{\mu}\,.\,\mathbf{H} = -\tfrac{1}{2}\gamma\hbar\boldsymbol{\sigma}\,.\,\mathbf{H}, \tag{38.8}$$

so that equation (38.7) becomes

$$\frac{dP_\alpha}{dt} = -\frac{1}{2}\gamma\sum_{\beta,\gamma,\delta=1}^{3} \varepsilon_{\beta\alpha\gamma}\,\mathrm{Tr}\,\{\sigma_\delta\sigma_\gamma\}H_\delta P_\beta. \tag{38.9}$$

Taking (38.2) into account, and using the obvious relation $\varepsilon_{\alpha\beta\gamma}A_\beta B_\gamma = (\mathbf{A}\times\mathbf{B})_\alpha$ for the Cartesian components of two arbitrary vectors **A** and **B** in a right-handed system of axes, it can be seen that (38a) then follows from (38.9). Note that equation (38a) has the same form as the classical equation for Larmor precession.

39. Denoting by H the operator representing the potential energy of interaction between a spin and the magnetic fields $\mathbf{H}_0$ and $\mathbf{H}_1$, and using the notation of problem 7 of Chapter IX, we have that

$$\begin{aligned} H &= -\gamma\hbar[H_0 s_z + H_1(s_x \cos\omega t + s_y \sin\omega t)] \\ &= \hbar[\omega_0 s_z + \tfrac{1}{2}\omega_1(s_+ e^{-i\omega t} + s_- e^{i\omega t})]. \end{aligned} \tag{39.1}$$

As in the preceding problem, we will use the density operator formalism. Now the thermal equilibrium (Boltzmann) distribution of the spins (brought about by the weak interactions between the spins and the thermal vibrations of the crystal lattice) is given by the equilibrium density operator:

$$\varrho_t^{(e)} = \frac{\exp\left(-\dfrac{H}{kT}\right)}{\mathrm{Tr}\left[\exp\left(-\dfrac{H}{kT}\right)\right]}. \tag{39.2}$$

If this thermal equilibrium is disturbed, we assume that the "relaxation" processes which tend to restore the equilibrium can be characterized by a relaxation time τ such that the rate of change with time of the spin system density operator ϱ_t, owing to these processes alone, is equal to $\dfrac{\varrho_t - \varrho_t^{(e)}}{\tau}$. Thus, bearing in mind equation (V.16), one can see that, in the presence of the rotating field, the equation of time-evolution of the density operator becomes in fact

$$\frac{\partial \varrho_t}{\partial t}+\frac{\varrho_t-\varrho_t^{(e)}}{\tau} = -\frac{i}{\hbar}[H, \varrho_t], \tag{39.3}$$

i.e.

$$\frac{\partial \Delta(t)}{\partial t} = -\frac{i}{\hbar}[H, \Delta(t)]-\frac{\Delta(t)}{\tau}-\frac{\partial \varrho_t^{(e)}}{\partial t}, \tag{39.4}$$

where $\Delta(t) = \varrho_t^{(t)}-\varrho_t^{(e)}$.

Assuming that the energy of a spin in the external magnetic fields is $\ll kT$, and expanding the exponential function in (39.2) as a power series, equation (39.4) can be written, in terms of the matrix elements of Δ, as

$$\left(\frac{\partial}{\partial t}-i\omega_0+\frac{1}{\tau}\right)\Delta_{12} = -\frac{1}{2}\,i\omega_1 e^{-i\omega t}(\Delta_{22}-\Delta_{11})+\frac{\hbar}{2kT}\,\omega\omega_1 e^{-i\omega t},$$

$$\left(\frac{\partial}{\partial t}+\frac{1}{\tau}\right)\Delta_{11} = \frac{1}{2}\,\omega_1(e^{-i\omega t}\Delta_{21}-e^{i\omega t}\Delta_{12}),$$

$$\Delta_{22} = \Delta_{11}, \quad \Delta_{21} = \Delta_{12}^*.$$

We now seek solutions of these equations in the form

$$\Delta_{11} = R_{11}, \quad \Delta_{12} = P_{12}e^{-i\omega t}, \quad \Delta_{21} = Q_{21}e^{i\omega t}, \quad (Q_{21} = P_{12}^*).$$

After some elementary calculation we find that

$$\begin{aligned} R_{11} &= \frac{\dfrac{\hbar}{2kT}\,\omega\omega_1^2\tau^2}{1+(\omega_0+\omega)^2\tau^2+\omega_1^2\tau^2}, \\ P_{12} &= \frac{\hbar}{2kT}\,i\omega\omega_1\tau\,\frac{1+i(\omega_0+\omega)\tau}{1+(\omega_0+\omega)^2\tau^2+\omega_1^2\tau^2}. \end{aligned} \tag{39.5}$$

In order to determine the real and the imaginary parts of the susceptibility χ, let us calculate

$$\langle M_x\rangle = -N\gamma\hbar\,\mathrm{Tr}\,(\varrho s_x) = -N\gamma\hbar\,\mathrm{Tr}\,[(\varrho_0+\Delta)s_x].$$

We find after some algebra that

$$\langle M_x\rangle = \chi_0 H_1 \cos\omega t+\chi_0\omega\tau H_1\,\frac{\sin\omega t-(\omega_0+\omega)\tau\cos\omega t}{1+(\omega_0+\omega)^2\tau^2+\omega_1^2\tau^2}. \tag{39.6}$$

On the other hand

$$\begin{aligned}(M_x+iM_y) &= \chi(H_x+iH_y) = (\chi'+i\chi'')(H_1\cos\omega t+iH_1\sin\omega t)\\ &= H_1(\chi'\cos\omega t-\chi''\sin\omega t)+iH_1(\chi'\sin\omega t+\chi''\cos\omega t)\end{aligned}$$

whence

$$M_x = H_1(\chi'\cos\omega t-\chi''\sin\omega t). \tag{39.7}$$

By comparing (39.6) with (39.7) we immediately obtain Bloch's relations (39a).

CHAPTER VII

Systems of Identical Particles. Second Quantization

1. Symmetry and Anti-symmetry of State Vectors

Particles of a system are said to be "identical" if the dynamical properties of the system are not changed by any permutation of these particles. Consider a complete set of observables for any one of a system of N identical particles and let $|\psi_{p_i}\rangle$ denote the simultaneous orthonormal eigenvectors of these observables, $\langle\psi_{p_i}|\psi_{p_j}\rangle = \delta_{p_i p_j}$.

Now consider the Hilbert space $\mathcal{E}$ which consists of all linear combinations of the basis vectors†

$$|\psi_{p_1}^{(1)}\psi_{p_2}^{(2)} \ldots \psi_{p_N}^{(N)}\rangle \equiv |\psi_{p_1}\rangle^{(1)}|\psi_{p_2}\rangle^{(2)} \ldots |\psi_{p_N}\rangle^{(N)}. \tag{VII.1}$$

In the state (VII.1) the particle 1 is in the state $|\psi_{p_1}\rangle$, the particle 2 in the state $|\psi_{p_2}\rangle$, and so on. A permutation of the N particles modifies their distribution among the N states $|\psi_{p_1}\rangle, \ldots, |\psi_{p_N}\rangle$. Note that any permutation can be written as a sequence of transpositions of pairs of particles.

Let $|u\rangle$ be a vector of $\mathcal{E}$, and P an operator which performs a permutation of the N particles. The vector $|u\rangle$ is said to be symmetrical if, for any arbitrary P, we have $P|u\rangle = |u\rangle$, and antisymmetrical if $P|u\rangle = (-1)^p |u\rangle$, where p is the number of transpositions equivalent to P. The symmetrical and the anti-symmetrical vectors form mutually orthogonal subspaces in $\mathcal{E}$, the corresponding projection operators being (Fig. VII.1),

$$S = \frac{1}{N!}\sum P, \quad A = \frac{1}{N!}\sum(-1)^p P, \tag{VII.2}$$

where the sums are over all possible permutations P of the particles.

† The vectors of $\mathcal{E}$ are usually specified in practice in a representation in which the position operators $x_i y_i z_i$ and the spin z-component operators s_z^i $(i = 1, 2, \ldots)$ of the particles are all diagonal. In such a representation the vector (VII.1) is represented by a product of one-particle wavefunctions

$$\psi_{p_1}(\xi_1)\,\psi_{p_2}(\xi_2) \ldots \psi_{p_N}(\xi_N) \tag{VII.1'}$$

where ξ_i stands for the position and the spin eigenvalues of the ith particle.

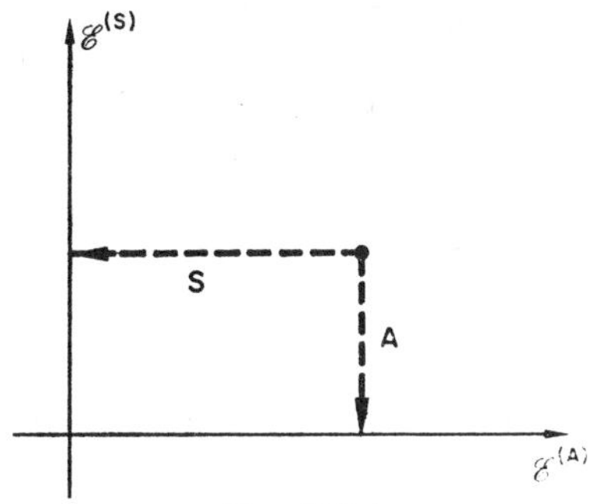

FIG. VII.1.

It follows from (VII.2) that the effect of A on a basis vector can be written in the form

$$A\left|\psi_{p_1}^{(1)}\psi_{p_2}^{(2)}\dots\psi_{p_N}^{(N)}\right\rangle \equiv \frac{1}{N!}\begin{vmatrix} |\psi_{p_1}\rangle^{(1)} & |\psi_{p_1}\rangle^{(2)} & \dots & |\psi_{p_1}\rangle^{(N)} \\ |\psi_{p_2}\rangle^{(1)} & |\psi_{p_2}\rangle^{(2)} & \dots & |\psi_{p_2}\rangle^{(N)} \\ |\psi_{p_N}\rangle^{(1)} & |\psi_{p_N}\rangle^{(2)} & \dots & |\psi_{p_N}\rangle^{(N)} \end{vmatrix} \tag{VII.3}$$

which is called a "Slater determinant".

Particle symmetry rules. By comparing theoretical predictions with experimental observations, it has been found that systems of particles which have spin $\frac{1}{2}$ (electrons, protons, neutrons, etc.) form totally anti-symmetrical states only ("fermions"), and that systems of particles with zero or integer spin (photons, π mesons, etc.) form totally symmetrical states only ("bosons"). No particles are known which can form both types of state, or states having no specified symmetry. Hence, the possible states of fermions, and of bosons, respectively, can be written as

$$\begin{aligned} |\psi\rangle^{(A)} &= A\left|\psi_{p_1}^{(1)}\psi_{p_2}^{(2)}\dots\psi_{p_N}^{(N)}\right\rangle, \\ |\psi\rangle^{(S)} &= S\left|\psi_{p_1}^{(1)}\psi_{p_2}^{(2)}\dots\psi_{p_N}^{(N)}\right\rangle. \end{aligned} \tag{VII.4}$$

The *Pauli exclusion principle* states that no two fermions can ever be simultaneously in the same quantum state; it can be seen to be a consequence of the above considerations.

2. Isotopic Spin

The charge independence of nuclear forces allows us to consider the proton and the neutron as different states of the same particle, or nucleon. Thus, in addition to position and spin variables, the state of a nucleon requires for its specification a "charge" variable which can take two values. Therefore in the study of nucleon systems a formalism similar to the one developed for the eigenstates of spin $\frac{1}{2}$ can be used. It is called the *isotopic spin formalism.*

Briefly, the base vectors of a two-dimensional "isotopic spin space" can be defined to be

$$\chi_p = \begin{pmatrix}1\\0\end{pmatrix}, \quad \chi_n = \begin{pmatrix}0\\1\end{pmatrix}, \tag{VII.5}$$

and the operators representing isotopic spin and nucleon charge are then

$$\mathbf{t} = \tfrac{1}{2}\boldsymbol{\tau}, \quad q = \tfrac{1}{2}(1+\tau_3)e, \tag{VII.6}$$

where the operation of $\boldsymbol{\tau}$ (τ_1, τ_2, τ_3) on the base vectors can be represented by Pauli matrices. For a system of N nucleons we can define a total isotopic spin operator and a total charge operator

$$\mathbf{T} = \sum_i \mathbf{t}^{(i)}, \quad Q = \left(\frac{N}{2}+T_3\right)e. \tag{VII.7}$$

The π-mesons, π^+, π^0, π^- can be considered as constituting three states of a particle having an isotopic spin equal to unity, with base vectors

$$\chi_{\pi^+} = \begin{pmatrix}1\\0\\0\end{pmatrix}, \quad \chi_{\pi^0} = \begin{pmatrix}0\\1\\0\end{pmatrix}, \quad \chi_{\pi^-} = \begin{pmatrix}0\\0\\1\end{pmatrix}. \tag{VII.8}$$

In this case the operations of the three components ($\theta_1, \theta_2, \theta_3$ say) of the isotopic spin operator are given by the matrices derived in problem 11 of Chapter VI.

For a system of π-mesons, we can then define a total isotopic spin operator and a total charge operator:

$$\boldsymbol{\Theta} = \sum_i \boldsymbol{\theta}^{(i)}, \quad Q = \Theta_3. \tag{VII.9}$$

Experimental data show that, in the interaction of π mesons with nucleons, the total isotopic spin $\mathbf{I} = \mathbf{T}+\boldsymbol{\Theta}$ is a constant of the motion, provided that Coulomb forces can be neglected.

3. Second Quantization

Let us denote by $n_1, n_2, \ldots$, the numbers of particles in the states

$$|\psi_{p_1}\rangle, \quad |\psi_{p_2}\rangle, \ldots; \qquad \sum_i n_i = N.$$

In the method of calculation known as *second quantization*, the basis vectors (VII.1) are specified by the corresponding *occupation numbers* $n_1, n_2, \ldots$

$$|\psi_{p_1}^{(1)}\psi_{p_2}^{(2)} \ldots \psi_{p_N}^{(N)}\rangle \to |n_1 n_2 \ldots\rangle. \tag{VII.10}$$

For a system of bosons the appropriate state vectors are the symmetrical linear combinations of basis vectors

$$|n_1 n_2 \ldots\rangle^{(S)} \equiv S|n_1 n_2 \ldots\rangle = \left(\frac{n_1!\,n_2!\ldots}{N!}\right)^{1/2} \sum |\psi_{p_1}^{(1)}\psi_{p_2}^{(2)} \ldots \psi_{p_N}^{(N)}\rangle. \tag{VII.11}$$

Here, in contrast to (VII.2), the summation is over all possible distinct permutations of the indices $p_1, p_2, \ldots, p_N$ (note that some of these may have equal values, in which case no new state is obtained by permuting them).

It is convenient to define the following operators:

(1) the "annihilation" operator

$$a_i|n_1n_2 \ldots n_i \ldots\rangle = \sqrt{n_i}\,|n_1n_2 \ldots n_{i-1} \ldots\rangle, \tag{VII.12}$$

(2) the "creation" operator

$$a_i^+|n_1n_2 \ldots n_i \ldots\rangle = \sqrt{n_i+1}\,|n_1n_2 \ldots n_{i+1} \ldots\rangle, \tag{VII.12$'$}$$

and (3) the "particle number" operator

$$n_i = a_i^+ a_i. \tag{VII.13}$$

The following commutation rules are then valid

$$[a_i, a_i^+] = 1, \quad [a_i, a_k] = 0, \quad [a_i^+, a_k^+] = 0, \quad i \neq k. \tag{VII.14}$$

For a system of fermions, on the other hand, the appropriate state vectors are the anti-symmetrical linear combinations of the basis vectors:

$$|n_1n_2 \ldots\rangle^{(A)} \equiv A|n_1n_2 \ldots\rangle. \tag{VII.15}$$

"Annihilation" and "creation" operators can again be defined, although not quite so straightforwardly as for boson states; they obey the "anti-commutation" rules:

$$\{a_i, a_i^+\} = 1, \quad \{a_i, a_k\} = 0, \quad \{a_i^+, a_k^+\} = 0, \quad i \neq k \tag{VII.16}$$

where

$$\{A, B\} = (AB+BA).$$

The method of second quantization is useful in the study of systems of identical particles in interaction (see problems 20, 21), and is essential in the treatment of systems in which particles can be created or annihilated, e.g. in the proper treatment of electromagnetic emission and absorption (of photons) by charged particles, of electron–positron pair creation and annihilation, of β-decay, etc.

Problems

1. Consider a system consisting of two identical particles, each of which has one-particle states represented in coordinate representation by the wavefunctions $\psi_\alpha(\mathbf{r})$, $\psi_\beta(\mathbf{r})$.

Let us define

$$\begin{aligned} \psi^{(A)}(\mathbf{r}_1, \mathbf{r}_2) &= \frac{1}{\sqrt{2}}[\psi_\alpha(\mathbf{r}_1)\,\psi_\beta(\mathbf{r}_2) - \psi_\alpha(\mathbf{r}_2)\,\psi_\beta(\mathbf{r}_1)], \\ \psi^{(S)}(\mathbf{r}_1, \mathbf{r}_2) &= \frac{1}{\sqrt{2}}[\psi_\alpha(\mathbf{r}_1)\,\psi_\beta(\mathbf{r}_2) + \psi_\alpha(\mathbf{r}_2)\,\psi_\beta(\mathbf{r}_1)], \end{aligned} \tag{1a}$$

i.e. the corresponding anti-symmetrical and symmetrical wavefunctions, respectively, of the system.

Now if the symmetrization rule for identical particles were ignored, the system would in general have the following wavefunction

$$\Psi = \lambda\psi^{(A)}+\mu\psi^{(S)}, \quad |\lambda|^2+|\mu|^2 = 1. \tag{1b}$$

Show that in this case the probability per unit volume of finding a particle at $\mathbf{r}_1$ and another at $\mathbf{r}_2$ depends on λ and on μ, and discuss this result.

2. Show that the symmetrization and the anti-symmetrization operators are orthogonal projection operators, i.e. that $S^2 = S$, $A^2 = A$, $SA = AS = 0$.

3. Find the state vectors of a system of two spin-$\frac{1}{2}$ particles which are simultaneous eigenvectors of the operators $\mathbf{S}^2$ and S_z, where $\mathbf{S}$ is the total spin operator of the system. Discuss the symmetry of these state vectors.

4. Solve the preceding problem in the case of two particles, each of spin unity.

5. Write down the normalized wavefunctions of a system of three identical bosons, which are in given one-particle states.

6. Show that for a system of two identical particles, each of spin s, the ratio of the number of symmetrical to the number of anti-symmetrical spin states is $(s+1)/s$.

7. Show that if the wavefunction of a system of two identical spinless particles is an eigenfunction of the orbital angular momentum of relative motion of the two particles, then the quantum number l necessarily has an even (or zero) value.

8. Use the result of the preceding problem to study the possibility of the decay

$$\mathrm{Be}_4^8 \to \mathrm{He}_2^4+\mathrm{He}_2^4,$$

if the Be nucleus is in an excited state, with a total angular momentum equal to unity.

9. Find the eigenvalues and the eigenvectors of the isotopic spin operator for a system of two nucleons. Knowing that the bound state of the deuteron consists of a mixture of $^3S_1$ and $^3D_1$ states only, deduce that its isotopic spin is zero.

10. Show that if we neglect the difference in mass between the proton and the neutron, and the Coulomb interaction between protons, then the Hamiltonian of a system of two nucleons commutes with the total isotopic spin.

11. Show that, assuming charge independence of nuclear forces to hold, the isotopic spin formalism is equivalent to the formalism which describes a system of nucleons as consisting of protons and of neutrons, treated as two distinct sets of identical particles.

12. Consider the following two possible nuclear reactions:

$$p+d \to d+n+\pi^+,$$
$$p+d \to d+p+\pi^0.$$

Show that the branching ratio of the two reactions is equal to two.

13. Show that the branching ratio of the reactions

$$n+p \rightarrow p+p+\pi^-,$$
$$n+p \rightarrow n+n+\pi^+$$

is unity.

14. Show that the elastic scattering process

$$\pi^+ + p \rightarrow \pi^+ + p$$

can take place only in the isotopic spin state with $T = \frac{3}{2}$, while

$$\pi^+ + n \rightarrow \pi^+ + n$$

can take place in both of the isotopic spin states, with $T = \frac{3}{2}$ and $T = \frac{1}{2}$.

15. The Hamiltonian of the harmonic oscillator (see (17a) Chapter II) can be written as

$$H(\xi) = \frac{\hbar\omega}{2}\left(\xi^2 - \frac{d}{d\xi^2}\right), \quad \xi = x\sqrt{\frac{m\omega}{\hbar}}. \tag{15a}$$

Show that the operators

$$a = \frac{1}{\sqrt{2}}\left(\xi + \frac{d}{d\xi}\right) = \frac{1}{\sqrt{2}}(\xi + ip_\xi), \tag{15b}$$

$$a^+ = \frac{1}{\sqrt{2}}\left(\xi - \frac{d}{d\xi}\right) = \frac{1}{\sqrt{2}}(\xi - ip_\xi) \tag{15b'}$$

can be treated as annihilation and creation operators respectively, and that

$$H = \frac{\hbar\omega}{2}(aa^+ + a^+a) = \hbar\omega\left(a^+a + \tfrac{1}{2}\right). \tag{15c}$$

16. Consider a system of two interacting harmonic oscillators of the same natural frequency. If the interaction energy is proportional to the product of the displacements of the oscillators from their equilibrium positions, study this system in second quantization.

17. Consider a system of N identical bosons, and let L_i be an operator which operates on the variables of the ith particle only. Find the expression for an operator of the form $L = \sum_{i=1}^{N} L_i$ in second quantization.

18. Let the complete set of observables ξ which define the one-particle states of a system of identical particles consist of the positions and the spins of the particles, and let us define the operators

$$\psi(\xi) = \sum_i \psi_i(\xi)a_i, \quad \psi^+(\xi) = \sum_i \psi_i^*(\xi)a_i^+. \tag{18a}$$

Show

(1) that the operators $\psi^+(\xi_0)$ and $\psi(\xi_0)$ create and annihilate respectively a particle at the point ξ_0,

(2) that the following commutation rules are valid if the particles are bosons:

$$\begin{aligned}[\psi(\xi), \psi(\xi')] &= [\psi^+(\xi), \psi^+(\xi')] = 0,\\ [\psi(\xi), \psi^+(\xi')] &= \delta(\xi-\xi') = \delta(\mathbf{r}-\mathbf{r}')\delta_{s'}\end{aligned} \tag{18b}$$

and that the following anti-commutation rules are valid if the particles are fermions:

$$\begin{aligned}\{\psi(\xi), \psi(\xi')\} &= \{\psi^+(\xi), \psi^+(\xi')\} = 0,\\ \{\psi(\xi), \psi^+(\xi')\} &= \delta(\xi-\xi').\end{aligned} \tag{18c}$$

19. Let us suppose that the Hamiltonian of a system of bosons has the form

$$H = \sum_{i=1}^{N}\left(-\frac{\hbar^2}{2m}\nabla_i^2+V(\xi_i)\right)+\tfrac{1}{2}\sum_{i,j=1}^{N} W(\xi_i, \xi_j), \tag{19a}$$

where ξ stands as before for the position and spin variables. By using the operators $\psi(\xi)$, $\psi^+(\xi)$ defined in the preceding problem, show that the Hamiltonian (19a) can be written in the form

$$\begin{aligned}H = &\int\left[\frac{\hbar^2}{2m}\nabla\psi^+(\xi).\nabla\psi(\xi)+V(\xi)\psi^+(\xi)\psi(\xi)\right]d\xi\\ &+\tfrac{1}{2}\int\int\psi^+(\xi)\psi^+(\xi')\, W(\xi, \xi')\,\psi(\xi')\,\psi(\xi)\, d\xi\, d\xi'\end{aligned} \tag{19b}$$

where the integrations include sums over the spin variables.

20. Consider a system of bosons whose Hamiltonian is of the form

$$H = \sum_{\alpha=1}^{N}\left[-\frac{\hbar^2}{2m}\nabla_\alpha^2+\tfrac{1}{2}\sum_{\beta=1}^{N} V(|\mathbf{r}_\alpha-\mathbf{r}_\beta|)\right] \tag{20a}$$

where $V(|\mathbf{r}_\alpha-\mathbf{r}_\beta|)$ is the two-particle interaction energy, and depends only on the distance between the particles. Show that, by choosing as one-particle functions the eigenfunctions of linear momentum with eigenvalue $\mathbf{p}$:

$$\psi_{\mathbf{p}}(\mathbf{r}) = L^{-3/2}e^{\frac{i}{\hbar}\mathbf{p}\cdot\mathbf{r}}, \tag{20b}$$

which are periodic and normalized in a cube of side L, the Hamiltonian (20a) can be written in second quantization in the form

$$H = \sum_k \frac{p_k^2}{2m}a_{\mathbf{p}_k}^+a_{\mathbf{p}_k}+\tfrac{1}{2}\sum\frac{1}{L^3}\nu(\mathbf{p}_l-\mathbf{p}_i)a_{\mathbf{p}_l}^+a_{\mathbf{p}_m}^+a_{\mathbf{p}_i}a_{\mathbf{p}_k} \tag{20c}$$

where $\nu(\mathbf{p}) = \int V(|\mathbf{q}|)e^{-i\mathbf{p}\cdot\mathbf{q}}\, d\mathbf{q}$, and the second summation in (20c) is carried out subject to the condition $\mathbf{p}_l+\mathbf{p}_m = \mathbf{p}_i+\mathbf{p}_k$.

21. Show that, if the interaction between the particles of the system described in the preceding problem is a weak one, the eigenvalues of the Hamiltonian (20c) are given approximately by the expression

$$E = E_0 + \sum_{p \neq 0} \varepsilon(p) \Lambda_{\mathbf{p}}, \tag{21a}$$

where

$$\varepsilon(p) = \sqrt{\frac{N}{L^3} \frac{p^2 \nu(p)}{m} + \frac{p^4}{4m^2}}, \tag{21b}$$

and $\Lambda_{\mathbf{p}}$ is a positive integer or zero for each value of $\mathbf{p}$. Deduce from this result that a non-ideal Bose gas can have superfluid properties.

Solutions

1. The probability per unit volume of finding a particle at $\mathbf{r}_1$ and another at $\mathbf{r}_2$ is a sum of two quantities

$$\begin{aligned} P(\mathbf{r}_1, \mathbf{r}_2) &= |\psi(\mathbf{r}_1, \mathbf{r}_2)|^2 + |\psi(\mathbf{r}_2, \mathbf{r}_1)|^2 \\ &= 2[|\lambda|^2 |\psi^{(A)}(\mathbf{r}_1, \mathbf{r}_2)|^2 + |\mu|^2 |\psi^{(S)}(\mathbf{r}_1, \mathbf{r}_2)|^2], \end{aligned} \tag{1.1}$$

and hence it depends on λ and μ. Thus, although the wavefunction (1b) represents the same physical state for arbitrary λ and μ, the probability density (1.1) depends on these parameters. To avoid this difficulty we could postulate that

$$|\psi^{(A)}(\mathbf{r}_1, \mathbf{r}_2)|^2 = |\psi^{(S)}(\mathbf{r}_1, \mathbf{r}_2)|^2, \tag{1.2}$$

for all $\mathbf{r}_1$, $\mathbf{r}_2$.

But this is impossible since we can write

$$|\psi^{(S)}(\mathbf{r}, \mathbf{r})| = \sqrt{2} |\psi_\alpha(\mathbf{r}) \psi_\beta(\mathbf{r})| \neq 0, \quad \psi^{(A)}(\mathbf{r}, \mathbf{r}) = 0, \tag{1.3}$$

which contradicts (1.2).

The difficulty can be avoided only by resorting to the symmetrization rule according to which the values of the parameters λ and μ are fixed at $\lambda = 0$, $\mu = 1$, or at $\lambda = 1$, $\mu = 0$, according to the type of particle being considered.

2. According to the definition (VII.2)

$$S = \frac{1}{N!} \sum P, \quad A = \frac{1}{N!} \sum (-1)^p P, \tag{2.1}$$

where the summation is to be taken over all $N!$ permutation operators. Let us suppose that these operators have been placed in some particular order and let us multiply them on the right (or on the left) by a given permutation operator P'. This merely changes the order in the arrangement of the $N!$ permutation operators. Owing to the summation in (2.1), it follows that

$$P'S = SP' = S, \quad P'A = AP' = (-1)^{p'}A. \tag{2.2}$$

From (2.1) and (2.2) the relations stated in the problem can easily be obtained.

3. This problem is essentially one of addition of angular momenta, and we can use the Clebsch–Gordan coefficients given in Table VI.2. Let the two one-particle states ($s_z = \pm\frac{1}{2}$) be denoted by

$$\chi_\alpha = \begin{bmatrix} 1 \\ 0 \end{bmatrix}, \quad \chi_\beta = \begin{bmatrix} 0 \\ 1 \end{bmatrix}. \tag{3.1}$$

The solution of the problem is then given in Table VII.1.

TABLE VII.1

State vectors of the system	Total spin S	S_z	Symmetry
$\chi_\alpha^{(1)}\chi_\alpha^{(2)}$	1	+1	Symmetrical
$\frac{1}{\sqrt{2}}(\chi_\alpha^{(1)}\chi_\beta^{(2)} + \chi_\beta^{(1)}\chi_\alpha^{(2)})$	1	0	Symmetrical
$\chi_\beta^{(1)}\chi_\beta^{(2)}$	1	−1	Symmetrical
$\frac{1}{\sqrt{2}}(\chi_\alpha^{(1)}\chi_\beta^{(2)} - \chi_\beta^{(1)}\chi_\alpha^{(2)})$	0	0	Anti-symmetrical

We shall also give an elementary solution of the problem. We rely on the fact that the total spin can have only the values $S = 1$ (with $S_z = -1, 0, 1$) and $S = 0$ (with $S_z = 0$). To show how the state vectors of the system can be determined, consider the case of the two states for which $S_z = 0$. The general form of the state vectors is then

$$a\chi_\alpha^{(1)}\chi_\beta^{(2)} + b\chi_\beta^{(1)}\chi_\alpha^{(2)} = a\begin{bmatrix} 1 \\ 0 \end{bmatrix}_1\begin{bmatrix} 0 \\ 1 \end{bmatrix}_2 + b\begin{bmatrix} 0 \\ 1 \end{bmatrix}_1\begin{bmatrix} 1 \\ 0 \end{bmatrix}_2 \tag{3.2}$$

where a and b are (for the moment) unknown coefficients. By imposing on (3.2) the condition that it be an eigenvector of the operator

$$\mathbf{S}^2 = \mathbf{s}_1^2 + \mathbf{s}_2^2 + 2\mathbf{s}_1.\mathbf{s}_2 = \left\{\frac{3}{2}\begin{bmatrix} 1 & 0 \\ 0 & 1 \end{bmatrix}_1 + \begin{bmatrix} 1 & 0 \\ 0 & 1 \end{bmatrix}_2\right\} + \frac{1}{2}\left\{\begin{bmatrix} 0 & 1 \\ 1 & 0 \end{bmatrix}_1\begin{bmatrix} 0 & 1 \\ 1 & 0 \end{bmatrix}_2 + \begin{bmatrix} 0 & -i \\ i & 0 \end{bmatrix}_1\begin{bmatrix} 0 & -i \\ i & 0 \end{bmatrix}_2 + \begin{bmatrix} 1 & 0 \\ 0 & -1 \end{bmatrix}_1\begin{bmatrix} 1 & 0 \\ 0 & -1 \end{bmatrix}_2\right\} \tag{3.3}$$

with the eigenvalue λ, we find that

$$(\lambda-1)a-b = 0, \quad -a+(\lambda-1)b = 0 \tag{3.4}$$

whence two possible values for λ can be found: $\lambda = 2$ (with $a = b$), and $\lambda = 0$ (with $a = -b$). Taking into account the normalization condition $|a|^2+|b|^2 = 1$, two of the required eigenvectors have thus been obtained. The others can be calculated in a similar way. Note that the vectors given in Table VII.1 are also eigenvectors of the product $\mathbf{s}_1 . \mathbf{s}_2$.

4. To solve this problem we can, as in the preceding problem, either use the Clebsch–Gordan coefficients given in Table VI.1, or proceed in a more elementary way. The result is summarized in Table VII.2, in which χ_α, χ_β, χ_γ are the one-particle states corresponding to $S_z = 1$, 0 and -1 respectively.

TABLE VII.2

State vectors of the system	Total spin S	Total spin S_z	Symmetry
$\chi_\alpha^{(1)}\chi_\alpha^{(2)}$	2	2	Symmetrical
$\frac{1}{\sqrt{2}}(\chi_\alpha^{(1)}\chi_\beta^{(2)}+\chi_\beta^{(1)}\chi_\alpha^{(2)})$	2	1	Symmetrical
$\sqrt{\frac{2}{3}}\left(\chi_\beta^{(1)}\chi_\beta^{(2)}+\frac{1}{2}\chi_\alpha^{(1)}\chi_\gamma^{(2)}+\frac{1}{2}\chi_\gamma^{(1)}\chi_\alpha^{(2)}\right)$	2	0	Symmetrical
$\frac{1}{\sqrt{2}}(\chi_\beta^{(1)}\chi_\gamma^{(2)}+\chi_\gamma^{(1)}\chi_\beta^{(2)})$	2	-1	Symmetrical
$\chi_\gamma^{(1)}\chi_\gamma^{(2)}$	2	-2	Symmetrical
$\frac{1}{\sqrt{2}}(\chi_\alpha^{(1)}\chi_\beta^{(2)}-\chi_\beta^{(1)}\chi_\alpha^{(2)})$	1	1	Anti-symmetrical
$\frac{1}{\sqrt{2}}(\chi_\alpha^{(1)}\chi_\gamma^{(2)}-\chi_\gamma^{(1)}\chi_\alpha^{(2)})$	1	0	Anti-symmetrical
$\frac{1}{\sqrt{2}}(\chi_\beta^{(1)}\chi_\gamma^{(2)}-\chi_\gamma^{(1)}\chi_\beta^{(2)})$	1	-1	Anti-symmetrical
$\frac{1}{\sqrt{3}}(\chi_\beta^{(1)}\chi_\beta^{(2)}-\chi_\alpha^{(1)}\chi_\gamma^{(2)}-\chi_\gamma^{(1)}\chi_\alpha^{(2)})$	0	0	Symmetrical

5. Let $|\psi_{p_1}\rangle$, $|\psi_{p_2}\rangle$, $|\psi_{p_3}\rangle$ be the normalized one-particle states. We have now to distinguish three cases:

(a) All the three occupied states are different, i.e. $p_1 \neq p_2 \neq p_3$. The state vector of the system will then be

$$\begin{aligned}|\psi\rangle^{(S)} = S\,|\psi_{p_1}^{(1)}\psi_{p_2}^{(2)}\psi_{p_3}^{(3)}\rangle = \frac{1}{\sqrt{3!}}\Big[|\psi_{p_1}^{(1)}\psi_{p_2}^{(2)}\psi_{p_3}^{(3)}\rangle \\ + |\psi_{p_1}^{(1)}\psi_{p_3}^{(2)}\psi_{p_2}^{(3)}\rangle + |\psi_{p_3}^{(1)}\psi_{p_2}^{(2)}\psi_{p_1}^{(3)}\rangle + |\psi_{p_3}^{(1)}\psi_{p_1}^{(2)}\psi_{p_2}^{(3)}\rangle \\ + |\psi_{p_2}^{(1)}\psi_{p_1}^{(2)}\psi_{p_3}^{(3)}\rangle + |\psi_{p_2}^{(1)}\psi_{p_3}^{(2)}\psi_{p_1}^{(3)}\rangle\Big]\end{aligned} \tag{5.1}$$

(b) Two of the three filled states are identical, e.g. $p_1 \neq p_2 = p_3$. Then

$$|\psi\rangle^{(S)} = \sqrt{\frac{2!}{3!}}\,\big[|\psi_{p_1}^{(1)}\psi_{p_2}^{(2)}\psi_{p_2}^{(3)}\rangle + |\psi_{p_2}^{(1)}\psi_{p_1}^{(2)}\psi_{p_2}^{(3)}\rangle + |\psi_{p_2}^{(1)}\psi_{p_2}^{(2)}\psi_{p_1}^{(3)}\rangle\big]. \tag{5.2}$$

The factor in front of this expression ensures the normalization of $|\psi\rangle^{(S)}$.

(c) All three particles are in the same state, i.e. $p_1 = p_2 = p_3$. Then

$$|\psi\rangle^{(S)} = |\psi_{p_1}^{(1)}\psi_{p_1}^{(2)}\psi_{p_1}^{(3)}\rangle. \tag{5.3}$$

6. Let χ_i $(i = -s,\ -s+1,\ \ldots,\ s)$ stand for the $(2s+1)$ possible one-particle spin states. A possible set of basis vectors in the space of the total spin will then consist of all vectors of the form $\chi_i^{(1)}\chi_k^{(2)}$, where $i,\ k = -s,\ -s+1,\ \ldots,\ s$; hence their number is $(2s+1)^2$. These vectors do not all have a well-defined symmetry under permutation of the spins, but a set of symmetrical and of anti-symmetrical basis vectors can easily be obtained. Thus, for $i = k$ the $(2s+1)$ vectors $\chi_i^{(1)}\chi_i^{(2)}$ are clearly symmetrical. For $i \neq k$, we can construct $2s(2s+1)$ symmetrical (or anti-symmetrical) combinations $\chi_i^{(1)}\chi_k^{(2)} \pm \chi_k^{(1)}\chi_i^{(2)}$. Since $(2s+1)+2s(2s+1) = (2s+1)^2$, it follows that the vectors

$$\chi_i^{(1)}\chi_i^{(2)}, \quad \chi_i^{(1)}\chi_k^{(2)} \pm \chi_k^{(1)}\chi_i^{(2)}; \qquad i \neq k = -s,\ -s+1,\ \ldots,\ s, \tag{6.1}$$

form a complete set of basis vectors having well-defined symmetry properties. The ratio of the number of symmetrical to the number of anti-symmetrical states is then

$$\frac{(2s+1)+s(2s+1)}{s(2s+1)} = \frac{s+1}{s}.$$

7. Let $\mathbf{r}_1$ and $\mathbf{r}_2$ be the position vectors of the two particles. Let us choose a new system of coordinates with the origin at the point $\frac{\mathbf{r}_1+\mathbf{r}_2}{2}$. We observe then that the operation of inversion of the coordinates in this system corresponds to a permutation of the two particles. Further, if we denote the orbital angular momentum of relative motion by l, then, under inversion of the coordinates, the wavefunction is multiplied by $(-1)^l$ (see problem 8, Chapter VI). Since the particles of the system are spinless and hence are bosons, the wavefunction depends only on the position coordinates of the two particles and must be symmetrical under their permutation. It follows that l must be even or zero.

8. Since α-particles are spinless, it follows from the preceding problem that the two-particle system can be found only in an even state of orbital angular momentum. Therefore, for reasons of parity conservation, the decay is not possible.

9. Since the isotopic spin of each nucleon is $\frac{1}{2}$, this problem is formally equivalent to problem 3; *mutatis mutandis*, Table VII.1 becomes Table VII.3:

TABLE VII.3

State vectors of the system	Total isotopic spin T	T_3	Symmetry	Physical interpretation
$\chi_1^{(1)} = \chi_p^{(1)}\chi_p^{(2)}$	1	+1	Symmetrical	2 protons
$\chi_0^{(1)} = \frac{1}{\sqrt{2}}(\chi_p^{(1)}\chi_n^{(2)} + \chi_n^{(1)}\chi_p^{(2)})$	1	0	Symmetrical	proton + neutron
$\chi_{-1}^{(1)} = \chi_n^{(1)}\chi_n^{(2)}$	1	−1	Symmetrical	2 neutrons
$\chi_0^{(0)} = \frac{1}{\sqrt{2}}(\chi_p^{(1)}\chi_n^{(2)} - \chi_n^{(1)}\chi_p^{(2)})$	0	0	Anti-symmetrical	proton + neutron

The component states of the bound deuteron, ${}^3S_1$ and ${}^3D_1$, correspond to orbital angular momenta $l = 0$ and $l = 2$ respectively, and, in consequence, the deuteron wavefunction is even in the coordinate variables (see problem 7). In addition, since the deuteron spin is equal to unity, the spin-dependent part of its wavefunction is symmetrical in the variables. In order to have a totally anti-symmetrical wavefunction, the isotopic-spin-dependent part has to be anti-symmetrical, which is the case if $T = 0$.

10. The conditions of the problem suggest that the forces between two nucleons are the same (for a given two-particle state) whether these nucleons are in fact neutrons and/or protons. Let us see how this hypothesis can be expressed in terms of isotopic spin. Systems consisting of two protons (or two neutrons) are described in isotopic spin space by the symmetrical state vectors $\chi_1^{(1)}$ or $\chi_{-1}^{(1)}$ (see Table VII.3). Systems consisting of a proton and a neutron are described by the state vectors $\chi_0^{(1)}$, $\chi_0^{(0)}$. All but $\chi_0^{(0)}$ are symmetrical, the latter being anti-symmetrical. Each will appear in wavefunctions multiplied by a position and spin function having the opposite symmetry, to preserve overall anti-symmetry. Now if we postulate that the Hamiltonian of the system is to be invariant under rotation of axes in isotopic spin space, it follows that the potential energy of interaction of the nucleons must be the same in the three states $\chi_1^{(1)}$, $\chi_{-1}^{(1)}$ and $\chi_0^{(1)}$, but may have a different value in the state $\chi_0^{(0)}$. This is equivalent to the "charge independence" hypothesis made above, which can be seen to be the same as the hypothesis that the purely nuclear forces may depend on the total isotopic spin but that they do not depend on its "z-component" projection T_3. The interaction Hamiltonian between nucleons therefore contains the isotopic spin operators only in the form of $\mathbf{T}^2$. However, since $\mathbf{T}^2 = (\mathbf{t}_1+\mathbf{t}_2)^2 = \mathbf{t}_1^2+\mathbf{t}_2^2+2\mathbf{t}_1.\mathbf{t}_2$, and since the eigenvalues of $\mathbf{t}_1^2$ and of $\mathbf{t}_2^2$ are equal to $\frac{3}{4}$, the dependence of the interaction Hamiltonian on the isotopic spin variables can be written in the form

$$H' = U_1+(\mathbf{t}_1.\mathbf{t}_2)U_2 \tag{10.1}$$

where U_1 and U_2 do not depend on the isotopic spin variables.†

Since, in the expression (10.1), there appears only the scalar product $\mathbf{t}_1.\mathbf{t}_2$, H' is clearly invariant under any rotation in isotopic spin space. Now the operator for such a rotation involves in general all the components of the total isotopic spin operator, and since H' commutes with this operator it follows that H' commutes with all the components of $\mathbf{T}$, i.e. that $[H', T_1] = [H', T_2] = [H', T_3] = [H', \mathbf{T}^2] = 0$. Note that the above arguments are no longer valid if we cannot neglect terms in the Hamiltonian which depend on the Coulomb interaction between protons and on the proton–neutron mass difference. However, even in this case, we still have $[H', T_3] = 0$, since this is in fact simply an expression of the law of conservation of (electric) charge.

11. Consider a system of N nucleons, of which N_p are protons and N_n are neutrons, and suppose that the protons have been labelled with the numbers 1 to N_p and the neutrons with the numbers N_p+1 to N_p+N_n $(= N)$. Let us denote the anti-symmetrizing operator of the N_p protons by A_{N_p}, the anti-symmetrizing operator of the N_n neutrons by A_{N_n}, and let the state vector of the whole system be $|\phi\rangle$. We have then that

$$A_{N_p}|\phi\rangle = |\phi\rangle, \quad A_{N_n}|\phi\rangle = |\phi\rangle. \tag{11.1}$$

Passing to the isotopic spin formalism, we denote by $|\boldsymbol{\Phi}\rangle$ the state vector of the system of nucleons, and by A the total anti-symmetrizing operator which, in contrast to A_{N_p} and A_{N_n}, can include permutations which change protons into neutrons and vice versa. From (VII.2) and (VII.4) we have that

$$A|\boldsymbol{\Phi}\rangle = |\boldsymbol{\Phi}\rangle, \quad A = \frac{1}{N!}\sum_P (-1)^p P. \tag{11.2}$$

Since any permutation P is performed both on the charge variables and on the other variables, it can be written in the form $P = P_o P_c$, where P_c is a permutation of the charge variables

† It is true that H' could depend on higher powers of $\mathbf{T}^2$. Such a dependence would add to (10.1) higher powers of the product $\mathbf{t}_1.\mathbf{t}_2$, which, however, can be expressed linearly in terms of $\mathbf{t}_1.\mathbf{t}_2$ (see problem 22, Chapter VI).

and P_o is the same permutation applied to the other variables. Hence

$$A = \frac{1}{N!} \sum_P (-1)^p P_o P_c \tag{11.2'}$$

We have now to show that a one-to-one correspondence can be established between the state vectors $|\phi\rangle$ and $|\Phi\rangle$ which leaves scalar products unchanged. Consider a state vector $|\phi\rangle$ which satisfies condition (11.1), and let $|\zeta\rangle$ be a vector which represents a charge state in which the first N_p particles are protons and the following N_n are neutrons, thus

$$|\zeta\rangle = |p^{(1)}p^{(2)} \dots p^{(N_p)};\ n^{(N_p+1)} \dots n^{(N)}\rangle. \tag{11.3}$$

The vector given (in the isotopic spin formalism) by

$$|\boldsymbol{\Phi}\rangle = \sqrt{\frac{N!}{N_p!N_n!}}\, A\,|\phi\rangle|\zeta\rangle \tag{11.4}$$

hen satisfies the first relation (11.2). Conversely, for a given $|\boldsymbol{\Phi}\rangle$, we can show that

$$|\phi\rangle = \sqrt{\frac{N!}{N_p!N_n!}}\,\langle\zeta|\boldsymbol{\Phi}\rangle. \tag{11.5}$$

Thus, by (11.4)

$$\sqrt{\frac{N!}{N_p!N_n!}}\,\langle\zeta|\boldsymbol{\Phi}\rangle = \sqrt{\frac{N!}{N_p!N_n!}}\,\langle\zeta|A|\boldsymbol{\Phi}\rangle = \frac{1}{N_p!N_n!}\sum_P (-1)^p P_o|\phi\rangle\langle\zeta|P_c|\zeta\rangle. \tag{11.6}$$

Now the $N!$ permutations in (11.6) can be divided into two classes. The first class contains permutations, which we denote by F, say, which do not exchange nucleons between the proton and the neutron states. The $N_p!N_n!$ "F-permutations" therefore leave the vector $|\zeta\rangle$ unchanged, and transform $|\phi\rangle$ into $(-1)^f|\phi\rangle$, and hence, writing $F = F_o F_c$,

$$\langle\zeta|F_c|\zeta\rangle = 1, \quad F_o|\phi\rangle = (-1)^f|\phi\rangle. \tag{11.7}$$

The permutations in the second class, which we denote by $G = G_o G_c$ say, exchange nucleons between the proton and the neutron states, so that $\langle\zeta|G_c|\zeta\rangle = 0$. In consequence the summation in (11.6) contains only $N_p!N_n!$ non-vanishing terms, all equal to $|\phi\rangle$. Hence (11.5) is verified.

Finally, let us show that scalar products are unchanged. By analogy with (11.4), we have

$$|\boldsymbol{\Omega}\rangle = \sqrt{\frac{N!}{N_p!N_n!}}\, A\,|\omega\rangle|\zeta\rangle$$

and using (11.5) it follows that

$$\langle\boldsymbol{\Omega}|\boldsymbol{\Phi}\rangle = \sqrt{\frac{N!}{N_p!N_n!}}\,\langle\omega|\langle\zeta|\boldsymbol{\Phi}\rangle = \langle\omega|\phi\rangle. \tag{11.8}$$

12. Since $T = T_3 = 0$ for the deuteron which appears in the initial and in the final states (see problem 9), it can be ignored when conservation of isotopic spin is being considered. Hence, when calculating the consequence of this conservation, the reaction can be written schematically as

$$\text{proton } (T = \tfrac{1}{2}) \rightarrow \text{nucleon } (T = \tfrac{1}{2}) + \pi \text{ meson } (T = 1). \tag{12.1}$$

For the left-hand side of (12.1) we have $T = T_3 = \frac{1}{2}$, so that for the right-hand side we must also have $T = T_3 = \frac{1}{2}$. Using the notation of section 2 and the Clebsch–Gordan coefficients given in Table VI.2, it can be seen that the wavefunction of the nucleon–π meson system has to have the form

$$-\sqrt{\tfrac{1}{3}}\chi_p\chi_{\pi^0} + \sqrt{\tfrac{2}{3}}\chi_n\chi_{\pi^+}. \tag{12.2}$$

The squares of the coefficients in (12.2) give the relative probabilities of the reactions $p \rightarrow p+\pi^0$ and $p \rightarrow n+\pi^+$. The required result follows immediately.

13. Since the wavefunction of a system of two protons (or of two neutrons) is antisymmetrical with respect to simultaneous position and spin permutation, it has to be symmetrical with respect to the isotopic spin variable. Therefore, since T can have only the values 0 or 1, and $T \geqslant |T_3|$, and, for the system $p+p$, $T_3 = 1$ (and $T_3 = -1$ for the system $n+n$), it follows that the only possible value of T is unity. Thus, the wavefunctions of the $p+p$ and of the π^+ meson systems have the same isotopic spin quantum numbers. A similar situation occurs for the system $n+n$ and the π^- meson: the right-hand sides of the two reactions are described by the same function of the isotopic spin variables; hence the two reactions which can result from the same initial state $\chi_p\chi_n$, will have equal probability. The equality of the cross-sections expresses in fact the charge symmetry of the nucleon–nucleon and nucleon–pion interactions, i.e. the invariance of these interactions under the simultaneous changes $n \rightarrow p$, $p \rightarrow n$, $\pi^+ \rightarrow \pi^-$, $\pi^- \rightarrow \pi^+$.

14. For the first process $T_3 = \frac{3}{2}$ and, since $|T_3| \leqslant T$, from the possible values $T = \frac{3}{2}$ and $T = \frac{1}{2}$ we can have only $T = \frac{3}{2}$. For the second process $T_3 = \frac{1}{2}$ and therefore, for T, both of the values $T = \frac{3}{2}$ and $T = \frac{1}{2}$ are possible.

15. The Hamiltonian can easily be put in the form (15c). For the remainder of the problem we require the relation (A.4) between Hermite polynomials:

$$\frac{dH_n}{d\xi} = 2nH_{n-1}(\xi). \tag{15.1}$$

Now the wavefunction $\psi_n(\xi)$ for the state of energy $E_n = \hbar\omega(n+\frac{1}{2})$ is given by

$$\psi_n(\xi) = N_n e^{-(\xi^2/2)} H_n(\xi) \tag{II.17.11}$$

and hence

$$\frac{\partial \psi_n}{\partial \xi} = 2\sqrt{\frac{n}{2}}\,\psi_{n-1} - \xi\psi_n. \tag{15.2}$$

Using now the relation

$$\xi\psi_n = \sqrt{\frac{n}{2}}\,\psi_{n-1} + \sqrt{\frac{n+1}{2}}\,\psi_{n+1} \tag{15.3}$$

it follows that

$$\frac{1}{\sqrt{2}}\left(\xi + \frac{\partial}{\partial \xi}\right)\psi_n = \sqrt{n}\,\psi_{n-1}, \quad \frac{1}{\sqrt{2}}\left(\xi - \frac{\partial}{\partial \xi}\right)\psi_n = \sqrt{n+1}\,\psi_{n+1}, \tag{15.4}$$

and hence that

$$a\psi_n = \sqrt{n}\,\psi_{n-1}, \quad a^+\psi_n = \sqrt{n+1}\,\psi_{n+1}. \tag{15.5}$$

Let us show now that, in the formalism of second quantization, a and a^+ can be regarded as annihilation and creation operators. ξ is now of course not related to the coordinate of a particle, but to the amplitude, e.g., of a plane wave forming part of the decomposition of some field, say a "phonon" field consisting of all elastic waves in a solid. The function $\psi_n(\xi)$ of this partial wave amplitude, with $n = 1$, is the "wavefunction" of a "state" of the partial wave of energy $E_1 = \hbar\omega(1+\frac{1}{2}) = E_0+\hbar\omega$, which differs from E_0 by a quantum of energy $\hbar\omega$. This we can assign to the existence of a phonon of energy $\hbar\omega$ in the partial wave considered. Similarly the state with $n = 2$ corresponds to $E_2 = E_0+2\hbar\omega$, which requires the existence of two phonons each of energy $\hbar\omega$. In general, the state ψ_n, with $E_n = E_0+n\hbar\omega$, corresponds to the existence of n phonons of energy $\hbar\omega$ in the partial wave considered and, consequently, in the second quantization formalism, it is written as $|n\rangle$. The relations (15.5) then become

$$a|n\rangle = \sqrt{n}\,|n-1\rangle, \quad a^+|n\rangle = \sqrt{n+1}\,|n+1\rangle, \quad [a, a^+] = 1 \tag{15.5'}$$

i.e. a and a^+ are annihilation and creation operators of phonons of energy $\hbar\omega$. It can be verified directly that a^+a is the phonon number operator. The state which does not contain any phonon is called the vacuum state and is denoted by $|0\rangle$. It can be defined also through the relation $a|0\rangle = 0$. It can be observed that, by starting from the vacuum state and using creation operators, we can construct any other state, thus:

$$|n\rangle = \frac{1}{\sqrt{n!}}(a^+)^n|0\rangle. \tag{15.6}$$

Phonons are in fact Bose particles of spin 1, and hence the values of n are not restricted. Note that, in this representation, "wave-packet" states of the partial wave amplitude ξ, in which the mean value of ξ oscillates with the classical frequency ω, describe the classical limit in which all field amplitudes are well-defined periodic functions of time. There is no such representation, and no corresponding classical limit, for fermion fields.

16. The Hamiltonian of the system of two oscillators can be written as:

$$H(x_1, x_2) = H(x_1)+H(x_2)+\lambda x_1 x_2 \tag{16.1}$$

where $H(x_1)$ and $H(x_2)$ are the Hamiltonians of the two oscillators. By introducing the notation

$$\xi_1 = \sqrt{\frac{m\omega}{\hbar}}\, x_1, \quad \xi_2 = \sqrt{\frac{m\omega}{\hbar}}\, x_2,$$

(16.1) becomes

$$H(\xi_1, \xi_2) = H(\xi_1)+H(\xi_2)+\frac{\lambda\hbar}{m\omega}\,\xi_1\xi_2, \tag{16.2}$$

where

$$H(\xi_1) = \frac{\hbar\omega}{2}\left(\xi_1^2-\frac{\partial^2}{\partial\xi_1^2}\right), \quad H(\xi_2) = \frac{\hbar\omega}{2}\left(\xi_2^2-\frac{\partial^2}{\partial\xi_2^2}\right). \tag{16.3}$$

By a further change of variables to the "normal coordinates" η_1, η_2,

$$\eta_1 = \sqrt{\frac{\omega_1}{2\omega}}\,(\xi_1+\xi_2), \quad \eta_2 = \sqrt{\frac{\omega_2}{2\omega}}\,(\xi_1-\xi_2) \tag{16.4}$$

where $\omega_1^2 = \omega^2+\lambda/m$, $\omega_2^2 = \omega^2-\lambda/m$, the Hamiltonian (16.2) becomes

$$H(\eta_1, \eta_2) = \frac{\hbar\omega_1}{2}\left(\eta_1^2-\frac{\partial^2}{\partial\eta_1^2}\right)+\frac{\hbar\omega_2}{2}\left(\eta_2^2-\frac{\partial^2}{\partial\eta_2^2}\right). \tag{16.5}$$

To proceed further we follow the lines of the solution of the preceding problem. Define

$$A_1 = \frac{1}{\sqrt{2}}\left(\eta_1+\frac{\partial}{\partial\eta_1}\right), \quad A_1^+ = \frac{1}{\sqrt{2}}\left(\eta_1-\frac{\partial}{\partial\eta_1}\right) \tag{16.6}$$

$$A_2 = \frac{1}{\sqrt{2}}\left(\eta_2+\frac{\partial}{\partial\eta_2}\right), \quad A_2^+ = \frac{1}{\sqrt{2}}\left(\eta_2-\frac{\partial}{\partial\eta_2}\right). \tag{16.6'}$$

The operators A_1, A_1^+, A_2, A_2^+ are then the annihilation and the creation operators of phonons of energies $\hbar\omega_1$ and $\hbar\omega_2$. The Hamiltonian (16.1) becomes

$$H = \hbar\omega_1 A_1^+ A_1+\hbar\omega_2 A_2^+ A_2+\frac{\hbar}{2}(\omega_1+\omega_2), \tag{16.7}$$

and a state of the system can be represented by the vector $|N_1N_2\rangle$, with energy $E = \hbar\omega_1 N_1+ +\hbar\omega_2 N_2+\frac{\hbar}{2}(\omega_1+\omega_2)$, where N_1 and N_2 are the eigenvalues of the operators $A_1^+A_1$, $A_2^+A_2$, and represent the number of phonons of energy $\hbar\omega_1$ and $\hbar\omega_2$ respectively. The commutation rules

$$[A_1, A_2] = [A_1^+, A_2^+] = 0, \quad [A_i, A_j^+] = \delta_{ij}; \quad i, j = 1, 2$$

are satisfied.

17. Let us find the matrix elements

$$\langle n_1', n_2' \dots | L | n_1 n_2 \dots \rangle. \tag{17.1}$$

Since $L = \sum_{i=1}^{N} L_i$, we need only consider a term of the form

$$\langle n_1' n_2' \dots | L_i | n_1 n_2 \dots \rangle. \tag{17.2}$$

By using the relation (VII.11) we obtain

$$L_i | n_1 n_2 \dots \rangle = \left(\frac{n_1! \, n_2! \dots n_k! \dots}{N!} \right)^{1/2} L_i \sum | \psi_{p_1}^{(1)} \psi_{p_2}^{(2)} \dots \psi_{p_N}^{(N)} \rangle,$$

where the sum is over all possible permutations of the indices $p_1, p_2, \dots, p_N$. But we have also that

$$L_i \sum | \psi_{p_1}^{(1)} \psi_{p_2}^{(2)} \dots \psi_{p_N}^{(N)} \rangle = \sum | \psi_{p_1}^{(1)} \rangle | \psi_{p_2}^{(2)} \rangle \dots L_i | \psi_{p_i}^{(i)} \rangle \dots | \psi_{p_N}^{(N)} \rangle$$

and consequently, in the double sum implied by (17.2), the only non-vanishing terms are those which contain products of one-particle state vectors which correspond (except for the ith) to the same one-particle states. Neglecting for the moment the normalization factors, these terms have the form

$$(L_i)_{lk} = \langle \psi_{p_l}^{(i)} | L_i | \psi_{p_k}^{(i)} \rangle. \tag{17.3}$$

But these are, for $l \neq k$, the matrix elements for transitions of the particle from the kth to the lth state. Thus the number of particles in the kth state is decreased by one, and the number in the lth state is increased by one. Taking into account the normalization factors, the corresponding non-vanishing matrix element is

$$\begin{aligned} &\langle \dots n_k - 1 \dots n_l \dots | L_i | \dots n_k \dots n_l - 1 \dots \rangle \\ &= \left(\frac{n_1! \dots (n_k-1)! \dots n_l! \dots}{N!} \right)^{1/2} \left(\frac{n_1! \dots n_k! \dots (n_l-1)! \dots}{N!} \right)^{1/2} \times \\ &\times \frac{(N-1)!}{n_1! \dots (n_k-1)! \dots (n_l-1)! \dots} (L_i)_{lk} = \frac{\sqrt{n_k n_l}}{N} (L_i)_{lk}. \end{aligned} \tag{17.4}$$

In (17.4) all possible permutations of $(N-1)$ particles (i.e. of all particles except the ith particle) have been taken into account. From (17.4) we obtain immediately

$$\langle \dots n_k - 1 \dots n_l \dots | L | \dots n_k \dots n_l - 1 \dots \rangle = N \frac{\sqrt{n_k n_l}}{N} (L_i)_{lk} = \sqrt{n_k n_l} (L_i)_{lk}. \tag{17.5}$$

For $l = k$ we find in a similar manner that

$$\langle n_1 n_2 \dots | L | n_1 n_2 \dots \rangle = \sum_l n_l (L_i)_{ll}. \tag{17.6}$$

Remembering the properties of the operators a and a^+, we can see, from (17.5) and (17.6), that L can be written in the form

$$L = \sum_{k,l} (L_i)_{lk} a_l^+ a_k . \tag{17.7}$$

Remarks: In a similar manner it can be shown that, in the second quantization representation, an operator of the form $L = \sum_{i,j=1}^{N} L_{ij}$, where L_{ij} acts simultaneously on the ith and on the jth particles, can be written in the form

$$L = \sum_{k,p,l,m} (L_{ij})_{lm,\,kp} a_l^+ a_m^+ a_k a_p , \tag{17.8}$$

where

$$(L_{ij})_{lm,\,kp} = \langle \psi_l \psi_m | L_{ij} | \psi_k \psi_p \rangle .$$

In particular, if the Hamiltonian of the system is written in terms of the variables ξ_i (the position and spin of the ith particle) in the form

$$H = \sum_{i=1}^{N} \left[-\frac{\hbar^2}{2m} \nabla_i^2 + V(\xi_i) \right] + \frac{1}{2} \sum_{i,j=1}^{N} W(\xi_i, \xi_j) = \sum_{i=1}^{N} H_i + \frac{1}{2} \sum_{i,j=1}^{N} W(\xi_i, \xi_j), \tag{17.9}$$

then, by (17.7) and (17.8), we can obtain its form in the second quantization formalism:

$$H = \sum_{k,l} (H_i)_{lk} a_l^+ a_k + \tfrac{1}{2} \sum_{k,p,l,m} (W)_{lm,\,kp} a_l^+ a_m^+ a_k a_p . \tag{17.10}$$

If the states $|\psi_k\rangle$ are chosen to be the eigenvectors of H_i, then

$$H = \sum_k E_k a_k^+ a_k + \tfrac{1}{2} \sum_{k,p,l,m} (W)_{lm,\,kp} a_l^+ a_m^+ a_k a_p , \tag{17.11}$$

where E_k is the energy of the particle in the kth state, neglecting interactions except in so far as these can be incorporated in the "smoothed out" potential V.

18. Let us show that the operator $\psi^+(\xi_0)$ creates a particle at the point ξ_0. We already know that the operator a_i^+ creates a particle in the state represented by the wavefunction $\psi_i(\xi)$. It follows that the operator $\psi^+(\xi_0)$ creates a particle in a state which is represented by the wavefunction $\sum_i \psi_i^*(\xi_0)\psi_i(\xi)$. But, in accordance with the closure relation (6.1) of Chapter I, we have $\sum_i |\psi_i\rangle\langle\psi_i| = 1$ and so, if we multiply the latter on the right and on the left by the vectors $|\xi\rangle$, $|\xi_0\rangle$, respectively, we find that

$$\sum_i \psi_i^*(\xi_0)\,\psi_i(\xi) = \langle \xi_0 | \xi \rangle = \delta(\xi_0 - \xi) = \delta(\mathbf{r}_0 - \mathbf{r}) \delta_{s_z^0 s_z} , \tag{18.1}$$

which represents the wavefunction of a particle created at ξ_0 (with spin z-component s_z^0). Similarly it can be shown that $\psi(\xi_0)$ annihilates a particle at ξ_0. The commutation relations (18b) and (18c) follow immediately from (VII.14) and (VII.16), using (18.1).

19. Consider in general an operator of the form $L = \sum_{i=1}^{N} L_i$, where L_i acts only on the ith particle (cf. problem 17). From (17.7) and (18a) we have that

$$L = \sum_{k,l} (L)_{lk} a_l^+ a_k = \sum_{k,l} \int \psi_l^*(\xi)\, L\psi_k(\xi)\, d\xi a_l^+ a_k = \int \psi^+(\xi)\, L\psi(\xi)\, d\xi \tag{19.1}$$

and, similarly, if L has the form $L = \sum_{i,j=1}^{N} L_{ij}$, then

$$L = \iint \psi^+(\xi)\, \psi^+(\xi')\, L\psi(\xi')\, \psi(\xi)\, d\xi\, d\xi'\,. \tag{19.2}$$

It is interesting to note that, for any physical quantity $f(\xi)$ which is a function of ξ only, we can write

$$f = \int f(\xi)\, \psi^+(\xi)\, \psi(\xi)\, d\xi \tag{19.1'}$$

and thus the product $\psi^+(\xi)\,\psi(\xi)\, d\xi$ can be regarded as the number (or, more strictly, the number operator) of particles to be found in the "volume" $d\xi$. From (19.1) and (19.2), after an integration by parts, we can obtain (19b). Similar expressions can be found for systems of fermions.

20. We shall make use of expression (17.11). It is natural to choose for the one-particle states the eigenvectors of the operator $H_\alpha = -\frac{\hbar^2}{2m}\nabla_\alpha^2$. Because the sum in (17.11) is over a discrete spectrum, and the spectrum of the operator H_α is continuous, we shall resort to the usual quantization procedure in a cube of side L. The momentum of the particle can then take only the values

$$p_x = \frac{2\pi\hbar n_x}{L}, \quad p_y = \frac{2\pi\hbar n_y}{L}, \quad p_z = \frac{2\pi\hbar n_z}{L}; \qquad (n_x, n_y, n_z) = \text{whole numbers},$$

and the one-particle wavefunctions are given by (20b). The matrix elements of the interaction operator $W = \sum_{\alpha,\beta} V(|\mathbf{r}_\alpha - \mathbf{r}_\beta|)$ are given by

$$W_{lm,\,ik} = \frac{1}{V^2} \int e^{-\frac{i}{\hbar}(\mathbf{p}_l\cdot\mathbf{r}_1 + \mathbf{p}_m\cdot\mathbf{r}_2)} V(|\mathbf{r}_1 - \mathbf{r}_2|) e^{\frac{i}{\hbar}(\mathbf{p}_i\cdot\mathbf{r}_1 + \mathbf{p}_k\cdot\mathbf{r}_2)}\, d\mathbf{r}_1\, d\mathbf{r}_2\,,$$

where $V = L^3$.

By introducing the new variable $\mathbf{q} = \mathbf{r}_1 - \mathbf{r}_2$ and bearing in mind that

$$\int_V e^{\frac{i}{\hbar}\mathbf{p}\cdot\mathbf{r}}\, d\mathbf{r} = \begin{cases} V & \text{for} \quad \mathbf{p} = 0 \\ 0 & \text{for} \quad \mathbf{p} \neq 0 \end{cases}$$

we find that

$$W_{lm,\,ik} = \begin{cases} \dfrac{1}{V}\, \nu(\mathbf{p}_l - \mathbf{p}_i) & \text{for} \quad \mathbf{p}_l + \mathbf{p}_m = \mathbf{p}_i + \mathbf{p}_k \\ 0 & \text{for} \quad \mathbf{p}_l + \mathbf{p}_m \neq \mathbf{p}_i + \mathbf{p}_k \end{cases}$$

where $\nu(\mathbf{p}) = \int V(|\mathbf{q}|) e^{-i\mathbf{p}\mathbf{q}}\, d\mathbf{q}$. The relation (20c) then follows immediately.

21. Let us suppose for a start that there is no interaction between the particles. If we denote by $n_{\mathbf{p}}$ the number of particles having momentum $\mathbf{p}$, we have $n_{\mathbf{p}} = 0$ if $\mathbf{p} \neq 0$ and $n_{\mathbf{p}} = N$ if $\mathbf{p} = 0$. Therefore, even if there is a weak interaction between the particles, the greater part of them will be found in the state of zero momentum. Using annihilation and creation operators we can write $a_0^+ a_0 = n_0 \approx N$, $a_{\mathbf{p}}^+ a_{\mathbf{p}} = n_{\mathbf{p}} \ll N$. The operators a_0^+ and a_0 satisfy the commutation rule $a_0 a_0^+ - a_0^+ a_0 = 1$, but since $a_0^+ a_0 = n_0 \gg 1$, we can neglect the non-commutativity of these operators, and treat them as ordinary quantities. Also, since $n_{\mathbf{p}} \ll n_0$, we can neglect the terms in the Hamiltonian (20c) which contain products of three or more operators $a_{\mathbf{p}}$ (or $a_{\mathbf{p}}^+$) with $\mathbf{p} \neq 0$, and retain only terms of the form $a_0^+ a_0 a_{\mathbf{p}} a_{-\mathbf{p}}$.

Finally, since

$$\nu(\mathbf{p}) = \int_0^\infty V(|\mathbf{q}|) q^2\, dq \int_0^\pi e^{-ipq\cos\theta} \sin\theta\, d\theta \int_0^{2\pi} d\phi = 4\pi \int V(|\mathbf{q}|) p^{-1} q\, dq \equiv \nu(p),$$

we obtain

$$\begin{aligned} H = \sum_{\mathbf{p}} \frac{\mathbf{p}^2}{2m} a_{\mathbf{p}}^+ a_p + \nu(0) \left[\frac{1}{2} \frac{n_0^2}{V} + \frac{n_0}{V} \sum_{\mathbf{p} \neq 0} a_{\mathbf{p}}^+ a_{\mathbf{p}} \right] + \frac{n_0}{V} \sum_{\mathbf{p} \neq 0} \nu(p) a_{\mathbf{p}}^+ a_{\mathbf{p}} \\ + \frac{a_0^2}{2V} \sum_{\mathbf{p} \neq 0} \nu(p) a_{\mathbf{p}}^+ a_{-\mathbf{p}}^+ + \frac{(a_0^+)^2}{2V} \sum_{\mathbf{p} \neq 0} \nu(p) a_{\mathbf{p}} a_{-\mathbf{p}} . \end{aligned} \tag{21.1}$$

The Hamiltonian (21.1) can be put into a simpler form if we note that

$$\sum_{p \neq 0} a_{\mathbf{p}}^+ a_{\mathbf{p}} = N - n_0 \approx 0.$$

Then

$$\begin{aligned} H = \sum_{\mathbf{p} \neq 0} \left[\frac{\mathbf{p}^2}{2m} + \frac{n_0}{V} \nu(p) \right] a_{\mathbf{p}}^+ a_{\mathbf{p}} + \frac{1}{2} \frac{N^2}{V} \nu(0) + \frac{a_0^2}{2V} \sum_{\mathbf{p} \neq 0} \nu(p) a_{\mathbf{p}}^+ a_{-\mathbf{p}}^+ \\ + \frac{(a_0^+)^2}{2V} \sum_{\mathbf{p} \neq 0} \nu(p) a_{\mathbf{p}} a_{-\mathbf{p}} . \end{aligned} \tag{21.2}$$

Let us now introduce operators $b_{\mathbf{p}}$ and $b_{\mathbf{p}}^+$ related to $a_{\mathbf{p}}$ and $a_{\mathbf{p}}^+$ by the relations

$$\frac{a_0^+ a_{\mathbf{p}}}{\sqrt{n_0}} = u_{\mathbf{p}} b_{\mathbf{p}} + v_{\mathbf{p}} b_{-\mathbf{p}}^+ ; \quad \frac{a_0 a_{\mathbf{p}}^+}{\sqrt{n_0}} = u_{\mathbf{p}} b_{\mathbf{p}}^+ + v_{\mathbf{p}} b_{-\mathbf{p}} , \tag{21.3}$$

where, for the moment, $u_{\mathbf{p}}$ and $v_{\mathbf{p}}$ are unknown functions of $\mathbf{p}$.

We shall determine these functions in such a way that the following two conditions are satisfied:

(a) The operators $b_{\mathbf{p}}$ and $b_{\mathbf{p}}^+$ have the usual commutators

$$\begin{aligned} [b_{\mathbf{p}}, b_{\mathbf{p}'}^+] = \delta_{\mathbf{p}\mathbf{p}'} \\ [b_{\mathbf{p}}, b_{\mathbf{p}'}] = [b_{\mathbf{p}}^+, b_{\mathbf{p}'}^+] = 0 \end{aligned} \tag{21.4}$$

(b) The Hamiltonian (21.2) transforms into a diagonal form

$$H = H_0 + \sum_{\mathbf{p}\neq 0} \varepsilon(p) b_{\mathbf{p}}^{+} b_{\mathbf{p}} \tag{21.5}$$

where $H_0 = (1/2)\,[(\nu(0))/V]\,N^2$.

It can be shown very easily that, if the Hamiltonian has the form (21.5), the following commutation rules are valid

$$\begin{aligned} [b_{\mathbf{p}}, H'] &= \varepsilon(p) b_{\mathbf{p}} \\ [b_{\mathbf{p}}^{+}, H'] &= -\varepsilon(p) b_{\mathbf{p}}^{+} \end{aligned} \tag{21.6}$$

where $H' = H - H_0$.

From (21.6), by using (21.2) and (21.3), the following set of homogeneous algebraic equations is obtained:

$$\begin{aligned} \varepsilon(p) u_{\mathbf{p}} &= w_{\mathbf{p}} u_{\mathbf{p}} + \frac{\nu(p)}{V} n_0 v_{\mathbf{p}} \\ -\varepsilon(p) v_{\mathbf{p}} &= w_{\mathbf{p}} v_{\mathbf{p}} + \frac{\nu(p)}{V} n_0 u_{\mathbf{p}} \end{aligned} \tag{21.7}$$

where

$$w_{\mathbf{p}} = \left[\frac{p^2}{2m} + \frac{\nu(p)}{V} N\right].$$

By equating to zero the determinant of the coefficients of $u_{\mathbf{p}}$ and $v_{\mathbf{p}}$ in (21.7), we obtain for $\varepsilon(p)$ the expression

$$\varepsilon(p) = \sqrt{\frac{n_0}{V}\frac{p^2}{m}\nu(p) + \frac{p^4}{4m^2}}\,. \tag{21.8}$$

$u_{\mathbf{p}}$ and $v_{\mathbf{p}}$ are then easily obtained and have the form

$$u_{\mathbf{p}} = \frac{1}{\sqrt{1-A_p^2}}, \qquad v_{\mathbf{p}} = \frac{A_{\mathbf{p}}}{\sqrt{1-A_p^2}} \tag{21.9}$$

where

$$A_{\mathbf{p}} = \frac{V}{n_0 \nu(p)}\left[\varepsilon(p) - \frac{p^2}{2m} - \frac{n_0}{V}\nu(p)\right].$$

The eigenvalues of the diagonal Hamiltonian (21.5) are then

$$E = E_0 + \sum_{\mathbf{p}\neq 0} \varepsilon(p) \Lambda_{\mathbf{p}}, \tag{21.10}$$

where

$$E_0 = \frac{N^2}{2V}\nu(0) + \sum_{\mathbf{p}\neq 0} \frac{1}{2}\left[\varepsilon(p) - \frac{p^2}{2m} - \frac{n_0}{V}\nu(p)\right]$$

and $\Lambda_{\mathbf{p}}$ is a positive integer or zero for each value of p. Since $\Lambda_{\mathbf{p}}$ is an eigenvalue of the operator $b_{\mathbf{p}}^{+}b_{\mathbf{p}}$, it can be regarded as the number of elementary oscillations ("quasi-particles") in a state of energy $\varepsilon(p)$.†

Let us now show that the above system (a "non-ideal Bose gas") can have superfluid properties. Let us suppose that the whole gas is in motion with velocity $\mathbf{v}$ with respect to some fixed frame of reference (e.g. the wall of a container). If we denote by E the total kinetic energy of the gas as measured in a frame of reference moving with the gas, and by E' the same quantity as measured in the fixed frame of reference, we have that

$$E' = E+N\frac{mv^2}{2}+\mathbf{v}.\mathbf{P}$$

where $\mathbf{P}$ is the total momentum of the gas as measured in the fixed frame. Since $\mathbf{P} = \sum_{\mathbf{p}} \mathbf{p}\Lambda_{\mathbf{p}}$, we can write

$$E' = E_0+N\frac{mv^2}{2}+\sum_{\mathbf{p}} \Lambda_{\mathbf{p}}(\varepsilon(p)+\mathbf{v}.\mathbf{p}). \tag{21.11}$$

Any decrease in the velocity of the gas is equivalent to the creation of a quasi-particle having a momentum in a direction opposite to that of $\mathbf{v}$. By (21.11) the change in the energy of the gas on the appearance of such a quasi-particle is $\Delta E' = \varepsilon(p)-|\mathbf{v}|.|\mathbf{p}|$. We shall show, however, that if the dominant forces between the gas particles are repulsive, then $\Delta E' > 0$, which is inconsistent with a decrease in velocity of the gas. The appearance of such a quasi-particle is therefore impossible, and the gas will continue to move with velocity $\mathbf{v}$, i.e. it will show superfluid properties.

Let us show that if the dominant forces between the gas molecules are repulsive, then, as stated above, $\Delta E' > 0$. This is equivalent to writing

$$v^* = \min\left(\frac{\varepsilon(p)}{|\mathbf{p}|}\right) > 0. \tag{21.12}$$

Now

$$\min\left(\frac{\varepsilon(p)}{|\mathbf{p}|}\right) = \left(\sqrt{\frac{n_0}{Vm}\nu(p)+\frac{p^2}{4m}}\right)_{p=0} = \sqrt{\frac{n_0}{mV}\nu(0)}$$

and, when repulsive forces are dominant, $\nu(0) = \int V(|\mathbf{q}|)\, d\mathbf{q} > 0$, so that relation (21.12) is verified.

† The reader can easily show that the momentum of these quasi-particles is equal to $\mathbf{p}$.

CHAPTER VIII

Perturbation Theory. The Variational Method†

1. Stationary State Perturbation Theory

Let $H = H_0 + H'$ be the Hamiltonian operator of a physical system, where H_0 is a time-independent operator (whose eigenfunctions are known), and H' is also a time-independent operator, called the perturbation. Through perturbation theory the energy spectrum and the eigenfunctions of the Hamiltonian H can be determined to various orders of useful approximation, provided that certain conditions are satisfied.

1. The spectrum of H_0 is discrete and non-degenerate.

Let ϕ_n and $E_n^{(0)}$ be the wavefunction and the energy respectively of a stationary state of the "unperturbed Hamiltonian" H_0, and ψ_n, E_n the corresponding quantities of the Hamiltonian H. To the first order of approximation of perturbation theory we have then that

$$E_n = E_n^{(0)} + H'_{nn}, \quad \psi_n = \phi_n + \sum_{m \neq n} \frac{H'_{mn}}{E_n^{(0)} - E_m^{(0)}} \phi_m \tag{VIII.1}$$

where $H'_{mn} \equiv \langle m | H' | n \rangle = \int \phi_m^* H' \phi_n \, dV$ are the matrix elements of the perturbation.

To the second order of approximation

$$E_n = E_n^{(0)} + H'_{nn} + \sum_{m \neq n} \frac{|H'_{mn}|^2}{E_n^{(0)} - E_m^{(0)}}. \tag{VIII.2}$$

A necessary condition for these results to be valid is that

$$|H'_{mn}| \ll |E_n^{(0)} - E_m^{(0)}| \tag{VIII.3}$$

for any m and n.

2. The spectrum of the Hamiltonian H_0 is discrete and degenerate.

Let us suppose that the energy level $E_n^{(0)}$ is f-fold degenerate. In general, the introduction of a perturbation removes or partly removes the degeneracy of degenerate energy levels.

† For other problems in perturbation theory and other methods of approximation, see Chapter XI.

The distinct energies which result from the initial level $E_n^{(0)}$ through the introduction of the perturbation are the solutions of the "secular equation":

$$\begin{vmatrix} H_{11}^{(n)}-E & H_{12}^{(n)} & H_{13}^{(n)}\dots \\ H_{21}^{(n)} & H_{22}^{(n)}-E & H_{23}^{(n)}\dots \\ H_{31}^{(n)} & H_{32}^{(n)} & H_{33}^{(n)}-E\dots \end{vmatrix} = 0. \tag{VIII.4}$$

where
$$H_{lk}^{(n)} = \langle nl\,|\,H'\,|\,nk\rangle = \int \phi_{nl}^{*} H' \phi_{nk}\,dV.$$

If E is any solution of (VIII.4), then the corresponding wavefunction in zero-order approximation is $\psi_n = \sum_{k=1}^{f} a_k \phi_{nk}$, where $\phi_{n1}, \phi_{n2}, \dots, \phi_{nf}$ is a set of f distinct (degenerate) eigenfunctions of H_0 with energy $E_n^{(0)}$, and the coefficients a_k are the solutions of the set of f homogeneous algebraic equations†

$$\sum_{k=1}^{f} (H_{lk}^{(n)} - E\delta_{lk}) a_k = 0, \quad 1 = 1, 2, \dots, f. \tag{VIII.5}$$

2. The Variational Method

By means of the "variational method", the lowest eigenvalues and the corresponding eigenfunctions of any Hamiltonian H can be found approximately without having to solve the Schrödinger equation. Thus the ground state energy E_0 is determined from the fact that

$$E_0 = \min \int \psi^* H \psi\, dv \tag{VIII.6}$$

where the range of ψ includes all arbitrary functions subject only to the normalization condition $\int |\psi|^2\, dV = 1$. Similarly the energy of the first excited level is given by

$$E_1 = \min \int \overline{\psi}^* H \overline{\psi}\, dv \tag{VIII.7}$$

where the range of $\overline{\psi}$ includes all arbitrary functions subject to the normalization condition and to the condition of being orthogonal to the ground state wavefunction:

$$\int |\overline{\psi}|^2\, dv = 0, \qquad \int \overline{\psi}^* \psi\, dv = 0. \tag{VIII.7'}$$

The relations (VIII.6) and (VIII.7) are exact, but, in practice, trial wavefunctions containing parameters have to be used, the latter being varied to give the required minima. If the choice of trial wavefunction is a good one, this procedure yields a good value for the energy.

† For a rigorous discussion of the validity of eigenfunction exansions and of perturbation theory, see *Eigenfunction Expansions* by E. C. Titchmarsh, Clarendon Press, Oxford, 1962; in particular, chapters XIX and XX of vol. II.

Problems

1. To the first order of approximation of perturbation theory, calculate the correction to the ground state of a hydrogen-like atom due to the finite spatial extension of the nucleus. For simplicity assume that the nucleus is spherical, of radius R, and that its charge Ze is uniformly distributed throughout its volume.

2. Solve the preceding problem with the hypothesis that the nuclear charge is uniformly distributed over the surface of a sphere of radius R.

3. Find the energy spectrum of a system whose Hamiltonian is

$$H = H_0 + H' = -\frac{\hbar^2}{2m}\frac{d^2}{dx^2} + \frac{1}{2}m\omega^2x^2 + ax^3 + bx^4 \tag{3a}$$

where a and b are small constants (the "anharmonic oscillator").

4. A plane rigid rotator having a moment of inertia I and an electric dipole moment $\mathbf{d}$ is placed in a homogeneous electric field $\mathbf{E}$. By considering the electric field as a perturbation, determine the first non-vanishing correction to the energy levels of the rotator.

5. Solve the previous problem for a rigid rotator which is not restricted to moving in a plane.

6. Calculate the perturbation of the first two energy levels of a hydrogen atom placed in an electric field $\mathbf{E}$ (the Stark effect).

7. Find the ground state energy of the He atom by treating the coulomb interaction between the two electrons as a perturbation (the spins of the electrons, and the symmetry or antisymmetry of the spatial wavefunction, may be ignored).

8. Inside an atomic nucleus the distance between nucleons is of the order of 10^{-13} cm. At such a distance the nuclear forces are much stronger than the Coulomb force between protons. Estimate the correction to the energy of a stationary state of a nucleus, due to Coulomb interaction between the protons.

9. Relation (VIII.3) gives a necessary condition for the validity of perturbation theory. Show, by taking as an example the case of a particle moving in the potential $V(x) = \frac{1}{2}m\omega^2x^2 + \lambda x^3$, that this condition is not sufficient.

10. Find, to the first order of perturbation, the changes in the energy levels of a hydrogen-like atom produced by a unit increase of the nuclear charge (due, for example, to a beta-emission). Using the exact energies, discuss the validity of the approximation used.

11. The states of a paramagnetic ion in a crystalline lattice are, according to the theory of paramagnetic resonance, the eigenfunctions of a "spin Hamiltonian":

$$H_S = a\mathcal{H}S_z + b\mathcal{H}I_z + \tfrac{1}{2}D(3\cos^2\theta - 1)\,[S_z^2 - \tfrac{1}{3}S^2] + \tfrac{1}{2}D\sin 2\theta \times$$
$$\times\left[S_+\left(S_z + \frac{1}{2}\right) + S_-\left(S_z - \frac{1}{2}\right)\right] + \frac{1}{4}D\sin^2\theta(S_+^2 + S_-^2) + AS_zI_z + \frac{A}{2}(S_+I_- + S_-I_+)$$

where S and I are the spin operators of the electrons and of the nucleus respectively, a, b, D, A are constants, $a \ll b$, and θ is the angle between the axis of symmetry of the crystal and the direction of the applied magnetic field $\mathcal{H}$.

Denoting by $|M_S M_I\rangle$, $(M_S = -S, -S+1, \ldots, S;\ M_I = -I, -I+1, \ldots, I)$, the eigenvectors of the "unperturbed" Hamiltonian $H_0 = a\mathcal{H}S_z + b\mathcal{H}I_z$, find the perturbation of the energy levels of H_0 which occurs when the terms in D and A are added to it to form H_S, to the second order of perturbation theory.†

12. Find the eigenvectors of the Hamiltonian H_S of the preceding problem, to the first order of perturbation theory, in terms of the eigenvectors of H_0.

13. Let $H = H_0 + H'$ be the time-independent Hamiltonian of a system. The operator

$$R(E) = \frac{1}{H-E}, \tag{13a}$$

considered as a function of the complex variable E, is called the "resolvent" of the Schrödinger equation $H|\psi\rangle = E|\psi\rangle$.

Show that if the Hamiltonian H_0 has a discrete non-degenerate spectrum $H_0|\phi_n\rangle = E_n^{(0)}|\phi_n\rangle$, then the energy of the perturbed level E_n is a pole of the diagonal matrix element $D_n(E) = \langle\phi_n|R(E)|\phi_n\rangle$. Deduce from this the expressions which give E_n in the usual formulation of perturbation theory.

14. Show that, in the preceding problem, the perturbed state vector is given by

$$|\psi_n\rangle = -N_n^{-1/2} \sum_m \text{Res}\,[\langle\phi_m|R(E)|\phi_n\rangle|\phi_m\rangle]_{E=E_n} \tag{14a}$$

where Res denotes the residue of the pole at $E = E_n$, and $N_n = -\text{Res}\,[D_n(E)]_{E=E_n}$ (assuming that $\langle\phi_n|\phi_n\rangle = 1$).

15. By using the variational method, find the ground state energy of an atom with two electrons and a nuclear charge Ze, using a trial wavefunction of the form

$$\psi(r_1, r_2) = \left(\frac{Z'^3}{\pi a^3}\right) e^{-Z'r_1/a} e^{-Z'r_2/a} \tag{15a}$$

where r_1 and r_2 are the distances of the two electrons from the nucleus, $a = \hbar^2/me^2$, and Z' is an adjustable parameter.

16. Since the harmonic oscillator wavefunction must be finite for any x, and must vanish as $x \to \pm\infty$, it seems reasonable to take as a trial function for the ground state wavefunction the form

$$\psi(x) = A \exp(-Bx^2), \quad B > 0. \tag{16a}$$

† F. Waldner, *Helv. Phys. Acta*, **35**, 756 (1962); I. Ursu, A. Nicula and S. Nistor, *Rev. Roum. Phys.* **10**, 229 (1965).

Find, by means of the variational method, the "best" value of B and of the ground state energy.

17. Using the results of the preceding problem, find by means of the variational method, an approximate wavefunction and energy for the first excited level of the harmonic oscillator.

18. Apply the variational method to the determination of the ground state of the hydrogen atom, using as trial wavefunctions the following expressions, which all have spherical symmetry,

$$\psi_1 = A_1 e^{-(b/a)r}, \quad \psi_2 = A_2 \frac{1}{b^2+\left(\frac{r}{a}\right)^2}, \quad \psi_3 = A_3 \frac{r}{a} e^{-(b/a)r} \tag{18a}$$

where a is the radius of the first Bohr orbit and b is an arbitrary constant. Perform the numerical calculations and discuss the results obtained.

19. Using the variational method, find an approximate energy and wavefunction for he $2s$ state of the hydrogen atom.

Solutions

1. The potential energy of the electron is given by

$$V(r) = \begin{cases} -\dfrac{Ze^2}{R}\left(\dfrac{3}{2}-\dfrac{1}{2}\dfrac{r^2}{R^2}\right), & 0 \leqslant r \leqslant R \\ -\dfrac{Ze^2}{r}, & r \geqslant R \end{cases} \tag{1.1}$$

In Fig. VIII.1, the potential energy $V(r)$ is shown as a continuous line, and the potential energy for a point nucleus as a broken line.

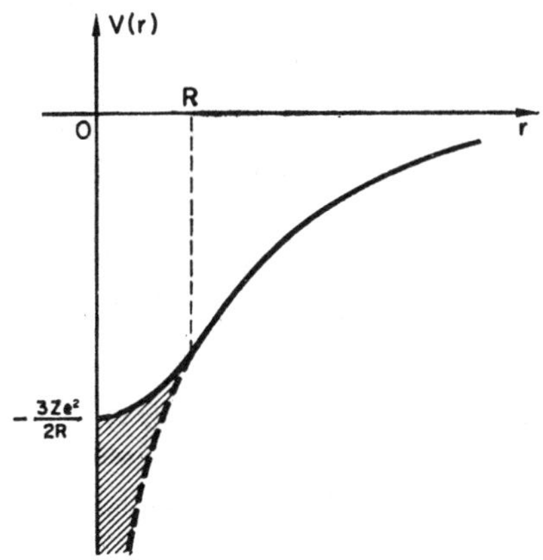

FIG. VIII.1.

Using perturbation theory, we take as the unperturbed Hamiltonian that of the hydrogen atom with a point nucleus ($V = -Ze^2/r$), and as the perturbation the following differences:

$$H' = \begin{cases} -\dfrac{Ze^2}{R}\left(\dfrac{3}{2}-\dfrac{1}{2}\dfrac{r^2}{R^2}\right)-\left(-\dfrac{Ze^2}{r}\right), & 0 \leqslant r \leqslant R \\ 0, & r \geqslant R. \end{cases} \tag{1.2}$$

The first-order correction to the ground state energy is then

$$E_1^{(1)} = \int \psi_{100}^*(\mathbf{r})\, H' \psi_{100}(\mathbf{r})\, dv \tag{1.3}$$

where, from Table II.1, we have that

$$\psi_{100}(\mathbf{r}) = \frac{1}{\sqrt{\pi a^3}} e^{-r/a}, \quad a = \frac{\hbar^2}{Zme^2}. \tag{1.4}$$

Bearing (1.2) in mind, it can be seen that the integration in (1.3) is to be carried out only within the volume of the sphere of radius a. Since the quantity a in (1.4) is the radius of the first Bohr orbit, and R is the radius of the nucleus, $r \ll a$, and the exponential $e^{-r/a}$ ist approximately equal to unity. Thus, for $R \sim 10^{-12}$ cm, $Z = Z_{\max} \sim 100$ we have that $a \sim 5 \times 10^{-11}$ cm, and so $a/R \approx 50$. Approximately, we have then that

$$E_1^{(1)} = \frac{1}{\pi a^3} \int_0^R \left[\frac{Ze^2}{r} - \frac{Ze^2}{R}\left(\frac{3}{2}-\frac{1}{2}\frac{r^2}{R^2}\right)\right] 4\pi r^2\, dr = \frac{2}{5}\frac{Ze^2}{a}\left(\frac{R}{a}\right)^2. \tag{1.5}$$

The fact that $E_1^{(1)} > 0$ could have been foreseen from the beginning, since, as Fig. VIII.1 shows, the depth of the potential well in which the electron moves is decreased when one takes into account the finite spatial extension of the nucleus. Since the unperturbed ground state energy is $E_1^{(0)} = -Z^2me^4/2\hbar^2$ (see (33.12) Chapter II), we have, to first order of perturbation theory,

$$E_1 \approx E_1^{(0)} + E_1^{(1)} = E_1^{(0)}\left[1 - \frac{4}{5}\left(\frac{R}{a}\right)^2\right]. \tag{1.6}$$

Remembering that the nuclear radius is approximately proportional to $Z^{1/3}$, and the first Bohr orbit radius to Z^{-1}, we arrive at the conclusion that the importance of the perturbation increases as $Z^{8/3}$.

The corrections to the energy levels of the atom due to the spatial extension of the nucleus are called "isotopic" corrections, since they vary from one isotope to another of the same element.

2. As in the preceding problem, the perturbation is different from zero only inside a sphere of radius R, in which it has the value

$$H' = Ze^2\left(\frac{1}{r} - \frac{1}{R}\right). \tag{2.1}$$

Then

$$E_1^{(1)} = Ze^2 \int_0^R \psi_{100}^*(r)\left(\frac{1}{r}-\frac{1}{R}\right)\psi_{100}(r)\,dv. \tag{2.2}$$

By the same arguments as in the preceding problem, we find that

$$E_1^{(1)} \approx \frac{2}{3}\frac{Ze^2}{a}\left(\frac{R}{a}\right)^2, \quad E_1 = E_1^{(0)}\left[1-\frac{4}{3}\left(\frac{R}{a}\right)^2\right]. \tag{2.3}$$

By comparing (2.3) with (1.6) we can see that, at least to a first approximation, the isotopic corrections do not depend on the distribution within the nucleus of the nuclear charge.

3. If $a = b = 0$, the Hamiltonian H reduces to that of a linear harmonic oscillator, H_0 say. We shall accordingly use perturbation theory, with $H' = ax^3+bx^4$. The following integrals must then be calculated:

$$H'_{nn} = a\int_{-\infty}^{+\infty} \psi_n x^3 \psi_n\,dx + b\int_{-\infty}^{+\infty} \psi_n x^4 \psi_n\,dx \tag{3.1}$$

where ψ_n denotes the nth harmonic oscillator eigenfunction ((17.11) Chapter II). Since x^3 is an odd function and ψ_n^2 is an even one, the first integral in (3.1) vanishes. The second integral has already been calculated in problem 5 of Chapter III, thus

$$\int_{-\infty}^{+\infty} \psi_n x^4 \psi_n\,dx = \frac{3}{4}x_0^4(2n^2+2n+1), \quad x_0^2 = \frac{\hbar}{m\omega}. \tag{3.2}$$

The first-order perturbation of the energy level $E_n^{(0)}$ is thus given by

$$H'_{nn} = \tfrac{3}{4}bx_0^4(2n^2+2n+1), \tag{3.3}$$

so that, to this approximation, the energy spectrum of the oscillator is

$$E_n = E_n^{(0)}+H'_{nn} = (n+\tfrac{1}{2})\,\hbar\omega+\tfrac{3}{4}bx_0^4(2n^2+2n+1). \tag{3.4}$$

4. The Schrödinger equation of a plane rigid rotator is

$$-\frac{\hbar^2}{2I}\frac{d^2\psi}{d\phi^2} = E\psi \tag{4.1}$$

where ϕ is the angle of rotation about the z-axis. The energies and the normalized wave-functions are readily found to be:

$$E_m^{(0)} = \frac{\hbar^2 m^2}{2I}, \quad \psi_m^{(0)}(\phi) = \frac{1}{\sqrt{2\pi}}e^{im\phi}, \quad m = 0, \pm 1, \pm 2, \ldots, \tag{4.2}$$

the levels for $m \neq 0$ being doubly degenerate, since states with $+m$ and with $-m$ have the same energy.

Treating the electric field **E** as a perturbation, the Hamiltonian of the system becomes

$$H = H_0+H' = -\frac{\hbar^2}{2I}\frac{d^2}{d\phi^2}-Ed\cos\phi. \tag{4.3}$$

Now, since the "parity" operator P defined by $Pf(\phi) = f(-\phi)$ (reflection in the z-axis) commutes with both H and H', there will be no mixing of a state with $-m$, say, into a (perturbed) state with $+m$; thus the perturbation theory for non-degenerate levels can be used. We have, then, that

$$H'_{mm'} = \int_0^{2\pi} \psi_m^{(0)*}H'\psi_{m'}^{(0)}\,d\phi = -\frac{Ed}{2\pi}\int_0^{2\pi} e^{i(m'-m)\phi}\cos\phi\,d\phi$$

$$= \begin{cases} 0, & m' \neq m\pm 1 \\ -\dfrac{Ed}{2}, & m' = m\pm 1. \end{cases} \tag{4.4}$$

From (VIII.1), (VIII.2) and (4.4) it follows immediately that

$$E_m^{(1)} = H'_{mm} = 0,$$

$$E_m^{(2)} = \frac{|H'_{m,m-1}|^2}{E_m^{(0)}-E_{m-1}^{(0)}}+\frac{|H'_{m,m+1}|^2}{E_m^{(0)}-E_{m+1}^{(0)}} = \frac{IE^2d^2}{\hbar^2(4m^2-1)}$$

and hence, to the second order of perturbation, we have

$$E_m = E_m^{(0)}+E_m^{(1)}+E_m^{(2)} = \frac{\hbar^2m^2}{2I}+\frac{IE^2d^2}{\hbar^2(4m^2-1)}. \tag{4.5}$$

5. The eigenfunctions and eigenvalues of the unperturbed problem (see problem 34, Chapter II) are

$$\langle\theta,\phi\,|\,l,m\rangle = Y_l^m(\theta,\phi), \quad E_l^{(0)} = \frac{\hbar^2}{2I}l(l+1). \tag{5.1}$$

In the presence of an electric field **E** in the direction of the z-axis, the Hamiltonian of the system contains the additional perturbation term

$$H' = -dE\cos\theta. \tag{5.2}$$

To solve the problem we have to evaluate the matrix elements $\langle l_1, m_1\,|\,H'\,|\,l_2, m_2\rangle$. This can be done directly, by using relation (A.28), or by using the results obtained in problem 30 of Chapter VI. The only non-vanishing matrix elements are those in which $m_1 = m_2$ and $l_1 = l_2\pm 1$.

For these we find the expressions:

$$\langle l,m\,|\cos\theta|\,l-1,m\rangle = \langle l-1,m\,|\cos\theta|\,l,m\rangle = \left(\frac{l^2-m^2}{4l^2-1}\right)^{1/2}. \tag{5.3}$$

Note that the energy spectrum of the rotator is degenerate except for the level $l = 0$. For this reason, it might seem at first sight that we have to use the perturbation theory applicable to degenerate levels. This can be avoided, however, since the free Hamiltonian $H_0 = \frac{\mathbf{l}^2}{2I}$, and the perturbed Hamiltonian $H = H_0+H'$ both commute with l_z. It follows that a (perturbed) state $|l, m\rangle$ will contain only unperturbed states $|l', m\rangle$ with the same quantum number m, and hence the problem can be solved by using only the perturbation theory for non-degenerate levels. Since $\langle l, m\,|H'|\,l, m\rangle = 0$, the first-order energy corrections are zero. In second order we have

$$E_{lm}^{(2)} = \frac{2I(Ed)^2}{\hbar^2} \sum_{l'\neq l} \frac{|\langle l, m|\cos\theta\,|l', m\rangle|^2}{l(l+1)-l'(l'+1)}. \tag{5.4}$$

Bearing in mind the selection rule $l' = l\pm 1$, and the relation (5.3), we find

$$\begin{aligned} E_{lm}^{(2)} &= \frac{2I(Ed)^2}{\hbar^2}\left[\frac{|\langle l, m|\cos\theta\,|l+1, m\rangle|^2}{l(l+1)-(l+2)(l+1)}+\frac{|\langle l, m|\cos\theta\,|l-1, m\rangle|^2}{l(l+1)-l(l-1)}\right] \\ &= \frac{(Ed)^2}{E_l^{(0)}}\,\frac{l(l+1)-3m^2}{2(2l-1)(2l+3)}. \end{aligned} \tag{5.5}$$

The energy levels in second-order approximation will thus be

$$E_{lm} \approx E_l^{(0)}\left[1+\left(\frac{Ed}{E_l^{(0)}}\right)^2 \frac{l(l+1)-3m^2}{2(2l-1)(2l+3)}\right]. \tag{5.6}$$

Note that the initial degeneracy is only partly removed, since the energy depends on l and on m^2, but not on m. States with m and $-m$ therefore have the same energy. It can be shown that this remaining degeneracy is retained in higher orders of perturbation theory. The problem we have just solved is relevant to the study of the polarization of diatomic molecules in weak electric fields.

6. In non-relativistic approximation, the electric field $\mathbf{E}$ does not interact with the electron's magnetic moment and for this reason the spin, and the fine structure resulting from spin–orbit interaction, can be neglected.† By choosing the z-axis along the direction of $\mathbf{E}$, the perturbation operator becomes

$$H' = -\mathbf{d}.\mathbf{E} = -e\mathbf{r}.\mathbf{E} = -ezE. \tag{6.1}$$

The ground state $|100\rangle$ of the hydrogen atom is non-degenerate. Because the corresponding wavefunction is even under the transformation $z \to -z$, it follows that $\langle 100\,|z|\,100\rangle = 0$, i.e. that the energy of the ground state is not perturbed. For $n = 2$, we have to remember

† Such an approximation is valid for electric fields weaker than about 10^3 V/cm.

that $E_2^{(0)}$ is fourfold degenerate. To study the perturbation of this level in first-order approximation, the following linear combination must be formed

$$\psi^{(0)} = \sum_{k=1}^{4} a_k \psi_k \tag{6.2}$$

where $\psi_1 = |200\rangle$, $\psi_2 = |210\rangle$, $\psi_3 = |211\rangle$, $\psi_4 = |21-1\rangle$.

By using the hydrogen atom wavefunctions (Table II.1) the only non-vanishing matrix elements of the perturbation are found to be

$$H'_{12} = H'_{21} = -eE\langle 200\,|z|\,210\rangle = -3eaE, \qquad a = \frac{\hbar^2}{me^2}. \tag{6.3}$$

By writing $\varepsilon = E_2 - E_2^{(0)}$, the secular equation (VIII.4) becomes $(\varepsilon^2 - 9e^2a^2E^2)\varepsilon^2 = 0$, and has the roots

$$\varepsilon_1 = 3eaE, \quad \varepsilon_2 = -3eaE, \quad \varepsilon_3 = \varepsilon_4 = 0. \tag{6.4}$$

The corresponding correct wavefunctions in zero-order approximation can be obtained by determining the coefficient a_k in (6.2) with the aid of the equations (VIII.5).

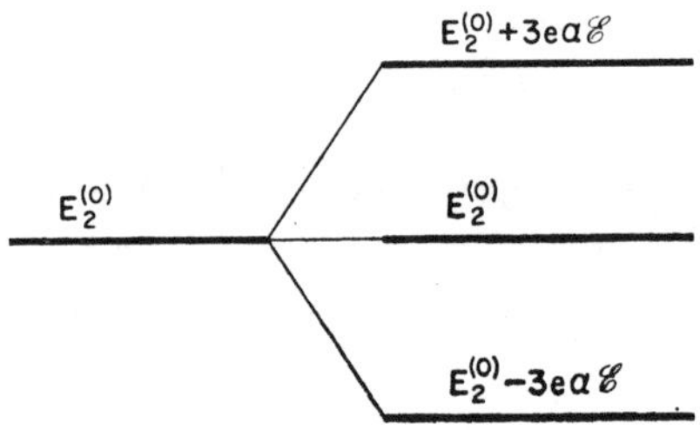

FIG. VIII.2.

Note that on applying the external electric field, the fourfold degenerate level $E_2^{(0)}$ is split into three levels, one of which is doubly degenerate, with $m = \pm 1$ (Fig. VIII.2). The energy splittings are proportional to the electric field. For this reason we say that this is an instance of the "linear" Stark effect. It can be shown that the linear Stark effect appears (in systems of particles moving under Coulomb interactions) only if there is degeneracy in zero-order with respect to the orbital quantum number l. If there is no such degeneracy, the first-order corrections vanish, and the Stark effect is then proportional to the square of the electric field (cf. the rotator in problems 4 and 5).

7. Consider in general the problem of two electrons in the Coulomb field of a nucleus of charge Ze. Examples are given by He, Li^+, Be^{++}, etc., for which $Z = 2, 3, 4, \ldots$. The potential energy of the system is then

$$V = -\frac{Ze^2}{r_1} - \frac{Ze^2}{r_2} + \frac{e^2}{r_{12}} \tag{7.1}$$

where r_1 and r_2 are the distances of the two electrons from the nucleus, and r_{12} is the distance between the electrons. If we assume a fixed nucleus (which is very nearly the case), the Schrödinger equation of the system is

$$H\psi = -\frac{\hbar^2}{2m}\left(\frac{\partial^2}{\partial x_1^2}+\frac{\partial^2}{\partial y_1^2}+\frac{\partial^2}{\partial z_1^2}+\frac{\partial^2}{\partial x_2^2}+\frac{\partial^2}{\partial y_2^2}+\frac{\partial^2}{\partial z_2^2}\right) + \left(-\frac{Ze^2}{r_1}-\frac{Ze^2}{r_2}+\frac{e^2}{r_{12}}\right)\psi = E\psi \tag{7.2}$$

where x_1, y_1, z_1 and x_2, y_2, z_2 are the Cartesian coordinates of the two electrons respectively, and m is the electron mass. Since, in the absence of the term

$$H' = \frac{e^2}{r_{12}}, \tag{7.3}$$

equation (7.2) can be solved exactly, we shall treat H' as a perturbation. The unperturbed wavefunctions $\psi^{(0)}$ can be written in the form

$$\psi^{(0)}(x_1, y_1, z_1, x_2, y_2, z_2) = u_1^{(0)}(x_1, y_1, z_1)\, u_2^{(0)}(x_2, y_2, z_2) \tag{7.4}$$

where $u_i^{(0)}(x_i, y_i, z_i)$, $i = 1, 2$, satisfy the equations

$$\frac{\partial^2 u_i^{(0)}}{\partial x_i^2}+\frac{\partial^2 u_i^{(0)}}{\partial y_i^2}+\frac{\partial^2 u_i^{(0)}}{\partial z_i^2}+\frac{2m}{\hbar^2}\left(E_i^{(0)}+\frac{Ze^2}{r_i}\right)u_i^{(0)} = 0, \quad E^{(0)} = E_1^{(0)}+E_2^{(0)}. \tag{7.5}$$

Equation (7.5) is the Schrödinger equation of the hydrogen-like atom. By using the expression (33.12) of Chapter II for the unperturbed energy of the ground state of a two-electron atom we have

$$E^{(0)} = -2Z^2E_H, \quad E_H = -E_1 = \frac{me^4}{2\hbar^2} = 13{\cdot}5 \text{ eV}.$$

The first-order correction due to the interaction between the electrons is

$$E^{(1)} = \int \psi^{(0)*}H'\psi^{(0)}\,dv = \int \frac{e^2}{r_{12}}\psi^{(0)2}\,dv \tag{7.6}$$

where $$\psi^{(0)} = \psi_{100} = \sqrt{\frac{Z^3}{\pi a^3}}\exp\left(-\frac{\varrho}{2}\right), \quad \varrho = \frac{2Zr}{a},$$

is the ground state wavefunction of a hydrogen-like atom (see Table II.1), and a is the radius of the first Bohr orbit.

Using spherical coordinates,

$$E^{(1)} = \frac{Ze^2}{2^5\pi^2 a}\iint \frac{e^{-\varrho_1-\varrho_2}}{\varrho_{12}}\,dv_1\,dv_2 = \frac{Ze^2}{2^5\pi^2 a}\int_0^{2\pi}\int_0^{\pi}\int_0^{\infty}\int_0^{2\pi}\int_0^{\pi}\int_0^{\infty}\frac{e^{-\varrho_1-\varrho_2}}{\varrho_{12}}\times$$
$$\times\varrho_1^2\,d\varrho_1\sin\theta_1\,d\theta_1\,d\phi_1\varrho_2^2\,d\varrho_2\sin\theta_2\,d\theta_2\,d\phi_2\,. \tag{7.7}$$

The integral

$$I = \iint \frac{e^{-\varrho_1-\varrho_2}}{\varrho_{12}}\, dv_1\, dv_2 \tag{7.8}$$

can be evaluated by using an electrostatic analogy. Thus I is the potential energy of the forces between two charge distributions of densities $e^{-\varrho_1}$ and $e^{-\varrho_2}$ respectively (Fig. VIII.3). To evaluate I, we can calculate first the electrostatic potential at any point due to the first charge distribution, and then calculate the potential energy of the second charge distribution in the field of the first one. Now, the potential at a distance r from the centre due to a spherical shell of charge density $e^{-\varrho_1}$ enclosed between the radii ϱ_1 and $\varrho_1+d\varrho_1$ is

$$d\Phi(r) = \begin{cases} 4\pi\varrho_1^2 e^{-\varrho_1}\, d\varrho_1 \cdot \dfrac{1}{\varrho_1} & \text{for} \quad r < \varrho_1 \\ 4\pi\varrho_1^2 e^{-\varrho_1}\, d\varrho_1 \cdot \dfrac{1}{r} & \text{for} \quad r > \varrho_1. \end{cases}$$

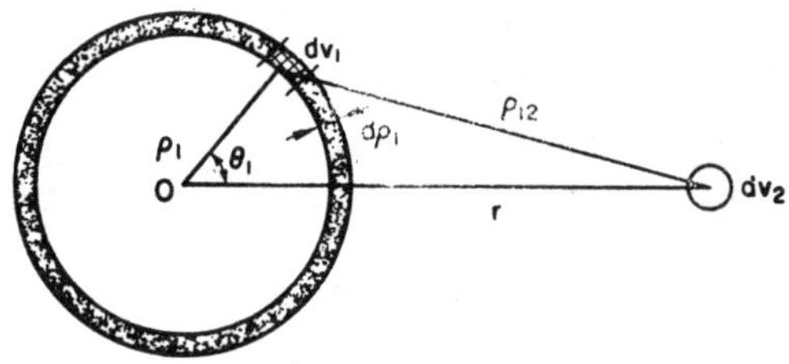

FIG. VIII. 3.

The electrostatic potential of the first distribution at a distance r from its centre is therefore

$$\Phi(r) = \frac{4\pi}{r}\int_0^r e^{-\varrho_1}\varrho_1^2\, d\varrho_1 + 4\pi\int_r^\infty e^{-\varrho_1}\varrho_1^2\, d\varrho_1 = \frac{4\pi}{r}[2-e^{-r}(r+2)],$$

and hence

$$I = \int \Phi(\varrho_2)e^{-\varrho_2}\, dv_2 = 16\pi^2\int_0^\infty [2-e^{-\varrho_2}(\varrho_2+2)e^{-\varrho_2}\varrho_2\, d\varrho_2] = 20\pi^2. \tag{7.9}$$

Finally,

$$E^{(1)} = \frac{5Ze^2}{8a} = \frac{5}{4}ZE_H, \quad E = E^{(0)}+E^{(1)} = -\left(2Z^2-\frac{5}{4}Z\right)E_H. \tag{7.10}$$

Numerical values calculated from (7.10) agree quite well with the experimental values obtained by adding the two ionization energies. As Z increases, the interaction between the two electrons (i.e. the perturbation) becomes smaller in comparison with their interaction with the nucleus, which is why the values calculated from (7.10) for B^{3+} and C^{4+} agree better with the experimental data than those for He or Li^+.

8. The unperturbed Hamiltonian H_0 consists of the kinetic energy operator of the nucleons and of the nuclear potential energy. The perturbation H' is a sum of $\frac{1}{2}Z(Z-1)$ terms, and has the form

$$H' = \sum_{i<j\leq Z} \frac{e^2}{r_{ij}}, \qquad r_{ij} = |\mathbf{r}_i - \mathbf{r}_j| \tag{8.1}$$

where $\mathbf{r}_i$ and $\mathbf{r}_j$ are the position vectors of two nucleons.

If we suppose that the potential energy of the nuclear forces has the form (24.4) of Chapter VI, we can choose, as unperturbed state vectors, the vectors $|\psi_{JM}\rangle$ which correspond to well defined values of $\mathbf{J}^2$ and J_z. We can then apply the perturbation theory for non-degenerate levels, since $\mathbf{J}^2$ and J_z commute both with H_0 and with H, so that states having different values of J and of M cannot be mixed together by the perturbation.†

We have then that

$$E^{(1)} = \langle\psi_{JM}|H'|\psi_{JM}\rangle. \tag{8.2}$$

The wavefunctions corresponding to the state vectors $|\psi_{JM}\rangle$ depend on the spatial coordinates and on the spins of the protons and of the neutrons. Because these vectors are antisymmetrical with respect to permutations of protons, the $\frac{1}{2}Z(Z-1)$ terms of H' give equal contributions to the integral (8.2), and hence

$$E^{(1)} = \tfrac{1}{2}Z(Z-1)\varepsilon, \tag{8.3}$$

$$\varepsilon = \left\langle \psi_{JM} \left| \frac{e^2}{r_{12}} \right| \psi_{JM} \right\rangle = \sum_{\text{spins}} \int |\psi_{JM}|^2 \frac{e^2}{r_{12}} d\mathbf{r}_1 \ldots d\mathbf{r}_Z \, d\mathbf{r}_{Z+1} \ldots d\mathbf{r}_A$$

where A (the "mass number") is the total number of protons and neutrons in the nucleus, and the summation is over the spin variables of all the nucleons.

Let

$$\varrho(\mathbf{r}_1, \mathbf{r}_2)\, d\mathbf{r}_1 \, d\mathbf{r}_2 = \sum_{\text{spins}} \left[\int |\psi_{JM}|^2 \, d\mathbf{r}_3 \ldots d\mathbf{r}_N\right] d\mathbf{r}_1 \, d\mathbf{r}_2$$

denote the probability of finding simultaneously the first proton in the volume element $d\mathbf{r}_1$ about the point $\mathbf{r}_1$ and the second in $d\mathbf{r}_2$ about $\mathbf{r}_2$. Then

$$\varepsilon = \int \frac{e^2}{r_{12}} \varrho(\mathbf{r}_1, \mathbf{r}_2)\, d\mathbf{r}_1 \, d\mathbf{r}_2. \tag{8.4}$$

The integral (8.4) can easily be evaluated approximately by considering a simplified model, in which all correlations between protons 1 and 2 are neglected (which involves neglecting, among other things, the Pauli principle). In this approximation we can write $\varrho(\mathbf{r}_1, \mathbf{r}_2) \approx$

† See also problems 1, 3, 4 of Chapter XI.

$\varrho(\mathbf{r}_1)\,\varrho(\mathbf{r}_2)$. Assuming further that the proton charge is uniformly distributed inside the nucleus, which we take to be a sphere of radius R, we have

$$\varrho(\mathbf{r}) = \begin{cases} 3/(4\pi R^3), & r < R \\ 0, & r > R. \end{cases}$$

The integral (8.4) can then be calculated as in the preceding problem. Finally, we have that

$$\varepsilon = \frac{6}{5}\frac{e^2}{R}, \quad E^{(1)} = \frac{3}{5}Z(Z-1)\frac{e^2}{R}. \tag{8.5}$$

Although the above approximations are quite rough, the expression (8.5) can be used to confirm the charge independence of nuclear forces.

9. For $\lambda = 0$, the Hamiltonian $H = -(\hbar^2/2m)\,d^2/dx^2 + V$ reduces to the Hamiltonian of a harmonic oscillator and hence possesses a discrete spectrum of energies $E_n = \hbar\omega(n+\frac{1}{2})$. For small values of λ, the conditions (VIII.3) become

$$\lambda|\langle m|x^3|n\rangle| \ll |E_m - E_n| = \hbar\omega|m-n| \tag{9.1}$$

and are always satisfied. However, for any $\lambda > 0$, however small, the Hamiltonian H has a continuous energy spectrum, since for large negative values of x the potential energy becomes smaller than the total energy of the particle. The particle can thus go through the potential barrier shown in Fig. VIII.4 along the negative direction of the x-axis, and so escape to infinity.

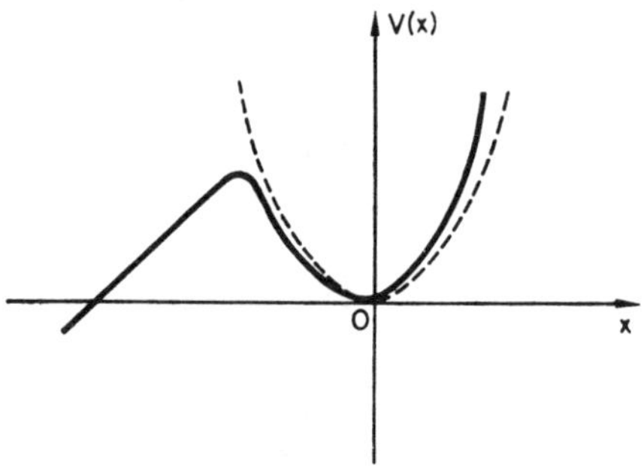

FIG. VIII.4.

We have here an example in which the perturbed quantities do not pass continuously into the unperturbed ones as $\lambda \to 0$ so that, although the conditions (VIII.3) are satisfied, perturbation theory is not, strictly speaking, applicable. Nevertheless, it can be seen that as $\lambda \to 0$, the probability that the particle will pass through the potential barrier also tends to zero, and the states are very nearly bound, or *quasi-bound*, for small λ. Due to this fact, calculations based on perturbation theory yield bound states which are good approximations to the correct quasi-bound states (see p. 204, footnote).

10. The perturbation operator is $H' = e^2/r$. We have then that

$$E_n^{(1)} = \left\langle nlm \left| \frac{1}{r} \right| nlm \right\rangle = \left\langle n \left| \frac{1}{r} \right| n \right\rangle .$$

Taking into account (7.5) of Chapter III, it follows that

$$E_n^{(1)} = -\frac{mZe^4}{\hbar^2 n^2} \tag{10.1}$$

The correction to the level $E_n^{(0)}$ can be obtained exactly in this case from (33.12) Chapter II, thus

$$\Delta E_n^{(0)} = -\frac{me^4}{2\hbar^2 n^2}[(Z+1)^2 - Z^2] = -\frac{me^4}{\hbar^2 n^2}(Z+\tfrac{1}{2}). \tag{10.2}$$

By comparing (10.2) with (10.1), we see that $E_n^{(1)} \approx \Delta E_n^{(0)}$ only in the limit of large Z, a conclusion which could have been drawn from the conditions of applicability of perturbation theory, i.e. from (VIII.3).

11. The unperturbed spectrum is given by the expression

$$E_{M_S M_I}^{(0)} = a\mathcal{H}M_S + b\mathcal{H}M_I. \tag{11.1}$$

Applying perturbation theory, we find that

$$E_{M_S M_I}^{(1)} = \langle M_S M_I | H' | M_S M_I \rangle = \tfrac{1}{2}D(3\cos^2\theta - 1)[M_S^2 - \tfrac{1}{3}S(S+1)] + AM_S M_I, \tag{11.2}$$

$$E_{M_S M_I}^{(2)} = \sum_{M_S', M_I'} \frac{|\langle M_S M_I | H' | M_S' M_I' \rangle|^2}{E_{M_S M_I}^{(0)} - E_{M_S' M_I'}^{(0)}} \approx \sum_{M_S', M_I'} \frac{|\langle M_S M_I | H' | M_S' M_I' \rangle|^2}{a\mathcal{H}(M_S - M_S')}. \tag{11.3}$$

The only non-vanishing matrix elements in (11.3) are

$$\langle M_S M_I | H' | M_S+1, M_I \rangle = \frac{D}{2}\sin 2\theta \left(M_S + \frac{1}{2}\right)[S(S+1) - M_S(M_S+1)]^{1/2}$$

$$\langle M_S M_I | H' | M_S-1, M_I \rangle = \frac{D}{2}\sin 2\theta \left(M_S - \frac{1}{2}\right)[S(S+1) - M_S(M_S-1)]^{1/2}$$

$$\langle M_S M_I | H' | M_S+2, M_I \rangle = \frac{D}{4}\sin^2\theta\,[S(S+1) - M_S(M_S+1)]^{1/2} \times$$
$$\times [S(S+1) - (M_S+1)(M_S+2)]^{1/2}$$

$$\langle M_S M_I | H' | M_S-2, M_I \rangle = \frac{D}{4}\sin^2\theta\,[S(S+1) - M_S(M_S-1)]^{1/2} \times \tag{11.4}$$
$$\times [S(S+1) - (M_S-1)(M_S-2)]^{1/2}$$

$$\langle M_S M_I | H' | M_S+1, M_I-1\rangle = \frac{A}{2}[S(S+1)-M_S(M_S+1)]^{1/2}\times$$
$$\times [I(I+1)-M_I(M_I-1)]^{1/2}$$
$$\langle M_S M_I | H' | M_S-1, M_I+1\rangle = \frac{A}{2}[S(S+1)-M_S(M_S-1)]^{1/2}\times$$
$$\times [I(I+1)-M_I(M_I+1)]^{1/2}.$$

Substituting for these in (11.3), we obtain

$$E^{(2)}_{M_S M_I} = \frac{D^2 \sin^2 2\theta}{4a\mathcal{H}}\left\{\left(M_S-\frac{1}{2}\right)^2[S(S+1)-M_S(M_S-1)]\right.$$
$$\left.-\left(M_S+\frac{1}{2}\right)^2[S(S+1)-M_S(M_S+1)]\right\}$$
$$+\frac{D^2\sin^4\theta}{32a\mathcal{H}}\{[S(S+1)-M_S(M_S-1)][S(S+1)-(M_S-1)(M_S-2)]$$
$$-[S(S+1)-M_S(M_S+1)][S(S+1)+(M_S+1)(M_S+2)]\}$$
$$+\frac{A^2}{2a\mathcal{H}}\{M_I[M_S^2-S(S+1)]+M_S[I(I+1)-M_I^2]\}. \tag{11.5}$$

12. Using (11.4) and denoting the state vectors in the zero order approximation by $|M_S M_I\rangle^{(1)}$, we find from (VIII.1) that

$$|M_S M_I\rangle^{(1)} = \frac{D}{2}\frac{\sin 2\theta}{a\mathcal{H}}\left(M_S-\frac{1}{2}\right)[S(S+1)-M_S(M_S-1)]^{1/2}|M_S-1, M_I\rangle$$
$$-\frac{D}{2}\frac{\sin 2\theta}{a\mathcal{H}}\left(M_S+\frac{1}{2}\right)[S(S+1)-M_S(M_S+1)]^{1/2}|M_S+1, M_I\rangle$$
$$+\frac{D}{8}\frac{\sin^2\theta}{a\mathcal{H}}[S(S+1)-(M_S-1)(M_S-2)]^{1/2}[S(S+1)-M_S(M_S-1)]^{1/2}|M_S-2, M_I\rangle$$
$$-\frac{D}{8}\frac{\sin^2\theta}{a\mathcal{H}}[S(S+1)-(M_S+1)(M_S+2)]^{1/2}[S(S+1)-M_S(M_S+1)]^{1/2}|M_S+2, M_I\rangle$$
$$+\frac{A}{2a\mathcal{H}}[S(S+1)-M_S(M_S-1)]^{1/2}[I(I+1)-M_I(M_I+1)]^{1/2}|M_S-1, M_I+1\rangle$$
$$+\frac{A}{2a\mathcal{H}}[S(S+1)-M_S(M_S+1)]^{1/2}[I(I+1)-M_I(M_I-1)]^{1/2}|M_S+1, M_I-1\rangle.$$

13. The resolvent $R(E)$ can be written in the form

$$R(E) = \frac{1}{H_0-E}-\frac{1}{H_0-E}H'R(E) \tag{13.1}$$

whence

$$\langle\phi_n|R(E)|\phi_n\rangle = \frac{1}{E_n^{(0)}-E} - \frac{1}{E_n^{(0)}-E}\sum_m \langle\phi_n|H'|\phi_m\rangle\langle\phi_m|R(E)|\phi_n\rangle$$

$$= \frac{1}{E_n^{(0)}-E} - \frac{1}{E_n^{(0)}-E}\langle\phi_n|H'|\phi_n\rangle\langle\phi_n|R(E)|\phi_n\rangle$$

$$- \frac{1}{E_n^{(0)}-E}\sum_{m\neq n}\langle\phi_n|H'|\phi_m\rangle\langle\phi_m|R(E)|\phi_n\rangle. \qquad (13.2)$$

Similarly for $\phi_m \neq \phi_n$ we have

$$\langle\phi_m|R(E)|\phi_n\rangle = -\frac{1}{E_m^{(0)}-E}\langle\phi_m|H'|\phi_n\rangle\langle\phi_n|R(E)|\phi_n\rangle$$

$$- \frac{1}{E_m^{(0)}-E}\sum_{r\neq n}\langle\phi_m|H'|\phi_r\rangle\langle\phi_r|R(E)|\phi_n\rangle. \qquad (13.3)$$

By iterating (13.3) we obtain

$$\langle\phi_m|R(E)|\phi_n\rangle = \frac{1}{E_m^{(0)}-E_0}\Big\{-\langle\phi_m|H'|\phi_n\rangle + \sum_{r\neq n}\langle\phi_m|H'|\phi_r\rangle\frac{1}{E_r^{(0)}-E}\times$$

$$\times\langle\phi_r|H'|\phi_n\rangle - \dots\Big\}D_n(E). \qquad (13.4)$$

Bearing in mind that

$$|\phi_r\rangle\frac{1}{E_r^{(0)}-E}\langle\phi_r| = \frac{1}{H_0-E}|\phi_r\rangle\langle\phi_r|, \qquad (13.5)$$

(13.4) can be written in the form

$$\langle\phi_m|R(E)|\phi_n\rangle = \frac{1}{E_m^{(0)}-E}\langle\phi_m|\Big\{-H'+H'\frac{1}{H_0-E}H' - \dots\Big\}'|\phi_n\rangle D_n(E), \qquad (13.6)$$

where $\{\dots\}'$ indicates that the quantities in brackets are to be calculated by introducing, where necessary, the operator $\sum_{r\neq n}|\phi_r\rangle\langle\phi_r|$. By substituting (13.6) into (13.2) we obtain

$$D_n(E) = \frac{1}{E_n^{(0)}-E} - \frac{1}{E_n^{(0)}-E}\langle\phi_n|H'|\phi_n\rangle D_n(E) - \frac{1}{E_n^{(0)}-E}\sum_{m\neq n}\langle\phi_n|H'|\phi_m\rangle$$

$$\frac{1}{E_m^{(0)}-E}\langle\phi_m|\Big\{-H'+H'\frac{1}{H_0-E}H' - \dots\Big\}'|\phi_n\rangle D_n(E). \qquad (13.7)$$

Using once more the identity (13.5), we see from (13.7) that

$$D_n(E) = \frac{1}{E_n^{(0)}-E} + \frac{1}{E_n^{(0)}-E}\langle\phi_n|\Big\{-H'+H'\frac{1}{H_0-E}H' - \dots\Big\}'|\phi_n\rangle D_n(E), \qquad (13.8)$$

i.e. that

$$D_n(E) = \frac{1}{E_n^{(0)}-E} + \frac{1}{E_n^{(0)}-E} F_n(E) D_n(E) \tag{13.9}$$

where

$$F_n(E) = \langle \phi_n | \left\{ -H' + H' \frac{1}{H_0-E} H' - \dots \right\}' | \phi_n \rangle. \tag{13.10}$$

From (13.9) it follows that

$$D_n(E) = \langle \phi_n | R(E) | \phi_n \rangle = \frac{1}{E_n^{(0)}-E-F_n(E)} \, . \tag{13.11}$$

Let us denote by E_n an energy value for which the denominator in (13.11) vanishes. Then, from (13.4), we see that $E = E_n$ is a pole of the function $\langle \phi_m | R(E) | \phi_n \rangle$ for any m, and we can therefore construct a vector

$$|\psi_n\rangle = C \lim_{E \to E_n} (E_n - E) R(E) | \phi_n \rangle \tag{13.12}$$

where C is a normalization constant which we leave undetermined. Writing

$$H - E_n = (H-E) + (E-E_n)$$

and operating with both sides of this operator relation on $C(E_n - E)R(E)|\phi_n$, we obtain

$$(H-E_n)C(E_n-E)R(E)|\phi_n\rangle = C(H-E)(E_n-E)\frac{1}{H-E}|\phi_n\rangle - C(E-E_n)^2 R(E)|\phi\rangle.$$

Finally, taking the limit $E \to E_n$, we obtain $(H-E_n)|\psi_n\rangle = 0$, i.e. E_n is an eigenvalue of H, as required.

For the second part of the problem, we substitute the expression (13.10) into $E_n^{(0)} - E_n - F_n(E_n) = 0$, and obtain

$$E_n = E_n^{(0)} + \langle \phi_n | H' | \phi_n \rangle + \sum_{m \neq n}' \frac{\langle \phi_n | H' | \phi_m \rangle \langle \phi_m | H' | \phi_n \rangle}{E_n - E_m^{(0)}}$$
$$+ \sum_{\substack{m \neq n \\ r \neq n}}' \frac{\langle \phi_n | H' | \phi_m \rangle \langle \phi_m | H' | \phi_r \rangle \langle \phi_r | H' | \phi_n \rangle}{(E_n - E_m^{(0)})(E_n - E_r^{(0)})} + \dots \tag{13.13}$$

This expression is known as the Wigner–Brillouin formula. Its disadvantage consists in the appearance of the unknown quantity E_n in the second and higher terms. However, by iterating we can obtain the usual results of (Rayleigh–Schrödinger) perturbation theory, thus

$$E_n^{(1)} = E_n^{(0)} + \langle \phi_n | H' | \phi_n \rangle,$$
$$E_n^{(2)} = E_n^{(0)} + \langle \phi_n | H' | \phi_n \rangle + \sum_{m \neq n} \frac{\langle \phi_n | H' | \phi_m \rangle \langle \phi_m | H' | \phi_n \rangle}{E_n^{(1)} - E_m^{(0)}} \, .$$

14. From (13.12), if we multiply by $\langle \phi_n |$, we have that

$$\langle \phi_n | \psi_n \rangle = C \lim_{E \to E_n} (E_n - E) D_n(E) = C N_n$$

where

$$N_n = -\lim_{E\to E_n}(E-E_n)D_n(E)$$

is the residue of $D_n(E)$ at the pole $E = E_n$. On the other hand, consider the identity

$$\langle\phi_n|R(E)|\psi_n\rangle = \left\langle\phi_n\left|\frac{1}{H-E}\right|\psi_n\right\rangle = \frac{1}{E_n-E}\langle\phi_n|\psi_n\rangle,$$

multiply both sides by $C(E_n-E)$, and take the limit $E\to E_n$. By (13.12), we find that $\langle\psi_n|\psi_n\rangle = C\langle\phi_n|\psi_n\rangle$. Since $\langle\psi_n|\psi_n\rangle = 1$, we obtain the result $C = N_n^{-1/2}$. Using (13.12) once more, we find that

$$|\psi_n\rangle = N_n^{-1/2}\lim_{E\to E_n}(E_n-E)R(E)|\phi_n\rangle = N_n^{-1/2}\lim_{E\to E_n}(E_n-E)\sum_m|\phi_m\rangle\langle\phi_n|R(E)|\phi_n\rangle$$
$$\equiv -N_n^{-1/2}\sum_m \text{Res}\,[\langle\phi_m|R(E)|\phi_n\rangle]_{E=E_n}|\phi_m\rangle.$$

The reader can verify that, as in the preceding problem, the usual formulae for the state vector in different orders of approximation of perturbation theory can be obtained from (14a).

It should be carefully noted that the normalization condition used above is $\langle\psi_n|\psi_n\rangle = 1$, whereas the normalization condition $\langle\phi_n|\psi_n\rangle = 1$ is often used in perturbation theory expressions.

15. From (15a) it can be seen that ψ is the product of two ground state wavefunctions of a hydrogen-like atom with nuclear charge $Z'e$:

$$\psi \equiv \psi_1\psi_2 = \sqrt{\frac{Z'^3}{\pi a^3}}\,e^{-\frac{Z'r_1}{a}}\cdot\sqrt{\frac{Z'^3}{\pi a^3}}\,e^{-\frac{Z'r_2}{a}} \tag{15.1}$$

Then, with the notation

$$H_i = -\frac{\hbar^2}{2m}\left(\frac{\partial^2}{\partial x_i^2}+\frac{\partial^2}{\partial y_i^2}+\frac{\partial^2}{\partial z_i^2}\right)-\frac{Z'e^2}{r_i}, \qquad i = 1, 2 \tag{15.2}$$

we have that

$$H_i\psi_i = -(Z')^2E_H, \qquad E_H = \frac{me^4}{2\hbar^2} = 13{\cdot}53 \text{ eV}. \tag{15.3}$$

From (VIII.6) it follows that

$$E_0 = \min\int\psi^*\left[-\frac{\hbar^2}{2m}(\Delta_1+\Delta_2)-Ze^2\left(\frac{1}{r_1}+\frac{1}{r_2}\right)+\frac{e^2}{r_{12}}\right]\psi\,dv$$
$$= \min\left[-2Z'^2E_H+(Z'-Z)e^2\int\psi^*\left(\frac{1}{r_1}+\frac{1}{r_2}\right)\psi\,dv+\int\psi^*\frac{e^2}{r_{12}}\psi\,dv\right]. \tag{15.4}$$

The first integral in (15.4) can be written as

$$\int\psi^*\left(\frac{1}{r_1}+\frac{1}{r_2}\right)\psi\,dv = 2\int\frac{\psi_1^2}{r_1}dv_1 = \frac{4Z'E_H}{e^2},$$

and the second integral is of the same form as (7.7), with Z' instead of Z.

Thus,

$$\int \psi^* \frac{e^2}{r_{12}} \psi \, dv = \frac{5}{4} Z' E_H$$

and hence

$$E_0 = \min \left[-2Z'^2 + \frac{5}{4} Z' + 4Z'(Z' - Z) \right] E_H.$$

The required minimum is attained when $Z' = Z - (5/16)$, so that

$$E_0 = -2\left(Z - \frac{5}{16}\right)^2 E_H. \tag{15.5}$$

Comparison of numerical results obtained from (15.5) with experimental data shows that in this case the variational method gives better results than the first order of perturbation theory.

16. To normalize $\psi(x, B)$ we require $A = (2B/\pi)^{1/4}$. By (II.17a), the ground state energy is then

$$E_0 = \min \int_{-\infty}^{+\infty} \psi^* H \psi \, dx = \min \left(\frac{\hbar^2 B}{2m} + \frac{m\omega^2}{8B} \right). \tag{16.1}$$

The minimum is attained when $B = m\omega/2\hbar$, so that

$$\psi_0(x) \equiv \psi\left(x, \frac{m\omega}{2\hbar}\right) = \left(\frac{m\omega}{\pi\hbar}\right)^{1/4} \exp\left(-\frac{m\omega x^2}{\hbar}\right), \quad E_0 = \frac{\hbar\omega}{2}. \tag{16.2}$$

Note that these results are in fact the accurate solutions of the problem (see problem 17 of Chapter II).

17. We have to choose a simple trial function, depending on a parameter, which is finite at all points of the x-axis, which vanishes as $x \to \pm\infty$, and which is orthogonal to the function ψ_0 given by the expression (16.2). It is clear that all these conditions are satisfied by

$$\psi_1(x, D) = Cx \exp(-Dx^2), \quad D > 0. \tag{17.1}$$

The normalization condition gives $C = 4(2\pi^{-1})^{1/2} D^{3/2}$. Then, by (VIII.7), we have

$$E_1 = \min \int \psi_1^* H \psi_1 \, dv = \min \left(\frac{3\hbar^2 D}{2m} + \frac{3mw^2}{8D} \right). \tag{17.2}$$

The minimum is attained when $D = m\omega/2\hbar$, so that

$$\psi_1(x) = \psi_1\left(x, \frac{m\omega}{2\hbar}\right) = \left(\frac{2}{\sqrt{\pi}}\right)^{1/2} \left(\frac{m\omega}{\hbar}\right)^{3/4} x \exp\left(-\frac{m\omega x^2}{2\hbar}\right), \quad E_1 = \frac{3}{2}\hbar\omega. \tag{17.3}$$

18. From the normalization condition we can determine the values of the constants A_1, A_2, A_3, thus:

$$A_1 = \left(\frac{b^3}{\pi a^3}\right)^{1/2}, \quad A_2 = \frac{1}{\pi}\left(\frac{b}{a^3}\right)^{1/2}, \quad A_3 = \left(\frac{b^5}{3\pi a^3}\right)^{1/2}. \tag{18.1}$$

For ψ_1 we have

$$E_0 = \min \int \psi_1^* \left(-\frac{\hbar^2}{2m} \Delta - \frac{e^2}{r} \right) \psi_1 \, dv$$
$$= \min \left\{ \frac{2\hbar^2 b^3}{ma^3} \int_0^\infty e^{-(b/a)r} \Delta e^{-(b/a)r} r^2 \, dr - \frac{4e^2 b^2}{a^3} \int_0^\infty e^{-(2b/a)r} r \, dr \right\}. \quad (18.2)$$

The two integrals immediately yield

$$\int_0^\infty e^{-(b/a)r} \Delta e^{-(b/a)r} r^2 \, dr = -\int_0^\infty \left(\frac{\partial}{\partial r} e^{-(b/a)r} \right)^2 r^2 \, dr = -\frac{a}{4b},$$

$$\int_0^\infty e^{-(2b/a)r} r \, dr = \frac{a^2}{4b^2}.$$

By substituting in (18.2) we then find that

$$E_0 = \min \left(\frac{\hbar^2 b^2}{2ma^2} - \frac{e^2 b}{a} \right) = -\frac{e^2}{2a}, \quad \psi_1 = \frac{1}{\sqrt{\pi a^3}} e^{-(r/a)}, \quad (18.3)$$

the minimum being attained when $b = 1$. By performing analogous calculations for the other two cases we obtain the results given in Table VIII.1.

TABLE VIII.1

$\psi =$	ψ_1	ψ_2	ψ_3
$b =$	1	$\frac{\pi}{4}$	$\frac{3}{2}$
$E_0 =$	$-E_H$	$-0{\cdot}81E_H$	$-0{\cdot}75E_H$

Note that the first set of results is accurate, while the other two are approximate only; this is related to the fact that ψ_1 behaves correctly at the origin and at infinity, ψ_2 behaves correctly at the origin but its asymptotic behaviour is different from that of the exact solution, while ψ_3 is incorrect at the origin but has a correct (exponentially decreasing) behaviour at infinity.

19. Since the $2s$ state wavefunction has spherical symmetry and vanishes as $r \to \infty$, we may suppose that it contains a factor $e^{-(b/a)r}$, where b is an adjustable parameter and a is the radius of the first Bohr orbit. We need another parameter in order to be able to apply the variational method to the first excited level, and this can be introduced by multiplying $e^{-(b/a)r}$ by a factor $((1+\gamma(r/a))$. We therefore take as a reasonable trial function for the $2s$ state wavefunction the form

$$\psi_{2s} = A\left(1+\frac{\gamma r}{a}\right)e^{-(b/a)r}. \tag{19.1}$$

From (VIII.7) and (VIII.7′) we obtain

$$\gamma = \frac{1}{3}(1+b), \quad A = \left[\frac{3b^5}{\pi a^3(a^2-ab+b^2)}\right]^{1/2}, \tag{19.2}$$

$$E_{2s} = \min \int \psi_{2s}^*\left(-\frac{\hbar^2}{2m}\nabla^2-\frac{e^2}{r}\right)\psi_{2s}\,d\mathbf{r} = \min\left\{\frac{e^2}{a}\left[-\frac{b}{2}+\frac{7b^2}{6}-\frac{b^2}{2(b^2-b+1)}\right]\right\}. \tag{19.3}$$

The minimum is attained when $b = \frac{1}{2}$, and thus

$$E_{2s} = -\frac{e^2}{8a}, \quad \psi_{2s} = (8\pi a^3)^{-1/2}\left(1-\frac{r}{2a}\right)e^{-(r/2a)}. \tag{19.4}$$

The expressions (19.4) are in fact exact (see problem 33, Chapter II). We could also have tried other forms for ψ_{2s}, for example, those given in problem 18, each multiplied by a factor $(1+\gamma(r/a))$. The results obtained would have been approximate only.

CHAPTER IX

Time-dependent Perturbations. Radiation Theory

1. Time-dependent Perturbations

Let us suppose that a system having a time-independent Hamiltonian H_0 is in the steady state m, at time $t = 0$, and that small, time-dependent, forces then begin to act on the system and continue to do so until a time $t = \tau$. If H' is the contribution to the Hamiltonian of the system due to these forces, then

$$H' = \begin{cases} V(t), & \text{for} \quad 0 \leqslant t \leqslant \tau \\ 0, & \text{for} \quad t < 0,\, t > \tau \end{cases} \tag{IX.1}$$

where $V(t)$ is the effective potential energy of the forces.

The probability, at time τ, that a transition $m \to n$ has occurred, due to the perturbation H', is given, to a first approximation, by

$$\omega_{nm}(\tau) = \frac{1}{\hbar^2} \left| \int_0^\tau \langle n|V(t)|m\rangle e^{i\omega_{nm}t} \right|^2, \qquad \hbar\omega_{nm} = E_n - E_m. \tag{IX.2}$$

If it happens that V is independent of t in the interval $0 \leqslant t \leqslant \tau$, then, by (IX.2),

$$\omega_{nm}(\tau) = \frac{4|\langle n|V|m\rangle|^2 \sin^2(\frac{1}{2}\omega_{nm}\tau)}{\hbar^2\omega_{nm}^2}. \tag{IX.3}$$

For any given $E_n (\neq E_m)$, this is a periodic function of τ, with a peak for $E_n \sim E_m$. In practice, the final states often have a continuous (or nearly continuous) energy spectrum. Then, if $\varrho(E)$ is such that the number of final states having energies in any range E to $E+dE$ is $\varrho(E)\,dE$, the total probability of transition summed over all final states (the main contribution coming, of course, from those whose energies are close to that of the initial state) is given by "Fermi's Golden Rule":

$$\omega_{nm}(\tau) = \frac{2\pi}{\hbar} |\langle n|V|m\rangle|^2 \varrho(E_n)\tau \tag{IX.4}$$

in which the matrix element is to be calculated for a typical state n of the set of final states having $E_m \sim E_n$. Since this result is proportional to τ, it follows that a transition probability per unit time, $P_{nm} = \omega_{nm}(T)/\tau$, can be defined. If the perturbation is periodic in time $V(t) = \sum_{\pm} V_0^{\pm} e^{\pm i\omega t}$, the transition probability has a form similar to that of (IX.4), and leads to a probability per unit time

$$P_{nm} = \frac{2\pi}{\hbar} |\langle n | V_0^{\pm} | m \rangle|^2 \varrho(E), \quad E_n = E_m \mp \hbar\omega. \tag{IX.5}$$

Note that $V(t)$ must be Hermitian, i.e. real, unless the $V_0^{\pm}$ are operators, in which case V_0^- must be the Hermitian adjoint of V_0^+.

2. Radiation

In classical theory,† it is shown that the vector potential of the electromagnetic radiation field can be expanded in plane waves, thus:

$$\mathbf{A} = \sum_{\lambda} (b_\lambda \mathbf{A}_\lambda + b_\lambda^* \mathbf{A}_\lambda^*), \quad \mathbf{A}_\lambda = \mathbf{e}_\lambda \sqrt{\frac{4\pi c^2}{V}} e^{i\mathbf{k}_\lambda \cdot \mathbf{r}}, \quad k_\lambda = \frac{\omega_\lambda}{c}, \tag{IX.6}$$

where $\mathbf{e}_\lambda$ includes (for each value of $\mathbf{k}_\lambda$) the two possible directions of polarization, V is the volume of the normalization cube and the sum is taken over all possible plane waves which are periodic in this cube. By taking

$$q_\lambda = b_\lambda + b_\lambda^*, \quad p_\lambda = -i\omega_\lambda (b_\lambda - b_\lambda^*),$$

the energy of the radiation field can be written in the Hamiltonian form

$$H_{\text{rad}} = \sum_{\lambda} \tfrac{1}{2}(p_\lambda^2 + \omega_\lambda^2 q_\lambda^2). \tag{IX.7}$$

In the quantum theory of radiation, the Hamiltonian H has the form (IX.7), but the quantities p_λ, q_λ, are operators which satisfy the commutation rules

$$[p_\lambda, q_\mu] = -i\hbar\delta_{\lambda\mu}, \quad [q_\lambda, q_\mu] = [p_\lambda, p_\mu] = 0. \tag{IX.8}$$

The vector potential (operator) and the Hamiltonian (operator) of the radiation field are then:

$$\mathbf{A} = \sum_{\lambda} (b_\lambda \mathbf{A}_\lambda + b_\lambda^+ \mathbf{A}_\lambda^*),$$
$$b_\lambda = \frac{1}{2\omega_\lambda}(\omega_\lambda q_\lambda + ip_\lambda), \quad b_\lambda^+ = \frac{1}{2\omega_\lambda}(\omega_\lambda q_\lambda - ip_\lambda); \tag{IX.9}$$

$$H_{\text{rad}} = \sum_{\lambda} \hbar\omega_\lambda (a_\lambda a_\lambda^+ + \tfrac{1}{2}),$$
$$a_\lambda = \sqrt{\frac{2\omega_\lambda}{\hbar}}\, b_\lambda, \quad a_\lambda^+ = \sqrt{\frac{2\omega_\lambda}{\hbar}}\, b_\lambda^+, \tag{IX.10}$$

† L. D. Landau and E. M. Lifshitz, *Classical Theory of Fields*, Addison–Wesley, Reading, Massachusetts, 1951.

where a_λ^+ and a_λ are creation and annihilation operators respectively (see section 3 and problem 15, Chapter VII), and satisfy the commutation rules for Bose particles (which are, in this case, the radiation field quanta or "photons"):

$$[a_\lambda, a_\mu^+] = \delta_{\lambda\mu}, \quad [a_\lambda, a_\mu] = [a_\lambda^+, a_\mu^+] = 0. \tag{IX.11}$$

The total field energy eigenvalues can be written in the form

$$E = \sum_\lambda E_\lambda n_\lambda + \sum_\lambda \frac{\hbar}{2}\omega_\lambda = \sum_\lambda E_\lambda n_\lambda + E_0, \tag{IX.12}$$

where $E_\lambda = \hbar\omega_\lambda$, and n_λ is the number of photons of energy E_λ. If all n_λ are equal to zero, the field has an energy E_0, called the vacuum energy. The presence of the (infinite) constant E_0 in the expression for the field energy does not affect the emission, absorption and scattering of light, since in these processes E_0 does not change, any change in field energy being due solely to the first term of (IX.12).

3. The Interaction of Radiation with Atomic Systems

A system which consists of an electron in a radiation field is described in non-relativistic quantum mechanics by the Hamiltonian (cf. problem 32, Chapter VI),

$$H = \frac{1}{2m}\left(\mathbf{p} - \frac{e}{c}\mathbf{A}\right)^2 + V + H_{\text{rad}} = \frac{\mathbf{p}^2}{2m} + V - \frac{e}{mc}(\mathbf{A}.\mathbf{p}) + \frac{e^2}{2mc^2}\mathbf{A}^2 + H_{\text{rad}}, \tag{IX.13}$$

where the potential ϕ of the radiation field is taken to be equal to zero,† and V is, for example, the electrostatic field of a nucleus. The part of the Hamiltonian which represents the interaction between the electron and the radiation is thus

$$H' = -\frac{e}{mc}(\mathbf{A}.\mathbf{p}) + \frac{e^2}{2mc^2}\mathbf{A}^2. \tag{IX.14}$$

By treating H' as a perturbation, it follows from (IX.14) that the probability per unit time for the electron to make a transition (due to the absorption of a photon of energy $E_\lambda = \hbar\omega_\lambda$) from a steady state 1 to a steady state 2 of the unperturbed electronic Hamiltonian, is given by

$$dP_{21} = \frac{2\pi}{\hbar}|\langle 2; \ldots n_\lambda - 1, \ldots | H' | 1; \ldots n_\lambda, \ldots \rangle|^2 \varrho(E)\, d\Omega \tag{IX.15}$$

where $d\Omega$ is an element of solid angle about the direction of propagation of the photon, and

$$\varrho(E) = \frac{\omega^2 V}{(2\pi c)^3 \hbar} \tag{IX.16}$$

is the number of states of the radiation field in the volume V per unit energy interval and per unit solid angle.

† For a pure radiation field, the gauge can be chosen to be such that $\phi = 0$, $\nabla.\mathbf{A} = 0$.

Using the properties of the creation and annihilation operators, one finds that

$$\langle 2; \ldots, n_\lambda - 1, \ldots |H'| 1; \ldots, n_\lambda, \ldots\rangle = -\frac{e}{m}\sqrt{\frac{2\pi\hbar n_\lambda}{V\omega_\lambda}}\langle 2 | e^{i\mathbf{k}_\lambda \cdot \mathbf{r}}(\mathbf{p}\cdot\mathbf{e}_\lambda)| 1\rangle. \quad \text{(IX.17)}$$

Writing, for simplicity, $\mathbf{k}_\lambda = \mathbf{k}$, $\mathbf{e}_\lambda = \mathbf{e}$, we have that

$$dP_{21} = \frac{4\pi^2 e^2 n_\lambda}{V\omega m^2} |\langle 2 | e^{i\mathbf{k}\cdot\mathbf{r}}(\mathbf{p}\cdot\mathbf{e})| 1\rangle|^2 \varrho(E)\, d\Omega. \quad \text{(IX.18)}$$

Similarly, for emission, one obtains the result

$$dP_{12} = \frac{4\pi e^2(n_\lambda + 1)}{V\omega m^2} |\langle 1 | e^{-i\mathbf{k}\cdot\mathbf{r}}(\mathbf{p}\cdot\mathbf{e})| 2\rangle|^2 \varrho(E)\, d\Omega. \quad \text{(IX.18')}$$

It follows from (IX.18′) that the transition probability dP_{12} consists of two terms. The first is proportional to the radiation field intensity (i.e. to the number n_λ of photons present) and corresponds to "stimulated" emission. The second term is independent of n_λ and corresponds to "spontaneous" emission. It very often happens that the wavelength of the absorbed (or of the emitted) photon is much greater than the linear dimensions of the region in which the probability of finding the electron is appreciably different from zero. In such cases the exponential factor can be put equal to unity, and, for emission, we then have the "dipole" transition probability:

$$dP_{12} = \frac{e^2\omega}{m^2 2\pi\hbar c^3} |(\mathbf{p}\cdot\mathbf{e})_{12}|^2 (n_\lambda + 1)\, d\Omega = \frac{\omega^3}{2\pi\hbar c^3} |(\mathbf{d}\cdot\mathbf{e})_{12}|^2 (n_\lambda + 1)\, d\Omega, \quad \text{(IX.19)}$$

where $\mathbf{d} = e\mathbf{r}$ is the electric dipole moment operator of the electron. Choosing the states of plane-polarization of the photon to be in the plane $(\mathbf{d}, \mathbf{k})$, and perpendicular to it, respectively, we can obtain from (IX.19)

$$dP_{12} = \frac{\omega^3}{2\pi\hbar c^3} |\mathbf{d}_{12}|^2 (n_\lambda + 1) \sin^2\theta\, d\Omega \quad \text{(IX.20)}$$

and, similarly,

$$dP_{21} = \frac{\omega^3}{2\pi\hbar c^3} |\mathbf{d}_{12}|^2 n_\lambda \sin^2\theta\, d\Omega \quad \text{(IX.20')}$$

where θ is the angle between the vectors $\mathbf{d}_{12}$ and $\mathbf{k}$. By integrating over the angles, we find finally from (IX.20) that the total energy of spontaneous radiation per excited atom per unit time is $4\omega^4/3c^3\, |\mathbf{d}_{12}|^2$. This expression is the same as the classical one if we make the equation $|\mathbf{d}_{12}|^2 \leftrightarrow \frac{1}{2}\mathbf{d}^2$, where $\mathbf{d}^2$ is the mean square classical dipole moment of the radiating system. The above relations, derived for one electron, are also valid for a system of electrons if $\mathbf{d}$ now stands for the sum of the dipole moments of all the electrons, and states 1 and 2 now refer, of course, to many-particle states.

If the dipole transition probability vanishes, the transition is said to be "forbidden". By expanding the exponential factor in (IX.18′) in a power series, expressions for the probabilities of the transitions known as "magnetic dipole", "electric quadrupole", and so on, can be obtained.

Problems

1. A charged-particle linear harmonic oscillator is in a time-dependent homogeneous electric field given by

$$\mathcal{E}(t) = \frac{A}{\sqrt{\pi}\,\tau} e^{-(t/\tau)^2}, \tag{1a}$$

where A and τ are constants. If, at $t = -\infty$, the oscillator is in its ground state, find, to a first approximation, the probability that it will be in its first excited state at $t = +\infty$.

2. A hydrogen atom is placed in a time-dependent homogeneous electric field given by

$$\mathcal{E}(t) = \frac{B\tau}{e\pi}\,\frac{1}{\tau^2+t^2} \tag{2a}$$

where B and τ are constants. If, at $t = -\infty$, the atom is in its ground state, calculate the probability that it will be in a $2p$ state at $t = +\infty$.

3. Calculate the probability that an electron in an atom will make a transition from the stationary state m to the stationary state n under the influence of a heavy charged particle which passes near the atom. Discuss the result obtained.

4. Let m and n be two stationary states of an atom on which the perturbation (IX.1) acts. Show that if the "adiabatic approximation" condition is satisfied:

$$\omega_{nm}^{-1} \frac{d}{dt} \langle n|V(t)|m\rangle \ll |E_n - E_m|, \quad \omega_{nm} = \frac{E_n - E_m}{\hbar} \tag{4a}$$

(i.e. if the change in time of the interaction energy which occurs during one period of oscillation of the atomic system is much smaller than the difference between the energies of the initial and the final states), then the probability of the transition $m \to n$ is very small. Find an approximate expression for the probability of the transition $m \to n$ in the opposite case, in which

$$\omega_{nm}^{-1} \frac{d}{dt} \langle n|V(t)|m\rangle \gg |E_n - E_m|, \tag{4b}$$

during the time in which the perturbation increases from zero to a maximum value V, i.e. the perturbation is "suddenly switched on".

5. A charged-particle linear harmonic oscillator is in its ground state when a time-dependent electric field is suddenly switched on. Calculate the probability of excitation of the nth level of the oscillator, assuming that the conditions under which perturbation theory can be used are not satisfied.

6. Calculate the probability of excitation of the $2s$ level of a hydrogen-like atom if a sudden change of the nuclear charge, $Z \to Z \pm 1$, occurs (e.g., due to $\beta^{\mp}$ decay).

7. "Magnetic resonance" phenomena are due to the absorption of energy by a system (having a spin and a magnetic moment) from a rotating magnetic field $\mathbf{H}_1 \equiv (H_{1x} = H_1 \cos \omega t,\ H_{1y} = H_1 \sin \omega t,\ H_{1z} = 0)$, which results in a change in the orientation of the spin of the system relative to a constant homogeneous magnetic field $\mathbf{H}_0 \equiv (0, 0, H_0)$. Denoting by γ the gyromagnetic ratio of the system, the rate of change with time of the spin wavefunction χ of the system is given by

$$i\hbar \frac{\partial \chi}{\partial t} = H\chi = -\gamma\hbar[H_0 s_z + H_1(s_x \cos \omega t + s_y \sin \omega t)]\,\chi. \tag{7a}$$

Suppose that, at $t = 0$, the spin is in the state m_s. Calculate the probability of finding it at time t in a state m_s' ($m_s, m_s' = -s, -s+1, \ldots, s-1, +s$). Study in detail the case $s = \frac{1}{2}$.

8. Obtain, to the first order of approximation of time-dependent perturbation theory, the differential cross-section for scattering by a constant potential (the "first Born approximation", see (X.10)).

9. Find the differential cross-section for inelastic scattering of a fast electron by a hydrogen atom in its ground state, which, as a result of the collision, is excited to the $2s$ state. Any exchange effects between the colliding electron and the bound one may be neglected.

10. Derive expressions for the transition probability due to the perturbation (IX.1), to arbitrarily high orders of perturbation theory.

11. Find the probability that the oscillator of problem 1 will make a transition to an excited state $n > 1$ under the influence of the perturbation (1a).

12. Let $H = H_0 + V$ be the Hamiltonian of two conservative systems in interaction. In the Schrödinger picture H_0, V and H are time independent.† Show that:

1. The matrix elements of the S-matrix defined in problem 10 can be written as

$$\langle n | S | m \rangle = \delta_{nm} - 2\pi i\, \delta(E_n - E_m) \langle n | T | m \rangle, \tag{12a}$$

where

$$\langle n | T | m \rangle = \langle n | V | m \rangle + \sum_l \frac{\langle n | V | l \rangle \langle l | V | m \rangle}{E_m - E_l - i\eta} + \sum_{l, l'} \frac{\langle n | V | l \rangle \langle l | V | l' \rangle \langle l' | V | m \rangle}{(E_m - E_l + i\eta)(E_m + E_{l'} + i\eta)} + \ldots, \tag{12b}$$

η being a small positive number.

† See the note on page 245.

2. The operator T satisfies the equation

$$T = V+V(E_m-H_0+i\eta)^{-1}T. \tag{12c}$$

13. A hydrogen atom in its ground state is placed in an electric field $\mathcal{E}(t) = \mathcal{E}_0 \sin \omega t$ of angular frequency $\omega > me^4/2\hbar^3$. Find the probability per unit time that the atom will be ionized. The wavefunctions of the electron in the ionized states may be taken to be plane waves.

14. Show that the term in $\mathbf{A}^2$ in the interaction Hamiltonian of an electron in a radiation field

$$H' = -\frac{e}{mc}(\mathbf{p}\,.\,\mathbf{A})+\frac{e^2}{2mc^2}\mathbf{A}^2 \tag{IX.14}$$

gives a contribution to the transition probability of a process only if two photons take part in that process.

15. Show that free electrons cannot absorb or emit photons.

16. Show that the total probability of spontaneous emission of an excited atom per unit time is of the order of $\omega/(137)^3$.

17. The intensity of magnetic dipole, and of electric quadrupole, radiation is proportional to the square of the matrix element of the first-order term in the expansion in series of the exponential in (IX.18′):

$$\langle 1\,|(\mathbf{k}\,.\,\mathbf{e})(\mathbf{p}\,.\,\mathbf{e})|\,2\rangle. \tag{17a}$$

Separate, in (17a), the part which represents electric quadrupole radiation from that which represents magnetic dipole radiation.

18. Find selection rules for the emission and absorption of electric dipole, electric quadrupole and magnetic dipole radiation, by an electron moving in a central field.

19. Show that, in quantum theory (as well as in classical theory), a charged linear harmonic oscillator can absorb or emit radiation of frequency equal to its own classical frequency only.

20. Using the expressions (IX.18) and (IX.18′) for the absorption and emission probabilities, find the spectral distribution of black-body radiation (Planck's formula).

21. Find the probabilities of the spontaneous transitions $(n, l) \to (n', l+1)$ and $(n, l) \to (n', l-1)$ for an electron in a central field.

22. Find the ratio of the intensities of the first two lines of the "Balmer series" of hydrogen.

23. The process of absorption of a photon by a bound particle, when the photon energy is greater than the binding energy of the particle, is called a "photoelectric" process. Find the differential cross-section of this process for a hydrogen-like atom in its ground state.

24. Consider a system of charged particles (atom, molecule, etc.) placed in the field of an electromagnetic plane wave whose wavelength is much greater than the linear dimensions of the system, and which induces in the system an electric dipole moment $\mathbf{d} = \beta\mathcal{E}$, proportional to the electric field amplitude $\mathcal{E} = \mathcal{E}_0 \cos \omega t$. Show how the quantity β, called the "polarization tensor" of the system, may be calculated, assuming the eigenstates of the unperturbed system to be brown.

It may be supposed that there are no eigenfrequencies of the system near the frequency of the external field.

Solutions

1. Note that the "characteristic time" τ enters into the expression for the perturbation in such a way that the total linear momentum P given (classically) by the field to the oscillator does not depend on τ,

$$P = \int_{-\infty}^{+\infty} e\mathcal{E}(t)\, dt = \frac{eA}{\sqrt{\pi}\,\tau} \int_{-\infty}^{+\infty} e^{-(t/\tau)^2}\, dt = eA.$$

The transition probability from the state n to the state k is given by (IX.2) to be

$$\omega_{kn} = \frac{1}{\hbar^2} \left| \int_{-\infty}^{+\infty} \langle k \,|\, W \,|\, n \rangle e^{i\omega_{kn}t}\, dt \right|^2, \tag{1.1}$$

where $W = -x\mathcal{E}(t)e$ is the effective (time-dependent) potential energy of the perturbing force. Knowing the matrix elements of x for the oscillator states (see (21.4) Chapter V),

$$\langle n \,|\, x \,|\, n+1 \rangle = \langle n+1 \,|\, x \,|\, n \rangle = \sqrt{\frac{(n+1)\hbar}{2m\omega}}, \tag{1.2}$$

we have that

$$\langle 0 \,|\, W \,|\, 1 \rangle = \langle 1 \,|\, W \,|\, 0 \rangle = \frac{P}{\sqrt{\pi}\tau} \sqrt{\frac{\hbar}{2m\omega}}\, e^{-(t/\tau)^2}. \tag{1.3}$$

Substituting (1.3) in (1.1) and remembering that $\omega_{10} = \omega$, we obtain

$$\omega_{10} = \frac{P^2}{2\pi\tau^2 m\hbar\omega} \int_{-\infty}^{+\infty} e^{i\omega t-(t/\tau)^2}. \tag{1.4}$$

Now, since

$$\int_{-\infty}^{+\infty} e^{i\beta x - \alpha x^2}\, dx = \sqrt{\frac{\pi}{\alpha}}\, e^{-(\beta^2/4\alpha)},$$

we find finally that

$$\omega_{10} = \frac{P^2}{2m\hbar\omega} e^{-(1/2)(\omega\tau)^2}. \tag{1.5}$$

Transition probabilities to levels with $n > 1$ are obtained only in higher orders of perturbation theory (see problem 11).

Let us discuss (1.5) as a function of the characteristic time τ. It can be seen that if $\tau \gg 1/\omega$ (i.e. if the time during which the perturbation has an appreciable value is much longer than the period of a classical oscillation), then the probability ω_{10} is very small. This is the "adiabatic" case (cf. problem 4). At the other extreme, if $\tau \ll 1/\omega$, then the probability ω_{10} is approximately constant. As $\tau \to 0$ the perturbation tends towards a δ-function of t, and the probability ω_{10} then appears as the ratio of the (classical) energy $P^2/2m$ given to the oscillator to the energy quantum $\hbar\omega$. For perturbation theory to be valid, the probability of a transition to an excited state has to be much smaller than the probability of remaining in the ground state, i.e. $\omega_{10} \ll (1-\omega_{10})$, whence $\omega_{10} \ll 1$. From (1.5) it can be seen that a sufficient condition for this is that

$$\frac{P^2}{2m} = \frac{(eA)^2}{2m} \ll \hbar\omega. \tag{1.6}$$

It is interesting to note that, if (1.6) holds, the possibility of excitation of the oscillator is a purely quantum-mechanical effect, since, from a semi-classical point of view, the energy $P^2/2m$ received from the field would be insufficient to allow the energy jump $\omega\hbar$ to occur. In the quantized system, however, the quantity $P^2/2m$ does not in fact represent the energy received by the oscillator from the field, so that the paradox is only an apparent one.

2. Choosing the z-axis in the direction of the electric field, it follows from (IX.2) that the required probability is

$$\omega = \frac{1}{\hbar^2} \left| \int_{-\infty}^{\infty} \langle 210 | V | 100 \rangle e^{i\omega t}\, dt \right|^2 \tag{2.1}$$

where $V = -er\mathscr{E}(t) \cos\theta$ and $\omega = 3me^4/8\hbar^2$. From Table 1 of Chapter II we see that the only possible transition from the ground state to the $2p$ state corresponds to the matrix element

$$\langle 210 | r\cos\theta | 100 \rangle = \frac{1}{4\sqrt{2}} \frac{1}{\pi a^3} \int \frac{r}{a} e^{-(r/2a)} \cos\theta (r\cos\theta) e^{-(r/a)}\, d\mathbf{r} = \frac{2^7\sqrt{2}a}{3^5}. \tag{2.2}$$

Substituting (2a) and (2.2) in (2.1), we obtain

$$\omega = \frac{2^{15}B^2\tau^2a^2}{3^{10}\hbar^2\pi^2} \left| \int_{-\infty}^{+\infty} \frac{e^{i\omega t}}{\tau^2+t^2}\, dt \right|^2. \tag{2.3}$$

The integral in (2.3) can be calculated using the theory of residues. The contour of integration in the plane of (complex) t is along the real axis, and can be closed by a semicircle of infinite radius in the upper half-plane (Fig. IX.1). On this semicircle the integral vanishes on account of the factor $e^{i\omega t}$.

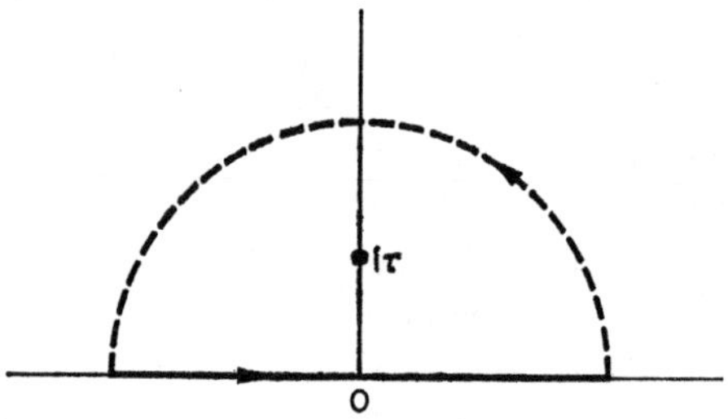

FIG. IX.1.

The only pole inside the contour of integration is at $t = +i\tau$, and hence

$$\int_{-\infty}^{+\infty} \frac{e^{i\omega t}}{t^2+\tau^2}\,dt = 2\pi i \operatorname{Res}\left(\frac{e^{i\omega t}}{t^2+\tau^2}\right)_{t=i\tau} = \frac{\pi}{\tau}\,e^{-\omega\tau}. \tag{2.4}$$

Finally, the required probability is

$$\omega = \frac{2^{15}B^2a^2}{3^{10}\hbar^2}\,e^{-2\omega\tau}. \tag{2.5}$$

This result can be discussed as in the preceding problem; the total linear momentum given by the electric field to the hydrogen atom is equal to B.

3. Since the particle is heavy, we suppose that its motion is not changed by any interaction with the atom. We suppose also that the nucleus of the atom remains at rest. Let us choose a coordinate system with its origin at the nucleus, and with its x-axis parallel to the direction of the particle's motion. Let v be the (constant) velocity of the latter and $\mathbf{R} \equiv (vt, D, 0)$ its position vector at time t. D is the distance of nearest approach to the atom, attained at time $t = 0$. If $\mathbf{r} \equiv (x, y, z)$ is the position vector of the electron, the potential energy of particle-electron interaction is given by

$$W(t) = -\frac{Ze^2}{|\mathbf{R}-\mathbf{r}|} = -\frac{Ze^2}{R} - \frac{Ze^2(xvt+Dy)}{R^3} + \ldots \tag{3.1}$$

where $R = \sqrt{(vt)^2+D^2}$ and Ze is the electric charge of the particle. Since $x, y \ll R$, in (3.1) we can restrict ourselves to the first two terms only, and we have then that

$$\langle n|W(t)|m\rangle = -\frac{Ze^2}{R^3}(x_{nm}vt+Dy_{nm}), \tag{3.2}$$

where x_{nm} and y_{nm} are the matrix elements of the coordinates x and y of the electron.

Substituting (3.2) in (IX.2) we obtain the transition probability from the state m to the state n of the atom:

$$\omega_{nm} = \frac{Z^2 e^4}{\hbar^2} \left| \int_{-\infty}^{+\infty} \frac{x_{nm} vt + D y_{nm}}{[(vt)^2 + D^2]^{3/2}} e^{i\omega_{nm} t}\, dt \right|^2 . \tag{3.3}$$

The integrand decreases rapidly with R. We can say that the effective duration of the interaction, or "collision", between the particle and the atom is of order D/v. Note that, as in problem 1, if the adiabatic condition $(\omega_{nm} D)/v \gg 1$ holds (i.e. if the effective duration of the collision is much greater than the characteristic period, ω_{nm}^{-1}, of the quantized system), then the value of the integral is almost zero and the collision cannot excite the atom. In the opposite limiting case, $(\omega_{nm} D)/v \ll 1$, the exponential in (3.3) can be put equal to unity, and the integral reduces to

$$\int_{-\infty}^{+\infty} \frac{x_{nm} vt + D y_{nm}}{[(vt)^2 + D^2]^{3/2}}\, dt = \int_{-\infty}^{+\infty} \frac{D y_{nm}}{[(vt)^2 + D^2]^{3/2}}\, dt = \frac{2 y_{nm}}{vD}$$

(note that the substitution $vt = D \tan \theta$ enables the integral to be evaluated) and, finally,

$$\omega_{nm} = \frac{4 Z^2 e^4 |y_{nm}|^2}{\hbar^2 D^2 v^2} . \tag{3.4}$$

4. Integrating by parts, we find that

$$\int_0^\tau \langle n | V(t) | m \rangle e^{i\omega_{nm} t}\, dt = \frac{1}{i\omega_{nm}} \left[\langle n | V(t) | m \rangle e^{i\omega_{nm} t} \right]_0^\tau - \frac{1}{i\omega_{nm}} \int_0^\tau e^{i\omega_{nm} t} \frac{d}{dt} \langle n | V(t) | m \rangle\, dt. \tag{4.1}$$

Substituting (4.1) in (IX.2), and remembering that $\langle n | V(t) | m \rangle$ vanishes for $t = 0$ and for $t = \tau$, we obtain

$$\omega_{nm}(\tau) = \frac{1}{\hbar^2 \omega_{nm}^2} \left| \int_0^\tau e^{i\omega_{nm} t} \frac{d}{dt} \langle n | V(t) | m \rangle\, dt \right|^2 . \tag{4.2}$$

For an adiabatic variation of the perturbation (4a) the second factor in the integral changes very slowly in comparison with the first, so that it can be taken outside the integration, with the result that

$$\omega_{nm}(\tau) = \frac{4}{\hbar^2 \omega_{nm}^4} \left| \frac{d}{dt} \langle n | V(t) | m \rangle \right|^2 \sin^2 \left(\frac{\tau}{2} \omega_{nm} \right). \tag{4.3}$$

Remembering (4a), it can be seen immediately that the probability $\omega_{nm}(\tau)$ is very small. Note that (4.3) can also be obtained without using perturbation theory, and that it is known as the "adiabatic" approximation.†

If the perturbation is applied suddenly, i.e. if $V(t)$ increases from zero to a maximum value V in a time Δt which is short in comparison with the periodic time ω_{nm}^{-1} (so that (4b) holds), and, subsequently, it varies and/or vanishes adiabatically, then the major contribution to the integral in (4.2) occurs in the interval Δt in which the perturbation is applied. In this time interval the factor $e^{i\omega_{nm}t}$ changes very little, and therefore can be regarded as a constant and taken outside the integral. One obtains then

$$\omega_{nm} \approx \frac{|\langle n\,|\,V\,|\,m\rangle|^2}{\hbar^2\omega_{nm}^2}, \tag{4.4}$$

where V is the maximum of the perturbation. Instances of (4a) and of (4b) have both been met already in problems 1, 2 and 3. Note that (4.4) can be used to calculate transition probabilities due to a sudden perturbation only if the maximum V is small enough for perturbation theory to be valid. Otherwise we have to use another method (see problems 5 and 6).

5. If the field is sufficiently weak for perturbation theory to be valid, expression (4.4) can be used. There is, however, another way of calculating the required probabilities which is independent of perturbation theory, and which is valid even if the field $\mathscr{E}$ is a strong one.

Let us suppose that the field is applied suddenly at $t = 0$. For $t \leqslant 0$ the oscillator is in the ground state $\psi_0(x)$. Since the perturbation is applied suddenly (i.e. in a time much shorter than the period of oscillation $1/\omega$), the wavefunction at time $t = +0$ will be the same as it was at $t = -0$ (see (17.12), of Chapter II), viz.,

$$\psi_0(x) = x_0^{-1/2}\pi^{-1/4}\exp\left[-\frac{1}{2}\left(\frac{x}{x_0}\right)^2\right], \tag{5.1}$$

where $x_0 = (\hbar/m\omega)^{1/2}$. However, for $t = +0$ the Hamiltonian of the system is

$$H = -\frac{\hbar^2}{2m}\frac{d^2}{dx^2} + \frac{m\omega^2x^2}{2} - e\mathscr{E}x, \tag{5.2}$$

or, in terms of a new variable $x_1 = x - e\mathscr{E}/m\omega^2$,

$$H = -\frac{\hbar^2}{2m}\frac{d^2}{dx_1^2} + \frac{m\omega^2}{2}x_1^2 - \frac{(e\mathscr{E})^2}{2m\omega^2}. \tag{5.3}$$

Now the Hamiltonian (5.3) is, except for a constant, that of an oscillator of the same frequency as the unperturbed one, but oscillating about the point $x = a$, where $a = e\mathscr{E}/m\omega^2$. The corresponding wavefunctions $\phi_n(x)$ can therefore be obtained from (17.12) of Chapter II by making the substitution $x \to x - a$, thus,

$$\phi_n(x) \equiv \psi_n(x-a) = \pi^{-1/4}(2^n n!\,x_0)^{-1/2}\,e^{-(x-a)^2/2x_0^2}\,H_n\left(\frac{x-a}{x_0}\right). \tag{5.4}$$

† M. Born and V. Fock, *Zs. f. Phys.* **51**, 165 (1928).

The initial wavefunction $\psi_0(x)$ is not an eigenfunction of the new Hamiltonian (5.2), but it can be written as a superposition of the new eigenfunctions,

$$\psi_0(x) = \sum_{n=0}^{\infty} c_n \phi_n(x) = \sum_{n=0}^{\infty} c_n \psi_n(x-a), \tag{5.5}$$

the c_n being constants. The time-dependent wavefunction of the oscillator in the electric field ($t \geqslant 0$) is then

$$\psi(x, t) = \sum_{n=0}^{\infty} c_n \psi_n(x-a) e^{-\frac{i}{\hbar} E_n t}, \tag{5.6}$$

and the probability of finding the oscillator at a time $t > 0$ in the state ϕ_n of energy $E_n = \hbar\omega\left(n+\frac{1}{2}\right) - \frac{(e\mathscr{E})^2}{2m\omega^2}$ is equal to $|c_n|^2$. Now

$$c_n = \int_{-\infty}^{+\infty} \psi_0(x)\, \psi_n^*(x-a)\, dx$$

$$= x_0^{-1}(\pi 2^n n!)^{-1/2} \int_{-\infty}^{+\infty} \exp\left[-\frac{1}{2}\left(\frac{x}{x_0}\right)^2 - \frac{1}{2}\left(\frac{x-a}{x_0}\right)^2\right] H_n\left(\frac{x-a}{x_0}\right) dx. \tag{5.7}$$

The integral in (5.7) is the same as in (11.2) of Chapter V if the change of variable $\frac{x-a}{x_0} = \xi$ is made. Hence

$$c_n = (-1)^n \frac{\left(\frac{a}{x_0}\right)^n \exp\left(-\frac{1}{4}\frac{a^2}{x_0^2}\right)}{\sqrt{2^n n!}}, \tag{5.8}$$

and, finally,

$$\omega_n = |c_n|^2 = \frac{\left(\frac{a}{x_0}\right)^{2n} \exp\left(-\frac{1}{2}\frac{a^2}{x_0^2}\right)}{2^n n!}. \tag{5.9}$$

Note that this expression has the form of a Poisson distribution $\frac{1}{n!}(\bar{n})^n e^{-\bar{n}}$, with the mean value

$$\bar{n} = \frac{1}{2}\left(\frac{a}{x_0}\right)^2 = \frac{e^2\mathscr{E}^2}{2mn\omega^3}. \tag{5.10}$$

The method given above is valid whatever the strength of the field. If $\mathscr{E} \gg (m\hbar\omega^3)^{1/2}/e$, i.e. if $\bar{n} \gg 1$, we have $\omega_0 = e^{-\bar{n}} \ll 1$. In other words the probability that the oscillator will remain in its ground state is very small. Since ω_n reaches its maximum value for $n = \bar{n}$, the most probable excited level is the one which has energy $E_{\bar{n}}$. The opposite limiting case,

$\bar{n} \ll 1$, corresponds to a weak perturbing field. In this case, from (5.10), we find that $\omega_n \approx \frac{1}{n!}(\bar{n})^n$. Since this expression decreases very rapidly with n, it has a negligible value except for $n = 1$, in which case

$$\omega_1 \approx \bar{n} = \frac{e^2\mathcal{E}^2}{2m\hbar\omega^3} \ll 1. \tag{5.11}$$

In this case perturbation theory is valid, and, as is to be expected, the result (4.4) obtained by means of perturbation theory (for a sudden switching on of the perturbation) is the same as (5.11).

6. The initial state of the atom has the wavefunction

$$\psi_{100} = \sqrt{\frac{Z^3}{\pi a^3}} \exp\left(-\frac{Zr}{a}\right), \quad a = \frac{\hbar}{me^2}. \tag{6.1}$$

Using the same method as in the preceding problem, the required probability is found to be

$$\omega(1s \to 2s) = \left|\int \phi^*_{200}\psi_{100}\, dV\right|^2 \tag{6.2}$$

where ϕ_{200} is the wavefunction of the $2s$ state of a hydrogen-like atom having a nuclear charge $(Z\pm 1)e$. From Table II.1 of Chapter II it can be seen that

$$\int \phi^*_{200}\psi_{100}\, dV = \int \frac{1}{4}\sqrt{\frac{(Z\pm 1)^3}{2\pi a^3}}\left[2-\frac{(Z\pm 1)\,r}{a}\right]\exp\left[-\frac{(Z\pm 1)\,r}{2a}\right]\times$$
$$\times\sqrt{\frac{Z^3}{\pi a^3}}\exp\left(-\frac{Zr}{a}\right)dV = (\mp 2)\frac{[2^3Z(Z\pm 1)]^{3/2}}{(3Z\pm 1)^4}.$$

The required probability is thus

$$\omega(1s \to 2s) = \frac{2^{11}Z^3(Z\pm 1)^3}{(3Z\pm 1)^8}. \tag{6.3}$$

For large values of Z the perturbation $V = \pm e^2/r$ is small. Hence in this case perturbation theory is also valid. Since $E_{2s} - E_{1s} = 3Z^2e^2/8a$, from (4.4) we have that

$$\omega(1s \to 2s) \approx \frac{16a^2}{9Z^4e^4}\left|\langle 2s\,|V|\,1s\rangle\right|^2. \tag{6.4}$$

A familiar calculation gives for the matrix element of (6.4)

$$\langle 2s\,|V|\,1s\rangle = \pm\frac{8Z^2e^2}{27a},$$

so that

$$\omega(1s \to 2s) \approx 2^{11}9^{-4}Z^{-2} \approx 0{\cdot}312Z^{-2}. \tag{6.5}$$

It can be seen that for large Z the expressions (6.3) and (6.5) are nearly equal.

7. Using the notation $\omega_0 = -\gamma\mathscr{H}_0$ (= the "Larmor precession frequency" of the system in the constant magnetic field $\mathscr{H}_0$) and $\omega_1 = -\gamma\mathscr{H}_1$, equation (7a) becomes

$$i\frac{\partial\chi}{\partial t} = \left[\omega_0 s_z + \frac{1}{2}\omega_1(s_+ e^{-i\omega t} + s_- e^{i\omega t})\right]\chi. \tag{7.1}$$

We first solve the problem in a way which gives the exact result. We change to a frame of reference which is rotating with the field $\mathscr{H}_1$, by the unitary transformation

$$\chi = U\chi_e = \exp(-i\omega s_z t)\,\chi_e.$$

Equation (7.1) then becomes

$$i\dot{\chi}_e = \{U^{-1}[\omega_0 s_z + \tfrac{1}{2}\omega_1(s_+ e^{-i\omega t} + s_- e^{i\omega t})]\,U - iU^{-1}\dot{U}\chi_e\}. \tag{7.2}$$

We have, however, that

$$\left.\begin{aligned} U^{-1}s_z U = e^{i\omega t s_z} s_z\, e^{-i\omega t s_z} &= s_z \\ e^{+i\omega s_z t} s_+\, e^{-i\omega s_z t} &= e^{i\omega t} s_+ \\ e^{+i\omega s_z t} s_-\, e^{-i\omega s_z t} &= e^{-i\omega t} s_- \end{aligned}\right\} \tag{7.3}$$

and hence

$$i\dot{\chi}_e = [(\omega_0 - \omega)\, s_z + \omega_1 s_x]\,\chi_e. \tag{7.4}$$

Equation (7.4) can be integrated immediately, to give

$$\chi_e = \chi_e(0)\exp\{-i[(\omega_0-\omega)\, s_z + \omega_1 s_x]\, t\}, \tag{7.5}$$

where $\chi_e(0) = \chi(0)$. By introducing the notation

$$a = [(\omega_0 - \omega)^2 + \omega_1^2]^{1/2}, \quad \sin\theta = \omega_1/a, \quad \cos\theta = (\omega_0 - \omega)/a, \tag{7.6}$$

the spin wavefunction can be written in the form

$$\chi = \exp(-i\omega s_z t)\exp[-ia(\mathbf{n}.\mathbf{s})]\,\chi(0), \tag{7.7}$$

where $\mathbf{n}$ is a unit vector having components $n_x = \sin\theta$, $n_y = 0$ $n_z = \cos\theta$. Since, initially, the spin is in the state $|m_s\rangle$, we have that

$$|\chi\rangle = \exp(-i\omega s_z t)\exp[-ia(\mathbf{n}.\mathbf{s})]\,|m_s\rangle. \tag{7.8}$$

From (7.8) it follows immediately that the probability of finding the spin in the state $|m_s'\rangle$ at time t is

$$P_{m_s' m_s} = |\langle m_s'|\exp(-i\omega s_z t)\exp[ia(\mathbf{n}.\mathbf{s})\, t]\,|\, m_s\rangle|^2 = |\langle m_s'|\exp[ia(\mathbf{n}.\mathbf{s})\, t]\,|\, m_s\rangle|^2. \tag{7.9}$$

In the special case $s = \frac{1}{2}$, (7.9) can be written easily in an explicit form. Writing $\mathbf{s} = \frac{1}{2}\boldsymbol{\sigma}$, where σ_x, σ_y, σ_z are the Pauli matrices, the following relations can easily be verified:

$$(\mathbf{n}.\boldsymbol{\sigma})^{2p} = (\mathbf{n})^{2p} = 1, \quad (\mathbf{n}.\boldsymbol{\sigma})^{2p+1} = (\mathbf{n}.\boldsymbol{\sigma}). \tag{7.10}$$

From (7.10) one finds that

$$\exp\left[ia(\mathbf{n}.\mathbf{s})\,t\right] = \exp\left[\frac{1}{2}\,ia(\mathbf{n}.\boldsymbol{\sigma})\,t\right] = \cos\frac{a}{2}\,t + i(\mathbf{n}.\boldsymbol{\sigma})\sin\frac{a}{2}\,t, \tag{7.11}$$

and, by substituting in (7.9),

$$P_{-1/2,\,1/2} = \left|\left\langle -\frac{1}{2}\right|\cos\frac{a}{2}\,t + i\sin\frac{a}{2}\,t\,(\cos\theta\sigma_z + \sin\theta\sigma_x)\left|\frac{1}{2}\right\rangle\right|^2$$

$$= \sin^2\theta\,\sin^2\frac{a}{2}\,t = \frac{\omega_1^2}{(\omega_0-\omega)^2+\omega_1^2}\sin^2\frac{1}{2}\left[(\omega_0-\omega)^2+\omega_1^2\right]^{1/2} t. \tag{7.12}$$

Let us try now to solve the problem by using perturbation theory. Since the rotating field is usually much weaker than the constant field, it follows from (IX.2), after some simple calculation, that, to first order in perturbation theory,

$$P_{-1/2,\,1/2} = \left(\frac{\omega_1}{\omega_0-\omega}\right)^2 \sin^2\frac{1}{2}\,(\omega_0-\omega)\,t. \tag{7.13}$$

It can be seen that (7.13) gives a good approximation to the exact result (7.12) only if $|\omega_0-\omega| \gg \omega_1$, i.e. if the rotating field has a frequency far enough removed from the resonance frequency $\omega = \omega_0$. From this an important fact emerges, viz., that, in the vicinity of a resonance, perturbation theory cannot be applied, no matter how small the perturbation may be. Thus, for $\omega = \omega_0$, (7.13) becomes

$$P_{-1/2,\,1/2} = \tfrac{1}{4}(\omega_1 t)^2, \tag{7.14}$$

an expression which would lead to probabilities greater than unity for $t > 2/\omega_1$. Similar contradictions occur in other situations, usually when the frequency of a perturbation approaches one of the eigenfrequencies of the perturbed system.

If the direction of rotation of the field $\mathcal{H}_1$ is reversed (i.e. if $\mathcal{H}_{1x} = \mathcal{H}_1\cos\omega t$, $\mathcal{H}_{1y} = -\mathcal{H}_1\sin\omega t$) the corresponding probability is obtained from (7.12) by replacing ω by $-\omega$, thus

$$P_{-1/2,\,1/2} = \frac{\omega_1^2}{(\omega_0+\omega)^2+\omega_1^2}\sin^2\frac{1}{2}\left[(\omega_0+\omega)^2+\omega_1^2\right]^{1/2} t. \tag{7.12'}$$

It can be seen that if ω is in the vicinity of the frequency $-\omega_0$ (7.13) gives a good approximation to the exact result (7.12′).

8. In order to solve this problem we use Fermi's Golden Rule (IX.4). The unperturbed Hamiltonian $H_0 = \mathbf{p}^2/2m$ is the free-particle Hamiltonian, and we shall treat the scattering potential $V(\mathbf{r})$ as a time-independent perturbation. In the usual cube of side L, the orthonormal eigenstates of H_0 are the plane waves

$$\psi_k = L^{-3/2}\,e^{i\mathbf{k}.\mathbf{r}}, \quad \mathbf{p} = \hbar\mathbf{k}. \tag{8.1}$$

If we denote by $\varrho(E)\,dE\,d\Omega$ the number of states (8.1) having an energy $E = p^2/2m$ in the interval $(E, E+dE)$, and a wave vector $\mathbf{k}$ in the solid angle $d\Omega$, then

$$\varrho(E)\,dE\,d\Omega = \left(\frac{L}{2\pi}\right)^3 dk_x\,dk_y\,dk_z = \left(\frac{L}{2\pi}\right)^3 k^2\,dk\,d\Omega, \tag{8.2}$$

whence

$$\varrho(E) = \left(\frac{L}{2\pi}\right)^3 k^2\frac{dk}{dE} = \left(\frac{L}{2\pi}\right)^3 \frac{mp}{\hbar^3}. \tag{8.3}$$

Consider now an incident flux of particles of energy $E = \frac{1}{2}mv_a^2$, and let $\hbar\mathbf{k}_a$ and $\hbar\mathbf{k}_b$ be the momenta (corresponding to the same energy) of the incident and of the scattered particles. From (IX.4), it follows that the transition probability per unit time from the state a to the state b is

$$\omega_{ba} = \frac{2\pi}{\hbar}\,|\langle\mathbf{k}_b\,|\,V\,|\,\mathbf{k}_a\rangle|^2\,\varrho(E). \tag{8.4}$$

Let $d\sigma(\theta, \phi)$ be the differential cross-section of the scattering process, i.e. the number of particles scattered into a solid angle $d\Omega$ about the direction (θ, ϕ) of $\mathbf{k}_b$, per unit time and per unit incident flux. We can thus write

$$\frac{d\sigma}{d\Omega} = \left(\frac{v_a}{L^3}\right)^{-1}\omega_{ba} = \frac{mL^3}{\hbar k_a}\,\omega_{ba}. \tag{8.5}$$

From (8.4), (8.3) and (8.5) it follows that

$$\frac{d\sigma}{d\Omega} = \frac{m^2}{4\pi^2\hbar^4}\left|\int e^{-i\mathbf{q}\cdot\mathbf{r}}V(\mathbf{r})\,d\mathbf{r}\right|^2, \quad \mathbf{q} = \mathbf{k}_b - \mathbf{k}_a \tag{8.6}$$

which is in fact the "first Born approximation" of collision theory (see X.10).

Remarks: In order to determine $\varrho(E)$, eigenstates of H_0 which are periodic and normalized in a cube of side L were used. We can alternatively work with eigenstates of H_0 which are "normalized to a δ-function", viz.,

$$\psi_{\mathbf{k}} = (2\pi)^{-3/2}e^{i\mathbf{k}\cdot\mathbf{r}}. \tag{8.1'}$$

In accordance with the closure relation we have (see problem 8, Chapter I) $\int|\mathbf{k}\rangle\,d\mathbf{k}\,\langle\mathbf{k}| = 1$, whence it follows that the number of states (8.1′) with wave vectors in the range $(\mathbf{k}, \mathbf{k}+d\mathbf{k})$ is equal to $d\mathbf{k}$. Hence we have immediately that

$$\varrho(E)\,dE\,d\Omega = d\mathbf{k} = k^2\,dk\,d\Omega, \tag{8.2'}$$

and thus, for these states,

$$\varrho(E) = \frac{mk}{\hbar^2} = \frac{mp}{\hbar^3}. \tag{8.3'}$$

9. Choosing the origin of coordinates at the nucleus (assumed to remain stationary owing to its large mass), and denoting by $\mathbf{r}_1$ and $\mathbf{r}_2$ the position vectors of the colliding electron and of the bound electron respectively, and by $\mathbf{p}_1$ and $\mathbf{p}_2$ the corresponding momentum operators, the unperturbed Hamiltonian is

$$H_0 = \frac{\mathbf{p}_1^2}{2m} + \frac{\mathbf{p}_2^2}{2m} - \frac{e^2}{r_2}. \tag{9.1}$$

The perturbation is the interaction energy between the colliding electron and the bound electron and its nucleus:

$$V = \frac{e^2}{r_{12}} - \frac{e^2}{r_1}. \tag{9.2}$$

The unperturbed initial and final wavefunctions are

$$L^{-3/2} e^{i\mathbf{k}_a \cdot \mathbf{r}_1} \psi_{100}(\mathbf{r}_2), \quad L^{-3/2} e^{i\mathbf{k}_b \cdot \mathbf{r}_1} \psi_{200}(\mathbf{r}_2),$$

where L is the side of the normalization cube, $k_a\hbar$ and $k_b\hbar$ are the initial and the final momenta of the colliding electron, and $\psi_{100}(\mathbf{r}_2)$ and $\psi_{200}(\mathbf{r}_2)$ are the initial and final wavefunctions of the hydrogen atom (see problem 8). From the law of conservation of energy, we find that

$$k_b^2 = k_a^2 - \frac{2m}{\hbar^2} \frac{3e^2}{8a}. \tag{9.3}$$

The differential cross-section is found from (IX.4) in the same way as in the preceding problem, except that now the magnitudes of $\mathbf{k}_a$ and $\mathbf{k}_b$ are not equal, but are related by (9.3). We find then that

$$\frac{d\sigma(\theta)}{d\Omega} = \frac{k_b}{k_a} \left(\frac{m}{2\pi\hbar^2}\right)^2 L^6 \, |\langle 200 \,|\, V \,|\, 100 \rangle|^2, \tag{9.4}$$

$$\langle 200 \,|\, V \,|\, 100 \rangle = L^{-3} \iint e^{i\mathbf{q} \cdot \mathbf{r}_1} \psi_{200}^*(r_2) \left(\frac{e^2}{r_{12}} - \frac{e^2}{r_1}\right) \psi_{100}(r_2) \, d\mathbf{r}_1 \, d\mathbf{r}_2, \quad \mathbf{q} = \mathbf{k}_b - \mathbf{k}_a. \tag{9.5}$$

The term e^2/r_1 does not yield any contribution to the integral because of the orthogonality of the wavefunctions $\psi_{100}(r_2)$ and $\psi_{200}(r_2)$. This result is to be expected, since any interaction between the colliding electron and the nucleus (which is assumed to remain at rest) cannot excite the atom. To calculate the other term we first carry out the integration over the variable $\mathbf{r}_1$:

$$\int \frac{e^{i\mathbf{q} \cdot \mathbf{r}}}{r_{12}} \, d\mathbf{r}_1. \tag{9.6}$$

This can be done as follows. Note that, in electrostatics, the quantity

$$\phi_\mathbf{q}(\mathbf{r}) = \int \frac{e^{-i\mathbf{q} \cdot \mathbf{r}'}}{|\mathbf{r}' - \mathbf{r}|} \, d\mathbf{r}' \tag{9.7}$$

is the potential at $\mathbf{r}$ due to an electric charge distribution of density $\varrho(\mathbf{r}') = e^{-i\mathbf{q} \cdot \mathbf{r}'}$. This potential therefore satisfies Poisson's equation,

$$\nabla^2 \phi_\mathbf{q}(\mathbf{r}) = -4\pi\varrho(\mathbf{r}) = 4\pi e^{-i\mathbf{q} \cdot \mathbf{r}},$$

whence, by inspection,

$$\phi_{\mathbf{q}}(\mathbf{r}) = \frac{4\pi}{q^2} e^{-i\mathbf{q}\cdot\mathbf{r}}. \tag{9.8}$$

With the help of this result and the use of the hydrogen atom wavefunctions (33.14), Chapter II, we obtain from (9.5), after integrating over $\mathbf{r}_2$,

$$\langle 200 | V | 100 \rangle = L^{-3} \frac{16\sqrt{2}\pi a^2 e^2}{(q^2a^2+\frac{9}{4})^3}, \quad a = \frac{\hbar^2}{me^2}, \tag{9.9}$$

$$\frac{d\sigma(\theta)}{d\Omega} = \frac{k_b}{k_a} \frac{128q^2}{(q^2a^2+\frac{9}{4})^6}, \quad q^2 = k_a^2+k_b^2-2k_ak_b \cos\theta. \tag{9.10}$$

From (9.9) we see that, in order to be able to apply perturbation theory, it is necessary that $k_a a \gg 1$, i.e. that the colliding electron should have a high enough energy. k_b is then nearly equal to k_a, and, by (9.3),

$$\begin{aligned} (k_a-k_b)(k_a+k_b) &= \frac{2m}{\hbar^2}\frac{3e^2}{8a} = \frac{3}{4a^2} \\ k_a+k_b \approx 2k_a, \quad & k_a-k_b \approx \frac{3}{8k_a a^2}. \end{aligned} \tag{9.11}$$

Thus for q^2 we can write

$$q^2 \approx \left(4k_a^2 - \frac{3}{2a^2}\right)\sin^2\frac{\theta}{2} \approx \left(2k_a \sin\frac{\theta}{2}\right)^2. \tag{9.12}$$

From (9.9) we see also that the main contribution to the total inelastic scattering cross-section comes from the region in which $qa \leqslant 1$, i.e. in which $\theta \leqslant (k_a a)^{-1}$. Outside this region, $\frac{d\sigma(\theta)}{d\Omega}$ decreases rapidly with increasing θ $\left(\text{approximately as } \operatorname{cosec}^{12}\frac{\theta}{2}\right)$, as opposed to the elastic collision cross-section (which decreases as $\operatorname{cosec}^4\theta/2$ according to Rutherford's formula). This strong forward peaking of the inelastic scattering cross-section is characteristic of inelastic processes in general. To calculate explicitly the total cross-section, we observe from (9.10) that the required integration over the solid angle $2\pi \sin\theta\, d\theta$ can be replaced by an integration over $2\pi q\, dq/k_ak_b$, between the limits k_a-k_b and k_a+k_b. For high energies, the main contribution to the integral comes from values of q near the lower limit. From (9.11) it follows that $(k_a-k_b)^2a^2 \approx (9/64k_a^2a^2) \ll (\frac{9}{4})$. We can therefore extend the integration over q from 0 to ∞, and we obtain finally

$$\sigma \approx \frac{128\pi}{5k_a^2}\left(\frac{2}{3}\right)^{10}. \tag{9.13}$$

10. Denoting by ϕ_n and E_n the eigenfunctions and the eigenvalues respectively of the time-independent Hamiltonian H_0, the solution of the Schrödinger equation,

$$i\hbar\frac{\partial\psi}{\partial t} = [H_0+V(t)]\psi, \tag{10.1}$$

can be written in the form

$$\psi = \sum_n a_n(t)\phi_n \exp\left(-\frac{iE_n t}{\hbar}\right). \tag{10.2}$$

Since, up to the moment when the perturbation is applied, the system is in the state m, we have

$$a_n(t) = \delta_{nm}, \qquad t \leqslant 0.$$

After the perturbation has ceased, i.e. for $t \geqslant \tau$, the coefficients a_n again have constant values, $a_{nm}(\tau)$ say, and the wavefunction becomes

$$\psi = \sum_n a_{nm}(\tau)\phi_n \exp\left(-\frac{i}{\hbar}E_n t\right). \tag{10.4}$$

The probability of the transition $m \to n$ will then be

$$\omega_{nm}(\tau) = |a_{nm}(\tau)|^2.$$

We have now to calculate the coefficients $a_{nm}(\tau)$. Substituting (10.2) in (10.1) an d performing some elementary calculations we obtain for the $a_n(t)$ a system of differential equations,

$$i\hbar\frac{d}{dt}a_n(t) = \sum_l \langle n|V(t)|l\rangle e^{i\omega_{nl}t}a_l(t). \tag{10.5}$$

We can solve (10.5) by successive approximations and obtain

$$a_{nm}^{(1)} = \frac{1}{i\hbar}\int_0^t \langle n|V(t')|m\rangle e^{i\omega_{nm}t'}\,dt', \tag{10.6}$$

$$\begin{aligned} a_{nm}^{(2)} &= \frac{1}{i\hbar}\int_0^t \langle n|V(t')|m\rangle e^{i\omega_{nm}t'}\,dt' \\ &+\left(\frac{1}{i\hbar}\right)^2 \sum_{n'}\int_0^t \langle n|V(t')|n'\rangle e^{i\omega_{nn'}t'}\int_0^{t'}\langle n'|V(t'')|m\rangle e^{i\omega_{n'm}t''}\,dt''\,dt'. \end{aligned} \tag{10.7}$$

In general (and ignoring questions of convergence) we can write

$$\begin{aligned} a_{nm}(t) = \Bigg\langle n \Bigg| \Bigg\{ 1 &+ \frac{1}{i\hbar}\int_0^t V_I(t')\,dt' + \left(\frac{1}{i\hbar}\right)^2\int_0^t V_I(t')\int_0^{t'} V_I(t'')\,dt''\,dt' \\ &+ \left(\frac{1}{i\hbar}\right)^3\int_0^t V_I(t')\int_0^{t'} V_I(t'')\int_0^{t''} V_I(t''')\,dt'''\,dt''\,dt' + \dots \Bigg\} \Bigg| m \Bigg\rangle, \end{aligned} \tag{10.8}$$

where $V_I(t)$ is the expression for the perturbation in the interaction picture. As is to be expected, the quantity $|a_{nm}^{(1)}(\tau)|^2$ is equal to $\omega_{nm}(\tau)$ in (IX.2).

The problem can be solved in a more elegant and general way, as follows. Consider a quantum-mechanical system consisting of two interacting sub-systems, whose total Hamiltonian $H = H_0+V$ consists of the sum H_0 of the Hamiltonians of the two systems (neglecting any interaction between them), and the time-independent interaction Hamiltonian† V. The time evolution of the wavefunction of the system in the Schrödinger picture is given by

$$\psi(t) = S(t, t_0)\,\psi(t_0) \tag{10.9}$$

where the operator $S(t, t_0)$ satisfies the differential equation (6a), Chapter V,

$$i\hbar \frac{\partial S(t, t_0)}{\partial t} = V_I(t)\, S(t, t_0), \tag{10.10}$$

with the initial condition $S(t_0, t_0) = 1$.

We can solve (10.10) to successive approximations by taking as the zero-order approximation $S(t, t_0) = 1$. Without concerning ourselves with questions of convergence, the solution can be written as

$$\begin{aligned} S(t, t_0) &= \sum_{k=0}^{\infty} S_k(t, t_0), \\ S_k(t, t_0) &= \left(\frac{1}{i\hbar}\right)^k \int_{t_0}^{t} dt_1 \int_{t_0}^{t_1} dt_2 \ldots \int_{t_0}^{t_{k-1}} dt_k V_I(t_1)\, V_I(t_2) \ldots V_I(t_k). \end{aligned} \tag{10.11}$$

The repeated integrals in (10.11) can be rewritten as follows. Consider, e.g., the term

$$S_2(t, t_0) = -\int_{t_0}^{t} dt_1 \int_{t_0}^{t_1} dt_2 V_I(t_1) V_I(t_2). \tag{10.12}$$

In the plane (t_1, t_2), (10.12) is a surface integral over the triangle shown in Fig. IX.2 below the diagonal.

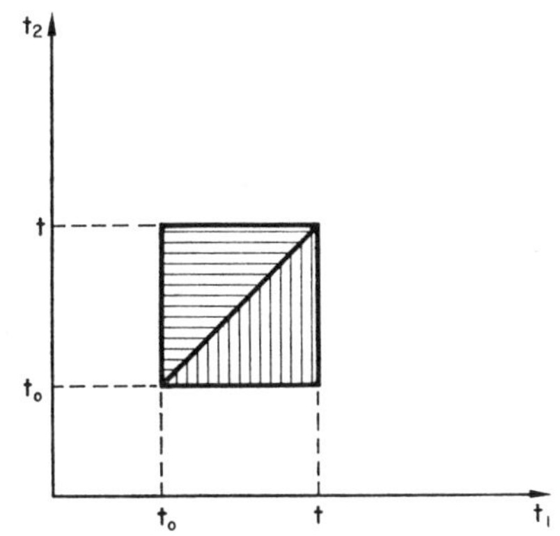

FIG. IX.2.

† In the preceding sections it was always supposed that H referred to a single system only. V is then a time-dependent *external perturbation*, whereas in the present section it is internal and time-independent.

Let us now change the order of integration in (10.12), so that

$$S_2(t, t_0) = -\int_{t_0}^{t} dt_2 \int_{t_0}^{t_2} dt_1 V_I(t_2)V_I(t_1). \tag{10.12'}$$

In the plane (t_1, t_2) the surface integral is now over the triangle above the diagonal. If the operators $V_I(t_1)$ and $V_I(t_2)$ commute, the expressions under the integral signs in (10.12) and in (10.12′) are equal to each other and $S_2(t, t_0)$ can be written as half the integral calculated over the whole of the square shown in the diagram.

If the operators $V_I(t_1)$ and $V_I(t_2)$ do not commute, we can still write

$$S(t, t_0) = -\tfrac{1}{2}\int_{t_0}^{t} dt_1 \int_{t_0}^{t} dt_2 \begin{cases} V_I(t_1)V_I(t_2), & t_2 < t_1 \\ V_I(t_2)V_I(t_1), & t_1 < t_2, \end{cases} \tag{10.13}$$

i.e.

$$S(t, t_0) = -\tfrac{1}{2}\int_{t_0}^{t} dt_1 \int_{t_0}^{t} dt_2 T(V_I(t_1)V_I(t_2)), \tag{10.13'}$$

where

$$T(V_I(t_1)V_I(t_2)) = \begin{cases} V_I(t_1)V_I(t_2), & t_2 < t_1 \\ V_I(t_2)V_I(t_1), & t_1 < t_2, \end{cases} \tag{10.14}$$

and, in general,

$$S_k(t, t_0) = \frac{1}{k!}\left(\frac{1}{i\hbar}\right)^k \int_{t_0}^{t} dt_1 \int_{t_0}^{t} dt_2 \ldots \int_{t_0}^{t} dt_k T(V_I(t_1)V_I(t_2) \ldots V_I(t_k)), \tag{10.15}$$

where T is called the "time-ordering operator" and has the property that it rearranges the factors which follow it in such a way that, for any given values of $t_1 \ldots t_k$, the order in which the operators $V_I(t)$ occur is such that the values of their arguments decrease from left to right. Formally we can write

$$S(t, t_0) = \exp\left(\frac{1}{i\hbar}\int_{t_0}^{t} V_I(t)\, dt\right), \tag{10.16}$$

provided the exponential is suitably interpreted.

The probability of the transition $m \to n$ is equal to the square of the matrix element

$$\langle n|S(t, t_0)|m\rangle = \langle n|\exp\left(\frac{1}{i\hbar}\int_{t_0}^{t} V_I(t)\, dt\right)|m\rangle. \tag{10.17}$$

If we let $t_0 \to -\infty$, and $t \to +\infty$, the resulting operator

$$S = S(-\infty, +\infty) = \exp\left(\frac{1}{i\hbar}\int_{-\infty}^{+\infty} V_I(t)\, dt\right) \tag{10.18}$$

is called the S-matrix (see also Chapter X).

11. Using the results obtained in the preceding problem, we have $\omega_{n0} = |a_{n0}|^2$, where

$$a_{n0} = \langle n|\exp\left(\frac{1}{i\hbar}\int_{-\infty}^{+\infty} V_I(t)\,dt\right)|0\rangle. \tag{11.1}$$

Since $V(t) = -ex\mathscr{E}(t)$, and the matrix elements of the oscillator coordinate are non-vanishing only between states whose quantum numbers differ by unity, it can be seen that

$$a_{n0} = \frac{1}{n!}\left(\frac{1}{i\hbar}\right)^n \int_{-\infty}^{+\infty} \langle n|V_I(t)|n-1\rangle\,dt \int_{-\infty}^{+\infty} \langle n-1|V_I(t)|n-2\rangle\,dt \ldots$$
$$\ldots \int_{-\infty}^{+\infty} \langle 1|V_I(t)|0\rangle\,dt. \tag{11.2}$$

Following now the solution given for problem 1, we obtain the required probability

$$\omega_{n0} = \frac{1}{n!}\left(\frac{P^2}{2m\hbar\omega}\right)^n e^{-(n/2)\omega^2\tau^2}. \tag{11.3}$$

A discussion of the physical significance of this result would be similar to that given for the solution of problem 1.

12. From (10.18) and (10.11) we have that

$$\langle n|S|m\rangle = \sum_{\alpha=0}^{\infty} \langle n|S^{(\alpha)}|m\rangle, \tag{12.1}$$

where

$$\begin{aligned}
\langle n|S^{(0)}|m\rangle &= \langle n|m\rangle = \delta_{nm}\\
\langle n|S^{(1)}|m\rangle &= \frac{1}{i\hbar}\langle n|\int_{-\infty}^{+\infty} V_I(t)\,dt|m\rangle\\
\langle n|S^{(2)}|m\rangle &= \left(\frac{1}{i\hbar}\right)^2\langle n|\int_{-\infty}^{+\infty} dt_1 \int_{-\infty}^{t_1} dt_2 V_I(t_1)V_I(t_2)|m\rangle\\
&\ldots\ldots\ldots\ldots\ldots\ldots\ldots\ldots
\end{aligned} \tag{12.2}$$

These matrix elements can be transformed in the following way

$$\langle n|S^{(1)}|m\rangle = -\frac{i}{\hbar}\langle n|V|m\rangle \int_{-\infty}^{\infty} e^{\frac{i}{\hbar}(E_n-E_m)t}\,dt = -2\pi i\delta(E_n-E_m)\langle n|V|m\rangle, \tag{12.3}$$

since

$$\int_{-\infty}^{\infty} e^{ixt}\,dt = 2\pi\delta(x). \tag{12.4}$$

Further, we can write

$$\langle n|S^{(2)}|m\rangle = \left(\frac{1}{i\hbar}\right)^2 \sum_l \int_{-\infty}^{\infty} dt_1 \langle n|V_I|l\rangle \int_{-\infty}^{t_1} dt_2 \langle l|V_I|m\rangle$$

$$= \left(\frac{1}{i\hbar}\right)^2 \sum_l \langle n|V|l\rangle\langle l|V|m\rangle \int_{-\infty}^{\infty} e^{\frac{i}{\hbar}(E_n-E_l)t_1}\, dt_1 \int_{-\infty}^{t_1} e^{\frac{i}{\hbar}(E_l-E_m)t_2}\, dt_2 .$$

In order to perform the second integration we make the change $E_l-E_m \to E_l-E_m-i\eta$ where η is a small positive real number which ensures the convergence of the integral at its lower limit. Then, in the results so obtained, η is made to tend to zero. Thus

$$\int_{-\infty}^{t_1} e^{\frac{i}{\hbar}(E_l-E_m)t_2}\, dt_2 \to \int_{-\infty}^{t_1} e^{\frac{i}{\hbar}(E_l-E_m-i\eta)t_2}\, dt_2 = i\hbar\, \frac{e^{\frac{i}{\hbar}(E_l-E_m-i\eta)t_2}}{E_m-E_l+i\eta} .$$

Hence

$$\langle n|S^{(2)}|m\rangle = \frac{1}{i\hbar} \sum \frac{\langle n|V|l\rangle\langle l|V|n\rangle}{E_m-E_l+i\eta} \int_{-\infty}^{+\infty} e^{\frac{i}{\hbar}(E_n-E_m-i\eta)t_1}\, dt_1$$

$$= -2\pi i\delta(E_n-E_m) \sum_l \frac{\langle n|V|l\rangle\langle l|V|m\rangle}{E_m-E_l+i\eta} . \qquad (12.5)$$

Similarly the matrix elements for higher orders can be obtained. Substituting these results into (12.1), the expression (12a) follows immediately. Note that the intermediate states l are eigenfunctions of the Hamiltonian H_0. Then

$$\frac{\langle n|V|l\rangle\langle l|V|m\rangle}{E_m-E_n+i\eta} = \langle n|V|l\rangle\langle l|(E_m-H_0+i\eta)^{-1}|l\rangle\langle l|V|n\rangle$$

and (12b) can be written as

$$T = V+V(E_m-H_0+i\eta)^{-1}V+V(E_m-H_0+i\eta)^{-1}V(E_m-H_0+i\eta)^{-1}V+\ \ldots \qquad (12.6)$$

Thus the operator T can be regarded as a solution (obtainable by successive approximation) of the operational equation (12c).

It is interesting to note that the first approximation to the operator T, $T = V$, leads to the expression (IX.4). Thus, it follows from problem 10 that the probability of the transition $m(-\infty) \to n(+\infty)$ is

$$\omega_{nm}(\infty) = |\langle n|S|m\rangle|^2 = 4\pi^2\delta^2(E_n-E_m)\,|\langle n|T|m\rangle|^2. \qquad (12.7)$$

On the other hand

$$\delta^2(x) = \frac{\delta(x)}{2\pi} \lim_{T\to\infty} \int_{-T}^{T} e^{ixt}\, dt = \frac{\delta(x)}{2\pi} \lim_{T\to\infty} \int_{-T}^{T} dt. \tag{12.8}$$

For the last equality in (12.8) we have used the identity $f(x)\,\delta(x-a) = f(a)\,\delta(x-a)$. The transition probability per unit time is then

$$\bar{P}_{nm} = \frac{\omega_{nm}(\infty)}{\lim\limits_{T\to\infty} \int_{-T}^{T} dt} = \frac{2\pi}{\hbar}\, \delta(E_n - E_m)\, |\langle n\,|T|\, m\rangle|^2. \tag{12.9}$$

It we take $T = V$, (12.9) becomes (IX.4).

13. Denoting by $\mathbf{r}$ the radius vector of the electron, and by e the electronic charge, the perturbation operator takes the form

$$V = -e\mathbf{r}\,.\,\mathcal{E}(t) = -e\mathbf{r}\,.\,\mathcal{E}_0 \sin \omega t = V_0\, e^{-i\omega t} + V_0^* e^{i\omega t},$$

$$V_0 = \frac{e}{2i}\, \mathbf{r}\,.\,\mathcal{E}_0. \tag{13.1}$$

It should be noted that, to ionize the hydrogen atom from the ground state, the minimum energy required is $-E_1 = me^4/2\hbar^2$, whence, for this problem, we obtain $\omega > me^4/2\hbar^3$.† In order to find the probability that ionization will occur under the influence of the perturbation (13.2), we shall apply formula (IX.5). We have then to calculate the matrix element

$$V_0^{\mathbf{k}1} = \int \psi_{\mathbf{k}}^* \left(\frac{e}{2i}\, \mathcal{E}_0 \,.\, \mathbf{r}\right) \psi_{100}\, d\mathbf{r}, \tag{13.2}$$

where

$$\psi_{100} = (\pi a^3)^{-1/2}\, e^{-r/a}, \quad \psi_{\mathbf{k}} = (2\pi)^{-3/2}\, e^{i\mathbf{k}.\mathbf{r}}, \tag{13.3}$$

and hence

$$V_0^{\mathbf{k}1} = \frac{e}{2i}\, (2\pi)^{-3/2}\, (\pi a^3)^{-1/2} \int e^{-i\mathbf{k}.\mathbf{r} - r/a} (\mathcal{E}_0 \,.\, \mathbf{r})\, d\mathbf{r}. \tag{13.4}$$

In order to evaluate the integral we introduce spherical coordinates (r, θ, ϕ) with the polar axis directed along $\mathbf{k}$. We have then that

$$\mathcal{E}_0 \,.\, \mathbf{r} = \mathcal{E}_0 r[\cos \Theta \cos \theta + \sin \Theta \sin \theta \cos (\phi - \phi_0)], \tag{13.5}$$

where Θ and ϕ_0 are the polar angles of $\mathcal{E}_0$. By substituting (13.5) in (13.4) and noting that

† Note that the approximation made by taking the final state of the electron to be a plane wave is accurate only if $\omega \gg me^4/2\hbar^3$ (in which case the ejected electron has a large kinetic energy).

the second term of (13.5) makes no contribution owing to the integration over ϕ, we obtain

$$V_0^{\mathbf{k}1} = \frac{ie}{2^{5/2}\pi^2 a^{3/2}} 2\pi\varepsilon_0 \cos\Theta \int_{-1}^{+1} \left(\int_0^\infty e^{-ikrx - r/a} r^2\, dr \right) x\, dx, \tag{13.6}$$

where $x = \cos\theta$. After performing the integration in (13.6) one finds

$$V_0^{\mathbf{k}1} = \frac{e\mathcal{E}_0 \cos\theta}{\pi(2a)^{3/2}} \frac{16ka^5}{(1+k^2a^2)^3}. \tag{13.7}$$

Taking into account (IX.5), (8.3′), and the relation $E_k = E_n + \hbar\omega$, we find the probability per unit time of ionization with ejection of an electron into the element of solid angle $d\Omega$ to be

$$dP_{\mathbf{k}1} = \frac{2^6 m a^7 e^2}{\pi\hbar^3} \frac{\mathcal{E}_0^2 k^3 \cos^2\Theta}{(1+k^2a^2)^6} d\Omega_{\mathbf{k}}. \tag{13.8}$$

With the notation $E_1 = -me^4/2\hbar^2 = -\hbar\omega_0$, the equation $E_k = E_n + \hbar\omega$ shows that $1+k^2a^2 = \omega/\omega_0$, and hence that

$$dP_{\mathbf{k}1} = \frac{64a^3\mathcal{E}_0^2}{\pi\hbar} \left(\frac{\omega_0}{\omega}\right)^6 \left(\frac{\omega}{\omega_0} - 1\right)^{3/2} \cos^2\Theta\, d\Omega_{\mathbf{k}}. \tag{13.8′}$$

Note that the probability (13.8′) is symmetrical about the direction of the applied field and vanishes along any direction perpendicular to it. By integrating (13.8′) over all directions of emission of the electron we obtain the total probability of ionization of the atom:

$$P =_{k1} \frac{256a^3\mathcal{E}_0^2}{3\hbar} \left(\frac{\omega_0}{\omega}\right)^2 \left(\frac{\omega}{\omega_0} - 1\right)^{5/2}. \tag{13.9}$$

For $\omega \approx \omega_0$ (i.e. near the ionization threshold) this probability increases from zero as $(\omega-\omega_0)^{3/2}$. For $\omega \gg \omega_0$ it decreases as $\omega^{-9/2}$. The maximum occurs when $\omega = \frac{4}{3}\omega_0$.

14. From (IX.9) we have

$$\mathbf{A}^2 = \sum_{\lambda,\lambda'} \{b_\lambda b_{\lambda'}(\mathbf{A}_\lambda . \mathbf{A}_{\lambda'}) + b_\lambda b_{\lambda'}^+(\mathbf{A}_\lambda . \mathbf{A}_{\lambda'}^*) + b_\lambda^+ b_{\lambda'}(\mathbf{A}_\lambda^* . \mathbf{A}_{\lambda'}) + b_\lambda^+ b_{\lambda'}^+(\mathbf{A}_\lambda^* . \mathbf{A}_{\lambda'}^*)\}. \tag{14.1}$$

Bearing in mind the properties of creation and annihilation operators, it follows from (IX.10) that the only non-vanishing matrix elements of the operators b_λ and b_λ^+ are

$$\begin{aligned} \langle n_1, \ldots, n_\lambda, \ldots | b_\lambda | n_1, \ldots, n_\lambda + 1, \ldots \rangle &= \sqrt{\frac{\hbar}{2\omega_\lambda}(n_\lambda + 1)}, \\ \langle n_1, \ldots, n_\lambda, \ldots | b_\lambda^+ | n_1, \ldots, n_\lambda - 1, \ldots \rangle &= \sqrt{\frac{\hbar}{2\omega_\lambda} n_\lambda}. \end{aligned} \tag{14.2}$$

The solution of the problem then follows from (14.1) and (14.2).

15. Let ψ_1 and ψ_2 be the wavefunctions of the initial and the final states of the electron. If the electron is free we can write ψ_1 and ψ_2 in the form of plane waves,

$$\psi_1 = C\exp\left(\frac{i}{\hbar}\mathbf{p}_1.\mathbf{r}\right), \quad \psi_2 = C\exp\left(\frac{i}{\hbar}\mathbf{p}_2.\mathbf{r}\right), \tag{15.1}$$

where C is a normalization constant. Let us consider the probability of emission. The matrix element from (IX.18′) is

$$\int \psi_1^* e^{-i\mathbf{k}_\lambda\cdot\mathbf{r}}(\mathbf{p}.\mathbf{e}_\lambda)\psi_2\, d\mathbf{r} = |C|^2 \int e^{-\frac{i}{\hbar}(\mathbf{p}_1+\hbar\mathbf{k}_\lambda)\cdot\mathbf{r}}\left(\frac{\hbar}{i}\mathbf{e}_\lambda.\nabla\right)e^{\frac{i}{\hbar}\mathbf{p}_2\cdot\mathbf{r}}\, d\mathbf{r}. \tag{15.2}$$

Now the Dirac delta function is the Fourier transform of unity, i.e.

$$\left(\frac{1}{2\pi}\right)^3 \int_{-\infty}^{+\infty} 1.e^{i\mathbf{k}.\mathbf{r}}\, d\mathbf{r} = \delta(\mathbf{k}),$$

and it therefore follows from (15.2) that

$$\int \psi_1^* e^{-i\mathbf{k}_\lambda\cdot\mathbf{r}}(\mathbf{p}.\mathbf{e}_\lambda)\psi_2\, d\mathbf{r} \sim \delta(\mathbf{p}_2-\hbar\mathbf{k}_\lambda-\mathbf{p}_1). \tag{15.3}$$

Relation (15.3) expresses the law of conservation of momentum as applied to the emission of a photon by a free electron,

$$\mathbf{p}_2 = \mathbf{p}_1+\hbar\mathbf{k}. \tag{15.4}$$

Now the law of conservation of energy states that

$$E_2 = E_1+\hbar\omega \tag{15.5}$$

and it can easily be seen that (15.4) is not consistent with (15.5). A similar argument can be used to show the impossibility of absorption.

In order to satisfy both conservation laws, the electron must interact with at least one other particle which can take up the excess momentum, e.g. the nucleus in the case of an electron in a bound state in an atom.

16. From (IX.20), by integrating over all space, we obtain (with $n_\lambda = 0$),

$$P_{12} = \frac{4\omega^3}{3\hbar c^3}|\mathbf{d}_{12}|^2 = \frac{4}{3}\alpha\left(\frac{|\mathbf{r}_{12}|}{c}\right)^2\omega^3, \quad \alpha = \frac{e^2}{\hbar c} = \frac{1}{137}. \tag{16.1}$$

In view of the resemblance between (16.1) and the classical formula (p. 228), it is clear that we can put $|\mathbf{r}_{12}| \sim a$, where a is the atomic radius. Taking as an example the hydrogen atom, we can write, for the probability per unit time of spontaneous transition from the first excited level to the ground state, the expression

$$P_{12} \sim \alpha\omega\left(\frac{a\omega}{c}\right)^2, \quad a = \frac{\hbar^2}{me^2} \sim \frac{e^2}{\hbar\omega},$$

or

$$P_{12} \sim \frac{\omega}{(137)^3}. \tag{16.2}$$

From (16.2) it follows that, for visible radiation ($\omega \sim 10^{15}$ sec^{-1}), the order of magnitude of the transition probability per unit time is 10^9 sec^{-1}.

17. Let a system of Cartesian coordinates be chosen such that the y axis is along the vector **e**, and the x axis is along the vector **k**. The matrix element (18a) then becomes

$$\begin{aligned}\langle 1 \,|(\mathbf{k}.\mathbf{r})(\mathbf{p}.\mathbf{e})|\, 2\rangle &= -ik\hbar\left\langle 1 \left| x\frac{\partial}{\partial y}\right| 2\right\rangle \\ &= -\frac{i}{2}k\hbar\left[\left\langle 1\left| x\frac{\partial}{\partial y}+y\frac{\partial}{\partial x}\right| 2\right\rangle+\left\langle 1\left| x\frac{\partial}{\partial y}-y\frac{\partial}{\partial x}\right| 2\right\rangle\right].\end{aligned} \tag{17.1}$$

Denoting by H_0 the Hamiltonian of the system, the vectors $|1\rangle$ and $|2\rangle$ will be eigenvectors of H_0, i.e. $H_0|1\rangle = E_1|1\rangle$ and $H_0|2\rangle = E_2|2\rangle$. From the operator relation

$$xyH_0-H_0xy = \frac{\hbar^2}{m}\left(x\frac{\partial}{\partial y}+y\frac{\partial}{\partial x}\right), \tag{17.2}$$

we have that

$$\left\langle 1\left| x\frac{\partial}{\partial y}+y\frac{\partial}{\partial x}\right| 2\right\rangle = \frac{\hbar^2}{m}(E_2-E_1)\langle 1\,|xy|\,2\rangle. \tag{17.3}$$

Then, from (17.1), it follows that

$$\langle 1\,|(\mathbf{p}.\mathbf{e})(\mathbf{k}.\mathbf{r})|\,2\rangle = -\tfrac{1}{2}ik\omega_{21}m\langle 1\,|xy|\,2\rangle+\frac{k}{2}\langle 1\,|l_z|\,2\rangle, \tag{17.4}$$

where $l_z = -i\hbar\left(x\frac{\partial}{\partial y}-y\frac{\partial}{\partial x}\right)$ is the z-component of the orbital angular momentum operator. The second term in (17.4) clearly corresponds to a magnetic dipole transition, while the first term corresponds to an electric quadrupole transition. These expressions are in fact similar to those obtained in the classical theory of radiation.

For other orientations of **e** and of **k**, matrix elements like $\langle 1\,|yz|\,2\rangle$, $\langle 1\,|zx|\,2\rangle$, $\langle 1\,|ly|\,2\rangle$, $\langle 1\,|lx|\,2\rangle$, also appear.

18. In order to have electric dipole transitions, the matrix element of the projection of the radius vector along the direction of polarization must not vanish.

Consider first the case in which the direction of polarization is along the z-axis. In this case $(\mathbf{r}.\mathbf{e}) = r\cos\theta = z$. For a central field, the dipole radiation matrix element will be proportional to

$$\int_0^{\pi}\int_0^{2\pi} Y_{l_1}^{m_1*}\cos\theta Y_{l_2}^{m_2}\sin\theta\, d\theta\, d\phi. \tag{18.1}$$

Here the quantum numbers l_1, m_1; l_2, m_2 refer to the states of the system before and after emission.

Since

$$\cos\theta Y_{l_2}^{m_2} = \left[\frac{(l_2+1+m_2)(l_2+1-m_2)}{(2l_2+1)(2l_2+3)}\right]^{1/2} Y_{l_2+1}^{m_2} + \left[\frac{(l_2+m_2)(l_2-m_2)}{(2l_2+1)(2l_2-1)}\right]^{1/2} Y_{l_2-1}^{m_2}$$

(see A.28), it follows that the integral (18.1) will be different from zero only if $\Delta m = m_2-m_1 = 0$ and $\Delta l = l_2-l_1 = \pm 1$.

Suppose now that the direction of polarization is along the x- or the y-axis. In this case we have to calculate the matrix elements of the coordinates x and y. This can easily be done if, instead of x and y, we consider the quantities $x\pm iy = r \sin\theta e^{\pm i\phi}$, which are related to the two directions of circular polarization in the xy plane. The matrix element for dipole radiation is then proportional to

$$\int_0^\pi \int_0^{2\pi} Y_{l_1}^{m_1*} \sin\theta e^{\pm i\phi} Y_{l_2}^{m_2} \sin\theta \, d\theta \, d\phi. \tag{18.2}$$

According to the same relation (A.28) we have that

$$\sin\theta Y_{l_2}^{m_2} = \left\{-\left[\frac{(l_2+1-m_2)(l_2+2-m_2)}{(2l_2+1)(2l_2+3)}\right]^{1/2} Y_{l_2+1}^{m_2-1} + \left[\frac{(l_2+m_2)(l_2-1+m_2)}{(2l_2+1)(2l_2-1)}\right]^{1/2} Y_{l_2-1}^{m_2-1}\right\} e^{i\phi}$$

and hence the integral (18.2) will be different from zero only if $\Delta m = \pm 1$, $\Delta l = \pm 1$. The dipole transition selection rules are therefore $\Delta m = 0, \pm 1$, $\Delta l = \pm 1$. This result could have been obtained directly from problem 29, Chapter VI, by noting that the position vector **r** is a polar vector. For electric quadrupole radiation (see problem 17), we have to calculate matrix elements of the products xy, yz and zx. Using the relations (A.28), we obtain, by a calculation similar to the one given above, the following selection rules:† $\Delta l = 0, \pm 2$ (excluding the case $l_1 = l_2 = 0$) and $\Delta m = 0, \pm 1, \pm 2$. The selection rules for magnetic dipole radiation are easily found to be $\Delta l = 0$, $\Delta m = \pm 1$. It is worth noting that for a purely central field, states with the same principal and orbital quantum numbers (n, l) have the same energy for all m, so that magnetic dipole transitions cannot occur between such states, although they satisfy the rule $\Delta l = 0$. However, if the system is put in a magnetic field, the energies of the states will depend on m (Zeeman effect), and magnetic dipole transitions can then occur between such states, in accordance with the selection rules $\Delta l = 0$, $\Delta m = \pm 1$.

19. The problem is that of determining the selection rules for an oscillator. For this we require the matrix elements of the coordinate x. From (21.4), Chapter V, the only non-vanishing matrix elements are

$$\langle n|x|m\rangle = \begin{cases} x_0\left(\dfrac{n+1}{2}\right)^{1/2}, & m = n+1 \\ x_0\left(\dfrac{n}{2}\right)^{1/2}, & m = n-1. \end{cases}$$

† These selection rules can be obtained directly by applying the Wigner–Eckart theorem (see Chapter VI).

It follows that the selection rule for the quantum number n is $\Delta n = \pm 1$. Therefore, for the oscillator, only transitions between adjacent levels are possible, and the frequency of the radiation emitted or absorbed is then always equal to the classical frequency of the oscillator.

20. Let us assume that the energies of the atoms of a black-body, in thermodynamic equilibrium with its own radiation, obey a Boltzmann distribution. Denoting by N_1 and N_2 the numbers of atoms with energies E_1 and E_2, $E_1 < E_2$, we have that

$$N_1\, dP_{21} = N_2\, dP_{12}, \quad \frac{N_1}{N_2} = \exp\left(\frac{\hbar\omega}{kT}\right). \tag{20.1}$$

From (IX.18), (IX.18′) and (20.1), it follows that the average number of photons in the volume V having given momentum and polarization is

$$n_\lambda = \left(e^{\frac{\hbar\omega}{kT}} - 1\right)^{-1}. \tag{20.2}$$

The number of states of the radiation field (i.e. the number of "oscillators") per unit frequency range is (see IX.16)

$$g(\omega) = 2\,\frac{4\pi\omega^2 V}{(2\pi c)^3}. \tag{20.3}$$

The factor 2 in this relation appears here because for any given momentum there are two possible states of polarization. Then, in the range $(\omega, \omega+d\omega)$, there will be $dn_\lambda = n_\lambda\, g(\omega)\, d\omega$ photons and therefore the field energy in this range is $\hbar\omega\, dn_\lambda$. Since the photons occupy a volume V, the black body radiation energy density (the Planck formula) is given by

$$\varrho(\omega, T) = \frac{\hbar\omega\, dn_\lambda}{V} = \frac{\hbar}{\pi^2 c^3}\,\frac{\omega^3}{e^{\frac{\hbar\omega}{kT}} - 1}. \tag{20.4}$$

21. If we consider only the transition $2 \equiv (n, l) \to 1 \equiv (n', l+1)$, it follows from (IX.19) that

$$P_{12} = \frac{4\omega^3}{3\hbar c^3}\,|\mathbf{d}_{12}|^2. \tag{21.1}$$

Since in a central field the states are degenerate with respect to the magnetic quantum number, we have that

$$|\mathbf{d}_{12}|^2 = e^2 \sum_{m=-1}^{+1} |\mathbf{r}_{12}|^2 = e^2 \sum_m \left[\tfrac{1}{2}\,|x_{12}+iy_{12}|^2 + \tfrac{1}{2}\,|x_{12}-iy_{12}|^2 + |z_{12}|^2\right]. \tag{21.2}$$

The matrix elements $(x+iy)_{12}$, $(x-iy)_{12}$ can be calculated easily by changing to spherical polar coordinates, in which

$$x+iy = r\sin\theta e^{i\phi}, \quad x-iy = r\sin\theta e^{-i\phi}, \quad z = r\cos\theta.$$

Now the wavefunctions of an electron in a central field are of the form

$$\psi_{nlm} = \phi_{nl}(r)\, Y_l^m(\theta, \phi), \quad \int |\psi_{nlm}|^2\, dV = 1, \tag{21.3}$$

so that, if we define

$$I_{n'l',\, nl} = \int_0^\infty r^3 \phi^*_{n'l'}(r)\, \phi_{nl}(r)\, dr$$

we obtain the results

$$\langle n', l+1, m+1 \,|\, x+iy \,|\, n, l, m\rangle = I_{n'l+1,\, nl}\left[\frac{(l+m+1)(l+m+2)}{(2l+1)(2l+3)}\right]^{1/2},$$

$$\langle n', l+1, m-1 \,|\, x-iy \,|\, n, l, m\rangle = -I_{n'l+1,\, nl}\left[\frac{(l-m+1)(l-m+2)}{(2l+1)(2l+3)}\right]^{1/2},$$

$$\langle n', l+1, m \,|\, z \,|\, n, l, m\rangle = I_{n'l+1,\, nl}\left[\frac{(l+m+1)(l-m+1)}{(2l+1)(2l+3)}\right]^{1/2},$$

$$\langle n', l-1, m+1 \,|\, x+iy \,|\, n, l, m\rangle = -I_{n'l-1,\, nl}\left[\frac{(l-m-1)(l-m)}{(2l-1)(2l+1)}\right]^{1/2},$$

$$\langle n', l-1, m-1 \,|\, x-iy \,|\, n, l, m\rangle = I_{n'l-1,\, nl}\left[\frac{(l+m-1)(l+m)}{(2l-1)(2l+1)}\right]^{1/2},$$

$$\langle n', l-1, m \,|\, z \,|\, n, l, m\rangle = I_{n'l-1,\, nl}\left[\frac{(l+m)(l-m)}{(2l-1)(2l+1)}\right]^{1/2},$$

and hence

$$\sum_m |x_{12}+iy_{12}|^2 = |I_{12}|^2 \sum_m \frac{(l+m+1)(l+m+2)}{(2l+1)(2l+3)}. \tag{21.4}$$

Remembering that

$$\sum_m m^2 = \tfrac{1}{3} l(l+1)(2l+1), \quad \sum_m m = 0, \quad \sum_m 1 = 2l+1$$

we find that

$$\sum_m |x_{12}+iy_{12}| = \frac{|I_{12}|^2}{(2l+1)(2l+3)}[(l+1)(l+2)(2l+1)+\tfrac{1}{3} l(l+1)(2l+1)] = \tfrac{2}{3}(l+1)\,|I_{12}|^2. \tag{21.5}$$

Similarly one can obtain the result

$$\sum_m |x_{12}-iy_{12}|^2 = 2\sum_m |z_{12}|^2 = \tfrac{2}{3}(l+1)\,|I_{12}|^2, \tag{21.6}$$

so that

$$\sum_m |\mathbf{r}_{12}|^2 = (l+1)\,|I_{12}|^2. \tag{21.7}$$

For the transition $2 \equiv (n, l) \rightarrow 1 \equiv (n', l-1)$, one obtains by a similar calculation the result

$$\sum_m |\mathbf{r}_{12}|^2 = l\,|I_{12}|^2. \tag{21.8}$$

22. The first two spectral lines of the Balmer series correspond to the transitions $(n = 3) \rightarrow (n' = 2)$ and $(n = 4) \rightarrow (n' = 2)$ respectively. We can use the results of the preceding problem by taking the $\phi_{nl}(r)$ to be the radial parts of the hydrogen atom wavefunction. From (33.14–16) Chapter II, and (A.33) we have

$$\phi_{2,0} = \frac{1}{\sqrt{2a^3}}\left(1-\frac{r}{2a}\right)e^{-(r/2a)}, \quad \phi_{2,1} = \frac{2}{2\sqrt{6a^3}}\,\frac{r}{a}\,e^{-(r/2a)}$$

$$\phi_{3,0} = \frac{2}{3\sqrt{3a^3}}\left(1-\frac{2r}{3a}+\frac{2r^2}{27a^2}\right)e^{-(r/3a)}$$

$$\phi_{3,1} = \frac{8}{27\sqrt{6a^3}}\,\frac{r}{a}\left(1-\frac{r}{6a}\right)e^{-(r/3a)}, \quad \phi_{3,2} = \frac{4}{81\sqrt{30a^3}}\,\frac{r^2}{a^2}\,e^{-(r/3a)}$$

$$\phi_{4,0} = \frac{1}{4\sqrt{a^3}}\left(1-\frac{3r}{4a}+\frac{r^2}{8a^2}-\frac{r^3}{192a^3}\right)e^{-(r/4a)}$$

$$\phi_{4,1} = \frac{1}{16}\sqrt{\frac{5}{3}}\,\frac{r}{a}\left(1-\frac{r}{4a}+\frac{r^2}{80a^2}\right)e^{-(r/4a)}$$

$$\phi_{4,2} = \frac{1}{64\sqrt{5a^3}}\,\frac{r^2}{a^2}\left(1-\frac{r}{12a}\right)e^{-(r/4a)}, \quad \phi_{4,3} = \frac{1}{768\sqrt{35a^3}}\,\frac{r^3}{a^3}\,e^{-(r/4a)}.$$

The first line H_α of the Balmer series results, according to the selection rules, from the transitions $(3, 0) \rightarrow (2, 1)$, $(3, 1) \rightarrow (2, 0)$, and $(3, 2) \rightarrow (2, 1)$, for which we have the matrix elements

$$\begin{aligned}
\langle 2, 1\,|r|\,3, 0\rangle &= \int_0^\infty r^3\phi_{21}\phi_{30}\,dr = \frac{4a}{15\sqrt{2}}\left(\frac{6}{5}\right)^5 \\
\langle 2, 0\,|r|\,3, 1\rangle &= \int_0^\infty r^3\phi_{20}\phi_{31}\,dr = \frac{32a}{15\sqrt{3}}\left(\frac{6}{5}\right)^5 \\
\langle 2, 1\,|r|\,3, 2\rangle &= \int_0^\infty r^3\phi_{21}\phi_{32}\,dr = \frac{64a}{15\sqrt{5}}\left(\frac{6}{5}\right)^5.
\end{aligned} \tag{22.1}$$

The second line H_β results from the transitions (4, 0) → (2, 1), (4, 1) → (2, 0) and (4, 2) → (2, 1), for which

$$\begin{aligned}
\langle 2, 1\,|r|\,4, 0\rangle &= \int_0^\infty r^3\phi_{21}\phi_{40}\,dr = \frac{2a}{9\sqrt{6}}\left(\frac{4}{3}\right)^5\\
\langle 2, 0\,|r|\,4, 1\rangle &= \int_0^\infty r^3\phi_{20}\phi_{41}\,dr = \frac{a}{3}\sqrt{\frac{5}{6}}\left(\frac{4}{3}\right)^5\\
\langle 2, 1\,|r|\,4, 2\rangle &= \int_0^\infty r^3\phi_{21}\phi_{42}\,dr = \frac{20a}{9\sqrt{30}}\left(\frac{4}{3}\right)^5.
\end{aligned} \tag{22.2}$$

The frequencies of the two spectral lines are

$$\omega_\alpha = \frac{e^4m}{2\hbar^3}\left(\frac{1}{2^2}-\frac{1}{3^2}\right) = \frac{5}{36}\,\frac{e^4m}{2\hbar^3},$$

$$\omega_\beta = \frac{e^4m}{2\hbar^3}\left(\frac{1}{2^2}-\frac{1}{4^2}\right) = \frac{3}{16}\,\frac{e^4m}{2\hbar^3}.$$

From (21.1), (21.7), (21.8), (22.1) and (22.2) we find the transition probabilities per unit time (per excited atom)

$$P_\alpha = \frac{4}{3}\,\frac{e^2}{\hbar c}\,\frac{\omega_\alpha^2}{c^2}\,a^2\left\{\left[\frac{4}{15\sqrt{2}}\left(\frac{6}{5}\right)^5\right]^2+\left[\frac{32}{15\sqrt{3}}\left(\frac{6}{5}\right)^5\right]^2+2\left[\frac{64}{15\sqrt{5}}\left(\frac{6}{5}\right)^5\right]^2\right\},$$

$$P_\beta = \frac{4}{3}\,\frac{e^2}{\hbar c}\,\frac{\omega_\beta^2}{c^2}\,a^2\left\{\left[\frac{2}{9\sqrt{6}}\left(\frac{4}{3}\right)^5\right]^2+\left[\frac{1}{3}\sqrt{\frac{5}{6}}\left(\frac{4}{3}\right)^5\right]^2+2\left[\frac{20}{9\sqrt{30}}\left(\frac{4}{5}\right)^5\right]^2\right\}.$$

Numerical computation then gives the results

$$P_\alpha = \left(\frac{c^2}{\hbar c}\right)^4\frac{c}{a}\,\frac{2^6\times 3727}{5^{10}}, \qquad P_\beta = \left(\frac{e^2}{\hbar c}\right)^4\frac{c}{a}\,\frac{2^6\times 209}{3^{13}}. \tag{22.3}$$

The intensities of the lines are then found from (22.3) to be

$$\begin{aligned}
J_\alpha &= P_\alpha.\hbar\omega_\alpha = \left(\frac{e^2}{\hbar c}\right)^4 c\,\frac{e^2}{a^2} = \frac{2^3\times 3727}{3^2\times 5^9}\ \text{erg sec}^{-1},\\
J_\beta &= P_\beta.\hbar\omega_\beta = \left(\frac{e^2}{\hbar c}\right)^4 c\,\frac{e^2}{a^2} = \frac{2\times 209}{3^{12}}\ \text{erg sec}^{-1}.
\end{aligned} \tag{22.4}$$

For the ratio of the intensities of the first two lines of the Balmer series we find

$$\frac{J_\alpha}{J_\beta} = \frac{2^2\times 3^{10}}{5^9}\times\frac{3727}{209} = 2{\cdot}16, \tag{22.5}$$

assuming that equal numbers of atoms are excited initially to the $n = 3$ and to the $n = 4$ states.

23. The kinetic energy of the electron ejected from the atom is given by the Einstein relation

$$T = \hbar\omega - I, \quad I = -E_1 \tag{23.1}$$

where I is the ionization energy of the atom. If the photon energy is large in comparison with I, we can neglect the Coulomb field which acts on the emitted electron. Then the normalized wavefunctions ψ_1 and ψ_2, with which we calculate the transition matrix elements, correspond respectively to the ground state of the hydrogen-like atom and to a free electron, i.e.

$$\psi_1 = \sqrt{\frac{Z^3}{\pi a^3}}\, e^{-\frac{Zr}{a}}, \quad \psi_2 = \frac{1}{(2\pi)^{3/2}}\, e^{\frac{i}{\hbar}\mathbf{p}.\mathbf{r}}. \tag{23.2}$$

The total probability per unit time of a transition to a state ψ_2 whose momentum $\mathbf{p}$ lies in a solid angle $d\Omega$ is, by (IX.4),

$$dP = \frac{2\pi}{\hbar}\, |H'_{21}|^2\, \varrho(E)\, d\Omega, \tag{23.3}$$

where H'_{21} is the matrix element of the perturbation (IX.14).

The magnitude of the momentum $\mathbf{p}$ is determined from the relation (23.1). From (IX.17) and (23.2), putting $n_\lambda = 1$, we have

$$H'_{21} = -\frac{e}{m(2\pi\hbar)}\sqrt{\frac{Z^3}{\pi a^3 V\omega}} \int e^{\frac{i}{\hbar}\mathbf{q}.\mathbf{r}}\, \frac{\hbar}{i}(\mathbf{e}.\nabla)\, e^{-\frac{Zr}{a}}\, d\mathbf{r}, \tag{23.4}$$

where $\mathbf{e}$ is the polarization vector, and $\mathbf{q} = \hbar\mathbf{k} - \mathbf{p}$ is the momentum transferred to the nucleus. Integrating by parts, we obtain

$$H'_{21} = -\frac{e}{m(2\pi\hbar)}\sqrt{\frac{Z^3}{\pi a^3 V\omega}}\, (\mathbf{e}.\mathbf{p}) \int e^{-\frac{Zr}{a}}\, e^{+\frac{i}{\hbar}\mathbf{q}.\mathbf{r}}\, d\mathbf{r}.$$

By changing to spherical polar coordinates with the polar axis in the direction of $\mathbf{q}$, we have

$$\int e^{-\frac{Z}{a}r}\, e^{\frac{i}{\hbar}\mathbf{q}.\mathbf{r}}\, dV = \frac{2\pi\hbar}{iq} \int_0^\infty \left(e^{\frac{i}{\hbar}qr} - e^{-\frac{i}{\hbar}qr}\right) e^{-\frac{Z}{a}r}\, r\, dr = \frac{8\pi a^3}{Z^3\left(1+\dfrac{q^2a^2}{Z^2\hbar^2}\right)^2},$$

and hence

$$H'_{21} = -\frac{4e}{m\hbar}\sqrt{\frac{a^3}{\pi Z^3 V\omega}}\, (\mathbf{e}.\mathbf{p}) \left(1+\frac{q^2a^2}{\hbar^2 Z^2}\right)^{-2}. \tag{23.5}$$

The differential cross-section for the photoelectric effect is obtained by dividing the transition probability per unit time by the flux density of the incident photons. Since the normalization has been chosen in such a way that there is only one photon in the volume V, the flux density of incident photons will be c/V. Then, taking into account (8.3′), we have

$$d\sigma = \frac{32\times 137^4}{Z^3}\frac{p(\mathbf{p}.\mathbf{e})^2}{(mc^2)^2}\frac{c^3}{\hbar\omega}r_0^2\left(1+\frac{q^2a^2}{\hbar^2Z^2}\right)^{-4}d\Omega, \tag{23.6}$$

where we have written $e^2/\hbar c = \frac{1}{137}$ and $e^2/mc^2 = r_0$ (the "classical radius" of the electron). Denoting by θ the angle between the planes ($\mathbf{p}$, $\mathbf{k}$) and ($\mathbf{e}$, $\mathbf{k}$), we have that

$$\mathbf{p}.\mathbf{e} = p\sin\theta\cos\phi, \quad q^2 = p^2+s^2-2ps\cos\theta; \quad s = \hbar k, \tag{23.7}$$

$$1+\frac{q^2a^2}{\hbar^2Z^2} = \frac{a^2}{\hbar^2Z^2}\left(\frac{Z^2\hbar^2}{a^2}+q^2\right) = \frac{a^2}{\hbar^2Z^2}\left(\frac{Z^2m^2e^4}{\hbar^2}+p^2+s^2-2ps\cos\theta\right).$$

From (23.1), and remembering that $I = Z^2e^4m/2\hbar^2$, it follows that

$$\frac{Z^2m^2e^4}{\hbar^2}+p^2 = 2m\hbar\omega = 2msc,$$

and hence that

$$1+\frac{q^2a^2}{\hbar^2Z^2} = \frac{a^2}{\hbar^2Z^2}s(2mc-s-2p\cos\theta) = \frac{a^2}{\hbar^2Z^2}2m\hbar\omega(1-\beta\cos\theta), \tag{23.8}$$

where $\beta = v/c$. To obtain the last equality the incident photon energy was assumed to be much smaller than the rest energy† of the electron, i.e. $\hbar\omega \ll mc^2$.

From (23.1) and the fact that the kinetic energy of the ejected electron is much greater than the ionization energy, we can take $p = \sqrt{2m\hbar\omega}$. From (23.6), (23.7) and (23.8) we then obtain

$$d\sigma = 4\sqrt{2}\frac{Z^5}{137^4}r_0^2\left(\frac{mc^2}{\hbar\omega}\right)^{7/2}\frac{\sin^2\theta\cos^2\phi}{(1-\beta\cos\theta)^4}d\Omega. \tag{23.9}$$

Since (23.9) has been obtained in a non-relativistic approximation, it is correct only to first order in β, so that we can write finally

$$d\sigma = 4\sqrt{2}\frac{Z^2}{137^4}r_0^2\left(\frac{mc^2}{\hbar\omega}\right)^{7/2}\sin^2\theta\cos^2\phi(1+4\beta\cos\theta)\,d\Omega. \tag{23.10}$$

From (23.10) it follows that almost all photo-electrons are emitted in the direction of polarization of the photon ($\theta = \pi/2$, $\phi = 0$), and that no photo-electrons are emitted in the direction of motion of the photon ($\theta = 0$). This maximum in the differential cross-section of the photoelectric effect along the direction of polarization of the photon becomes more marked with increasing photon energy. We can obtain the total cross-section of the photo-

† This was of course implied from the start, since this is a non-relativistic treatment of the photoelectric effect.

electric effect for K shell electrons by integrating (23.10) over θ and ϕ and multiplying it by 2, since there are two electrons in the K shell; we obtain in this way

$$\sigma = \frac{32\sqrt{2}}{3}\pi\frac{Z^5}{134^4}r_0^2\left(\frac{mc^2}{\hbar\omega}\right)^{7/2}. \tag{23.11}$$

24. Let us suppose that, until the electromagnetic field is applied, the system is in the stationary state $\phi_n(\mathbf{r})$ which is an eigenfunction with eigenvalue E_n of the unperturbed Hamiltonian H_0. The interaction energy between the particle of charge e and the electromagnetic field is, in the dipole approximation, $-e\mathbf{r}.\mathscr{E}_0 \cos\omega t$, so that the Schrödinger equation of the system is

$$i\hbar\frac{\partial\psi}{\partial t} = (H_0 - e\mathbf{r}.\mathscr{E}_0\cos\omega t)\psi. \tag{24.1}$$

Let us try to determine the perturbed wavefunction in the form

$$\psi_n(\mathbf{r}, t) = \phi_n(\mathbf{r})e^{-i\omega_n t} + u_n(\mathbf{r})e^{-i(\omega_n-\omega)t} + v_n(\mathbf{r})e^{-i(\omega_n+\omega)t}, \tag{24.2}$$

where $\omega_n = E_n/\hbar$, and the terms in u_n and v_n are small. Substituting (24.2) into (24.1) and restricting ourselves to first-order terms, we obtain

$$\hbar(\omega_n-\omega)u_n e^{i\omega t} + \hbar(\omega_n+\omega)v_n e^{-i\omega t} = -H_0 u_n e^{i\omega t} + H_0 v_n e^{-i\omega t} - e(\mathbf{r}.\mathscr{E}_0)\phi_n\cos\omega t. \tag{24.3}$$

Equating the coefficients of the exponentials:

$$\begin{aligned}\hbar(\omega_n-\omega)u_n &= H_0 u_n - \frac{e}{2}(\mathbf{r}.\mathscr{E}_0)\phi_n\\ \hbar(\omega_n+\omega)v_n &= H_0 v_n - \frac{e}{2}(\mathbf{r}.\mathscr{E}_0)\phi_n.\end{aligned} \tag{24.4}$$

In order to solve these equations we expand u and v in terms of the orthonormal functions ϕ_n,

$$u_n = \sum_l A_{nl}\phi_l, \qquad v_n = \sum_l B_{nl}\phi_l. \tag{24.5}$$

Taking into account the fact that the ϕ_l are eigenfunctions of H_0, and substituting (24.5) into (24.4), we find that

$$\begin{aligned}\hbar\sum_l(\omega_{nl}-\omega)A_{nl}\phi_l &= -\frac{e}{2}(\mathbf{r}.\mathscr{E}_0)\phi_n\\ \hbar\sum_l(\omega_{nl}+\omega)B_{nl}\phi_l &= -\frac{e}{2}(\mathbf{r}.\mathscr{E}_0)\phi_n,\end{aligned} \tag{24.6}$$

whence

$$\begin{aligned}A_{nl} = -\frac{\mathscr{E}_0 \cdot \mathbf{p}_{ln}}{2\hbar(\omega_{nl}-\omega)}, \quad B_{nl} &= \frac{\mathscr{E}_0\cdot\mathbf{p}_{ln}}{2\hbar(\omega_{nl}+\omega)}\\ \mathbf{p}_{ln} &= \int\phi_l e\mathbf{r}\phi_n\,dV.\end{aligned} \tag{24.7}$$

Since the expressions (24.2), (24.5) and (24.7) are in first-order approximation only, it is necessary for their validity to assume that the electric field is sufficiently weak and that its frequency is far from any resonance frequency of the system.

Now the electric dipole moment of the system in the state ψ_n is

$$\mathbf{D}_n = \int \psi_n^* e\mathbf{r}\psi_n \, dr$$

and, using (24.2), (24.5) and (24.7), we obtain finally

$$\begin{aligned} \mathbf{D}_n = \mathbf{p}_{nn} - \frac{e^{i\omega t}}{2\hbar} \sum_l \left[\frac{(\mathscr{E}_0 \cdot \mathbf{p}_{ln})\mathbf{d}_{nl}}{\omega_{ln} - \omega} + \frac{(\mathscr{E}_0 \cdot \mathbf{p}_{ln}^*)\mathbf{d}_{ln}}{\omega_{ln} + \omega} \right] \\ - \frac{e^{-i\omega t}}{2\hbar} \sum_l \left[\frac{(\mathscr{E}_0 \cdot \mathbf{p}_{ln})\mathbf{p}_{nl}}{\omega_{nl} + \omega} + \frac{(\mathscr{E}_0 \cdot \mathbf{p}_{ln}^*)\mathbf{p}_{ln}}{\omega_{nl} - \omega} \right]. \end{aligned} \tag{24.8}$$

Note that the electric dipole moment consists of two parts: (1) the permanent dipole moment $\mathbf{p}_{nn}$ of the system in the state ϕ_n, and (2) an induced electric dipole moment which is periodic in time with the same frequency as the external field. This induced moment $\mathbf{d}_n = \mathbf{D}_n - \mathbf{p}_{nn}$ leads to coherent scattering and to dispersion. From (24.8) we can write

$$\mathbf{d}_n = \mathrm{Re}\,(\beta \mathscr{E}_0 e^{i\omega t}), \tag{24.9}$$

where Re means the real part of the expression in parentheses, and β is the tensor

$$\beta = \begin{pmatrix} \beta_{xx} & \beta_{xy} & \beta_{xz} \\ \beta_{yx} & \beta_{yy} & \beta_{yz} \\ \beta_{zx} & \beta_{zy} & \beta_{zz} \end{pmatrix} \tag{24.10}$$

which has components of the form

$$\beta_{xy} = -\frac{1}{\hbar} \sum_l \left[\frac{(\mathbf{p}_{ln})_y (\mathbf{p}_{nl}^*)_x}{\omega_{nl} - \omega} + \frac{(\mathbf{p}_{ln}^*)_y (\mathbf{p}_{ln})_x}{\omega_{nl} + \omega} \right]. \tag{24.11}$$

Since $\mathbf{p}_{ln} = \mathbf{p}_{nl}^*$, (24.10) is Hermitian, i.e. $\beta_{xy} = \beta_{yx}^*$, etc., and, in consequence, the diagonal terms are real. In the particular case in which all the components of the tensor are real (i.e. there is no phase difference between the field $\mathscr{E}(t)$ and the induced moment $\mathbf{d}_{nn}$), we have from (24.9) that

$$\mathbf{d}_n = \beta \mathscr{E}, \tag{24.9'}$$

where β is now simply a scalar, i.e. $\beta_{xy} = \beta_{xz} = \beta_{yz} = 0$ and $\beta_{xx} = \beta_{yy} = \beta_{zz} = \beta$. From (24.11), an expression similar to that obtained from the classical theory of dispersion can be found:

$$\beta = \frac{e^2}{m} \sum_l \frac{f_{ln}}{\omega_{ln}^2 - \omega^2}, \qquad f_{ln} = \frac{2m\omega_{ln}}{\hbar} |\langle l | x | n \rangle|^2. \tag{24.12}$$

The quantity f_{ln} is called the $n \to l$ dipole transition intensity.† The expressions (24.12) can

† The following identity (Thomas–Reiche–Kuhn) is valid $\sum_l f_{l,}^{(\alpha)} = 1$; $\alpha \equiv x, y, z$ (see problem 23, Chapter V).

be used to find the dielectric constant of a rarefied gas in accordance with the formula of classical electrodynamics:†

$$\varepsilon = 1+4\pi N_\beta = 1+4\pi N\sum_l \frac{\frac{e^2}{m}f_{ln}}{\omega_{ln}^2-\omega^2} \tag{24.13}$$

where N is the number of atoms (or molecules) per unit volume. If the atoms are in the ground state, ($n = 0$), the quantities f_{l0} are all positive and the refractive index $n = \sqrt{\varepsilon}$ increases with the frequency ω. If the atoms are in excited states, there will be ranges of frequencies for which the refractive index decreases with frequency. Such cases lead to anomalous dispersion.

† It was assumed above that all the atoms were initially in the same state n. For a mixed ensemble (see section 3, Chapter V), if we denote by N_n the number of atoms in the state n, (24.13) has to be replaced by

$$\varepsilon = 1+4\pi\sum_n\sum_l N_n \frac{\frac{e^2}{m}f_{ln}}{\omega_{ln}^2-\omega^2}. \tag{24.14}$$

CHAPTER X

Collision Theory

1. Potential Scattering

1.1. Scattering Amplitudes and Scattering Cross-sections

In non-relativistic quantum mechanics, the scattering of particles of mass m, momentum $\mathbf{k}\hbar$ and energy E by a potential $V(\mathbf{r})$ is determined by the asymptotic behaviour of the solution $\psi_{\mathbf{k}}^{+}(\mathbf{r})$, corresponding to out-going scattered waves, of the Schrödinger equation

$$(\Delta+k^2)\,\psi(\mathbf{r}) = U(\mathbf{r})\,\psi(\mathbf{r}) \tag{X.1}$$

$$k^2 = 2mE/\hbar^2, \qquad U(\mathbf{r}) = 2mV(\mathbf{r})/\hbar^2. \tag{X.2}$$

If the potential has a finite range (i.e. if $V(\mathbf{r})$ is appreciably different from zero only for $|\mathbf{r}| \leqslant r_0$ say), the problem of solving the differential equation (X.1) with the appropriate boundary conditions is equivalent to that of solving the Fredholm-type integral equation

$$\psi_{\mathbf{k}}^{+}(\mathbf{r}) = \phi_{\mathbf{k}}(\mathbf{r}) - \frac{1}{4\pi}\int \frac{\exp(ik\,|\mathbf{r}-\mathbf{r}'|)}{|\mathbf{r}-\mathbf{r}'|}\,U(\mathbf{r}')\,\psi_{\mathbf{k}}^{+}(\mathbf{r}')\,d\mathbf{r}', \tag{X.3}$$

where

$$\phi_{\mathbf{k}}(\mathbf{r}) = \exp(i\mathbf{k}.\mathbf{r}) \tag{X.4}$$

and satisfies the homogeneous equation

$$(\Delta+k^2)\,\phi_{\mathbf{k}}(\mathbf{r}) = 0. \tag{X.5}$$

At large distances from the scattering centre ($r \gg r_0$) we have

$$\psi_{\mathbf{k}}^{+}(\mathbf{r}) \sim \phi_{\mathbf{k}}(\mathbf{r}) + A(E, \theta, \phi)\frac{e^{ikr}}{r}, \qquad r \gg r_0 \tag{X.6}$$

where $\mathbf{r} = (r, \theta, \phi)$ in polar coordinates having $\mathbf{k}$ as polar axis, and

$$A(E, \theta, \phi) = -\frac{1}{4\pi}\langle \phi_{\mathbf{k}'}\,|\,U\,|\,\psi_{\mathbf{k}}^{+}\rangle; \quad \phi_{\mathbf{k}}(\mathbf{r}) = e^{i\mathbf{k}.\mathbf{r}}, \quad \mathbf{k}' = k\frac{\mathbf{r}}{r}. \tag{X.7}$$

$A(E, \theta, \phi)$ is called the "scattering amplitude".

The differential scattering cross-section is then

$$d\sigma(E, \theta, \phi) = |A(E, \theta, \phi)|^2 \, d\Omega, \quad d\Omega = \sin\theta \, d\theta \, d\phi. \tag{X.8}$$

Integrating this expression over all directions, we obtain the total scattering cross-section

$$\sigma(E) = \int \frac{d\sigma(E, \theta, \phi)}{d\Omega} \, d\Omega. \tag{X.9}$$

The scattering amplitude obtained from (X.7) by using the zero-order approximation $\psi_{\mathbf{k}}^{+}(\mathbf{r}) = \phi_{\mathbf{k}}(\mathbf{r})$ to the solution of the integral equation (X.3) is called the Born amplitude, and can be written in the form

$$A^{(B)} = -\frac{1}{4\pi} \langle \phi_{\mathbf{k}'} | U | \phi_{\mathbf{k}} \rangle. \tag{X.10}$$

Denoting by $\hbar\mathbf{q} = \hbar(\mathbf{k}-\mathbf{k}')$ the momentum transferred on scattering, we can write

$$A^{(B)}(\mathbf{q}) = -\frac{1}{4\pi} \int e^{i\mathbf{q}\cdot\mathbf{r}} U(\mathbf{r}) \, d\mathbf{r}. \tag{X.10'}$$

The ratio

$$F(\mathbf{q}) = \frac{A^{(B)}(\mathbf{q})}{A^{(B)}(0)} \tag{X.11}$$

is called the "form factor" and it characterizes the interference between waves scattered in different volume elements of the scattering field. The mutual elastic scattering of two particles with masses m_1 and m_2 and a potential energy of interaction $V(\mathbf{r}_2 - \mathbf{r}_1)$ can be expressed in terms of the motion of particle 1 relative to particle 2, which is formally identical with that of a single particle of "reduced" mass $m = m_1 m_2 (m_1 + m_2)^{-1}$ moving in a potential $V(\mathbf{r})$ referred to a fixed origin. Thus, in the centre of mass system (CMS) of the two particles, the direction of motion after scattering of particle 1, (θ, ϕ), coincides with that of the hypothetical single particle and is related to the directions of motion after scattering, (θ_1, ϕ_1) and (θ_2, ϕ_2), of particles 1 and 2 in the laboratory system (in which the second particle was at rest before the scattering occurred) by the relations

$$\tan\theta_1 = \frac{\sin\theta}{\gamma + \cos\theta}, \quad \gamma = \frac{m_1}{m_2}, \quad \theta_2 = \frac{1}{2}(\pi - \theta)$$
$$\phi_1 = \phi, \quad \phi_2 = \phi + \pi. \tag{X.12}$$

1.2. The Method of Partial Waves

For a central potential, $V(\mathbf{r}) = V(r)$, the direction of the momentum of the incident particle constitutes an axis of symmetry of the problem. Choosing the polar axis in this direction, the solution† of equation (X.1) will be independent of ϕ, and will have the form

$$\psi_{\mathbf{k}}^{+}(r, \theta) = \sum_{l=0}^{\infty} a_l \frac{R_{l,k}(r)}{r} P_l(\cos\theta) \tag{X.13}$$

where, for convenience, we have written $a_l = i^l(2l+1)e^{i\delta_l(k)}$.

The functions $R_{l,k}(r)$ are the solutions of the radial equation (II.15) which obey the condition (II.16) at the origin, and which (for potentials which vanish at infinity more rapidly than r^{-1} and are less singular at the origin than r^{-2}) have the following asymptotic behaviour for large r:

$$R_{l,k}(r) \sim \frac{1}{k} \sin\left(kr - \frac{l\pi}{2} - \delta_l(k)\right). \tag{X.14}$$

This asymptotic behaviour of $R_{l,k}(r)$ corresponds to the "normalization"

$$\int_0^{\infty} R_{l,k}(r)\, R_{l,k'}(r)\, dr = \frac{\pi}{2k^2}\, \delta(k-k').$$

For large r, it is convenient to write the "outgoing" parts of the partial waves of (X.13) in the form

$$e^{2i\delta_l(k)} \phi_{l,\mathbf{k}}^{+}(r, \theta) \tag{X.15}$$

where $\phi_{l,\mathbf{k}}^{+}(r, \theta)$ is the outgoing part of the l-state wavefunction for the limiting case of a zero scattering potential.

The effect of the non-zero potential then appears in the phase shift factor, and thus the scattering due to a central potential can be represented as a unitary transformation of "free" outgoing partial waves by a scattering operator, which, in the above angular momentum representation, takes the simple form of a diagonal "scattering matrix" whose elements are

$$S_{ll'} = \exp[2i\delta_l(k)]\delta_{ll'}. \tag{X.16}$$

The various functions associated with elastic scattering are thus

(a) The phase-shifts $\delta_l(k)$,

(b) The eigenvalues of the scattering matrix

$$S_l(k) = \exp[2i\delta_l(k)], \tag{X.17}$$

known as the "S-functions" for each value of the (orbital) angular momentum l of the partial waves,

† In this chapter we emphasize, for reasons which will soon be apparent, the k-dependence of the radial function, writing the latter as $R_{l,k}(r)$.

(c) The partial wave amplitudes

$$A_l(E) = \frac{1}{2ik}(S_l(k)-1) = \frac{1}{k} e^{i\delta_l(k)} \sin \delta_l(k), \tag{X.18}$$

(d) The scattering amplitude

$$A(E, \theta) = \sum_{l=0}^{\infty} (2l+1) A_l(E) P_l(\cos \theta), \tag{X.19}$$

(e) The differential cross-section for elastic scattering

$$d\sigma(E, \theta) = |A(E, \theta)|^2 d\Omega, \tag{X.20}$$

(f) The total cross-section for elastic scattering

$$\sigma(E) = \sum_{l=0}^{\infty} \sigma^{(l)}(E) = \frac{4\pi}{k^2} \sum_{l=0}^{\infty} (2l+1) \sin^2 \delta_l(k) = \frac{4\pi}{k} \operatorname{Im} A(E, 0). \tag{X.21}$$

To investigate the analytic properties of the scattering matrix, solutions $f(\pm k, l; r)$ of equation (II.15) which are regular at infinity, called Jost solutions, are introduced. These are chosen to satisfy the boundary condition

$$\lim_{r \to \infty} f(\pm k, l; r) e^{\pm ikr} = 1. \tag{X.22}$$

In general, the Jost solutions do not satisfy the condition (II.16); their values at $r = 0$ are called the Jost functions and are denoted by $F_l(\pm k)$. We have then the important relationship

$$S_l(k) = e^{i\pi l} \frac{F_l(k)}{F_l(-k)}. \tag{X.23}$$

When the scattering process is such that it can be accompanied by various internal changes in the scattered "particles", which can take up or give out energy (such as excitation or ionization of atoms, excitation of nuclei, etc.) one says that it can occur in a number of possible "channels". Any channel which results in a change in the internal energy of one or more particles is said to be "inelastic". The existence of inelastic channels has an influence on the properties of the elastic one studied above. In the asymptotic region $r \to \infty$, the radial part of the wavefunction in the elastic channel will still consist (as in the purely elastic case) of a sum of outgoing and of incoming spherical waves. But, because of the presence of the inelastic channels, the amplitude of the outgoing wave is attenuated by a factor $\exp[-2\gamma_l(k)]$, where γ_l is real and positive, so that now

$$R_{l,k}(r) \underset{r \to \infty}{\sim} \frac{1}{2ik} \left\{ e^{-2\gamma_l} \exp\left[i\left(kr - \frac{l\pi}{2} + \delta_l\right)\right] - \exp\left[-i\left(kr - \frac{l\pi}{2} + \delta_l\right)\right]\right\}. \tag{X.24}$$

Thus, when inelastic channels are present, we have the following expressions for the elastic scattering amplitude and for the total elastic scattering cross-section:

$$A(E, \theta) = \frac{1}{2ik} \sum_{l=0}^{\infty} (2l+1)(e^{2i\eta_l}-1) P_l(\cos\theta), \tag{X.25}$$

$$\sigma_{el}(E) = \sum_{l=0}^{\infty} \sigma_{el}^{(l)}(E) = \frac{\pi}{k^2} \sum_{l=0}^{\infty} (2l+1)|1-C_l|^2, \tag{X.26}$$

where $C_l = \exp(2i\eta_l)$, and $\eta_l = \delta_l + i\gamma_l$ (the "complex phase-shift").

The total inelastic scattering cross-section is given by

$$\sigma_{inel}(E) = \sum_{l=0}^{\infty} \sigma_{inel}^{(l)}(E) = \frac{\pi}{k^2} \sum_{l=0}^{\infty} (2l+1)(1-|C_l|^2), \tag{X.27}$$

and the total cross-section for all channels is then

$$\sigma_{tot}(E) = \sigma_{el}(E) + \sigma_{inel}(E). \tag{X.28}$$

1.3. The Regge Pole Method

The Regge pole method is the name given to a method of studying the analytic properties of the scattering amplitude in which the orbital angular momentum l is treated as a continuous complex variable. In order to extend the definition (X.19) of the scattering amplitude to allow for complex values of l, it is assumed that in the complex plane of l an analytic function

$$A(l, E) \equiv \frac{1}{2ik}[S(l, E)-1] \tag{X.29}$$

can be defined, whose values, for integer values of l, coincide with the corresponding values of the partial amplitude $A_l(E)$. The scattering amplitude $A(E, \theta)$ can then be represented by a Watson–Sommerfeld integral

$$A(E, \theta) = -\frac{i}{2} \int_C \frac{2l+1}{\sin \pi l} A(l, E) P_l(-\cos\theta)\, dl, \tag{X.30}$$

in which the contour of integration C is chosen so that the "accidental poles" of the function $A(l, E)$ on the axis Re (l), Fig. X.1, are excluded. For Re $(l) > -\frac{1}{2}$, and given energy, the quantity under the integral sign in (X.30) has a finite number of poles, N, which are all in the half-plane Im $(l) \geqslant 0$.

As the energy E changes, these poles move along curves called Regge trajectories (see problems 36, 37). Thus, for example, for a potential consisting of a superposition of Yukawa potentials,

$$V(r) = \int_{\mu}^{\infty} \varrho(\alpha) \frac{e^{-\alpha r}}{r}\, d\alpha, \qquad \int_{\mu}^{\infty} \varrho(\alpha)\, d\alpha < \infty,$$

the function $A(l, E)$ tends to zero as $l \to \infty$ faster than $P_l(-\cos\theta)/\sin \pi l$ diverges so that the product tends to zero. By closing the contour of integration C as shown in Fig. X.2 and using the theory of residues, (X.30) becomes

$$A(E, \theta) = \frac{i}{2} \int_{-1/2-i\infty}^{-1/2+i\infty} \frac{2l+1}{\sin \pi l} A(l, E)\, P_l(-\cos\theta)\, dl + \sum_{n=1}^{N} \frac{\beta_n(E)}{\sin \pi l_n} P_{l_n}(-\cos\theta). \qquad \text{(X.31)}$$

Here $l_n(E)$ is the position of the nth pole and $\beta_n(E)$ is the quantity

$$\beta_n(E) = -\pi(2l_n+1)\ \text{Residue}\ [A(l, E)]_{l=l_n}\,.$$

A useful feature of the Regge pole representation (X.31) is that it distinguishes the singular part of the amplitude from the non-singular (integral) part.

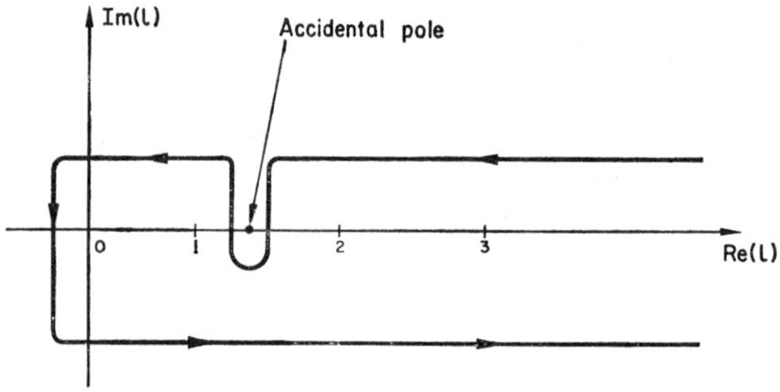

FIG. X.1.

2. The Lippmann–Schwinger Equations

Let H be the time-independent Hamiltonian of a quantized system consisting of several mutually interacting sub-systems. Let us make the following assumptions.

1. H can be expressed as the sum of two terms

$$H = H_0 + H', \qquad \text{(X.32)}$$

such that the eigenvalues and eigenfunctions of H_0 (called the "reference-system Hamiltonian") are known.

2. The spectrum of H_0 has a continuous part, which extends over the entire domain $E \geqslant 0$.

3. The "interaction Hamiltonian" H', no matter how it affects the discrete part of the spectrum of H_0, does not "displace" the continuous part, i.e. if $|\Phi_a\rangle$ is any eigenfunction of

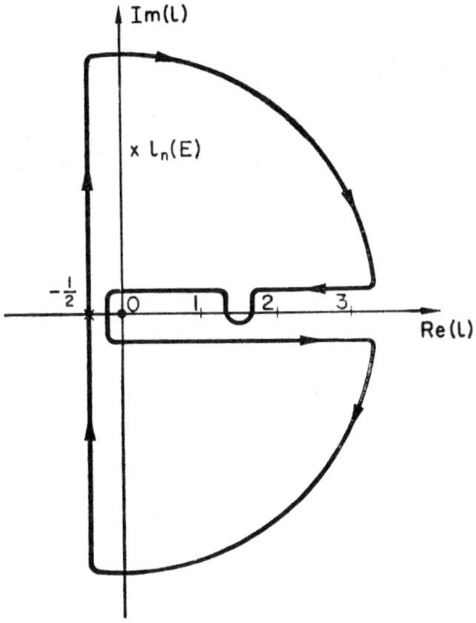

FIG. X.2.

H_0 with energy E_a in the continuous part of the spectrum of H_0, so that

$$H_0|\Phi_a\rangle = E_a|\Phi_a\rangle, \tag{X.33}$$

then there exists a corresponding eigenfunction $|\psi_a\rangle$ of H with the same energy E_a, belonging to the continuous part ($E \geqslant 0$) of the spectrum of H,

$$H\,|\psi_a\rangle = E_a|\,\psi_a\rangle. \tag{X.34}$$

Under these conditions, two solutions $|\psi^+\rangle$ and $|\psi^-\rangle$ of the Schrödinger equation (X.34) can be defined, called the "in" and the "out" solutions respectively, which satisfy the Lippmann–Schwinger equations

$$|\,\psi_a^{\pm}\rangle = |\Phi_a\rangle + \lim_{\eta\to+0} \frac{1}{E_a - H_0 \pm i\eta} H'\,|\,\psi_a^{\pm}\rangle. \tag{X.35}$$

These equations are valid in any representation, and once the representation has been chosen they become Fredholm-type integral equations. Thus, e.g., for a single particle in coordinate representation, $H_0 = -\frac{\hbar^2}{2m}\nabla^2$, $H' = V(\mathbf{r})$, and (X.35) becomes simply the integral equation (X.3) for potential scattering.

Problems

1. Using the Green's function method, deduce the integral equation for scattering (X.3).

2. Find the conditions of validity of the Born approximation.

3. Study the possibility of applying the Born approximation to the following potentials:

1. The "exponential" potential $V(r) = V_0 \exp\left(-\frac{r}{r_0}\right)$.
2. The "screened Coulomb" potential, or "Yukawa" potential, $V(r) = \frac{A}{r} \exp\left(-\frac{r}{r_0}\right)$.
3. The "spherical potential well" $V(r) = -V_0$ for $r < r_0$, $V(r) = 0$ for $r > r_0$.

4. Find the differential cross-section for scattering, in the Born approximation, for potentials with spherical symmetry, as a function of the angle of scattering.

Apply this expression to the following potentials:

$$1. \quad V(r) = \frac{A}{r} \exp\left(-\frac{r}{r_0}\right) \tag{4a}$$

$$2. \quad V(r) = V_0 \exp\left(-\frac{r^2}{2r_0^2}\right), \quad \text{the "Gaussian potential"} \tag{4b}$$

$$3. \quad V(r) = -V_0 \quad \text{for} \quad r < r_0, \quad V(r) = 0 \quad \text{for} \quad r > r_0. \tag{4c}$$

5. Using the results obtained in the preceding problem, calculate (in Born approximation) the total cross-section for scattering for the potentials (4a)–(4c).

6. In order to describe the interaction between two nucleons, Yukawa introduced the potential (4a).

Calculate the form factor and the total cross-section for neutron–proton scattering in Born approximation using the Yukawa potential. How does the total cross-section for scattering behave at high energies, where the use of the Born approximation is justified?

7. Using the transformation formulae (X.12), deduce the relation

$$|A_1(\theta_1, \phi_1)|^2 = \frac{(1+2\gamma\cos\theta+\gamma^2)^{3/2}}{|1+\gamma\cos\theta|} |A(\theta, \phi)|^2, \tag{7a}$$

where $A_1(\theta_1, \phi_1)$ and $A(\theta, \phi)$ are the scattering amplitudes in the laboratory and in the centre of mass systems respectively.

8. The wavefunction of a system of identical particles has to be symmetrical (for bosons) or anti-symmetrical (for fermions) with respect to the exchange of any two particles. Bearing this fact in mind, express the elastic scattering cross-section of two spinless bosons in terms of the scattering amplitude.

9. By generalizing the result of the preceding problem to include particles with spin, express the differential cross-section for proton–proton scattering in terms of the scattering amplitude, assuming that the interaction potential has spherical symmetry.

10. Knowing the Coulomb scattering amplitude

$$A(\theta) = -\frac{\lambda}{2k \sin^2 \dfrac{\theta}{2}} \exp\left[-i\lambda \ln\left(\sin^2 \frac{\theta}{2}\right) + 2i\eta\right], \tag{10a}$$

where

$$\lambda = \frac{Z_1 Z_2 e^2}{\hbar v}, \quad E = \frac{\hbar^2}{2m} k^2 = \frac{mv^2}{2}, \quad \eta = \arg \Gamma(1+i\lambda),$$

calculate the differential cross-section for elastic scattering of two identical particles of spin $\frac{1}{2}$ in Coulomb interaction, in the CMS and in the laboratory system.

11. Bearing in mind the asymptotic behaviour (X.14) of the radial equation, deduce the integral formula

$$\sin \delta_l(k) = -\frac{2m}{\hbar^2} k \int_0^\infty r j_l(kr)\, V(r)\, R_{l,\,k}(r)\, dr, \tag{11a}$$

which can be used to calculate the phase shift $\delta_l(k)$ for a state of orbital angular momentum l.

12. By using (11.6), study the dependence of the phase-shift $\delta_l(k)$ on small changes in the potential (k and l being kept constant).

13. Find the total cross-section for scattering of slow particles by the following spherical potential well:

$$V(r) = -V_0 \quad \text{if} \quad r < r_0 \quad \text{and} \quad V(r) = 0 \quad \text{if} \quad r > r_0.$$

14. Find the total cross-section for scattering of slow particles by the following spherical potential barrier:

$$V(r) = +V_0 \quad \text{if} \quad r < r_0 \quad \text{and} \quad V(r) = 0 \quad \text{if} \quad r > r_0.$$

15. Find the phase shifts and the differential cross-section for scattering of particles of mass m by the potential $V(r) = A/r^2$. Discuss the case in which $2mA\hbar^{-2} \ll 1$.

16. Show that, in the WKB approximation, the phase-shift $\delta_l(k)$ is given by

$$\delta_l(k) = \lim_{R\to\infty} \left[\int_a^R \sqrt{k^2 - U(r) - (l+\tfrac{1}{2})^2 r^{-2}}\, dr - \int_{a_0}^R \sqrt{k^2 - (l+\tfrac{1}{2})^2 r^{-2}}\, dr\right], \tag{16a}$$

where the lower limits of integration a and a_0 are the zeros of the functions under the integral signs (if $k^2 - U(r) - (l+\frac{1}{2})^2 r^{-2}$ has more than one zero, a is to be the largest of them).

17. Calculate, in the WKB approximation, the phase-shifts $\delta_l(k)$, for $l \gg 1$, for scattering by a potential which at large enough distances has the form $V(r) = Ar^{-n}$, $n > 2$.

18. Using the results of problem 1 show that the radial function $R_{l,k}(r)$ satisfies the integral equation

$$R_{l,k}(r) = rj_l(kr)e^{-\delta_l(k)} - krh_l^{(+)}(kr) \int_0^r r'j_l(kr')\, U(r')\, R_{l,k}(r')\, dr'$$

$$-ikrj_l(kr) \int_r^\infty r'h_l^{(+)}(kr')\, U(r')\, R_{l,k}(r')\, dr' \tag{18a}$$

where the $h_l^{(+)}(kr)$ are Hankel functions of the first kind.

19. Show that the Jost solutions $f(+k, l; r)$ and $f(-k, l; r)$ of the radial equation

$$\left[\frac{d^2}{dr^2} + k^2 - U(r) + \frac{l(l+1)}{r^2}\right] f(\pm k, l; r) = 0 \tag{19a}$$

are linearly independent.

Express the solution $R_{l,k}$, which is regular at the origin:

$$R_{l,k}(0) = 0, \quad R'_{l,k}(0) = 1 \tag{19b}$$

in terms of the Jost solutions, and hence deduce relation (X.23).

20. Find the Jost functions $F_0(\pm k)$, and the function $S_0(k)$, for the scattering of particles of energy E and orbital angular momentum $l = 0$ by a spherical potential well (4c). What properties do the poles of $S_0(k)$ have if $-V_0 < E < 0$?

21. Show that the Jost solution $f(k, l; r)$ satisfies the integral equation

$$f(k, l; r) = e^{-ikr} + \frac{1}{k} \int_r^\infty \sin k(r' - r)\, U_l(r') f(k, l; r')\, dr' , \tag{21a}$$

where

$$U_l(r) = U(r) + l(l+1)r^{-2} \tag{21b}$$

and that, in the complex k-plane, the corresponding Jost function $F_l(k)$ is holomorphic in the lower half-plane if

$$\int_r^\infty |U_l(r')|\, dr' < \infty. \tag{21c}$$

Show that $F_l(k)$ is also holomorphic in the region of the upper half-plane in which the condition

$$\int_r^\infty |U_l(r')|\, e^{2br'}\, dr' < \infty. \tag{21d}$$

is satisfied.

22. Using the integral equation (20a), show that the following relations hold:

$$S_l(k)S_l^*(k^*) = 1 \tag{22a}$$

$$S_l(k)S_l(-k) = 1. \tag{22b}$$

23. Determine the Bargmann strip† of the function $S_0(k)$ in the complex plane of k for the scattering of particles of orbital angular momentum $l = 0$ by the Yukawa potential (4a).

24. Study the arrangement and the physical meaning of the poles and the zeros of the function $S_l(k)$ in the Bargmann strip.

25. Using Cauchy's theorem

$$N = \frac{1}{2\pi i}\int\limits_C \frac{f'(z)}{f(z)}\,dz, \tag{25a}$$

which gives the number of zeros of the holomorphic function $f(z)$ inside the closed contour C, show that the following equality (Levinson's theorem) holds:

$$\delta_l(+0)-\delta_l(\infty) = n_l.\pi, \tag{25b}$$

where n_l is the number of bound states with orbital angular momentum l, provided that the Bargmann strip includes the whole complex plane of k.

26. Show that, for low energies, bound and virtual states of the same energy give equal contributions to the total cross-section for scattering.

27. Find the radial function $R_{l,\,k}(r)$, the partial amplitude $A_l(k)$, and the diagonal elements $S_l(k)$ of the matrix S, for scattering by a potential which is a δ-function on a sphere of radius a:

$$U(r) = -U_0\delta(r-a). \tag{27a}$$

28. Find the smallest value of U_0 which can lead to a bound state of orbital angular momentum l_0 in the potential (27a). Study the possibility of the existence of bound states with $l = 0$.

29. Show that the following relation holds between the elastic scattering amplitude (in the presence of inelastic channels) and the total cross-section:

$$\sigma_{\text{tot}}(E) = \frac{4\pi}{k}\,\text{Im}\,A(E, 0). \tag{29a}$$

30. Show that, for a given $\sigma_{\text{inel}}^{(l)}$, the elastic scattering cross-section $\sigma_{\text{el}}^{(l)}$ is such that

$$\sqrt{\sigma_0^{(l)}}-\sqrt{\sigma_0^{(l)}-\sigma_{\text{inel}}^{(l)}} \leqslant \sqrt{\sigma_{\text{el}}^{(l)}} \leqslant \sqrt{\sigma_0^{(l)}}+\sqrt{\sigma_0^{(l)}+\sigma_{\text{inel}}^{(l)}}, \tag{30a}$$

where $\sigma_0^{(l)} = \pi k^{-2}(2l+1)$.

† The Bargmann strip of $S_l(k)$ is the region of the complex plane of k in which $F_l(k)$ and $F_l(-k)$ are both holomorphic.

31. Assuming the existence and the uniqueness of the function (X.29), deduce relation (X.30), which gives the analytic continuation of the scattering amplitude into the complex l plane.

32. Show that, in the complex l plane, the solution $R_{l,k}(r)$ of the radial equation (II.15) is completely determined by the following condition:

$$R_{l,k} \sim r^{l+1} \quad \text{as} \quad r \to 0 \tag{32a}$$

provided

$$\text{Re}\,(l) > -\tfrac{1}{2}. \tag{32b}$$

33. Show that the solution $R_l(k, r)$ of the radial equation (II.15) which obeys (32a) is a holomorphic function in the complex l-plane if Re $(l) > -\frac{1}{2}$.

34. Show that, for positive energies ($E > 0$), all the poles of $S(l, k)$ are in the region Im $(l) > 0$, Re $(l) > -\frac{1}{2}$, of the complex l plane.

35. Show that, for negative energies ($k^2 < 0$), all the poles of $S(l, k)$ are on the real axis of the complex l plane.

36. Show that, for $E < 0$, the poles of $S(l, k)$ move along the axis Re(l) towards greater values of Re (l) as E increases, i.e. that

$$(dl_p(E)/dE) > 0. \tag{36a}$$

37. Study the motion of the poles of $S(l, k)$ for $E > 0$.

38. Show that a single Regge pole causes the appearance of bound states (or of resonances) whenever its trajectory for $E < 0$ (or $E > 0$) passes through the physical values $l = 0$, $1, 2, \ldots$ (or comes close to one of the physical values $l = 0, 1, 2, \ldots$).

39. Show that, in the Hilbert space of the eigenfunctions of the reference Hamiltonian H_0, the operators $(E_a - H_0 \pm i\eta)^{-1}$ are well defined, i.e. their eigenvalues and their eigenfunctions can be found.

40. Show that the solution of the Schrödinger-type equation

$$i\hbar \frac{\partial}{\partial t} |\psi_\eta(t)\rangle = e^{\frac{\eta}{\hbar}|t|} H_I' |\psi_\eta(t)\rangle, \qquad \eta > 0 \tag{40a}$$

with the initial condition $|\psi_\eta(-\infty)\rangle = |\Phi\rangle$, in which H_I' is the operator H' in the interaction picture, is equal in the limit $\eta \to 0$ to the "in" state vector $|\psi^+\rangle$, at $t = 0$, i.e. that

$$|\psi^+\rangle = \lim_{\eta \to 0} |\psi_\eta(0)\rangle. \tag{40b}$$

41. Let $\{\xi\}$ be a given representation. Using the completeness condition

$$\int |\xi\rangle\, d\xi \langle\xi| = 1 \tag{41a}$$

for the states $|\xi\rangle$ which form the basis of the representation, show that the Lippmann–Schwinger equations lead to Fredholm-type integral equations for the representations of states in the $\{\xi\}$ representation.

42. Show that, for the potential scattering of a particle, in which $H_0 = -\frac{\hbar^2}{2m}\nabla^2$, $H' = V(\mathbf{r})$, $E_k = \frac{\hbar^2 k^2}{2m}$, in coordinate representation, the Lippmann–Schwinger equation

$$|\psi_k^{\pm}\rangle = |\Phi_k\rangle + \lim_{\eta\to 0}\frac{1}{E_k - H_0 \pm i\eta} H' |\psi_k^{\pm}\rangle, \tag{42a}$$

leads to the integral equation (X.3) for the corresponding wavefunctions.

43. Starting from equations (X.35), show that

$$|\psi_a^{\pm}\rangle = |\Phi_a\rangle + \lim_{\eta\to +0}\frac{1}{E_a - H \pm i\eta} H' |\Phi_a\rangle. \tag{43a}$$

44. Let $|\psi_a^+\rangle$ and $|\psi_b^+\rangle$ be two different "in" states of a system. Show that the orthonormality relation

$$\langle \psi_a^+ | \psi_b^+\rangle = \delta_{ab} \tag{44a}$$

holds, where δ_{ab} is the Dirac delta-function, and the states are suitably normalized.

45. Using the results of the preceding problems deduce the relations

$$\langle \Phi_b | \psi_a^{\pm}\rangle = \delta_{ab} + \lim_{\eta\to +0}\frac{1}{E_a - E_b \pm i\eta}\langle \Phi_b | H' | \psi_a^{\pm}\rangle \tag{45a}$$

$$\langle \psi_\alpha^L | \Phi_b\rangle = \frac{1}{E_\alpha - E_b}\langle \psi_\alpha^L | H' | \Phi_b\rangle, \tag{45b}$$

in which the superscript L refers to bound states of H.

Solutions

1. Let us write the Schrödinger equation (X.1) in the form

$$L_k(\mathbf{r})\,\psi(\mathbf{r}) = U(\mathbf{r})\,\psi(\mathbf{r}) \tag{1.1}$$

where $L_k(\mathbf{r}) = \nabla^2 + k^2$.

Using the Dirac δ-function, (1.1) can be written in the form

$$L_k(\mathbf{r})\,\psi(\mathbf{r}) = \int U(\mathbf{r}')\,\psi(\mathbf{r})\,\delta(\mathbf{r}' - \mathbf{r})\,d\mathbf{r}.$$

Multiplying on the left by $L_k^{-1}(\mathbf{r})$ and adding the solution $\phi_{\mathbf{k}}(\mathbf{r}) = \exp(i\mathbf{kr})$ of the homogeneous equation:

$$L_k(\mathbf{r})\,\phi_{\mathbf{k}}(\mathbf{r}) = 0, \tag{1.2}$$

normalized in such a way that the probability current density $\mathbf{j}_\mathbf{k} = \text{Re}\left[\frac{\hbar}{im}\phi_\mathbf{k}^* \nabla\phi_\mathbf{k}\right]$ of the incident particles is numerically equal to their velocity $\hbar\mathbf{k}/m$, we obtain

$$\psi(\mathbf{r}) = \phi_\mathbf{k}(\mathbf{r}) + \int U(\mathbf{r}')\,\psi(\mathbf{r}')\,L_k^{-1}(\mathbf{r})\,\delta(\mathbf{r}'-\mathbf{r})\,d\mathbf{r}'.$$

With the notation

$$L_k^{-1}(\mathbf{r})\,\delta(\mathbf{r}'-\mathbf{r}) = G_k(\mathbf{r}-\mathbf{r}'), \tag{1.3}$$

we arrive at the integral equation

$$\psi(\mathbf{r}) = \phi_\mathbf{k}(\mathbf{r}) + \int G_k(\mathbf{r}-\mathbf{r}')\,U(\mathbf{r}')\,\psi(\mathbf{r}')\,d\mathbf{r}' \tag{1.4}$$

in which the Green's function $G_k(\mathbf{r}-\mathbf{r}')$ is the solution of the differential equation

$$L_k(\mathbf{r})\,G_k(\mathbf{r}-\mathbf{r}') = \delta(\mathbf{r}-\mathbf{r}'). \tag{1.5}$$

Remembering the well-known integral representation

$$\delta(\mathbf{r}-\mathbf{r}') = (2\pi)^{-3}\int \exp\,[i\mathbf{q}\,.\,(\mathbf{r}-\mathbf{r}')\,d\mathbf{q},$$

one obtains from (1.3)

$$G_k(|\mathbf{r}-\mathbf{r}'|) = (2\pi)^{-3}\int L_k^{-1}(\mathbf{r})\exp\,[i\mathbf{q}\,.\,(\mathbf{r}-\mathbf{r}')]\,d\mathbf{q}.$$

Since

$$L_k(\mathbf{r})\exp\,[i\mathbf{q}\,.\,(\mathbf{r}-\mathbf{r}')] = (k^2-q^2)\exp\,[i\mathbf{q}\,.\,(\mathbf{r}-\mathbf{r}')], \quad \text{and} \quad L_k^{-1}(\mathbf{r})\,L_k(\mathbf{r}) = 1$$

we have

$$L_k^{-1}(\mathbf{r})\exp\,[i\mathbf{q}\,.\,(\mathbf{r}-\mathbf{r}')] = (k^2-q^2)^{-1}\exp\,[i\mathbf{q}\,.\,(\mathbf{r}-\mathbf{r}')].$$

Therefore

$$G_k(|\mathbf{r}-\mathbf{r}'|) = \frac{1}{(2\pi)^3}\int \frac{\exp\,[i\mathbf{q}\,.\,(\mathbf{r}-\mathbf{r}')]}{k^2-q^2}\,d\mathbf{q}. \tag{1.6}$$

Changing to spherical coordinates with the polar axis along $\mathbf{R} = \mathbf{r}-\mathbf{r}'$, and integrating over the angular variables, we obtain

$$G_k(R) = \frac{1}{4i\pi^2 R}\int_{-\infty}^{+\infty} \frac{qe^{iqR}}{k^2-q^2}\,dq, \quad R = |\mathbf{r}-\mathbf{r}'|. \tag{1.7}$$

The integrand in (1.7) has two poles, at the points $q = +k$ and $q = -k$. Strictly speaking the integral does not exist. This difficulty can be overcome by considering q as a complex variable and changing the contour of integration in the complex q plane so as to avoid the poles. Different contours then define Green's functions which correspond to different boundary conditions. There are three basic ways of avoiding the poles (Fig. X.3). Any other possibility yields a linear combination of these. Since $R > 0$, in each case the contour of integration has to be closed in the upper half-plane by a semicircle of infinite radius.

In case (a), only the pole from $q = +k$ gives a contribution, and, using the theory of residues we obtain the corresponding Green's function:

$$G_k^+(|\mathbf{r}-\mathbf{r}'|) = -\frac{1}{4\pi}\frac{e^{ik|\mathbf{r}-\mathbf{r}'|}}{|\mathbf{r}-\mathbf{r}'|}. \tag{1.8}$$

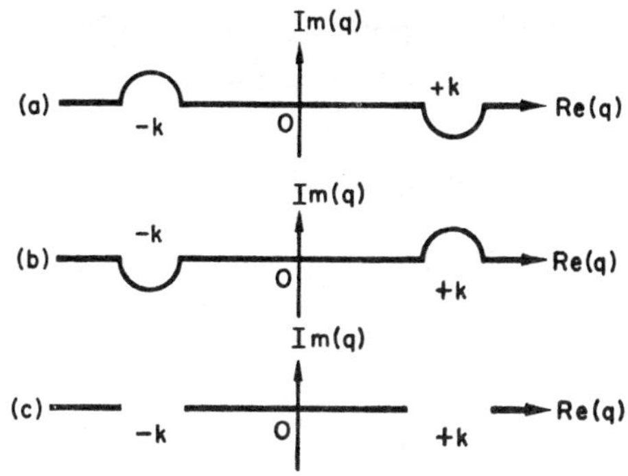

FIG. X.3.

Similarly, in case (b) we obtain:

$$G_k^-(|\mathbf{r}-\mathbf{r}'|) = -\frac{1}{4\pi}\frac{e^{-ik|\mathbf{r}-\mathbf{r}'|}}{|\mathbf{r}-\mathbf{r}'|}. \tag{1.9}$$

The contour (c) gives the Cauchy principal value of $G(R)$. It can be calculated easily by taking the arithmetical mean of the results obtained when both poles are included in the contour of integration, and when both are excluded from it, respectively.

The corresponding Green's function is found to be:

$$G_k^1(|\mathbf{r}-\mathbf{r}'|) = -\frac{1}{2\pi}\frac{\cos(k|\mathbf{r}-\mathbf{r}'|)}{|\mathbf{r}-\mathbf{r}'|}. \tag{1.10}$$

From the three solution $\psi_\mathbf{k}^+(\mathbf{r})$, $\psi_\mathbf{k}^-(\mathbf{r})$ and $\psi_\mathbf{k}^1(\mathbf{r})$ of the integral equation (1.4), only

$$\psi_\mathbf{k}^+(\mathbf{r}) = \phi_\mathbf{k}(\mathbf{r}) - \frac{1}{4\pi}\int \frac{\exp(ik|\mathbf{r}-\mathbf{r}'|)}{|\mathbf{r}-\mathbf{r}'|} U(\mathbf{r}')\,\psi_\mathbf{k}^+(\mathbf{r}')\,d\mathbf{r}' \tag{1.11}$$

represents the solution of the scattering problem, since in the asymptotic range $r \gg r_0$ only $\psi_\mathbf{k}^+(\mathbf{r})$ is a superposition of a plane wave and an outgoing wave (see X.6).

The solution $\psi_\mathbf{k}^-$, although it has no physical meaning, plays an important role in the formal theory of scattering.

2. If we compare the scattering amplitudes (X.7) and (X.10), it can be seen that the Born approximation consists in limiting the solution (obtained by iteration) of the integral equation (X.3):

$$\psi_\mathbf{k}^+(\mathbf{r}) = \phi_\mathbf{k}(\mathbf{r}) - \frac{1}{4\pi}\int \frac{e^{ik|\mathbf{r}-\mathbf{r}'|}}{|\mathbf{r}-\mathbf{r}'|} U(\mathbf{r}')\,\phi_\mathbf{k}(\mathbf{r}')\,d\mathbf{r}' + \dots$$

to its first term $\psi_{\mathbf{k}}^{+}(\mathbf{r}) = \phi_{\mathbf{k}}(\mathbf{r})$. This approximation is valid only if, within the range of the interaction, $r \leqslant r_0$, the following condition holds:

$$|\phi_{\mathbf{k}}(\mathbf{r})| \gg \left| \frac{1}{4\pi} \int \frac{e^{ik|\mathbf{r}-\mathbf{r}'|}}{|\mathbf{r}-\mathbf{r}'|} U(\mathbf{r}')\, \phi_{\mathbf{k}}(\mathbf{r}')\, d\mathbf{r}' \right|. \tag{2.1}$$

Since the potentials in common use all have their maximum value at the origin, we obtain, after making the substitutions $r = 0$ and $\phi_{\mathbf{k}}(\mathbf{r}) = \exp(i\mathbf{k}.\mathbf{r})$, the condition

$$\frac{1}{4\pi} \left| \int \frac{U(\mathbf{r})}{r} e^{i(kr+\mathbf{k}\cdot\mathbf{r})}\, d\mathbf{r} \right| \ll 1. \tag{2.2}$$

If the kinetic energy of the incident particle is sufficiently small ($kr_0 \ll 1$), the exponential in (2.2) is approximately equal to unity and we then have the condition

$$\bar{V} \ll \frac{\hbar^2}{2mr_0^2}, \tag{2.3}$$

where

$$\bar{V} = \frac{1}{4\pi r_0^2} \left| \int \frac{V(\mathbf{r})}{r}\, d\mathbf{r} \right|. \tag{2.4}$$

In accordance with the uncertainty relation between position and momentum, the quantity $\hbar^2/2mr_0^2$ is the uncertainty in the kinetic energy of a particle "localized" in a region of linear dimensions r_0. Thus (2.3) requires that, in the low energy range, the potential energy should be much smaller than the kinetic energy of the particles.

If $V(\mathbf{r}) = V(r)$, then, by choosing the direction of $\mathbf{k}$ as the polar axis of a polar coordinate system and performing the integration in (2.2) over the angular variables, we obtain the condition of validity of the Born approximation for potentials with spherical symmetry:

$$\frac{1}{2kr_0^2} \left| \int_0^\infty V(r)[e^{2ikr} - 1]\, dr \right| \ll \frac{\hbar^2}{2mr_0^2}. \tag{2.5}$$

If the kinetic energy of the particle is large ($kr_0 \gg 1$) then the contribution of the oscillating term can be neglected and (2.5) becomes

$$\frac{1}{2kr_0^2} \left| \int_0^\infty V(r)\, dr \right| \ll \frac{\hbar^2}{2mr_0^2}. \tag{2.6}$$

From this requirement it follows that for high kinetic energies the Born approximation is in general more useful than it is for low kinetic energies. For low energies ($kr_0 \ll 1$), the exponential in (2.5) can be expanded in a power series. Keeping the first two terms, we find condition (2.3) once again.

3. (1) For the exponential potential we have

$$\int_0^\infty V(r)\,(e^{2ikr}-1)\,dr = -\frac{2ikr_0^2V_0}{2ikr_0-1}. \tag{3.1}$$

Condition (2.3) then requires that

$$2mr_0^2V_0 \ll \hbar^2\sqrt{1+4k^2r_0^2}\,, \tag{3.2}$$

whence the following conditions can be obtained:

$$2mr_0^2V_0 \ll \hbar^2, \quad \text{for low energies} \quad (kr_0 \ll 1),$$
$$mr_0V_0 \ll k\hbar^2, \quad \text{for high energies} \quad (kr_0 \gg 1).$$

(2) For $V(r) = \dfrac{A}{r}\exp(-ar)$, where $a = 1/r_0$, we need to calculate the integral

$$I(a) = \int_0^\infty e^{-ar}(e^{2ikr}-1)\frac{dr}{r}. \tag{3.3}$$

Now, if we differentiate with respect to a, we find that

$$\frac{dI}{da} = -\int_0^\infty e^{-ar}(e^{2ikr}-1)\,dr = \frac{1}{a}-\frac{1}{a-2ik}.$$

From which

$$I(a) = \ln a - \ln(a-2ik)+C,$$

i.e.

$$I = -\ln(1-2ikr_0)+C = -\ln\sqrt{1+4k^2r_0^2}+i\tan^{-1}(2kr_0)+C.$$

Since for $r_0 = 0$ we have $I = 0$, it follows that $C = 0$, and thus the condition of validity of the Born approximation (2.5) becomes

$$mA\left[(\ln\sqrt{1+4k^2r_0^2})^2+\arctan g^2(2kr_0)\right]^{1/2} \ll k\hbar^2. \tag{3.4}$$

This condition is satisfied only for high energies.

(3) In this case (2.5) becomes

$$\frac{m}{k\hbar^2}\left|\int_0^{r_0} V_0(e^{2ikr}-1)\,dr\right| = \frac{mV_0}{k^2\hbar^2}\{\sin^2 kr_0+kr_0[kr_0-\sin(2kr_0)]\}^{1/2} \ll 1. \tag{3.5}$$

4. Note that the matrix element

$$\langle\phi_{\mathbf{k}'}|U|\phi_{\mathbf{k}}\rangle = \int U(\mathbf{r})e^{i\mathbf{q}\mathbf{r}}\,d\mathbf{r} = \mathcal{U}(\mathbf{q}), \tag{4.1}$$

say, which determines the Born amplitude (X.10), is the Fourier transform of the potential, for the momentum transfer $\hbar q$:

$$\hbar \mathbf{q} = \hbar(\mathbf{k}-\mathbf{k}'), \quad q = 2k \sin \frac{\theta}{2}, \tag{4.2}$$

where θ, the angle between $\mathbf{k}$ and $\mathbf{k}'$, is the angle of scattering. If the potential has spherical symmetry, then, from (4.1), we obtain

$$\mathcal{U}(\mathbf{q}) = \frac{4\pi}{q} \int_0^\infty rU(r) \sin qr \, dr, \tag{4.3}$$

whence it follows that the Fourier transform of the potential depends only on the magnitude of the momentum transfer,

$$\mathcal{U}(\mathbf{q}) = \mathcal{U}(q).$$

Thus

$$A^{(B)} = -\frac{1}{4\pi} \mathcal{U}\left(2k \sin \frac{\theta}{2}\right) \tag{4.4}$$

and

$$d\sigma = \frac{1}{16\pi^2} \left| \mathcal{U}\left(2k \sin \frac{\theta}{2}\right) \right|^2 d\Omega = \left(\frac{m}{2\pi\hbar^2}\right)^2 \left| \mathcal{V}\left(2k \sin \frac{\theta}{2}\right) \right|^2 d\Omega. \tag{4.5}$$

If the potential is an even function of r, then (4.3) becomes

$$\mathcal{U}(q) = \frac{2\pi}{iq} \int_{-\infty}^{+\infty} rU(r) e^{iqr} \, dr. \tag{4.6}$$

(1) For

$$V(r) = \frac{A}{r} \exp\left(-\frac{r}{r_0}\right), \tag{4.7}$$

we obtain from (4.3)

$$\mathcal{V}(q) = 4\pi A \left(q^2 + \frac{1}{r_0^2}\right)^{-1}. \tag{4.8}$$

And thus

$$d\sigma = \left(\frac{2mA}{4p^2 \sin^2 \frac{\theta}{2} + \frac{\hbar^2}{r_0^2}}\right)^2 d\Omega, \quad \mathbf{p} = \hbar \mathbf{k}. \tag{4.9}$$

As $r_0 \to \infty$ the screening disappears, and, by taking $A = Z_1 Z_1 e^2$, (4.7) becomes the Coulomb potential $V(r) = Z_1 Z_2 e^2/r$, and (4.9) becomes the well-known Rutherford formula for

Coulomb scattering:

$$\frac{d\sigma}{d\Omega} = \left(\frac{mZ_1Z_2e^2}{2p^2 \sin^2 \dfrac{\theta}{2}}\right)^2 = \left(\frac{Z_1Z_2e^2}{4E \sin^2 \dfrac{\theta}{2}}\right)^2, \tag{4.10}$$

where E is the kinetic energy of the incident particles. It should be mentioned that this expression, although it was obtained in Born approximation, is in fact valid for all energies. The following important features of Coulomb scattering follow from it:

(a) The cross-section depends only on the absolute value of the charges and not on their signs.

(b) The angular distribution is independent of the energy.

(c) For a given angle, the differential cross-section is inversely proportional to the square of the energy.

(d) The total cross-section is infinite, since the integral of $d\sigma(E, \theta)/d\Omega$ over all solid angles diverges at the lower limit of small angles of scattering.

(2) Since the Gaussian potential is an even function of r, we can use (4.6) to calculate the scattering cross-section and obtain

$$\mathcal{V}(q) = (2\pi r_0^2)^{3/2} V_0 \exp\left(-\tfrac{1}{2} r_0^2 q^2\right),$$

and hence

$$d\sigma = 2\pi\left(\frac{mr_0^3V_0}{\hbar^2}\right)^2 \exp\left(-4k^2r_0^2 \sin^2 \frac{\theta}{2}\right) d\Omega. \tag{4.11}$$

(3) The potential is in this case an even function also, so that we have, from (4.6), that

$$\mathcal{V}(q) = \frac{4\pi V_0}{q^2}\left(r_0 \cos qr_0 - \frac{\sin qr_0}{q}\right)$$

and hence

$$d\sigma = 4\left(\frac{mr_0^3V_0}{\hbar^2}\right)^2 \frac{(\sin qr_0 - qr_0 \cos qr_0)^2}{(qr_0)^6} d\Omega. \tag{4.12}$$

A common feature of the scattering differential cross-sections (4.9), (4.11) and (4.12) is that they become isotropic, i.e. independent of the scattering angle, at low energies ($kr_0 \ll 1$). This fact is characteristic of all potentials which have a finite range.

5. In order to obtain the total cross-sections we have to integrate the differential cross-sections (4.9), (4.11) and (4.12) over all solid angles. The calculations are much simplified if we return to the variable $q = 2k \sin \theta/2$ and replace the element of solid angle $d\Omega = \sin \theta \, d\theta \, d\phi$ by $2\pi q \, dq/k^2$ after performing the (trivial) integration over ϕ. The results are:

(1) For the Yukawa potential

$$\sigma = 16\pi\left(\frac{Amr_0^2}{\hbar^2}\right)^2 \frac{1}{4k^2r_0^2+1}. \tag{5.1}$$

(2) For the Gaussian potential

$$\sigma = \frac{2\pi^2}{k^2}\left(\frac{mr_0^2V_0}{\hbar^2}\right)^2 (1-e^{-4k^2r_0^2}). \tag{5.2}$$

(3) For the spherical potential well

$$\sigma = \frac{2\pi}{k^2}\left(\frac{mr_0^2V_0}{\hbar^2}\right)^2\left[1-\frac{1}{(2kr_0)^2}+\frac{\sin 4kr_0}{(2kr_0)^3}-\frac{\sin^2 2kr_0}{(2kr_0)^4}\right]. \tag{5.3}$$

It should be remembered that these expressions are valid only when the Born approximation is valid. In limiting cases (5.3) gives

$$\sigma = \frac{16\pi}{9} r_0^2\left(\frac{mr_0^2V_0}{\hbar^2}\right)^2 \quad \text{for } kr_0 \ll 1 \tag{5.4}$$

$$\sigma = \frac{2\pi}{k^2}\left(\frac{mr_0^2V_0}{\hbar^2}\right)^2 \quad \text{for } kr_0 \gg 1. \tag{5.5}$$

6. The neutron–proton elastic scattering problem can be reduced to the problem of the scattering of a hypothetical particle of mass $m = m_p m_n(m_p+m_n)^{-1} \approx m_p/2$ in the potential $V(r) = (A/r)\exp(-r/r_0)$.

Using (4.4) and (4.8) we find the Born amplitude:

$$A^{(B)}(q) = -\frac{m}{2\pi\hbar^2}\mathcal{V}(q) = \frac{m_p r_0^2 A}{\hbar^2(1+r_0^2q^2)}. \tag{6.1}$$

Hence the form factor can be found:

$$F(q) = \frac{A^{(B)}(q)}{A^{(B)}(0)} = (1+r_0^2q^2)^{-1} = \left(1+4r_0^2k^2\sin^2\frac{\theta}{2}\right)^{-1}. \tag{6.2}$$

From the definition of the Born amplitude (X.10′), it can be seen that each volume element of space $d\mathbf{r}$ in which the potential is not negligible makes a contribution $-(1/4\pi)e^{i\mathbf{q}\cdot\mathbf{r}}U(\mathbf{r})d\mathbf{r}$ to the amplitude. The factor e^{iqr} determines the phase shift between the wave scattered by each volume element $d\mathbf{r}$ and the wave scattered by the volume element at the origin. If $U(\mathbf{r})$ does not change its sign, then, for forward scattering ($q = 0$), the waves scattered by all volume elements will have the same phase and the Born amplitude will have a maximum value

$$A^{(B)}(0) = -\frac{1}{4\pi}\int U(\mathbf{r})d\mathbf{r}.$$

In all other directions the contributions of the different volume elements will have different phases, and thus the form factor defined by (X.11) characterizes the interference between waves scattered in the various volume elements. The total cross-section for n–p scattering in the CM system is found from (5.1):

$$\sigma(E) = 4\pi \left(\frac{m_p r_0^2 A}{\hbar^2}\right)^2 \frac{1}{1+\dfrac{4m_p r_0^2}{\hbar^2}E} = \frac{\sigma(0)}{1+\dfrac{4m_p r_0^2}{\hbar^2}E}. \tag{6.3}$$

From this it can be seen that, for high energies the total cross-section for n–p scattering is proportional to E^{-1}:

$$\sigma(E) \approx \frac{\hbar^2}{4m_p r_0^2} \cdot \frac{\sigma(0)}{E}. \tag{6.4}$$

7. By definition the differential cross-section for scattering, $d\sigma(\theta, \phi)$, in the CM system, is the number of particles of type 1 (say) scattered per unit time into an element of solid angle $d\Omega$ about the direction (θ, ϕ), when a beam of particles of type 1 collides with a beam of particles of type 2 travelling in the opposite direction, with equal and opposite momentum, the flux of particles of type 1 relative to each particle of type 2 being equal to unity. From this definition it follows that

$$d\sigma(\theta, \phi) = d\sigma_1(\theta_1, \phi_1), \tag{7.1}$$

i.e. that

$$|A(\theta, \phi)|^2 \, d\Omega(\theta, \phi) = |A_1(\theta_1, \phi_1)|^2 \, d\Omega_1(\theta_1, \phi_1), \tag{7.2}$$

where (θ_1, ϕ_1) is the direction of motion (after collision) of particle 1 in the laboratory coordinates which corresponds to a motion in the direction (θ, ϕ) in the CM system.

From (7.1) it follows that the total cross-sections for scattering as measured in the laboratory and in the CM systems are equal:

$$\sigma = \int d\sigma(\theta, \phi) = \int d\sigma_1(\theta_1, \phi_1) = \sigma_1. \tag{7.3}$$

Bearing in mind the transformation relations (X.12), we obtain immediately the result

$$\frac{d\Omega(\theta, \phi)}{d\Omega_1(\theta_1, \phi_1)} = \frac{\sin\theta \, d\theta \, d\phi}{\sin\theta_1 \, d\theta_1 \, d\phi_1} = \frac{(1+2\gamma\cos\theta+\gamma^2)^{3/2}}{|1+\gamma\cos\theta|} \tag{7.4}$$

and thus from (7.2) one obtains (7a).

8. Consider the elastic collision of two identical bosons. Let $\mathbf{R}, \mathbf{P}, \mathbf{r}$ and $\mathbf{p}$ denote respectively the position vector of the centre of mass of the system, the total linear momentum of the system, the relative position vector of the two particles, and the "relative" momentum; then

$$\mathbf{R} = \tfrac{1}{2}(\mathbf{r}_1+\mathbf{r}_2), \qquad \mathbf{r} = \mathbf{r}_1-\mathbf{r}_2$$
$$\mathbf{P} = \mathbf{p}_1+\mathbf{p}_2, \qquad \mathbf{p} = \tfrac{1}{2}(\mathbf{p}_1-\mathbf{p}_2).$$

Since **R** is invariant with respect to an exchange of the two particles, under which **r** becomes $-\mathbf{r}$, the wavefunction $\psi(\mathbf{R}, \mathbf{r}, t)$ which describes the state of the system has to satisfy the following symmetry condition:

$$\psi(\mathbf{R}, \mathbf{r}, t) = +\psi(\mathbf{R}, -\mathbf{r}, t). \tag{8.1}$$

Let us suppose for a moment that the two particles can be distinguished. In the CMS, the differential cross-section for particle 1 to be scattered into the direction (θ, ϕ) is then equal to the probability (per unit time, etc.) that the relative position vector **r** will end up pointing in the direction (θ, ϕ), i.e.

$$d\sigma_1(\theta, \phi) = |A(\theta, \phi)|^2 \, d\Omega. \tag{8.2}$$

Similarly, the differential cross-section for particle 2 to be scattered into the direction (θ, ϕ) is equal to the probability (per unit time, etc.) that the relative position vector **r** will end up pointing in the opposite direction to (θ, ϕ), i.e.

$$d\sigma_2(\theta, \phi) = |A(\pi-\theta, \phi+\pi)|^2 \, d\Omega. \tag{8.3}$$

However, if the two particles are not distinguishable, the above argument must be supplemented as follows:

(a) Since the observer cannot distinguish the particles after scattering, $d\sigma_1(\theta, \phi)$ and $d\sigma_2(\theta, \phi)$ cannot be measured separately.

(b) The function $\psi(\mathbf{R}, \mathbf{r}, t)$ has to be symmetrical (for bosons). Condition (a) does not lead to anything new from the quantum-mechanical point of view. We merely take the observed differential cross-section $d\sigma(\theta, \phi)$ to be the total number of particles (1 and 2) scattered into the element of solid angle $d\Omega$ about the direction of observation (θ, ϕ) per unit time per unit flux of incident particles, so that

$$d\sigma(\theta, \phi) = d\sigma_1(\theta, \phi) + d\sigma_2(\theta, \phi). \tag{8.4}$$

Note, however, that if we retain the definition of the *total* cross-section as being the number of particles scattered out of the incident beam per unit time per unit incident flux, we have

$$\sigma = \tfrac{1}{2} \int d\sigma(\theta, \phi). \tag{8.5}$$

The requirement (b) is a quantum-mechanical one. Let us introduce the symmetrized wavefunction

$$\hat{\psi}(\mathbf{R}, \mathbf{r}, t) = \frac{1}{\sqrt{2}} [\psi(\mathbf{R}, \mathbf{r}, t) + \psi(\mathbf{R}, -\mathbf{r}, t)]. \tag{8.6}$$

Since this symmetrization is valid in the asymptotic region also, it implies a corresponding symmetrization of the scattering amplitude, i.e.

$$\hat{A}(\theta, \phi) = \frac{1}{\sqrt{2}} [A(\theta, \phi) + A(\pi-\theta, \phi+\pi)]. \tag{8.7}$$

Then for $d\sigma_1$ and $d\sigma_2$ we obtain

$$d\sigma_1(\theta, \phi) = |\hat{A}(\theta, \phi)|^2 \, d\Omega = \tfrac{1}{2} \, |A(\theta, \phi) + A(\pi - \theta, \phi + \pi)|^2 \, d\Omega,$$
$$d\sigma_2(\theta, \phi) = |\hat{A}(\pi - \theta, \phi + \pi)|^2 \, d\Omega = d\sigma_1(\theta, \phi)$$

whence

$$d\sigma(\theta, \phi) = 2 \, |\hat{A}(\theta, \phi)|^2 \, d\Omega = |A(\theta, \phi) + A(\pi - \theta, \phi + \pi)|^2 \, d\Omega. \tag{8.8}$$

In conclusion, the symmetry of the wavefunction appropriate to indistinguishable bosons gives rise in scattering processes (and elsewhere) to a new quantum phenomenon called the "exchange effect". This leads to an appreciable increase in the differential cross-section; in particular, if the scattering amplitude does not depend on the angle ϕ, then, in the direction $\theta = \pi/2$, the differential cross-section (8.8) is increased by a factor of 4.

9. Although under the conditions of this problem the total spin of the two-proton system is a constant of the motion, the interactions of the two protons in the singlet and in the triplet spin states can be different. Let $A_s(\theta)$ and $A_t(\theta)$ be the corresponding singlet and triplet scattering amplitudes (as calculated without symmetrization). Since protons are fermions, the *p–p* system wavefunction must be anti-symmetrical with respect to the exchange of the two particles. If this function represents a triplet spin state, it will be symmetrical under a permutation of the spin variables, and hence it must be anti-symmetrical under a permutation of the coordinates $\mathbf{r}_1$ and $\mathbf{r}_2$. The correctly symmetrized triplet amplitude will then be

$$\hat{A}_t(\theta) = \frac{1}{\sqrt{2}} [A_t(\theta) - A_t(\pi - \theta)], \tag{9.1}$$

and the differential cross-section for *p–p* scattering in the triplet spin state becomes

$$d\sigma_t(\theta) = 2 \, |\hat{A}_t(\theta)|^2 \, d\Omega = |A_t(\theta) - A_t(\pi - \theta)|^2 \, d\Omega. \tag{9.2}$$

If, on the other hand, the wavefunction represents a singlet spin state, it will be anti-symmetrical under a permutation of the spin variables and hence must be symmetrical under a permutation of the coordinates. We then have

$$\hat{A}_s(\theta) = \frac{1}{\sqrt{2}} [A_s(\theta) + A_s(\pi - \theta)]^2 \, d\Omega, \tag{9.3}$$

and the scattering differential cross-section for the singlet spin state becomes

$$d\sigma_s(\theta) = 2 \, |\hat{A}_s(\theta)|^2 \, d\Omega = |A_s(\theta) + A_s(\pi - \theta)|^2 \, d\Omega. \tag{9.4}$$

If the target and the incident beam both consist of randomly oriented (non-polarized) protons, then at each collision the orientations of the spins of the colliding particles will be distributed at random. Now there are three independent triplet states and only one singlet state. It follows that, in any collision, the probability that the total spin $S = 1$ (triplet state), is $\frac{3}{4}$, while the probability that the total spin $S = 0$ (singlet state), is $\frac{1}{4}$. It follows that, for unpolarized beam and target protons,

$$d\sigma(\theta) = \tfrac{3}{4}\, d\sigma_t(\theta)+\tfrac{1}{4}\, d\sigma_s(\theta) = \left\{\tfrac{3}{4}\,|\,A_t(\theta)-A_t(\pi-\theta)\,|^2+\tfrac{1}{4}\,|\,A_s(\theta)+A_s(\pi-\theta)\,|^2\right\} d\Omega. \quad (9.5)$$

Note that if the forces between the particles were spin-independent, then we should have

$$A_s(\theta) = A_t(\theta) = A(\theta),$$

say, and

$$d\sigma(\theta) = \left\{|\,A(\theta)\,|^2+|\,A(\pi-\theta)\,|^2-\tfrac{1}{2}[A^*(\theta)\,A(\pi-\theta)+A(\theta)\,A^*(\pi-\theta)]\right\} d\Omega. \quad (9.6)$$

10. Since the Coulomb interaction does not depend on the spin state (singlet or triplet) of the system of two particles, the scattering differential cross-section in the CMS can be calculated by using (9.6).

After some simple calculation we obtain

$$d\sigma(\theta) = \left(\frac{Z^2e^2}{4E}\right)^2\left[\frac{1}{\sin^4\frac{\theta}{2}}+\frac{1}{\cos^4\frac{\theta}{2}}-\frac{\cos\left[\frac{Z^2e^2}{\hbar v}\ln\left(\tan^2\frac{\theta}{2}\right)\right]}{\sin^2\frac{\theta}{2}\cos^2\frac{\theta}{2}}\right] d\Omega, \quad (10.1)$$

where E is the energy of the system in the CMS, v is the relative velocity of the two particles, and $Z_1 = Z_2 = Z$. The expression (10.1) is called the Mott formula. It can be applied to electron–electron elastic scattering, and, for low energies, when nuclear forces can be neglected, it can be applied also to proton–proton elastic (Coulomb) scattering (crudely speaking, if the energy is low enough, Coulomb repulsion prevents the protons from coming close enough for nuclear forces to be effective).

At high energies the contribution of Coulomb forces to proton–proton scattering is negligible in comparison with that of the purely nuclear forces, and the Mott formula is no longer valid. A calculation of the electron–electron elastic scattering differential cross-section (neglecting the spin and the indistinguishability of the particles) would, in accordance with (8.2)–(8.4), have given the result

$$d\sigma_{\rm cl}(\theta) = \left(\frac{e^2}{4E}\right)^2\left[\sin^{-4}\frac{\theta}{2}+\cos^{-4}\frac{\theta}{2}\right] d\Omega. \quad (10.2)$$

The third term in (10.1) has thus a quantum-mechanical origin and is due to the exchange interaction between identical particles. Formally, it shows itself in the "interference" between the amplitudes $A(\theta)$ and $A(\pi-\theta)$ (cf. relation (9.6)).

In the limit $\hbar \to 0$, this term oscillates rapidly about the value zero, so that its mean value vanishes in any solid angle, however small, and the Mott formula passes into (10.2). The differential cross-section for scattering in the laboratory system can be obtained immediately by substituting $\theta = 2\theta_1$ and $d\Omega = \sin\theta\, d\theta\, d\phi = 4\cos\theta_1 \sin\theta_1\, d\theta_1\, d\phi_1 = 4\cos\theta_1\, d\Omega_1$ in (10.1), in accordance with the transformation relations (X.12).

The energy of relative motion $E = mv^2/2$, when expressed in terms of the velocity v_1 in the laboratory system and the mass m_0 of the incident particle, is $E = m_0v^2/4 = m_0v_1^2/4$.

11. The expression which will be deduced below (and from which (11a) follows as a particular case) is useful for comparing the phase shifts $\delta_l(k)$ and $\hat{\delta}_l(k)$ $\left(\text{for orbital angular momentum } l \text{ and energy } E = \frac{\hbar^2 k^2}{2m}\right)$ due to two potentials $V(r)$ and $\hat{V}(r)$ respectively. The relevant radial equations are

$$\left[\frac{d^2}{dr^2}+k^2-\left(U(r)+\frac{l(l+1)}{r^2}\right)\right]R_{l,k}(r)=0 \tag{11.1}$$

$$\left[\frac{d^2}{dr^2}+k^2-\left(\hat{U}(r)+\frac{l(l+1)}{r^2}\right)\right]\hat{R}_{l,k}(r)=0 \tag{11.1'}$$

with the condition at the origin that $R_{l,k}(0) = \hat{R}_{l,k}(0) = 0$, and the asymptotic behaviour for large r

$$\begin{aligned} R_{l,k}(r) &\sim \frac{1}{k}\sin\left(kr-\frac{1}{2}l\pi+\delta_l(k)\right) \\ \hat{R}_{l,k}(r) &\sim \frac{1}{k}\sin\left(kr-\frac{1}{2}l\pi+\hat{\delta}_l(k)\right). \end{aligned} \tag{11.2}$$

The Wronskian of the two solutions

$$W(R_{l,k}, \hat{R}_{l,k}) = R_{l,k}\hat{R}'_{l,k}-\hat{R}_{l,k}R'_{l,k} \tag{11.3}$$

vanishes at the origin, and

$$\lim_{r\to\infty} W(R_{l,k}, \hat{R}_{l,k}) = \frac{1}{k}\sin\left(\delta_l(k)-\hat{\delta}_l(k)\right). \tag{11.4}$$

On the other hand, from equations (11.1), we find that

$$W(R_{l,k}, \hat{R}_{l,k})\Big|_a^b = -\int_a^b \hat{R}_{l,k}(U-\hat{U})R_{l,k}\,dr. \tag{11.5}$$

By choosing $a = 0$ and $b = \infty$, we then find that

$$\frac{1}{k}\sin\left(\delta_l(k)-\hat{\delta}_l(k)\right) = -\frac{2m}{\hbar^2}\int_0^\infty \hat{R}_{l,k}(V-\hat{V})R_{l,k}\,dr. \tag{11.6}$$

This relation is valid for any two potentials $V(r)$ and $\hat{V}(r)$ provided only that they tend to zero more rapidly than $1/r$ at infinity and that, at the origin, they have a singularity (if any) not stronger than $1/r^2$. If $\hat{V}(r) \equiv 0$, then $\hat{\delta}_l(k) = 0$ and, in accordance with (A.46),

$$\hat{R}_{l,k}(r) = rj_l(kr).$$

The integral formula (11a) then follows immediately from (11.6).

Remarks: In order to calculate the phase shifts $\delta_l(k)$, the solution $R_{l,k}(r)$ of the radial equation (11.1) must be known. But, if $V(r)$ is small enough, $R_{l,k}(r)$ differs only slightly from the free spherical wave $rj_l(kr)$, and the phase shift is almost zero.

Thus, without too large an error, we can replace $R_{l,k}(r)$ by $rj_l(kr)$, and sin $\delta_l(k)$ by $\delta_l(k)$ in (11a) and obtain the following approximate formula

$$\delta_l(k) \approx -\frac{2m}{\hbar^2} k \int_0^\infty r^2 j_l^2(kr) V(r) \, dr. \tag{11.7}$$

The error in making this approximation is negligible if $V(r)$ is small enough in comparison with $E - l(l+1)\hbar^2/2mr^2$. This is true at high energies provided that, for large values of l, the potential decreases rapidly enough as $r \to \infty$. If the potential has a finite range, e.g. if $V(r) = 0$ for all $r > r_0$, then at low energies, $kr_0 \ll 1$, we have, by (A.49),

$$j_l(kr) \sim \frac{(kr)^l}{(2l+1)!!} \tag{11.8}$$

and, using (11.7), we obtain

$$\delta_l(k) \approx -\frac{2m(kr_0)^{2l+1}}{\hbar^2[(2l+1)!!]^2} \int_0^{r_0} rV(r) \left(\frac{r}{r_0}\right)^{2l+1} dr. \tag{11.9}$$

From this expression it can be seen that the phase shifts decrease rapidly with increasing l. If, for a given potential, the energy of the particles is such that $kr_0 \ll 1$, then the particles are said to be "slow", and from what has been said above we can conclude that in order to study their scattering it is sufficient to consider s-waves ($l = 0$) only.

Starting from expression (11.6), the following approximate formula can be derived:

$$\delta_l(k) - \hat{\delta}_l(k) \approx -\frac{2m}{\hbar^2} k \int_0^\infty \hat{R}_{l,k}^2 (V - \hat{V}) \, dr. \tag{11.10}$$

This relation is useful if the solutions $\hat{R}_{l,k}$ of (11.1′) are known for a potential $\hat{V}$ which differs only slightly from V. A quite good approximation to the $\delta_l(k)$ can then be obtained without having to solve equation (11.1) for V.

12. From (11.6), the following conclusions concerning changes in the phase shifts due to small changes in the potential can be drawn:

If $V(r) - \hat{V}(r) = \Delta V$, say, is a small quantity, then $\Delta\delta_l = \delta_l(k) - \hat{\delta}_l(k)$ will also be small, and the difference between $R_{l,k}$ and $\hat{R}_{l,k}$ will be negligible. It follows then that

$$\Delta\delta_l = -\frac{2m}{\hbar^2} k \int_0^\infty R_{l,k}^2 \Delta V \, dr. \tag{12.1}$$

If the variation ΔV in the potential has the same sign over the whole range 0 to ∞ of r, then the change in the phase shift $\Delta\delta_l$ has the opposite sign. Thus an increase in the potential (i.e. in the repulsive forces) leads to a decrease in the phase shift, and a decrease in the potential (i.e. in the attractive forces) leads to an increase in the phase shift.

The relation (X.14) defines the phase shift $\delta_l(k)$ to an additive constant $2n\pi$. To eliminate this ambiguity, let a continuous change of the potential from 0 to $V(r)$ be considered. The phase shift then changes from zero to some value $\delta_l(k)$ which does not depend on the way in which the potential was varied in going from zero to $V(r)$, and which therefore can be taken as a unique definition of the phase shift. Thus, if the potential $V(r)$ is everywhere repulsive, we can pass from zero to $V(r)$ by summing all the in finitesimal contributions. In accordance with (12.1) each such contribution decreases the phase shift and in consequence $\delta_l(k)$ will be negative. Similarly, if $V(r)$ is everywhere attractive, then $\delta_l(k)$ will be positive. More generally, if $V(r) > \hat{V}(r)$ for all r, then $\delta_l(k) < \hat{\delta}_l(k)$; if $V(r) < \hat{V}(r)$ for all r, then $\delta_l(k) > \hat{\delta}_l(k)$.

13. We have to find the phase shifts $\delta_l(k)$, i.e. to find the asymptotic forms of the radial functions which are the solutions of the equations

$$\left[\frac{d^2}{dr^2}+K^2-\frac{l(l+1)}{r^2}\right] R^{\text{int}}_{l,k}(r) = 0, \quad K^2 = \frac{2m}{\hbar^2}(E+V_0), \quad r < r_0$$

$$\left[\frac{d^2}{dr^2}+k^2-\frac{l(l+1)}{r^2}\right] R^{\text{ext}}_{l,k}(r) = 0, \quad k^2 = \frac{2m}{\hbar^2}E, \quad r > r_0$$

with the conditions at the origin $R^{\text{int}}_{l,k}(0) = 0$.

If the particles are "slow" ($kr_0 \ll 1$), the main contribution to the scattering cross-section is that due to the s-wave (see (11.9)). It is then sufficient to calculate δ_0.

The corresponding radial function satisfying the condition at the origin is

$$R_{0,k}(r) = \begin{cases} R^{\text{int}}_{0,k}(r) = A \sin Kr \\ R^{\text{ext}}_{0,k}(r) = B \sin (kr+\delta_0). \end{cases} \tag{13.1}$$

From the continuity of the logarithmic derivative at $r = r_0$,

$$\left.\frac{R'^{\text{int}}_{0,k}(r)}{R^{\text{int}}_{0,k}(r)}\right|_{r=r_0} = \left.\frac{R'^{\text{ext}}_{0,k}(r)}{R^{\text{ext}}_{0,k}(r)}\right|_{r=r_0}, \tag{13.2}$$

it follows that

$$\delta_0 = \tan^{-1}\left(\frac{k}{K}\tan Kr_0\right) - kr_0. \tag{13.3}$$

The partial scattering cross-section for $l = 0$ is then

$$\sigma_0 = \frac{4\pi}{k^2}\sin^2\delta_0 = \frac{4\pi}{k^2}\sin^2\left[\tan^{-1}\left(\frac{k}{K}\tan Kr_0\right) - kr_0\right], \tag{13.4}$$

and is very nearly equal to the total cross-section σ.

If the kinetic energy of the incident particles is very small ($k \to 0$), we have

$$\delta_0 \approx kr_0\left(\frac{\tan k_0 r_0}{k_0 r_0}-1\right), \quad k_0^2 = \frac{2m}{\hbar^2}V_0, \tag{13.5}$$

and therefore

$$\sigma \approx \sigma_0 \approx 4\pi r_0^2\left(\frac{\tan k_0 r_0}{k_0 r_0}-1\right)^2. \tag{13.6}$$

If the potential well is such that $k_0 r_0 \ll 1$, then

$$\sigma \approx 4\pi r_0^2 \frac{k_0^4 r_0^4}{9} = \frac{16\pi}{9} r_0^2 \left(\frac{m r_0^2 V_0}{\hbar^2}\right)^2. \tag{13.7}$$

This expression for the total cross-section is the same as the one found by means of the Born approximation (5.4).

From (13.6) it can be seen that, as V_0 increases, the scattering cross-section increases, and becomes infinite for $k_0 r_0 = \pi/2$. The condition $k_0 r_0 = \pi/2$ is also the condition for the appearance in the potential well of the first bound state (see problem (II.10) and (II.30)). A further increase in V_0 is then followed by a decrease in the scattering cross-section, until it vanishes when $\tan kr_0 = kr_0$. Further increases in V_0 lead to an oscillatory variation of the scattering cross-section between 0 and ∞. The infinite values of σ occur whenever, with increasing V_0, a new bound state becomes possible. This oscillatory behaviour of the scattering cross-section for slow particles explains in a qualitative way the experimental fact that the cross-sections for scattering of slow electrons by atoms bear no relation to the geometrical cross-sections of the atoms. It should be noted, however, that if the value of $k_0 r_0$ is nearly an odd multiple of $\pi/2$, then (13.5) and (13.6) cannot be used for quantitative calculations. Indeed, in this case $\tan Kr_0$ is very large and we cannot use a power series expansion in (13.4), as we did in (13.6). In the argument of the sine function of (13.4), only kr_0 (which by assumption is $\ll 1$) can be neglected.

Under these conditions we obtain instead

$$\delta_0 = \tan^{-1}\left(\frac{k}{K}\tan Kr_0\right), \quad \sigma = \frac{4\pi}{k^2+\xi^2} \tag{13.8}$$

where

$$\xi = \frac{K}{\tan Kr_0} \ll \frac{1}{r_0}. \tag{13.9}$$

(13.8) gives the "resonance scattering" cross-section for small values of k, when the potential well is such that a small variation of its parameters would lead to the appearance (or the disappearance) of a bound state.

14. The required results can be obtained immediately by substituting $-V_0$ for V_0 in the solution of the preceding problem. Thus we find

$$\delta_0 = \tan^{-1}\left(\frac{k}{K'}\tanh K'r_0\right) - kr_0, \quad K' = \left[\frac{2m}{\hbar^2}(V_0 - E)\right]^{1/2}, \tag{14.1}$$

whence, for very small kinetic energies ($k \to 0$),

$$\delta_0 \approx kr_0\left(\frac{\tanh k_0 r_0}{k_0 r_0} - 1\right), \quad k_0^2 = \frac{2m}{\hbar^2}V_0 \tag{14.2}$$

and hence

$$\sigma \approx \sigma_0 \approx 4\pi r_0^2\left(\frac{\tanh k_0 r_0}{k_0 r_0} - 1\right)^2. \tag{14.3}$$

If the well is such that $k_0 r_0 \ll 1$, then

$$\frac{\tanh k_0 r_0}{k_0 r_0} \approx 1 - \tfrac{1}{3}(k_0 r_0)^2$$

and the expression for the scattering cross-section, as is to be expected, is the same as (13.7).

As $V_0 \to \infty$, the scattering cross-section becomes

$$\sigma = 4\pi r_0^2. \tag{14.4}$$

This value is four times the classical scattering cross-section $\sigma_{cl} = \pi r_0^2$ for a hard sphere of radius r_0. The greater value of the total cross-section as calculated by quantum mechanics is due to the wave properties of particles, which in this case lead to diffraction in the neighbourhood of the sphere.

15. In order to find the phase shifts we have to determine the asymptotic form of the solution of the radial equation

$$\left[\frac{d^2}{dr^2} + k^2 - \left(\frac{2mA}{\hbar^2 r^2} + \frac{l(l+1)}{r^2}\right)\right] R_{l,k}(r) = 0, \tag{15.1}$$

which satisfies the condition $R_{l,k}(0) = 0$ and is bounded as $r \to \infty$. By defining a quantity λ to be such that

$$\lambda(\lambda+1) = l(l+1) + \frac{2mA}{\hbar^2}, \tag{15.2}$$

we can see that the function which satisfies the above conditions is (except for a constant factor)

$$R_{l,k}(r) = rj_\lambda(kr), \tag{15.3}$$

where the index λ of the spherical Bessel function is the positive root of (15.2), i.e.

$$\lambda = -\tfrac{1}{2} + \sqrt{(l+\tfrac{1}{2})^2 + \frac{2mA}{\hbar^2}}. \tag{15.4}$$

From (A.48) we have that, for large r,

$$R_{l,k}(r) \sim \frac{1}{k} \sin \left(kr - \frac{\pi}{2}\lambda\right).$$

Comparing this expression with (X.14) we find that

$$\delta_l = -\frac{\pi}{2}\left[\sqrt{(l+\tfrac{1}{2})^2 + \frac{2mA}{\hbar^2}} - (l+\tfrac{1}{2})\right]. \tag{15.5}$$

In this example we can easily verify the conclusions of problem 12. Thus if $V(r)$ is repulsive, $(A > 0)$, then $\delta_l < 0$, and if $V(r)$ is attractive, $(A < 0)$, then $\delta_l > 0$.

In accordance with (X.18) and (X.19), the fact that δ_l is independent of k means that the scattering amplitude can be written as

$$A(E, \theta) = \frac{1}{k} A_0(\theta), \tag{15.6}$$

where

$$A_0(\theta) = \sum_{l=0}^{\infty} (2l+1)\, P_l(\cos\theta) e^{i\delta_l} \sin \delta_l \tag{15.7}$$

and does not depend on the energy of the particles.

The differential cross-section for scattering is then

$$d\sigma(E, \theta) = k^{-2}\, |A_0(\theta)|^2\, d\Omega. \tag{15.8}$$

If $2mA\hbar^{-2} \ll 1$, then, from (15.5),

$$-\delta_l \approx \frac{\pi m A}{(2l+1)\hbar^2} \ll 1$$

and thus

$$A_0(\theta) \approx \sum_{l=0}^{\infty} (2l+1)\, P_l(\cos\theta)\delta_l \approx -\frac{\pi m A}{\hbar^2} \sum_{l=0}^{\infty} P_l(\cos\theta).$$

Using the generating function (A.13) of the Legendre polynomials, we observe that

$$\sum_{l=0}^{\infty} P_l(\cos\theta) = \left(2 \sin \frac{\theta}{2}\right)^{-1}$$

and therefore

$$d\sigma(E, \theta) = \frac{\pi^3 m A^2}{2\hbar^2 E} \cot \frac{\theta}{2}\, d\Omega. \tag{15.9}$$

If we attempt to calculate the total cross-section for scattering we obtain an infinite value, as in the case of the Coulomb potential.

16. The phase of the radial function $R_{l,k}(r)$ in WKB approximation (see IV.12.16)) is given by the integral

$$\int_a^r \sqrt{k^2 - U(r) - \left(l+\frac{1}{2}\right)^2 \Big/ r^2}\, dr + \frac{\pi}{4}, \tag{16.1}$$

where a is a zero of the integrand. Since the range in which the particle moves is beyond the classical turning point, $r > a$, we have to take a to be the largest zero of the integrand.

Subtracting from (16.1) the phase of the radial function corresponding to free motion:

$$\int_{a_0}^r \sqrt{k^2 - \left(l+\frac{1}{2}\right)^2 \Big/ r^2}\, dr + \frac{\pi}{2} \tag{16.2}$$

and proceeding to the limit $r \to \infty$, we obtain, in accordance with the definition, the phase shift

$$\delta_l(k) = \lim_{R\to\infty} \left[\int_a^R \sqrt{k^2 - U(r) - (l+\tfrac{1}{2})^2/r^2}\, dr - \int_{a_0}^R \sqrt{k^2 - (l+\tfrac{1}{2})^2/r^2}\, dr \right]. \tag{16.3}$$

Using the fact that

$$\lim_{r\to\infty} R_l(r) = 2D_- \sin\left[\frac{\pi}{4} + \int_a^\infty \left(\sqrt{k^2 - U(r) - \left(l+\frac{1}{2}\right)^2 \Big/ r^2} - k\right) dr + k(r-a)\right] \tag{IV.12.17}$$

another useful expression for the phase-shift can be obtained, which is equivalent to (16.3). Thus, by comparing (IV.12.17) with (X.14), we find that

$$\delta_l(k) = \left(l+\frac{1}{2}\right)\frac{\pi}{2} - ka + \int_a^\infty \left(\sqrt{k^2 - U(r) - \left(l+\frac{1}{2}\right)^2 \Big/ r^2} - k\right) dr. \tag{16.4}$$

(16.3) has the advantage that, for $l \gg 1$, it can be put into a simple form, since, if $l \gg 1$, a is also large; $V(r)$ is then small over all the range of integration, and hence $a \approx a_0$ and

$$\delta_l(k) = -\tfrac{1}{2} \int_a^\infty U(r) [k^2 - (l+\tfrac{1}{2})^2/r^2]^{-1/2}\, dr. \tag{16.5}$$

Since $l \gg 1$, the approximation $l+\frac{1}{2} \approx l$ can be used, and thus, finally,

$$\delta_l(k) = -\tfrac{1}{2} \int_{l/k}^\infty U(r) (k^2 - l^2/r^2)^{-1/2}\, dr. \tag{16.6}$$

17. The conditions of this problem are such that we can use (16.6). We have then that

$$\delta_l(k) = -\frac{mA}{\hbar^2} \int_{l/k}^\infty r^{-n} (k^2 - l^2/r^2)^{-1/2}\, dr.$$

Making the substitution $l^2/k^2r^2 = t$, we obtain Euler's "beta function", and hence

$$\delta_l(k) = -\frac{mAk^{n-2}}{2\hbar^2 l^{n-1}} \frac{\Gamma\left(\frac{1}{2}\right)\Gamma\left(\frac{n-1}{2}\right)}{\Gamma\left(\frac{n}{2}\right)}. \tag{17.1}$$

18. As we saw in problem 1, $\psi_{\mathbf{k}}^{+}(\mathbf{r})$ satisfies the integral equation

$$\psi_{\mathbf{k}}^{+}(\mathbf{r}) = \phi_{\mathbf{k}}(\mathbf{r}) + \int G_k(|\mathbf{r}-\mathbf{r}'|)\, U(\mathbf{r}')\, \psi_{\mathbf{k}}^{+}(\mathbf{r}')\, d\mathbf{r}', \tag{18.1}$$

where $\phi_{\mathbf{k}}(\mathbf{r}) = \exp(i\mathbf{k}\,.\,\mathbf{r})$ and

$$G_k(|(\mathbf{r}-\mathbf{r}'|) = \frac{1}{(2\pi)^3} \int \frac{e^{i\mathbf{q}\,.(\mathbf{r}-\mathbf{r}')}}{k^2-q^2}\, d\mathbf{q}. \tag{18.2}$$

The integral equation (18a) is obtained from (18.1) by making series expansions of $\phi_{\mathbf{k}}(\mathbf{r})$ and of $\psi_{\mathbf{k}}^{+}(\mathbf{r})$ in spherical harmonics.

In accordance with (A.50′) we have

$$\phi_{\mathbf{k}}(\mathbf{r}) = 4\pi \sum_{l=0}^{\infty} \sum_{m=-l}^{m=+l} i^l j_l(kr) Y_l^{m*}(\theta_{\mathbf{k}}, \phi_{\mathbf{k}}) Y_l^m(\theta_{\mathbf{r}}, \phi_{\mathbf{r}}).$$

Choosing the z-axis along $\mathbf{k}$, we have $\theta_{\mathbf{k}} = \phi_{\mathbf{k}} = 0$, and, owing to the spherical symmetry of the potential, $m = 0$. Therefore the expansion (X.13) can be written by means of (A.24) as

$$\psi_{\mathbf{k}}(r) = 4\pi \sum_{l=0}^{\infty} \sum_{m=-l}^{+l} i^l e^{i\delta_l(k)} \frac{R_{l,k}(r)}{r} Y_l^{m*}(\theta_{\mathbf{k}}, \phi_{\mathbf{k}}) Y_l^m(\theta_{\mathbf{r}}, \phi_{\mathbf{r}}). \tag{18.3}$$

The Green's function $G_k(|\mathbf{r}-\mathbf{r}'|)$, which depends on r, r' and $\cos\alpha = \mathbf{r}\,.\,\mathbf{r}'/rr'$, can be expanded in a series of Legendre polynomials $P_l(\cos\alpha)$. Thus, bearing in mind the addition theorem (A.30), the function $G_k(|\mathbf{r}-\mathbf{r}'|)$ can be written in the form

$$G_k(|\mathbf{r}-\mathbf{r}'|) = \sum_{l'=0}^{\infty} \sum_{m'=-l'}^{+l'} G_k^{(l')}(r, r') Y_{l'}^{m'*}(\theta_{\mathbf{r}'}, \phi_{\mathbf{r}'}) Y_{l'}^{m'}(\theta_{\mathbf{r}}, \phi_{\mathbf{r}}). \tag{18.4}$$

Now, using (18.1), (A.50′), (18.3), (18.4) and (A.27), after a little calculation we find that

$$R_{l,k}(r) = rj_l(kr)e^{-i\delta_l(k)} + r\int_0^{\infty} r' G_k^{(l)}(r; r')\, U(r')\, R_{l,k}(r')\, dr', \tag{18.5}$$

i.e.

$$R_{l,k}(r) = rj_l(kr)e^{-i\delta_l(k)} + r\int_0^{r} r' T_k^{(l)}(r, r')\, U(r')\, R_{l,k}(r')\, dr' + r\int_r^{\infty} r' T_k^{(l)}(r; r')\, U(r')\, R_{l,k}(r')\, dr'. \tag{18.5′}$$

In order to determine the functions $G_k^{(l)}(r; r')$ we expand the plane waves $e^{i\mathbf{q}\cdot\mathbf{r}}$ and $e^{-i\mathbf{q}\cdot\mathbf{r}}$ in spherical harmonics in accordance with (A.50′). Thus, using (A.27), (18.2) becomes

$$G_k(|\mathbf{r}-\mathbf{r}'|) = \frac{2}{\pi}\sum_{l=0}^{\infty}\sum_{m=-l}^{+l}\left(\int_0^{\infty} q^2\,\frac{j_l(qr)\,j_l(qr')}{k^2-q^2}\,dq\right) Y_l^{m*}(\theta_{\mathbf{r}'},\,\phi_{\mathbf{r}'})\,Y_l^m(\theta_{\mathbf{r}},\,\phi_{\mathbf{r}}). \tag{18.6}$$

By comparing (18.4) and (18.6), and remembering that the integrand is an even function of q, it can be seen that

$$G_k^{(l)}(r;\,r') = \frac{1}{\pi}\int_{-\infty}^{+\infty} q^2\,\frac{j_l(qr)\,j_l(qr')}{k^2-q^2}\,dq. \tag{18.7}$$

This integral can be evaluated easily by means of the theory of residues. Note first that $j_l(z)$ is a whole function of z, so that the poles of the integrand are simply the zeros of the denominator ($q = \pm k$).

Suppose now that $r > r'$ (as in the first integral of (18.5′)), and replace $j_l(qr)$ by $\frac{1}{2i}$ $[h_l^{(+)}(qr)-h_l^{(-)}(qr)]$. Then, since $h_l^{(\pm)}(z) \sim \frac{1}{z}\exp\left[\pm i\left(z-\frac{l\pi}{2}\right)\right]$ as $z \to \infty$, it follows that $h_l^{(+)}(qr)j_l(qr') \to 0$ as $|q| \to \infty$ in the upper half-plane, and that $h_l^{(-)}(qr)j_l(qr') \to 0$ as $|q| \to \infty$ in the lower half-plane. Thus the integrals in the expression

$$G_k^{(l)}(r;\,r') = \frac{1}{2\pi i}\int_{-\infty}^{+\infty} q^2\,\frac{h_l^{(+)}(qr)j_l(qr')}{k^2-q^2}\,dq - \frac{1}{2\pi i}\int_{-\infty}^{+\infty} q^2\,\frac{h_l^{(-)}(qr)j_l(qr')}{k^2-q^2}\,dq \tag{18.8}$$

can each be completed by a semicircle of infinite radius, in the appropriate half-plane. Remembering that, for a scattering problem, we must take for the first integral of (18.8) the residue of the pole at $q = +k$, and for the second integral the residue of the pole at $q = -k$, and bearing in mind that $h_l^{(-)}(-z) = (-1)^{l+1}h_l^{(+)}(z)$, we obtain after some simple calculation the correct Green's function

$$G_k^{(l)}(r;\,r') = -kh_l^{(+)}(kr)\,j_l(kr'), \qquad r > r'.$$

If $r < r'$ (as in the second integral of (18.5′)) we replace $j_l(qr')$ by $\frac{1}{2i}[h_l^{(+)}(qr')-h_l^{(-)}(qr')]$ and, after calculations similar to those given above, we obtain the result

$$G_k^{(l)}(r;\,r') = -kh_l^{(+)}(kr')\,j_l(kr), \qquad r < r'$$

and therefore, finally,

$$G_k^{(l)}(r;\,r') = -k\begin{cases} h_l^{(+)}(kr)\,j_l(kr'), & r > r' \\ h_l^{(+)}(kr')\,j_l(kr), & r < r'. \end{cases} \tag{18.9}$$

19. Since the two Jost solutions $f(\pm k,\, l;\, r)$ have the same value of k^2, their Wronskian does not depend on r. Also, from the conditions (X.22), it can be seen that, for large r, $f(\pm k,\, l;\, r) \sim \exp(\mp ikr)$.

The independence of r of the Wronskian enables us to calculate its value from these asymptotic forms, and the result

$$W[f(k, l; r),\ f(-k, l; r)] = 2ik \neq 0, \tag{19.1}$$

demonstrates the linear independence of the two solutions $f(\pm k, l; r)$.

Thus the required radial function which satisfies conditions (19b) can be expressed as a linear combination of $f(k, l; r)$ and of $f(-k, l; r)$, i.e.

$$R_{l,k}(r) = A(k, l)f(k, l; r)+B(k, l)f(-k, l; r). \tag{19.2}$$

The coefficients $A(k, l)$ and $B(k, l)$ can easily be found by noting that

$$\begin{aligned} W[f(k, l; r),\ R_{l,k}(r)] &= A(k, l)\, W[f(k, l; r),\ f(k, l; r)] \\ &+B(k, l)\, W[f(k, l; r),\ f(-k, l; r)] = 2ikB(k, l), \\ W[f(-k, l; r),\ R_{l,k}(r)] &= A(k, l)\, W[f(-k, l; r),\ f(k, l; r)] \\ &+B(k, l)\, W[f(-k, l; r),\ f(-k, l; r)] = -2ikA(k, l). \end{aligned} \tag{19.3}$$

On the other hand

$$\begin{aligned} W[f(k, l; 0), R_{l,k}(0)] &= f(k, l; 0)\, R'_{l,k}(0) - f'(k, l; 0)\, R_{l,k}(0) \equiv F_l(k) \\ W[f(-k, l; 0), R_{l,k}(0)] &= f(-k, l; 0)\, R'_{l,k}(0) - f'(k, -l; 0)\, R_{l,k}(0) \equiv F_l(-k), \end{aligned} \tag{19.4}$$

and, because $W[f(\pm k, l; r), R_{l,k}(r)]$ is independent of r, we obtain from (19.3) and (19.4) the results

$$A(k, l) = -\frac{F_l(-k)}{2ik}, \qquad B(k, l) = \frac{F_l(k)}{2ik}, \tag{19.5}$$

and thus

$$R_{l,k}(r) = \frac{1}{2ik}[F_l(k)f(-k, l; r) - F_l(-k)f(k, l; r)]. \tag{19.6}$$

Note that, in the asymptotic region ($r \to \infty$),

$$R_{l,k}(r) \sim \left[\frac{1}{2ik} F_l(k)e^{ikr} - F_l(-k)e^{-ikr}\right]. \tag{19.7}$$

Now from (X.14) we have that

$$R_{l,k}(r) \sim \frac{1}{2ik} e\left[^{i\left(\delta_l(k)-\frac{l\pi}{2}\right)} e^{ikr} - e^{-i\left(\delta_l(k)-\frac{l\pi}{2}\right)} e^{-ikr}\right]. \tag{19.8}$$

By comparing (19.7) with (19.8) we obtain the following relation between the Jost functions and the diagonal elements of the scattering matrix:

$$S_l(k) = e^{2i\delta_l(k)} = e^{i\pi l} \frac{F_l(k)}{F_l(-k)}. \tag{19.9}$$

20. Except for a multiplying constant which is included in the normalization factor of the wavefunction, the Jost solutions are

$$f(\pm k, 0; r) = A \sin Kr + B \cos Kr, \qquad r < r_0$$
$$f(\pm k, 0; r) = e^{\mp ikr}, \qquad r > r_0$$

where

$$K = \left[\frac{2m}{\hbar^2}(E+V_0)\right]^{1/2}.$$

A and B can be determined from the continuity of the solutions and of their derivatives at $r = r_0$. Thus the inner solution has the form

$$f(\pm k, 0; r) = \frac{e^{\mp ikr_0}}{K}[(K \sin Kr_0 \mp ik \cos Kr_0) \sin Kr + (K \cos Kr_0 \pm ik \sin Kr_0) \cos Kr], \qquad r < r_0.$$

The Jost functions follow immediately:

$$F_0(\pm k) = e^{\mp ikr_0}\left[\cos Kr_0 \pm i\frac{k}{K} \sin Kr_0\right], \tag{20.1}$$

and then

$$S_0(k) = \frac{F_0(k)}{F_0(-k)} = \frac{K \cos Kr_0 + ik \sin Kr_0}{K \cos Kr_0 - ik \sin Kr_0} e^{-2ikr_0}. \tag{20.2}$$

If the kinetic energy of the particles is less than the depth of the potential well, i.e. if $-V_0 < E < 0$, then $k = i\left(\frac{2m}{\hbar^2}|E|\right)^{1/2} = ik_0$, say, and the poles of the function $S_0(k)$ (i.e. the zeros of $F_0(-k)$) are determined from the condition

$$K \cos Kr_0 + k_0 \sin Kr_0 = 0,$$

i.e.

$$r_0 k_0 = -r_0 K \cot Kr_0. \tag{20.3}$$

By noting that (20.3) is the same as (30.3), Chapter II, we arrive at the conclusion that the energy values, corresponding to the poles of $S_0(k)$, coincide with the energies of bound states in the (spherical) potential well. This concrete example serves to demonstrate some properties which are valid in general for the functions $S_l(k)$ corresponding to an orbital angular momentum l. Since (20.2) is valid for complex values of k, it follows that $S_0(k)$, being the ratio of two holomorphic functions, is a meromorphic function of complex k. Note that

$$S_0(-k) = S_0^{-1}(k) \tag{20.4}$$

and hence that

$$\delta_0(-k) = -\delta_0(k) \tag{20.5}$$

and also that

$$[S_0(k^*)]^* = S_0^{-1}(k), \tag{20.6}$$

a property which is called "unitarity".

Finally, we can show, by using (20.2), that

$$\lim_{|k|\to\infty} S_0(k) = 1 \tag{20.7}$$

and that

$$\lim_{|k|\to\infty} \delta_0(k) = 0, \tag{20.8}$$

which express the fact that for very high energy particles no scattering occurs ($S_0(k) = 1$), i.e. the motion of the particles is not affected by any potentials they encounter.

21. Since, from (19a),

$$\frac{d^2 f(k, l; r)}{dr^2} = (U_l(r) - k^2)\, f(k, l; r)$$

and

$$\frac{d^2}{dr'^2}[\sin k(r'-r)] = -k^2 \sin k(r'-r),$$

we have that

$$\int_r^\infty \sin k(r'-r)\, U_l(r')\, f(k, l; r')\, dr'$$

$$= \int_r^\infty \left\{\sin k(r'-r)\frac{d^2 f(k, l; r')}{dr'^2} - f(k, l; r')\frac{d^2}{dr'^2}[\sin k(r'-r)]\right\} dr'$$

$$= \int_r^\infty \frac{d}{dr'}\left\{\sin k(r'-r)\frac{df(k, l; r')}{dr'} - f(k, l; r')\frac{d}{dr'}\sin k(r'-r)\right\} dr'.$$

By integrating and using the boundary condition (X.22), the integral equation (21a) follows directly. In order to simplify the calculation, consider the integral equation

$$g(k, l; r) = e^{ikr} f(k, l; r) = 1 + \frac{1}{k}\int_r^\infty \frac{1 - e^{-2ik(r'-r)}}{2i}\, U_l(r')\, g(k, l; r')\, dr', \tag{21.1}$$

whose solution can be written formally as

$$g(k, l; r) = \sum_{i=0}^{\infty} g_i(k, l; r) \tag{21.2}$$

where

$$g_0(k, l; r) = 1,$$

$$g_{n+1}(k, l; r) = \frac{1}{k}\int_r^\infty \frac{1-e^{2ik(r'-r)}}{2i} U_l(r') g_n(k, l; r')\, dr'. \tag{21.3}$$

This expansion in a series defines the solution $g(k, l; r)$ of the integral equation (21.1) only if the series converges. We shall prove that this is so provided $\text{Im}(k) = b < 0$, $(k \neq 0)$, and

$$\int_r^\infty |U_l(r')|\, dr' = M_l(r) \qquad \text{say, is finite.}$$

Thus, if $b < 0$, we have

$$\left|\frac{1-e^{-2ik(r'-r)}}{2ik}\right| < \left|\frac{1}{2ik}\right| + \left|\frac{e^{2b(r'-r)}}{2ik}\right| < \frac{1}{|k|},$$

and hence

$$|g_{n+1}| < \frac{1}{|k|}\int_r^\infty |U_l(r')| \,.\, |g_n(k, l; r')|\, dr'.$$

It follows that

$$|g_1(k, l; r)| < \frac{M_l(r)}{|k|},$$

$$|g_2(k, l; r)| < \frac{1}{|k|^2}\int_r^\infty |U_l(r')|\, M_l(r')\, dr' = -\frac{1}{|k|^2}\int_r^\infty \frac{dM_l(r')}{dr'} M_l(r')\, dr' = \frac{M_l^2(r)}{2|k|^2},$$

and, in general, that

$$|g_n(k, l; r)| < \frac{M_l^n(r)}{n!\,|k|^n}. \tag{21.4}$$

It is clear then that the series (21.2) converges, and hence that a solution exists for $b < 0$ and $k \neq 0$. In addition it follows that $g(k, l; r)$ is a holomorphic function, since the series is uniformly convergent. Further, it can be seen that

$$|g(k, l; r)-1| = \left|\sum_{n=0}^\infty g_n(k, l; r) - 1\right| = \left|\sum_{n=1}^\infty g_n(k, l; r)\right|$$

$$\leqslant \sum_{n=1}^\infty |g_n(k, l; r)| \leqslant \sum_{n=1}^\infty \frac{M_l^n(r)}{n!\,|k|^n} = \exp\left(\frac{M_l(r)}{|k|}\right) - 1,$$

and hence

$$|g(k, l; r)-1| \leq \exp\left(\frac{M_l(r)}{|k|}\right)-1, \tag{21.5}$$

whence

$$\lim_{|k|\to\infty} g(k, l; r) = 1,$$

for any direction of k in the lower half-plane.

Since $g(k, l; 0) = F_l(k)$ we can conclude that, if (21c) is valid, then the Jost functions $F_l(k)$ and $F_l(-k)$ are holomorphic in the half-planes Im $(k) < 0$ and Im $(k) > 0$ respectively, and that $\lim\limits_{|k|\to\infty} F_l(k) = 1$ and $\lim\limits_{|k|\to\infty} F_l(-k) = 1$, for any directions in the appropriate half-planes.

Consider now the case Im $(k) = b > 0$. Then

$$\left|\frac{1-e^{-2ik(r'-r)}}{2ik}\right| < \frac{e^{2b(r'-r)}}{|k|}.$$

and it follows that

$$|g_1(k, l; r)| < \frac{e^{2br}}{|k|}\int_r^\infty |U_l(r')|\, e^{2br'}\, dr' \equiv \frac{e^{2br}}{|k|} P_l(r)$$

and, in general, that

$$|g_n(k, l; r)| < P_l(r)\frac{M_l^{n-1}(r)}{(n-1)!}\frac{1}{|k|^n}e^{-2br}. \tag{21.6}$$

Thus the function $g(k, l; r)$ is holomorphic in the region of the upper half-plane in which

$$P_l(r) = \int_r^\infty U_l(r')e^{2br'}\, dr' \tag{21.7}$$

s finite. The Jost functions $F_l(k)$ and $F_l(-k)$ are therefore holomorphic in the region of the upper (or the lower) half-plane, in which

$$\int_r^\infty |U_l(r')|\, e^{2br'}\, dr' \tag{21.8}$$

is finite. Note that if $b > 0$, then $M_l(r)$ is finite if $P_l(r)$ is.

22. Since $f^*(-k^*, l; r)$ satisfies the same integral equation (21a) as $f(k, l; r)$, we have

$$f^*(-k^*, l; r) = f(k, l; r) \tag{22.1}$$

and hence

$$F_l^*(-k^*) = F_l(k). \tag{22.2}$$

Thus

$$S_l(k) = e^{i\pi l}\frac{F_l(k)}{F_l(-k)} = e^{i\pi l}\frac{F_l^*(-k^*)}{F_l^*(k^*)} = \frac{1}{S_l^*(k^*)} = S_l^*(-k^*), \tag{22.3}$$

whence the unitarity relation

$$S_l(k)\,S_l^*(k^*) = 1$$

and the reciprocity relation

$$S_l(k)\,S_l(-k) = 1$$

both follow.

The relations have already been obtained for $l = 0$ for the particular system specified in problem 20. The relation (22.1) is called the "crossing symmetry relation" of the Jost function.

23. Since, for the Yukawa potential,

$$M_0(r) = \int_r^\infty |U(r')|\,dr' = \frac{2m\,|A|}{\hbar^2}\int_r^\infty \frac{e^{-r'/r_0}}{r'}\,dr' < \infty,$$

it follows from problem 21 that $F_0(k)$ and $F_0(-k)$ are holomorphic in the whole of the lower (or, respectively, the upper) half-plane of k. However,

$$P_0(r) = \int_r^\infty |U(r')|\,e^{2br'}\,dr' = \frac{2m\,|A|}{\hbar^2}\int_r^\infty \frac{e^{(2br_0-1)r'/r_0}}{r'}\,dr',$$

converges only for $b < 1/(2r_0)$.

In conclusion, for Yukawa-type potentials, $F_0(k)$ and $F_0(-k)$ are holomorphic functions of k in the regions in which $b < 1/(2r_0)$ and $b > -1/(2r_0)$ respectively. Hence the Bargmann strip of the function $S_0(k)$, which is the region of the complex k-plane which is common to both the above-mentioned regions, (Fig. X.4), is the region defined by

$$-\frac{1}{2r_0} < \mathrm{Im}\,(k) < +\frac{1}{2r_0}. \tag{23.1}$$

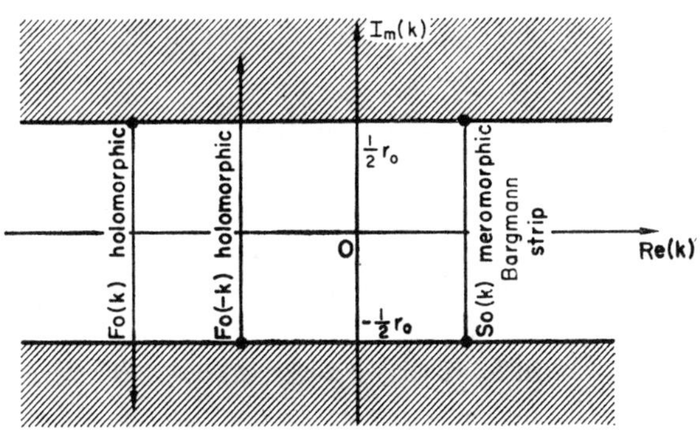

FIG. X.4.

The Bargmann strip is important in the study of $S_0(k)$ because in it $F_0(k)$ is holomorphic, and the poles of $S_0(k)$ are therefore due, not to singularities in $F_0(k)$, but to zeros of $F_0(-k)$.

24. Since, in the Bargmann strip, the Jost functions $F_l(\pm k)$ are holomorphic, the poles of $S_l(k)$ are given by the condition

$$F_l(-k) = 0. \tag{24.1}$$

We shall show first that, in the part of the Bargmann strip which lies in the upper half of the k-plane, all the poles of $S_l(k)$ lie on the imaginary axis. Suppose that there is a pole $k_0 = a+ib$, $(F_l(-k_0) = 0)$ with $b > 0$ and $a \neq 0$; then, from the relation

$$S_l(k) = S_l^*(-k^*) \tag{24.2}$$

it follows that there is a pole at $-k_0^*$ also.

Further, by (19.6),

$$R_{l,k_0}(r) = \frac{1}{2ik_0}[F_l(k_0)f(-k_0, l; r) - F_l(-k_0)f(k_0, l; r)] = \frac{1}{2ik_0}F_l(k_0)f(-k_0, l; r),$$

$$R_{l,-k_0^*}(r) = -\frac{1}{2ik_0^*}F_l(-k_0^*)f(k_0^*, l; r).$$

Thus there would exist two solutions, both regular at the origin ($r = 0$), viz., $f(-k_0, l; r) = f(-a-ib, l; r)$ and $f(k_0^*, l; r) = f(a-ib, l; r)$, which describe two states corresponding to the two eigenvalues $E_\pm = \frac{\hbar^2}{2m}(a \pm ib)^2$, i.e. two states which are mutually orthogonal, so that

$$\int_0^\infty f(a-ib, l; r)f(-a-ib, l; r)\,dr = 0. \tag{24.3}$$

This is, however, impossible, since, according to (22.1),

$$f(a-ib, l; r) = f^*(-a-ib, l; r) \tag{24.4}$$

so that, unless $f(-a-ib, l; r) \equiv 0$, the integral (24.3) cannot be zero. If $a = 0$, $b > 0$, then $E_+ = E_-$, and, instead of the orthogonality relation (24.3), a normalization condition is required

$$\int_0^\infty f^*(-ib, l; r)f(-ib, l; r)\,dr = N_f > 0,$$

which can evidently be satisfied.

Hence in the part of the Bargmann strip which lies in the upper half of the k-plane, $S_l(k)$ has no poles except on the imaginary axis. For $b < 0$ the argument given above is no longer valid, since, in accordance with (X.22), in the asymptotic region $r \to \infty$, $f(-a-ib, l; r) \sim \exp[(ia-b)r]$, and thus, unlike the case $b > 0$, the wavefunction is not square integrable.

In the part of the Bargmann strip which lies in the lower half of the k-plane, the poles of $S_l(k)$ do not have to be on the imaginary axis.

Let us consider the physical significance of the poles of $S_l(k)$. We start with those placed in the upper half of the k-plane. Let $k_0 = ib_0$, ($b_0 > 0$), be one of them. The corresponding solution

$$R_{l,\,k_0}(r) = -\frac{1}{2b_0} F_l(ib_0)\, f(-ib_0, l; r)$$

is, by definition, bounded at the origin (see (19b)), whence it follows that the Jost solution $f(-ib_0, l; r)$ is regular at $r = 0$. Because it behaves at infinity as $\exp(-b_0 r)$, it is thus a square integrable radial function and describes a bound state. It follows that any poles on the positive imaginary axis (within the Bargmann strip) describe bound states.

If $k_0 = ib_0$, where $b_0 < 0$ (within the limits of the Bargmann strip) is a pole of the function $S_l(k)$, then the corresponding state is called a "virtual state". Virtual states do not represent physical states of systems in the usual sense of quantum mechanics, since their radial wavefunctions are not square integrable. They do have an important observable effect, however, since their influence on the scattering cross-section is the same as would be that of a bound state of the same energy $\left(E_0 = -\dfrac{\hbar^2}{2m} b_0^2\right)$.

Those poles of $S_l(k)$ which occur in the lower half of the k-plane (inside the Bargmann strip) and which are not on the imaginary axis appear in complex conjugate pairs $(k_0, -k_0^*)$, as can be seen. The states corresponding to these poles are called "resonances". From the relations

$$S_l(k) = [S_l^*(k^*)]^{-1} \quad \text{and} \quad S_l(k) = [S_l(-k)]^{-1} \tag{24.5}$$

it can be seen that if $S_l(k)$ has a pole at k_0 then it has a zero at the conjugate point k_0^* and also at $-k_0$. It is thus clear that zeros in the lower half-plane can occur only on the imaginary axis.

Thus, to bound (or virtual) states there correspond zeros of $S_l(k)$ on the imaginary axis in the lower (or the upper) half of the k-plane, and to resonances there correspond zeros of $S_l(k)$ in the upper half of the k-plane, off the imaginary axis.

Figure X.5 shows how the poles and the zeros of the function $S_l(k)$ in the Bargmann strip are related as a direct consequence of (24.2) and (24.5). From these relations it follows

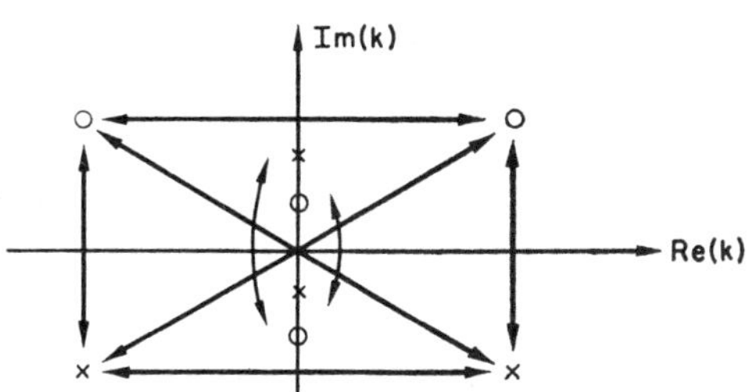

FIG. X.5.

that on the real axis the function $S_l(k)$ can have neither poles nor zeros. Remember that only real values of k correspond to real scattering states.

25. From Cauchy's theorem we have that

$$N = \frac{1}{2\pi i} \int_C [\ln f(z)]' \, dz = \frac{1}{2\pi i} \ln f(z) \Big|_C . \tag{25.1}$$

Since $F_l(-k)$ is a holomorphic function in the upper half of the k-plane, in which it has zeros only on the imaginary axis, and since between these zeros and the bound states there is a one-to-one correspondence, it follows that by choosing the contour $C = C_1+C_2+C_3$ shown in Fig. X.6 we obtain

$$n_l = \frac{1}{2\pi i} [\ln F_l(-k)]_C . \tag{25.2}$$

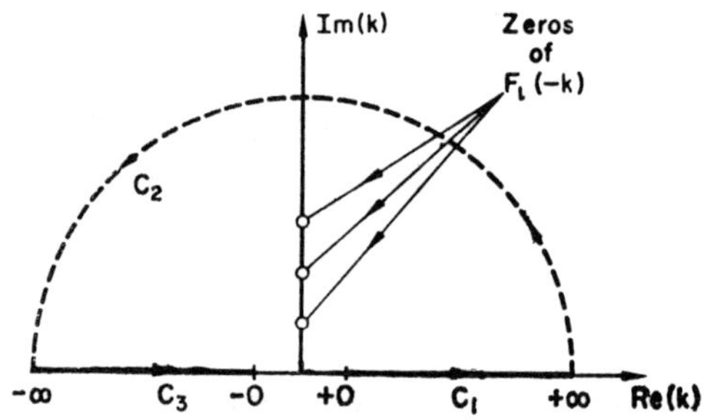

FIG. X.6.

As we saw in problem 21, $\lim_{|k|\to\infty} F_l(-k) = 1$, along any direction in the upper half of the k-plane. For this reason (25.2) becomes

$$n_l = \frac{1}{2\pi i} [\ln F_l(-k)]_{C_1+C_3} . \tag{25.3}$$

By comparing (X.17) with (X.23) and taking into account the crossing-symmetry (22.2) of the Jost functions, we can write on the Re (k) axis

$$F_l(-k) = |F_l(-k)| \exp\left[-i\left(\delta_l(k) - \frac{l\pi}{2}\right)\right], \tag{25.4}$$

and hence

$$n_l = \frac{1}{2\pi i}\left[\ln |F_l(-k)| - i\left(\delta_l(k) - \frac{l\pi}{2}\right)\right]_{C_1+C_3} . \tag{25.5}$$

Since, on C_1+C_3, the function $|F_l(-k)|$ has no zeros, and is equal to 1 at both ends, (25.5) becomes

$$n_l = \frac{1}{2\pi}\left[\frac{l\pi}{2} - \delta_l(k)\right]_{C_1+C_3}. \tag{25.6}$$

From the reciprocity relation (22b) we have

$$\delta_l(k) = -\delta_l(-k), \tag{25.7}$$

and from (25.6), Levinson's theorem follows immediately. Supposing that (20.8) holds for any value of l, Levinson's theorem can be written as

$$\delta_l(+0) = n_l\pi. \tag{25.8}$$

26. Let $k_0 = ib_0$, $b_0 < 0$, $|b_0| \ll 1$, be a virtual state, so that $F_l(-k_0) = 0$. By expanding the Jost function in a Taylor's series about k_0 and retaining only the first non-vanishing term, we have

$$F_l(-k) \approx F_l(-k_0) + \frac{k-k_0}{1!}\left.\frac{dF_l(-k)}{dk}\right|_{k=k_0} = (k-k_0)\left.\frac{dF_l(-k)}{dk}\right|_{k=k_0},$$

whence

$$S_l(k) = e^{2i\delta_l(k)} = e^{i\pi l}\frac{F_l(k)}{F_l(-k)} \approx -e^{i\pi l}\frac{k+ib_0}{k-ib_0}. \tag{26.1}$$

Since at low energies the principal contribution to the total scattering cross-section comes from the s-waves,

$$\sigma(E) \approx \frac{4\pi}{k^2}\sin^2\delta_0(k),$$

and from (26.1) we then find that

$$\sigma(E) \approx \frac{4\pi}{k^2+b_0^2}. \tag{26.2}$$

Thus it can be seen that, at low energies, poles corresponding to virtual states are associated with high values of the scattering cross-section, which depends as shown on the positions of the poles. Further, it can easily be seen that if there existed, instead of a given virtual state, a bound state of the same energy $E_0 = -\hbar^2 b_0^2/2m$, we would obtain the same expression for the scattering cross-section. This means that it is impossible, from measurements of the scattering cross-section alone, to determine the sign of b_0, in other words, that such measurements alone cannot distinguish between bound and virtual states. For example: the neutron–proton scattering cross-section both in the singlet and in the triplet states shows an appreciable increase at low energies of the form (26.2). This can be explained easily in the triplet state through the existence of the deuteron bound state. In the singlet state such a bound state does not exist. The increase in the scattering cross-section at low energies follows, however, from the existence of a virtual singlet state.

27. By substituting (27a) into the integral equation (18a) we obtain immediately

$$R_{l,k}(r) = rj_l(kr)e^{-i\delta_l(k)} + kaU_0 rR_{l,k}(a)\begin{cases} h_l^{(+)}(kr)\, j_l(ka), & a < r \\ j_l(kr)\, h_l^{(+)}(ka), & a > r \end{cases} \tag{27.1}$$

from which $R_{l,k}(a)$ can easily be determined by putting $r = a$.

Thus we find that

$$R_{l,k}(a) = \frac{aj_l(ka)e^{i\delta_l(k)}}{1-ka^2U_0 j_l(ka)\, h_l^{(+)}(ka)}. \tag{27.2}$$

Using now (11a) where, in our case, $\frac{2m}{\hbar^2} V(r) = U(r) = -U_0\, \delta(r-a)$, it follows that

$$\frac{1}{k} e^{i\delta_l(k)} \sin \delta_l(k) = \frac{a^2U_0[j_l(ka)]^2}{1-ka^2U_0 j_l(ka)\, h_l^{(+)}(ka)} \tag{27.3}$$

which, by (X.18), is precisely the scattering partial amplitude.

With the notation $x = ka$ and $g = aU_0$, the expression for the partial amplitude becomes

$$A_l(k) = \frac{ag[j_l(x)]^2}{1-gxj_l(x)\, h_l^{(+)}(x)} \tag{27.4}$$

and thus

$$S_l(k) = 1+\frac{2igx[j_l(x)]^2}{1-gxj_l(x)\, h_l^{(+)}(x)}. \tag{27.5}$$

28. As we saw in problem 24, bound states correspond to poles of the scattering matrix on the positive imaginary axis in the complex k-plane. The analytical continuation of (27.5) for complex values of k, is given by

$$S_l(k) = 1+\frac{2igz[j_l(z)]^2}{1-gzj_l(z)\, h_l^{(+)}(z)} \tag{28.1}$$

in which we have written $z = a[\text{Re}\,(k)+i\,\text{Im}\,(k)] = x+iy$. Since $j_l(z)$ is an entire function, the poles of S_l are due solely to the zeros of

$$D_l(g, z) \equiv 1-gzj_l(z)\, h_l^{(+)}(z). \tag{28.2}$$

The value of U_0 we are looking for corresponds to the bound state of zero energy, for which $x = y = 0$, $E = -\hbar^2y^2/2ma^2$, and the orbital angular momentum $l = l_0$.

Since, in accordance with (A.49),

$$\lim_{z\to 0} zj_{l_0}(z)\, h_{l_0}^{(+)}(z) = \frac{1}{2l_0+1},$$

it follows that

$$D_{l_0}(g, 0) = 1-\frac{g}{2l_0+1}, \quad g = aU_0.$$

Hence, if $U_0 = \frac{1}{a}(2l_0+1)$, there is a bound state in the potential (27a) with $l = l_0$ and zero energy.

For s-waves, $l = 0$, (28.2) becomes (see (A.47)),

$$D_0(g, z) = 1 - ge^{iz}\frac{\sin z}{z}, \tag{28.3}$$

and the condition for a bound state, $D_0(g, iy) = 0$, $y > 0$, leads to the equation

$$\frac{1}{g} = \frac{1-e^{-2y}}{y}. \tag{28.4}$$

The existence of bound states can be studied easily by means of the plot shown in Fig. X.7, in which the broken lines represent constant values of $1/g$ and the right-hand side of (28.4) is plotted as a continuous curve.

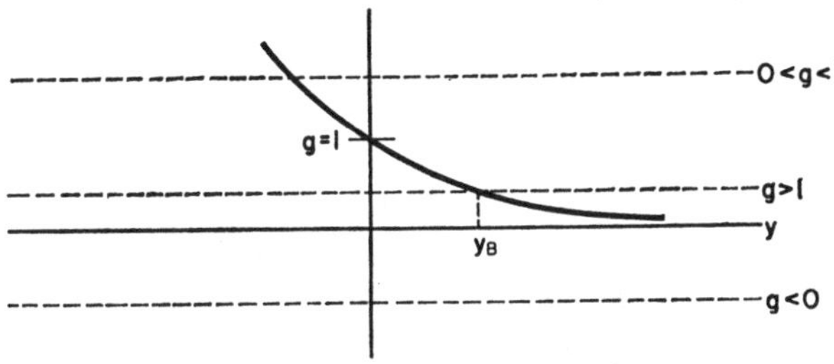

FIG. X.7.

It can be seen immediately that in the potential (27a) only one bound state can exist for $l = 0$, and this appears only if $g \geqslant 1$. The binding energy of this state is $E = -\hbar^2 y_B^2/2ma^2$. A bound state with $E = 0$ is, in agreement with what has been said above (in which, in this case, $l_0 = 0$), obtained if $g = 1$, i.e. if $U_0 = 1/a$.

29. From the relations (X.26)–(X.28) we have that

$$\sigma_{\text{tot}}(E) = \frac{2\pi}{k^2}\sum_{l=0}^{\infty}(2l+1)(1-\text{Re } C_l). \tag{29.1}$$

From (X.25), putting $\theta = 0$, and from (29.1), there follows the relation (29a) which is a generalization of the optical theorem (X.21). It expresses a very important fact, viz., that for any energy, the elastic scattering amplitude in the forward direction determines completely the total scattering cross-section (elastic and inelastic).

30. The relation (30a) follows from $\sigma_{\text{el}}^{(l)} = \sigma_0^{(l)}|1-C_l|^2$ and $\sigma_{\text{inel}}^{(l)} = \sigma_0^{(l)}(1-|C_l|^2)$ with the help of some elementary algebra. It is interesting that, if $\sigma_{\text{inel}}^{(l)} \neq 0$, then $\sigma_{\text{el}}^{(l)} \neq 0$ also, i.e. the existence of scattering in an inelastic channel implies the existence of elastic scattering.

31. The derivation of the Watson–Sommerfeld formula follows immediately. Thus, inside the C contour, the integrand has poles only where $\sin \pi l = 0$, i.e. only for integer values (= physical values) of l. Thus, the integral becomes a series consisting of the sum of the residues each multiplied by $2\pi i$.

Now the residue at $l = n$ is

$$-\frac{i}{2}\,\frac{(2n+1)\,A_n(E)\,P_n(-\cos\theta)}{\pi\cos n\pi} = \frac{1}{2\pi i(-1)^n}(2n+1)\,A_n(E)(-1)^n\,P_n(\cos\theta)$$

$$= \frac{1}{2\pi i}(2n+1)\,A_n(E)\,P_n(\cos\theta),$$

which, multiplied by $2\pi i$, becomes the nth term of the series (X.19) for the scattering amplitude $A(E, \theta)$.

32. The condition (32a) is fulfilled if, as $r \to 0$, the radial equation (II.15) passes into the equation

$$\left[\frac{d^2}{dr^2} - \frac{l(l+1)}{r^2}\right] R_{l,\,k}(r) = 0. \tag{32.1}$$

The two linearly independent solutions of this equation are r^{l+l} and r^{-l}. If Re $(l) > -\frac{1}{2}$, then as $r \to 0$ we have $|r^{l+1}| < |r^{-l}|$ and thus it is impossible to add the solution r^{-1} to the solution r^{l+1}, without violating the condition (32a).

If Re $(l) < -\frac{1}{2}$ then, as $r \to 0$, we have $|r^{l+1}| > |r^{l}|$ and it will always be possible to add the solution r^{-l} to the solution r^{l+1}, without changing its behaviour near the origin.

If Re $(l) = -\frac{1}{2}$, both solutions have the same behaviour near the origin. In conclusion, the condition (32a) determines uniquely the solution of the radial equation (II.15) in the complex l-plane only if Re $(l) > -\frac{1}{2}$.

33. The required solution of the radial equation (II.15) can be found from the integral equation:

$$R_{l,\,k}(r) = r^{l+1} + \int_0^r G_l(r, r')\,(U(r') - k^2)\,R_{l,\,k}(r')\,dr', \tag{33.1}$$

provided the Green's function satisfies the differential equation

$$\left[\frac{d^2}{dr^2} - \frac{l(l+1)}{r^2}\right] G_l(r, r') = \delta(r - r'). \tag{33.2}$$

Starting from the solutions of the homogeneous equation it can easily be verified that

$$G_l(r, r') = \begin{cases} \dfrac{1}{2l+1}\left(\dfrac{r'^{l+1}}{r^l} - \dfrac{r^{l+1}}{r'^l}\right) & \text{if} \quad r > r' \\ 0 & \text{if} \quad r < r'. \end{cases} \tag{33.3}$$

(33.3) in fact satisfies (33.2) for any r, except at the point $r = r'$, where its first derivative undergoes a jump equal to unity, and in consequence its second derivative behaves like $\delta(r-r')$.

Thus

$$R_{l,k} = r^{l+1} + \frac{1}{2l+1}\int_0^r \left(\frac{r'^{l+1}}{r^l} - \frac{r^{l+1}}{r'^l}\right)(U(r')-k^2)\,R_{l,k}(r')\,dr'. \tag{33.4}$$

Let us show that the solution of this equation satisfies condition (32a) at the origin. We have to show that, as $r \to 0$, the integral tends to zero faster than r^{l+1}. For this to be the case, it is sufficient that, as $r \to 0$,

$$\int_0^r \frac{R_{l,k}(r')}{r'^l}(U(r')-k^2)\,dr' < \infty, \tag{33.5}$$

and

$$\int_0^r r'^{l+1}(U(r')-k^2)\,R_{l,k}(r')\,dr' \sim r^s, \tag{33.6}$$

where

$$s > 2l+1. \tag{33.7}$$

If, as $r \to 0$, we have $U(r) \sim r^{-n}, n > 0$, and $R_{l,k}(r) \sim r^{l+1}$, then (33.5) converges if $-l-n+l+1 > -1$, i.e. if

$$n < 2. \tag{33.8}$$

Performing the integration in (33.6) we see that (33.7) becomes $2(l+1)-n+1 > 2l+1$, which is equivalent to (33.8). The convergence condition $(2l+1)-n > -1$ must thus be imposed, i.e.

$$l > \frac{n-3}{2}. \tag{33.9}$$

The inequalities (33.8) and (33.9) are the conditions under which the integral equation (33.4) defines the required solution $R_{l,k}(r)$.

The restriction $n < 2$ imposed on the potential is clearly necessary, since, for $n > 2$, the behaviour of the radial function at small distances is determined by the nature of the potential and not by the centrifugal term $l(l+1)/r^2$, as was tacitly assumed when (32.1) was derived. For attractive potentials, if $n > 2$, the particles "fall" into the centre of the field. For $n = 2$, this also occurs, but only in the lower states of orbital angular momentum. For $n < 2$ the restriction (33.9), imposed evidently on Re (l), has the same significance as condition (32b), and so it can be considered to be satisfied.

Finally, since (33.4) is an integral equation of the well-known Volterra type, its solution will be a holomorphic function of the parameter l in the domain Re $(l) > -\frac{1}{2}$.

Remarks: Since $R_{l,k}(r)$ is a holomorphic function in the domain Re $(l) > -\frac{1}{2}$, it follows from (19.7) that the functions $F(k, l)$ and $F(-k, l)$, which represent the analytic continuation of the Jost functions in the complex l-plane, will also be holomorphic (for Re $(l) > -\frac{1}{2}$). This means that the analytical continuation of $S_l(k)$,

$$S(l, k) = e^{2i\delta(l, k)} = e^{i\pi l}\frac{F(k, l)}{F(-k, l)}, \tag{33.10}$$

will be a meromorphic function in the domain Re $(l) > -\frac{1}{2}$.

Also, since $f^*(-k^*, l^*; r)$ and $f(k, l; r)$ satisfy the same integral equation (21a), we have that

$$f^*(-k^*, l^*; r) = f(k, l; r), \tag{33.11}$$

and

$$F^*(-k, l^*) = F(k, l). \tag{33.12}$$

Hence the unitarity condition (22a) becomes

$$S(l, k)\, S^*(l^*, k^*) = 1. \tag{33.13}$$

34. For the proof, we require equation (II.15) and the similar equation with l replaced by l'. Multiplying these equations by $R_{l', k}(r)$ and $R_{l, k}(r)$ respectively, and subtracting the results, we obtain

$$R_{l',k}\frac{d^2R_{l,k}}{dr^2} - R_{l,k}\frac{d^2R_{l',k}}{dr^2} = (l-l')(l+l'+1)\frac{R_{l,k}R_{l',k}}{r^2}. \tag{34.1}$$

By integrating from 0 to ∞ we find that

$$(l-l')(l+l'+1)\int_0^\infty \frac{R_{l,k}R_{l',k}}{r^2}\,dr = \left(R_{l',k}\frac{dR_{l,k}}{dr} - R_{l,k}\frac{dR_{l',k}}{dr}\right)\Bigg|_0^\infty. \tag{34.2}$$

Since Re $(l) > -\frac{1}{2}$ and Re $(l') > -\frac{1}{2}$, it follows by (32a) that the right-hand side of (34.2) vanishes for $r = 0$, and, as $r \to \infty$, we find from (19.7) that

$$(l-l')(l+l'+1)\int_0^\infty \frac{R_{l,k}R_{l',k}}{r^2}\,dr = \frac{1}{ik}[F(k, l')F(-k, l) - F(k, l)F(-k, l')]. \tag{34.3}$$

By putting $l' = l^*$ and using (33.12), we find that

$$4k\,\mathrm{Im}\,(l)\,\mathrm{Re}\left(l+\frac{1}{2}\right)\int_0^\infty \frac{|R_{l,k}|^2}{r^2}\,dr = |F(k, l)|^2 - |F(-k, l)|^2. \tag{34.4}$$

The point l_p at which $S(l, k)$ has a pole is a zero of $F(-k, l)$, i.e. $F(-k, l_p) = 0$. Since Re $(l_p) > -\frac{1}{2}$, it follows from (34.4) that

$$\mathrm{Im}\,(l_p) > 0 \tag{34.5}$$

which shows that the poles of $S(l, k)$ are to be found only in the upper half-plane of complex l.

Remarks: Another interesting result can be obtained from (34.3), by taking the limit $l \to l'$, treating l and l' as real variables. We then obtain

$$\int_0^\infty \frac{R_{l,k}^2}{r^2}\,dr = \frac{i}{(2l+1)k} \lim_{l\to l'} \frac{F(k,l)\,F(-k,l') - F(k,l')\,F(-k,l)}{l-l'}$$

$$= -\frac{i}{(2l+1)k} F(k,l)\,F(-k,l) \frac{d}{dl} \ln\left[\frac{F(-k,l)}{F(k,l)}\right]$$

$$= \frac{1}{(2l+1)k} F(k,l)\,F(-k,l)\left[\pi - 2\frac{d\delta(l,k)}{dl}\right]. \tag{34.6}$$

Since, for real l and k, $F(k,l)\,F(-k,l) = |F(k,l)|^2$, it follows from (34.6) that $\dfrac{d\delta_l(l)}{dl} < \dfrac{\pi}{2}$, i.e.

$$\delta_{l+1}(k) - \delta_l(k) < \frac{\pi}{2}. \tag{34.7}$$

Hence, for a given real and positive value of k, the phase shifts which correspond to two consecutive physical values of the orbital angular momentum differ by less than $\pi/2$.

35. If we have $k = ib$, $b > 0$, the poles of $S(l,k)$, i.e. the zeros of $F(-k,l)$, correspond to bound states, since, as $r \to \infty$,

$$R_{l,k}(r) \sim -\frac{F(ib,l)}{2b}\,e^{-br} \to 0.$$

These poles can occur only on the real axis Re (l), since their occurrence off the real axis would imply the existence of eigenfunctions with real energies of a non-Hermitian Hamiltonian (the centrifugal term $l(l+1)r^{-2}$ is not real for complex values of l). This fact can be proved also directly by using (34.3) with $k = ib$, $l = l_p$ and $l' = l_p^*$, where l_p is a pole of $S(l,k)$, so that $F(-k,l_p) = 0$. Taking into account (34.12), one finds then that Im $(l_p) = 0$.

Similar conclusions can be arrived at for virtual states, for which $k = ib$, $b > 0$, and, as $r \to \infty$,

$$R_{l,k}(r) \sim \frac{F(-ib,l_p)}{2b}\,e^{+br} \to \infty.$$

Remember that all the above results are valid only in the range Re $(l) > -\frac{1}{2}$.

36. Multiply (II.15), written for $R_{l_p,k}(r)$, by $R_{l_p,k}(r)$, and integrate from zero to infinity. Omitting terms which vanish at $r = 0$ and at $r = \infty$, we find that

$$J(R_{l_p,k}, E, l_p) \equiv -\int_0^\infty \left(\frac{dR_{l_p,k}}{dr}\right)^2 dr - \frac{2m}{\hbar^2}\int_0^\infty V(r)\,R_{l_p,k}^2\,dr$$

$$+\frac{2mE}{\hbar^2}\int_0^\infty R_{l_p,k}^2\,dr - l_p(l_p+1)\int_0^\infty r^{-2}R_{l_p,k}^2\,dr = 0. \tag{36.1}$$

The total variation of this expression is

$$\delta J = \frac{\partial J}{\partial R_{l_p, k}}\,\delta R_{l_p, k} + \frac{\partial J}{\partial E}\,\delta E + \frac{\partial J}{\partial l_p}\,\delta l_p = 0.$$

Because the term $(\partial J/\partial R_{l_p, k})$ leads precisely to the radial equation (II.15), it will be equal to zero (cf. the variational principle for this equation). Thus

$$\frac{\partial J}{\partial E}\,\delta E + \frac{\partial J}{\partial l_p}\,\delta l_p = \left(\frac{2m}{\hbar^2}\int_0^\infty R^2_{l_p, k}\,dr\right)\delta E - \left[(2l_p+1)\int_0^\infty r^{-2}R^2_{l_p, k}\,dr\right]\delta l_p = 0,$$

i.e.

$$\frac{\delta l_p}{\delta E} = \frac{2m}{\hbar^2}\,\frac{1}{2l_p+1}\,\frac{\int_0^\infty R^2_{l_p, k}\,dr}{\int_0^\infty r^{-2}R^2_{l_p, k}\,dr} > 0. \tag{36.2}$$

Thus, as the energy increases from negative values to positive ones, the poles of $S(l, k)$ move as follows: for $E < 0$ the poles are on the real axis ($\mathrm{Im}\,(l_p) = 0$) and move to the right as the energy increases; this displacement has an upper bound, since there cannot exist bound states for an arbitrarily large centrifugal barrier. For $E \geqslant 0$ all the poles are in the upper half-plane of l (Fig. X.8).

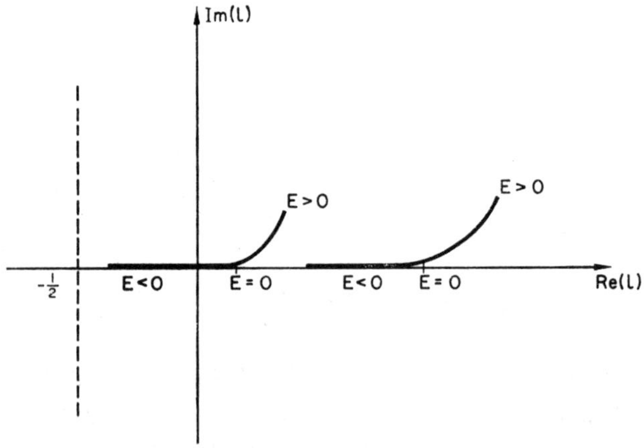

FIG. X.8.

These conclusions are of course valid only in the range $\mathrm{Re}\,(l) > -\frac{1}{2}$.

37. We have seen that, for negative energies, the motion of the poles along the axis Re (l) has an upper limit. Restrictions on the motion of the poles exist also for positive energies, although in this case l_p is complex. In order to study the trajectory of the poles in the range

$E > 0$, we formally suppose that r is a pure imaginary quantity, $r = iy$, $y > 0$. If l_p is a pole of $S(l, k)$, so that $F(-k, l_p) = 0$, then, in the asymptotic region, in accordance with (19.7), the function

$$R_{l_p, k}(r) = R_{l_p, k}(iy) = \phi_{l_p, k}(y)$$

say, becomes, as $y \to \infty$,

$$\phi_{l_p, k}(y) \sim \frac{F(k, l_p)}{2ik} e^{-ky}. \tag{37.1}$$

Now let the radial equations (II.15) for $\phi_{l_p, k}(y)$ and for $\phi^*_{l_p, k}(y)$, be multiplied by $\phi^*_{l_p, k}(y)$ and by $\phi_{l_p, k}(y)$, respectively and let the difference be integrated over y from zero to infinity. We find then that

$$\int_0^\infty \left[\phi^*_{l_p, k}\frac{d^2\phi_{l_p, k}}{dy^2} - \phi_{l_p, k}\frac{d^2\phi^*_{l_p, k}}{dy^2}\right] dy - \left[\left(l_p+\frac{1}{2}\right)^2 - \left(l_p^*+\frac{1}{2}\right)^2\right] \times$$

$$\times \int_0^\infty y^{-2}|\phi_{l_p, k}|^2\, dy + 2i\int_0^\infty |\phi_{l_p, k}|^2 \operatorname{Im}[U(iy)]\, dy = 0. \tag{37.2}$$

We suppose, of course, that the potential is such that $U(iy)$ has a meaning. The first integral on the left-hand side of (37.2) vanishes, as can be seen by integrating by parts using (37.1) and the condition at the origin (22a). Thus (37.2) becomes

$$2 \operatorname{Im}(l_p) \operatorname{Re}\left(l_p+\frac{1}{2}\right) = \frac{\int_0^\infty |\phi_{l_p, k}|^2 \operatorname{Im}[U(iy)]\, dy}{\int_0^\infty y^{-2}|\phi_{l_p, k}|^2\, dy} \tag{37.3}$$

If we suppose that $\operatorname{Im}[U(iy)] < My^{-2}$ for some constant M, it then follows that

$$2 \operatorname{Im}(l_p) \operatorname{Re}(l_p+\tfrac{1}{2}) < M, \tag{37.4}$$

which means that the product of the real and the imaginary parts of l_p is bounded, i.e. for $E > 0$ the poles of $S(l, k)$ cannot move away to infinity. As the energy increases, the poles either leave the range $\operatorname{Re}(l) > -\frac{1}{2}$, or come asymptotically close to the straight line $\operatorname{Re}(l) = -\frac{1}{2}$ while maintaining the inequality (37.4) (see Fig. X.9).

38. Let us suppose that the energy E is close to that of a bound state or of a resonance. This means that the scattering amplitude $A(E, \theta)$ will have a very large value, and that, in accordance with (X.31), its behaviour will be dominated by one of the Regge poles. Denoting the position of this pole by $l_n = \alpha(E)$, say, and the corresponding $\beta_n(E)$ by $\beta(E)$, the scattering amplitude is given approximately by the expression

$$A(E, \theta) \approx \frac{\beta(E)}{\sin \pi\alpha(E)} P_{\alpha(E)}(-z), \quad z = \cos\theta. \tag{38.1}$$

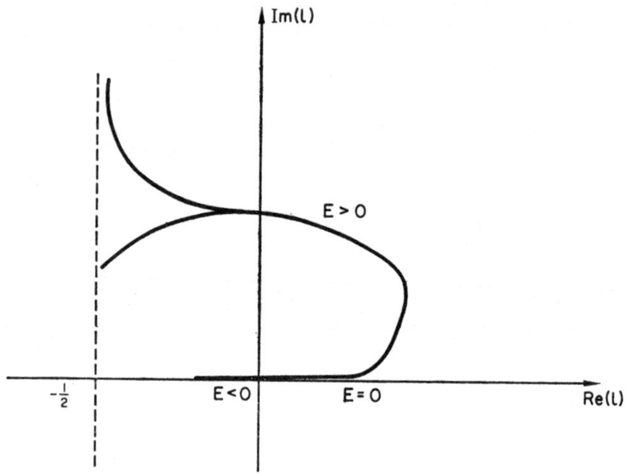

FIG. X.9.

Let us now find the projection of this expression onto the partial wave having orbital quantum number $l = 0, 1, 2, \ldots$. From (X.19) it follows that

$$A_l(E) = \tfrac{1}{2} \int_{-1}^{1} A(E, \theta)\, P_l(z)\, dz. \tag{38.2}$$

Using the formula

$$\frac{1}{2} \int_{-1}^{1} P_l(z)\, P_\alpha(-z)\, dz = \frac{1}{\pi} \frac{\sin \pi\alpha}{(\alpha - l)(\alpha + l + 1)} \tag{38.3}$$

we find from (38.1) and (38.2) that

$$A_l(E) \approx \frac{1}{\pi} \frac{\beta(E)}{(\alpha(E) - l)(\alpha(E) + l + 1)}, \qquad l = 0, 1, 2, \ldots \tag{38.4}$$

From this it can be seen immediately that for the energy E for which $\alpha(E) = 1$ (which, in accordance with the results of problem 35, implies that $E < 0$) the partial amplitude $A_l(E)$ has a pole. Since a pole of $A_l(E)$, with $E < 0$, represents a bound state with orbital angular momentum l, it follows that a single Regge pole trajectory can determine the appearance of bound states with various orbital angular momenta, since such a bound state occurs whenever the Regge pole trajectory passes through one of the "physical" values $l = 0, 1, 2, \ldots$

Let us examine now the relation between Regge poles and resonances. Suppose that, for a particular energy $E_r > 0$, the real part of the pole, $\alpha(E)$, is very close to a physical value of l. $\alpha(E)$ can then be expanded in a Taylor's series about the value $E = E_r$,

$$\alpha(E) \approx 1 + \left[\frac{d\,\mathrm{Re}\,(\alpha)}{dE}\right]_{E=E_r} (E - E_r) + i[\mathrm{Im}\,(\alpha)]_{E=E_r}.$$

By substituting this expression into (38.4) we find that

$$A_l(E) \approx \frac{\beta(E_r)}{\pi\{2l+1+i[\text{Im}\,(\alpha)]_{E=E_r}\}} \cdot \frac{1}{A\left(E-E_r+i\dfrac{\Gamma}{2}\right)} \tag{38.5}$$

where

$$A = \left[\frac{d\,\text{Re}\,(\alpha)}{dE}\right]_{E=E_r}, \quad \frac{\Gamma}{2} = \frac{[\text{Im}\,(\alpha)]_{E=E_r}}{(d\,\text{Re}\,(\alpha)/dE)_{E=E_r}}. \tag{38.6}$$

If $[\text{Im}\,(\alpha)]_{E=E_r} \ll 1$, the partial amplitude $A_l(E)$ shows a typical resonance behaviour for $E \sim E_r$, the half-width $\Gamma/2$ of the resonance being given by (38.6). Thus whenever the trajectory of a Regge pole, for $E > 0$, passes close to a physical value $l = 0, 1, 2, \ldots$, it produces a resonance in the corresponding partial amplitude. Evidently a single Regge pole can produce resonances in more than one partial wave. Thus a resonance in the amplitude of a partial wave is related directly to the behaviour of the amplitudes of other partial waves; a Regge pole of given energy E affects in theory all the partial amplitudes, but its influence is only slight on partial amplitudes whose value of l differs considerably from Re $[\alpha(E)]$.

39. Let us see first what happens when we operate on the base vectors with the operators $(E_a - H_0 \pm i\eta)^{-1}$.

If $|\Phi_E\rangle$ is an eigenvector with eigenvalue E of H_0, then

$$(E_a - H_0 \pm i\eta)\,|\Phi_E\rangle = (E_a - E \pm i\eta)\,|\Phi_E\rangle. \tag{39.1}$$

By operating on the left with $(E_a - H_0 \pm i\eta)^{-1}$, we obtain

$$(E_a - H_0 \pm i\eta)^{-1}\,|\Phi_E\rangle = (E_a - E \pm i\eta)^{-1}\,|\Phi_E\rangle \tag{39.2}$$

whence it follows that the eigenvector $|\Phi_E\rangle$ with eigenvalue E of H_0 is also an eigenvector of the operators $(E_a - H_0 \pm i\eta)^{-1}$, with the eigenvalues $(E_0 - E \pm i\eta)^{-1}$. From (39.1) and (39.2) it can easily be shown that the converse is also true.

Since the eigenvalues and eigenvectors of H_0 are known by hypothesis (see 1), it follows that they are known for the operators $(E_a - H_0 \pm i\eta)^{-1}$ too.

40. The differential equation (40a), together with the initial condition $|\psi_\eta(-\infty)\rangle = |\phi\rangle$, is equivalent to the integral equation

$$|\psi_\eta(t)\rangle = |\Phi\rangle - \frac{i}{\hbar}\int_{-\infty}^{t} e^{-\frac{\eta}{\hbar}|t'|} H_I'(t')\,|\psi_\eta(t')\rangle\,dt'. \tag{40.1}$$

In particular, using (V.6) and (V.8), we have that

$$|\psi_\eta(0)\rangle = |\Phi\rangle - \frac{i}{\hbar}\int_{-\infty}^{0} e^{+\frac{\eta}{\hbar}t'} e^{\frac{i}{\hbar}H_0t'} H' e^{-\frac{i}{\hbar}H_0t'}\,|\psi_\eta(t')\rangle\,dt'. \tag{40.2}$$

Let us substitute for $|\psi_\eta(t')\rangle$ in (40.2) its value given by (40.1), then

$$|\psi_\eta(0)\rangle = |\Phi\rangle + \frac{1}{E-H_0+i\eta} H'|\Phi\rangle$$

$$+\left(\frac{i}{\hbar}\right)^2 e^{\frac{\eta}{\hbar}t'} e^{\frac{i}{\hbar}H_0 t'} H' e^{-\frac{i}{\hbar}H_0 t'} dt' \int_{-\infty}^{t'} e^{\frac{\eta}{\hbar}t''} e^{\frac{i}{\hbar}H_0 t''} H' e^{-\frac{i}{\hbar}H_0 t''} |\psi_\eta(t'')\rangle \, dt''. \tag{40.3}$$

By substituting (40.1) once more into (40.3), we find that

$$|\psi_\eta(0)\rangle = |\Phi\rangle + \frac{1}{E-H_0+i\eta} H'|\Phi\rangle + \frac{1}{E-H_0+i\eta} H' \frac{1}{E-H_0+i\eta} H'|\Phi\rangle + \ldots$$

$$= |\Phi\rangle + \frac{1}{E-H_0+i\eta} H'|\psi_\eta(0)\rangle. \tag{40.4}$$

By comparing (40.4) with the Lippmann–Schwinger equation we see that $\lim_{\eta\to 0} |\psi_\eta(0)\rangle = |\psi^+\rangle$.

In other words, the time-dependent solution $|\psi_\eta(t)\rangle$ enables us to calculate $|\psi^+\rangle$. The introduction of the "adiabatic" potential $H'_{I\eta}(t) = \exp\left(-\frac{\eta}{\hbar}|t|\right) H'_I(t)$ enables us to avoid some divergences in the calculations, since we take the limit $\eta \to \pm 0$ only after the calculations have been otherwise completed.

41. By multiplying (X.35) on the left by the bra vector $\langle\xi|$, and introducing the integral (41a) between H' and $|\psi^\pm\rangle$, we find that

$$\langle\xi|\psi_a^\pm\rangle = \langle\xi|\Phi_a\rangle + \int \lim_{\eta\to+0} \langle\xi|(E_a - H_0 \pm i\eta)^{-1} H'|\xi'\rangle\langle\xi'|\psi_a^\pm\rangle \, d\xi'. \tag{41.1}$$

The projections $\langle\xi|\psi_a^\pm\rangle$ and $\langle\xi|\Phi_a\rangle$ are the wavefunctions in the $\{\xi\}$ representation, and we denote them by $\psi_a^\pm(\xi)$ and $\Phi_a(\xi)$ respectively. If we introduce the notation

$$\lim_{\eta\to+0} \langle\xi|(E_a - H_0 \pm i\eta)^{-1} H'|\xi'\rangle = K^\pm(\xi, \xi'; E_a), \tag{41.2}$$

then (41.1) takes the form of the well-known Fredholm-type integral equation

$$\psi_a^\pm(\xi) = \Phi_a(\xi) + \int K^\pm(\xi, \xi'; E_a)\, \psi_a^\pm(\xi')\, d\xi'. \tag{41.3}$$

42. A possible complete set of eigenstates of H_0 consists of the plane wave states of momentum $\hbar\mathbf{k}$ which, in coordinate representation, can be written as

$$\langle\mathbf{r}|\Phi_k\rangle = e^{i\mathbf{k}\cdot\mathbf{r}}. \tag{42.1}$$

On account of the completeness of this system (normalized to $(2\pi)^3\,\delta(\mathbf{r}-\mathbf{r}')$), and of (39.2), we can write

$$\frac{1}{E_k - H_0 \pm i\eta} = \frac{1}{(2\pi)^3} \int |\Phi_{k'}\rangle \frac{d\mathbf{k}'}{E_k - E_{k'} \pm i\eta} \langle\Phi_{k'}|. \tag{42.2}$$

The matrix elements of the operator $(E_k - H_0 \pm i\eta)^{-1}$ in coordinate representation are then

$$\left\langle \mathbf{r}' \left| \frac{1}{E_k + (\hbar^2/2m)\nabla^2 \pm i\eta} \right| \mathbf{r}'' \right\rangle = \int \frac{\langle \mathbf{r}' | \Phi_{k'} \rangle \langle \Phi_{k'} | \mathbf{r}'' \rangle}{E_k - E_{k'} \pm i\eta} \, d\mathbf{k}'$$

$$= \frac{2m}{(2\pi)^3 \hbar^2} \int \frac{\exp\,[i\mathbf{k}' \cdot (\mathbf{r}' - \mathbf{r}'')]}{k^2 - k'^2 \pm i\varepsilon} \, d\mathbf{k}', \tag{42.3}$$

where $\varepsilon = (2m/\hbar^2)\eta$.

Performing the integration over the angular variables, we obtain

$$\left\langle \mathbf{r}' \left| \frac{1}{E_k + (\hbar^2/2m)\nabla^2 \pm i\eta} \right| \mathbf{r}'' \right\rangle = \frac{2m}{\hbar^2} \frac{1}{4\pi^2 i} \frac{1}{|\mathbf{r}' - \mathbf{r}''|} \int_{-\infty}^{+\infty} \frac{k' \exp\,[ik' |\mathbf{r}' - \mathbf{r}''|]}{(k^2 \pm i\varepsilon) - k'^2} \, dk'. \tag{42.4}$$

If we compare this expression with (1.7)–(1.9) we arrive at the conclusion that

$$\lim_{\eta \to 0} \left\langle \mathbf{r}' \left| \frac{1}{E_k + (\hbar^2/2m)\nabla^2 \pm i\eta} \right| \mathbf{r}'' \right\rangle$$

$$= -\frac{2m}{\hbar^2} \frac{1}{4\pi} \frac{\exp\,(\pm ik\,|\mathbf{r}' - \mathbf{r}''|)}{|\mathbf{r}' - \mathbf{r}''|} \equiv \frac{2m}{\hbar^2} G^{\pm}(|\mathbf{r}' - \mathbf{r}''|). \tag{42.5}$$

If we now use (42.1), the Lippman–Schwinger equation (42a) can be written as

$$|\psi_k^+\rangle = |\Phi_k\rangle + \lim_{\eta \to 0} \frac{1}{(2\pi)^3} \int |\Phi_{k'}\rangle \frac{d\mathbf{k}'}{E_k - E_{k'} + i\eta} \langle \Phi_{k'} | H' | \psi_{k'}^+ \rangle. \tag{42.6}$$

Since in our case $H' = V(\mathbf{r})$, on account of the completeness of the set of coordinate operator eigenvectors, we can write

$$\langle \Phi_{k'} | H' | \psi_k^+ \rangle = \iint \langle \Phi_{k'} | \mathbf{r}'' \rangle \langle \mathbf{r}'' | V(\mathbf{r}) | \mathbf{r}' \rangle \langle \mathbf{r}' | \psi_k^+ \rangle \, d\mathbf{r}' \, d\mathbf{r}''. \tag{42.7}$$

The function $V(\mathbf{r})$ depends only on the coordinates, so that its matrix elements in coord nate representation are simply

$$\langle \mathbf{r}'' | V(\mathbf{r}) | \mathbf{r}' \rangle = V(\mathbf{r}'')\, \delta(\mathbf{r}'' - \mathbf{r}').$$

By performing the integration over $\mathbf{r}''$, we obtain from (42.7):

$$\langle \Phi_{k'} | H' | \psi_k^+ \rangle = \int \langle \Phi_k | \mathbf{r} \rangle V(\mathbf{r}') \langle \mathbf{r}' | \psi_k^+ \rangle \, d\mathbf{r}'. \tag{42.8}$$

Now $\langle \mathbf{r} | \Phi_k \rangle = \exp\,(i\mathbf{k} \cdot \mathbf{r})$, and $\langle \mathbf{r} | \psi_k^+ \rangle$ is the wavefunction of the "in" state $|\psi_k^+\rangle$ in coordinate representation (let us denote it by $\psi_{\mathbf{k}}^+(\mathbf{r})$). Thus (42.8) becomes

$$\langle \phi_{k'} | H' | \psi_k^+ \rangle = \int e^{-i\mathbf{k}' \cdot \mathbf{r}'} V(\mathbf{r}')\, \psi_{\mathbf{k}}^+(\mathbf{r}') \, d\mathbf{r}'. \tag{42.9}$$

Substituting this expression into (42.6), and multiplying by the bra vector $\langle \mathbf{r}|$, we obtain

$$\psi_{\mathbf{k}}^+(\mathbf{r}) = e^{i\mathbf{k} \cdot \mathbf{r}} + \frac{2m}{\hbar^2} \lim_{\varepsilon \to 0} \frac{1}{(2\pi)^3} \iint \frac{\exp i\mathbf{k}' \cdot (r - r')}{k^2 - k'^2 + i\varepsilon} V(\mathbf{r}')\, \psi_{\mathbf{k}}^+(\mathbf{r}') \, d\mathbf{k}' \, d\mathbf{r}'.$$

The integral over k' is the same as the one in (42.3), and hence, using (42.5) and denoting $(2m/\hbar^2)\, V(\mathbf{r})$ by $U(\mathbf{r})$, we have

$$\psi_{\mathbf{k}}^{+}(\mathbf{r}) = e^{i\mathbf{k}\cdot\mathbf{r}} - \frac{1}{4\pi}\int \frac{\exp(ik|\mathbf{r}-\mathbf{r}'|)}{|\mathbf{r}-\mathbf{r}'|}\, U(\mathbf{r}')\,\psi_{\mathbf{k}}^{+}(\mathbf{r}')\, d\mathbf{r}'$$

which is what we had to demonstrate.

43. Multiplying (X.35) on the left by H', we have

$$H'|\psi_a^{\pm}\rangle = H'|\Phi_a\rangle + H' \lim_{\eta\to+0} \frac{1}{E_a - H_0 \pm i\eta} H'|\psi_a^{\pm}\rangle. \tag{43.1}$$

With this relation we can write further

$$|\psi_a^{\pm}\rangle = |\Phi_a\rangle + \lim_{\eta\to+0} \frac{1}{E_a - H_0 \pm i\eta} H'|\Phi_a\rangle$$

$$+ \lim_{\eta\to+0} \frac{1}{E_a - H_0 \pm i\eta} H' \lim_{\eta\to+0} \frac{1}{E_a - H_0 \pm i\eta} H'|\psi_a^{\pm}\rangle.$$

Multiplying on the left by $\lim\limits_{\eta\to+0}(E_a - H_0 \pm i\eta)$, and reordering the terms, we obtain

$$H' \lim_{\eta\to+0} \frac{1}{E_a - H_0 \pm i\eta} H'|\psi_a^{\pm}\rangle = \lim_{\eta\to+0} (E_a - H_0 \pm i\eta)|\psi_a^{\pm}\rangle$$

$$- \lim_{\eta\to+0} (E_a - H_0 \pm i\eta)|\Phi_a\rangle - H'|\Phi_a\rangle$$

$$= \lim_{\eta\to+0} (E_a - H_0 \pm i\eta)|\psi_a^{\pm}\rangle - H'|\Phi_a\rangle \mp i\eta|\Phi_a\rangle.$$

Substituting in (43.1) we find that

$$H'|\psi_a^{\pm}\rangle - \lim_{\eta\to+0} (E_a - H_0 \pm i\eta)|\psi_a^{\pm}\rangle = \mp i\eta|\Phi_a\rangle.$$

Since $H = H_0 + H'$, after operating with the inverse operator $\lim\limits_{\eta\to+0}(E_a - H \pm i\eta)^{-1}$ we obtain

$$|\psi_a^{\pm}\rangle = \pm \lim_{\eta\to+0} \frac{i\eta}{E_a - H \pm i\eta}|\Phi_a\rangle. \tag{43.2}$$

Finally, since

$$\pm\frac{i\eta}{E_a - H \pm i\eta} = 1 - \frac{E_a - H}{E_a - H \pm i\eta}$$

and $H|\Phi_a\rangle = (H_0 + H')|\Phi_a\rangle = E_a|\Phi_a\rangle + H'|\Phi_a\rangle$, the required expression (43a) follows immediately from (43.2).

Remarks: Since

$$\frac{1}{E_a-H\pm i\eta}=\frac{1}{E_a-H_0\pm i\eta-H'},$$

we may use the operator identity

$$\frac{1}{A+B}=\frac{1}{A}\left(1-B\frac{1}{A+B}\right) \tag{43.3}$$

by taking $A = E_a-H_0\pm i\eta$ and $B = -H'$, to obtain

$$\frac{1}{E_a-H\pm i\eta}=\frac{1}{E_a-H_0\pm i\eta}+\frac{1}{E_a-H_0\pm i\eta}H'\frac{1}{E_a-H\pm i\eta}. \tag{43.4}$$

Hence, by successive approximation, one finds that

$$\frac{1}{E_a-H\pm i\eta}=\frac{1}{E_a-H_0\pm i\eta}+\frac{1}{E_a-H_0\pm i\eta}H'\frac{1}{E_a-H_0\pm i\eta}$$
$$+\frac{1}{E_a-H_0\pm i\eta}H'\frac{1}{E_a-H_0\pm i\eta}H'\frac{1}{E_a-H_0\pm i\eta}+\dots. \tag{43.5}$$

Thus we obtain for the states $|\psi_a^\pm\rangle$ the Born series

$$|\psi_a^\pm\rangle=|\Phi_a\rangle+\lim_{\eta\to+0}\frac{1}{E_a-H_0\pm i\eta}H'|\Phi_a\rangle$$
$$+\lim_{\eta\to+0}\frac{1}{E_a-H_0\pm i\eta}H'\frac{1}{E_a-H_0\pm i\eta}|\Phi_a\rangle+\dots. \tag{43.6}$$

44. Since the operators H and H' are Hermitian, by (43a) we have that

$$\langle\psi_a^+|\psi_b^+\rangle=\langle\Phi_a|\psi_b^+\rangle+\lim_{\eta\to+0}\langle\Phi_a|H'\frac{1}{E_a-H-i\eta}|\psi_b^+\rangle.$$

Further, since

$$\frac{1}{E_a-H-i\eta}|\psi_b^+\rangle=\frac{1}{E_a-E_b-i\eta}|\psi_b^+\rangle,$$

by using the Lippman–Schwinger equation we can write

$$\langle\psi_a^+|\psi_b^+\rangle=\langle\Phi_a|\Phi_b\rangle+\lim_{\eta\to+0}\langle\Phi_a|\frac{1}{E_b-H_0\pm i\eta}H'|\psi_b^+\rangle$$
$$+\lim_{\eta\to+0}\frac{1}{E_a-E_b-i\eta}\langle\Phi_a|H'|\psi_b^+\rangle.$$

Hence, using (39.2), we obtain

$$\langle\psi_a^+|\psi_b^+\rangle=\langle\Phi_a|\Phi_b\rangle+\lim_{\eta\to+0}\frac{1}{E_b-E_a+i\eta}\langle\Phi_a|H'|\psi_b^+\rangle$$
$$+\lim_{\eta\to+0}\frac{1}{E_a-E_b-i\eta}\langle\Phi_a|H'|\psi_b^+\rangle.$$

In the limit $\eta \to +0$ we have

$$\langle \psi_a^+ | \psi_b^+ \rangle = \langle \Phi_a | \Phi_b \rangle. \tag{44.1}$$

Since the eigenvectors which belong to the continuous spectrum of the reference system Hamiltonian are taken to be orthonormalized to a Dirac δ-function, relation (44a) follows immediately from (44.1).

Similarly we obtain

$$\langle \psi_a^- | \psi_b^- \rangle = \langle \Phi_a | \Phi_b \rangle \tag{44.2}$$

whence

$$\langle \psi_a^- | \psi_b^- \rangle = \delta_{ab}. \tag{44.3}$$

Remarks: If the spectrum of H, besides having a continuous part ($E \geqslant 0$), has also a discrete part (which by hypothesis is in the range $E < 0$), then the corresponding bound state $|\psi_\alpha^L\rangle$ will be orthogonal to the states $|\psi_a^\pm\rangle$, i.e.

$$\langle \psi_a^+ | \psi_\alpha^L \rangle = \langle \psi_a^- | \psi_\alpha^L \rangle = 0. \tag{44.4}$$

It is assumed, and in simple cases it can be shown explicitly, that the "in" states together with the bound states, and the "out" states together with the bound states, form complete sets,† so that

$$\begin{aligned} \int_0^\infty |\psi_E^+\rangle\, dE \langle \psi_E^+| + \sum_\alpha |\psi_\alpha^L\rangle \langle \psi_\alpha^L| &= 1, \\ \int_0^\infty |\psi_E^-\rangle\, dE \langle \psi_E^-| + \sum_\alpha |\psi_\alpha^L\rangle \langle \psi_\alpha^L| &= 1. \end{aligned} \tag{44.5}$$

45. By multiplying equation (X.35) by $\langle \Phi_b |$, and using the Hermitian property of H_0 and the relation (39.2), (45a) follows immediately.

Further, we have that

$$\langle \psi_\alpha^L | H' | \Phi_b \rangle = \langle \psi_\alpha^L | (H - H_0) | \Phi_b \rangle = \langle \psi_\alpha^L | H | \Phi_b \rangle - \langle \psi_\alpha^L | H_0 | \Phi_b \rangle = (E_\alpha - E_b) \langle \psi_\alpha^L | \Phi_b \rangle. \tag{45.1}$$

In accordance with the hypotheses made concerning the spectra of the operators H and H_0, we have $E_\alpha \neq E_b$, and thus, by dividing (45.1) by $E_\alpha - E_b$, (45b) follows directly.

† For a rigorous discussion of complete sets of the eigenfunctions of the type considered here, see chapters XI and XII of vol. II of the reference given in the footnote on p. 204.

CHAPTER XI

Atoms and Molecules

THE basic problem in the study of many-electron atoms is that of solving the Schrödinger equation with the following Hamiltonian:

$$H = \sum_i \left(-\frac{\hbar^2}{2m} \nabla_i^2 - \frac{Ze^2}{r_i} \right) + \sum_{i<j}\sum \frac{e^2}{r_{ij}} + W, \tag{XI.1}$$

in which W is the potential energy operator of the spin-dependent interactions (spin–orbit, spin–spin and hyperfine interactions). To find solutions, approximation methods, such as perturbation theory, the variational method (see Chapter VIII), the self-consistent field approximation of Hartree and Fock, the statistical model of Thomas and Fermi, etc., have to be used. In the self-consistent field approximation, the actual Coulomb interaction between the electrons is replaced by a mean field having spherical symmetry, together with small residual interactions. If W is much smaller than these residual interactions it can be neglected to a first approximation. This procedure leads to the *Russell–Saunders coupling* scheme for the electronic angular momenta $\mathbf{L} = \sum_i \mathbf{l}_i$, $\mathbf{S} = \sum_i \mathbf{s}_i$, $\mathbf{J} = \mathbf{L}+\mathbf{S}$. If the opposite is the case, then the angular momenta have to be coupled in accordance with the "j–j" *coupling* scheme: $\mathbf{j}_i = \mathbf{l}_i+\mathbf{s}_i$, $\mathbf{J} = \sum_i \mathbf{j}_i$. In the Fermi–Thomas statistical model, which is used for the study of heavy atoms, the point charges of the electrons are replaced by a continuous charge distribution with charge density $-e\varrho(r)$. The electrostatic potential $\Phi(r)$ due to the nucleus and to the distribution of electronic charge satisfies the classical Poisson equation

$$\nabla^2\Phi = 4\pi e\varrho \tag{XI.2}$$

with the boundary conditions

$$\lim_{r\to 0} r\Phi = Ze, \qquad \lim_{r\to\infty} \Phi = 0. \tag{XI.3}$$

The potential Φ can be expressed in terms of a dimensionless function ϕ through the relation

$$\Phi(r) = \frac{Ze}{b}\frac{\phi(x)}{x} \tag{XI.4}$$

where $x = r/b$, $b = 0{\cdot}885Z^{-1/3}a$, $a = \hbar^2/me^2$, and $\phi(x)$ satisfies the Thomas–Fermi equation

$$\frac{d^2\phi}{dx^2} = \frac{1}{\sqrt{x}}\phi^{3/2} \tag{XI.5}$$

with the boundary conditions $\phi(0) = 1$, $\phi(\infty) = 0$.

Problems

1. Find the corrections to the energy levels of the hydrogen atom due to spin–orbit coupling:

$$H' = \frac{\hbar^2 e^2}{2m^2c^2}\frac{1}{r^3}\mathbf{l}\cdot\mathbf{s}. \tag{1a}$$

2. Find the relativistic correction to the energy level E_n of a hydrogen atom, to order v^2/c^2 in its relativistic Hamiltonian.

3. Find the energy spectrum of an atom placed in a constant, homogenous magnetic field **H**, if the potential energy of the interaction between the electrons and the field is much smaller than the spin–orbit coupling energy ("Zeeman effect").

4. Find the energy spectrum of an atom placed in a constant, homogenous magnetic field **H**, if the potential energy of the interaction of the electrons with the field is much greater than the spin–orbit coupling energy ("Paschen–Back effect"). Show that, in states in which the total orbital angular momentum $\mathbf{L} = \sum_k \mathbf{l}_k$ and the total spin angular momentum $\mathbf{S} = \sum_k \mathbf{s}_k$ vanish, the atom has diamagnetic properties.

5. Find the energy spectrum of a hydrogen atom placed in a magnetic field **H** if the energy of interaction of the electron with the magnetic field is of the same order of magnitude as the spin–orbit coupling energy.

6. Two identical particles move in a potential well $V(r)$. Knowing the solutions of the Schrödinger equation for a single particle in this potential, solve the problem of the two particles, if their mutual interaction can be treated as a first-order perturbation.

7. Suppose that, at time $t = 0$, the first particle of the preceding problem is found to be in the rth (unperturbed) state and the second in the sth (unperturbed) state. At what time t will the occupation of the states be reversed?

8. Remembering that the Thomas–Fermi function $\phi(x)$ of a neutral atom has an exponential behaviour, determine by means of a trial function $\phi(x) = e^{-\alpha x}$ used in a variational principle, the best approximation of this type to the solution of equation (XI.5).

9. From the results of the preceding problem, find the nuclear charge Z' of an imaginary hydrogen-like atom which is such that its ground state energy is equal to the energy of an electron in the K shell of a Z-electron atom.

10. In the study of the vibrational spectra of molecules, the harmonic oscillator potential $V(x) = m\omega^2x^2/2$ is found to be valid only to a first approximation. More accurate results can be obtained by using the potential.

$$V(x) = V_0(1-e^{-\mu x})^2 \tag{10a}$$

proposed by Morse.†

Find the resulting energy levels of the molecular oscillator.

11. In the study of the rotational and vibrational spectra of diatomic molecules, it is found that the potential energy of interaction between the two atoms of the molecule can be written as

$$V(\varrho) = V_0\left(\frac{1}{\varrho^2}-\frac{2}{\varrho}\right), \quad V_0 > 0, \tag{11a}$$

where $\varrho = r/a$ is the distance between the two nuclei (Fig. XI.1). Find the corresponding energy levels.

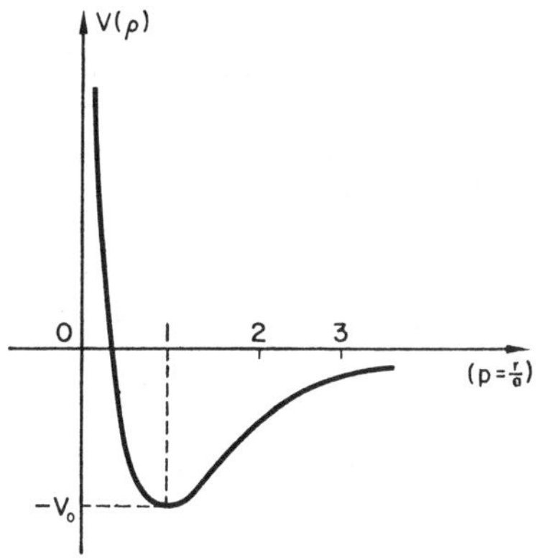

FIG. XI.1.

Solutions

1. We use perturbation theory. As unperturbed wavefunctions we could take the simultaneous eigenfunctions of the operators $\mathbf{l}^2$, $\mathbf{s}^2$, l_z and s_z. However, we would then be obliged to use the perturbation theory for degenerate levels. We can avoid this difficulty in the same way as in problem 18 of Chapter VI, i.e. by using the simultaneous eigenfunctions of the operators $\mathbf{l}^2$, $\mathbf{j}^2$ and j_z as unperturbed wavefunctions. Since $\mathbf{l}^2$, $\mathbf{j}^2$ and j_z commute with H', it follows that only states with the same quantum numbers l, j, m_j (but with different principal quantum numbers n) can contribute to the perturbation of an unperturbed state

† P. M. Morse, *Phys. Rev.* **34**, 57 (1928).

of given l, j, m_j. Thus we can apply non-degenerate perturbation theory to these states. To obtain the required answer it is sufficient to calculate the mean value

$$\langle H' \rangle = \left\langle \frac{\hbar^2 e^2}{2m^2c^2} \frac{1}{r^3} \mathbf{l}.\mathbf{s} \right\rangle \tag{1.1}$$

in a state with given l and j. Now $\langle \mathbf{j}^2 \rangle = j(j+1) = l(l+1)+s(s+1)+2\langle \mathbf{l}.\mathbf{s}\rangle$, whence

$$\langle H' \rangle = \frac{\hbar^2 e^2}{4m^2c^2} [j(j+1)-l(l+1)-s(s+1)] \left\langle \frac{1}{r^3} \right\rangle. \tag{1.2}$$

But

$$\left\langle \frac{1}{r^3} \right\rangle = \frac{1}{n^3(l+1)(l+\frac{1}{2})l} \left(\frac{me^2}{\hbar^2} \right)^3, \tag{III.7.6}$$

so that

$$\langle H' \rangle = \frac{me^4}{\hbar^2} \left(\frac{e^2}{\hbar c} \right)^2 \frac{j(j+1)-l(l+1)-s(s+1)}{4n^3 l(l+\frac{1}{2})(l+1)}. \tag{1.3}$$

Remembering that $s = \frac{1}{2}$ and $j = l \pm \frac{1}{2}$, we find after some simple calculation that

$$E^{(1)}_{nlj} = \langle H' \rangle = \frac{me^4}{\hbar^2} \left(\frac{e^2}{\hbar c} \right)^2 \frac{1}{2n^3} \left(\frac{1}{l+\frac{1}{2}} - \frac{1}{j+\frac{1}{2}} \right). \tag{1.4}$$

2. The relativistic Hamiltonian of a particle of rest mass m_0 and electric charge $-|e|$ moving in a Coulomb field $|e|/r$ is

$$H = c\sqrt{p^2+m_0^2c^2} - \frac{e^2}{r}. \tag{2.1}$$

If $v \ll c$, then, to the required approximation,

$$c\sqrt{p^2+m_0^2c^2} = m_0c^2\left(1+\frac{p^2}{m_0^2c^2}\right)^{1/2} = m_0c^2 + \frac{p^2}{2m_0} - \frac{p^4}{8m_0^3c^2}$$

and hence

$$H - m_0c^2 = \left(\frac{p^2}{2m_0} - \frac{e^2}{r}\right) - \frac{p^4}{8m_0^3c^2} = H_0 + H'. \tag{2.2}$$

If we take m_0 to be the reduced mass of the electron–proton system, and put $\mathbf{p} = -i\hbar \triangle$, H_0 becomes the non-relativistic Hamiltonian operator of the hydrogen atom, and H' can be treated as a perturbation.

Thus, using the eigenfunction equation $H_0\psi_{nlm} = E_n\psi_{nlm}$ and the relations (6.5) and (7.5) of Chapter III the relativistic correction to the energy of the state (nlm) of the hydrogen atom is, to the required approximation,

$$\Delta E_1 = \int \psi^*_{nlm} H' \psi_{nlm}\, d\mathbf{r} = -\frac{1}{8m_0^3c^2}\int \psi^*_{nlm} p^4 \psi_{nlm}\, d\mathbf{r}$$

$$= \frac{1}{2m_0c^2}\int \psi^*_{nlm}\left(E_n+\frac{e^2}{r}\right)^2 \psi_{nlm}\, d\mathbf{r} = \frac{3E_n^2}{2m_0c^2} - \frac{\left(\frac{m_0e^4}{\hbar^2}\right)}{n^3c^2m_0(2l+1)}$$

$$= \left[\frac{2}{8n^4} - \frac{1}{(2l+1)n^3}\right]\frac{m_0e^4}{\hbar^2}\left(\frac{e^2}{\hbar c}\right)^2. \tag{2.3}$$

Remarks: As was shown in the preceding problem, because of spin-orbit coupling the energy of the (nlj) state must be corrected by an amount

$$\Delta E_2 = \frac{m_0e^4}{\hbar^2}\left(\frac{e^2}{\hbar c}\right)^2 \frac{1}{2n^3}\left(\frac{1}{l+\frac{1}{2}} - \frac{1}{j+\frac{1}{2}}\right). \tag{2.4}$$

Hence, combining the correction (2.4) due to spin–orbit coupling with the first order relativistic correction (2.3), the energy of the hydrogen atom in the state (nj) becomes

$$E_{nj} = m_0c^2+E_n+\Delta E_1+\Delta E_2 = m_0c^2 - \frac{m_0e^4}{2\hbar^2n^2} + \frac{m_0e^8}{2\hbar^4c^3n^4}\left(\frac{3}{4} - \frac{n}{j+\frac{1}{2}}\right). \tag{2.5}$$

Since $m_0 \approx m_e$, the third term on the right-hand side of (2.5) is the same as that given by the fine-structure formula (cf. problem 32, Chapter XII).

3. In order to solve this problem we have to take into account the electron spin. For this reason we shall use the Pauli equation, (32.8) Chapter VI. The wavefunction $\psi(t)$ of a stationary state of the atom is written, as usual, in the form

$$\psi(t) = \psi e^{-\frac{iEt}{\hbar}}. \tag{3.1}$$

Substituting (3.1) into (32.8) Chapter VI we find

$$\left\{\frac{1}{2m}\sum_k \mathbf{p}_k^2 + \frac{|e|}{2mc}(\mathbf{L}+2\mathbf{S}).\mathbf{H} + \frac{e^2}{8mc^2}\sum_k (\mathbf{H}\times\mathbf{r}_k)^2 + U\right\}\psi = E\psi, \tag{3.2}$$

where U includes the spin–orbit interactions, the interactions between the electrons, and the interactions of the electrons with the nucleus. We assume that the external magnetic field is weak enough so that in (3.2) we can neglect terms proportional to the square of the field. We apply perturbation theory by taking as the perturbation

$$H' = \frac{|e|}{2mc}(\mathbf{L}+2\mathbf{S}).\mathbf{H} = -\frac{e}{2mc}(\mathbf{J}+\mathbf{S}).\mathbf{H} = -\frac{eH}{2mc}(J_z+S_z), \tag{3.3}$$

where $\mathbf{J} = \mathbf{L}+\mathbf{S}$ is the total angular momentum operator. $H_0 = \frac{1}{2m}\sum_k \mathbf{p}_k^2 + U$ is then the unperturbed Hamiltonian. Since U includes the spin–orbit interactions as well as the other time-independent interactions, the energy levels are only $(2J+1)$-fold degenerate (see problem 1), each state having a different value M of the component J_z of the total angular momentum operator. At first sight, it seems that we have to use the perturbation theory for degenerate levels, but this can be seen to be unnecessary if we note that H_0, as well as $H = H_0+H'$, commutes with J_z and, for that reason, H' does not mix states with different M.

In order to find the perturbation of the energy levels to first approximation, we need therefore calculate only the diagonal elements of the matrix of the operator H'. Now we have, for a start, that

$$\langle J, M \,|\, J_z \,|\, J, M\rangle = 0. \tag{3.4}$$

Let us then proceed to calculate $\langle J, M \,|S_z|\, J, M\rangle$. Note that the following commutation rules hold:

$$\begin{aligned} [J_x, S_x] &= 0, \quad [J_x, S_y] = i\hbar S_z, \quad [J_x, S_z] = -i\hbar S_y \\ [J_y, S_y] &= 0, \quad [J_y, S_z] = i\hbar S_x, \quad [J_y, S_x] = -i\hbar S_z \\ [J_z, S_z] &= 0, \quad [J_z, S_x] = i\hbar S_y, \quad [J_z, S_y] = -\, i\hbar S_x\,. \end{aligned} \tag{3.5}$$

From (3.5) we have

$$(J_x+iJ_y)(S_x+iS_y)-(S_x+iS_y)(J_x+iJ_y) = 0. \tag{3.6}$$

Let us determine the matrix elements of the two terms of (3.6) between the states $|J, M+1\rangle$ and $|J, M-1\rangle$.

Now

$$\langle J, M+1 \,|J_+S_+|\, J, M-1\rangle = \langle J, M+1 \,|S_+J_+|\, J, M-1\rangle,$$

and hence, by the multiplication rule for matrices,

$$\langle M+1 \,|J_+|\, M\rangle\langle M \,|S_+|\, M-1\rangle = \langle M+1 \,|S_+|\, M\rangle\langle M \,|J_+|\, M-1\rangle,$$

where, for simplicity, we have omitted the index J.

Using (VI.3), we find that

$$\frac{\langle M+1 \,|S_+|\, M\rangle}{\sqrt{(J+M+1)(J-M)}} = \frac{\langle M \,|S_+|\, M-1\rangle}{\sqrt{(J+M)(J-M+1)}} = A, \tag{3.7}$$

say. In an analogous manner we find that

$$\frac{\langle M+2 \,|S_+|\, M+1\rangle}{\sqrt{(J+M+2)(J-M-1)}} = A, \tag{3.8}$$

whence we see that the quantity A is independent of M. Hence we have

$$\langle M \,|S_+|\, M-1\rangle = A\sqrt{(J+M)(J-M+1)}. \tag{3.9}$$

The diagonal elements of the matrix of the operator S_z can now be obtained by using the commutation rule $[J_-, S_+] = -2\hbar S_z$. We have that

$$-2\hbar\langle M|S_z|M\rangle = \langle M|J_-|M+1\rangle\langle M+1|S_+|M\rangle - \langle M|S_+|M-1\rangle\langle M-1|J_-|M\rangle. \tag{3.10}$$

Using (VI.3) and (3.9), one obtains

$$\langle M|S_z|M\rangle = AM. \tag{3.11}$$

It remains for us to determine A. We have, on the one hand,

$$\langle M|\mathbf{J}.\mathbf{S}|M\rangle = \hbar\frac{2J(J+1)-L(L+1)+S(S+1)}{2}, \tag{3.12}$$

and, on the other hand,

$$\langle M|\mathbf{J}.\mathbf{S}|M\rangle = \tfrac{1}{2}\langle M|S_+|M-1\rangle\langle M-1|J_-|M\rangle + \tfrac{1}{2}\langle M|S_-|M+1\rangle\langle M+1|J_+|M\rangle + \hbar M\langle M|S_z|M\rangle. \tag{3.13}$$

Using (VI.3), (3.9), (3.11) and comparing (3.12) with (3.13) we find finally that

$$A = \hbar\frac{J(J+1)+S(S+1)-L(L+1)}{2J(J+1)}. \tag{3.14}$$

Thus an energy level with the quantum numbers J, L and S is split by the magnetic field into $2J+1$ equidistant levels:

$$E_M^{(1)} = \frac{|e|\,H\hbar M}{2mc}\left(1+\frac{J(J+1)+S(S+1)-L(L+1)}{2J(J+1)}\right) = \frac{|e|\,H\hbar M}{2mc}g, \tag{3.15}$$

say, where g is called the Landé factor.

In order to obtain correct results by using perturbation theory as above, $E_M^{(1)}$ must be small in comparison with the fine structure splitting of the atomic levels.

Note that in the above solution we have assumed that J, L and S are "good" quantum numbers, i.e. that the Russell–Saunders coupling scheme is valid for the atom in question.

4. If the energy of interaction of the atom and the magnetic field is much greater than the energy of the spin–orbit coupling, the latter can be neglected in the unperturbed Hamiltonian H_0 (see Problem 3). The unperturbed state vectors will, in this case, be $|LSM_LM_S\rangle = |LM_L\rangle\,|SM_S\rangle$, while the perturbation operator remains the same, i.e.

$$H' = \frac{|e|\,\hbar}{2mc}(J_z+S_z)\,H = \frac{|e|\,\hbar}{2mc}(L_z+2S_z)\,H. \tag{4.1}$$

Although the degeneracy of an energy level having the quantum numbers L and S is $(2L+1)(2S+1)$, we can still use non-degenerate perturbation theory for the same reasons as those given in the preceding problem. The first-order correction to the energy of a level with quantum numbers L and S is then

$$E^{(1)}_{M_L,\, M_S} = \frac{|e|\,\hbar H}{2mc}(M_L+2M_S). \tag{4.2}$$

Note that the initial degeneracy is in general lifted by the magnetic field. It can be shown that the perturbation H' of problems 3 and 4 determines the paramagnetic properties of the atom. Thus, denoting by m_z the z-component of the atom's magnetic moment, we have $E^{(1)} = -Hm_z$. On the other hand, from (3.15) and (4.2) we see that $E^{(1)}$ is a quantity which can take the same number of positive as of negative values. In an ensemble of atoms in thermodynamic equilibrium the negative values of $E^{(1)}$ (i.e. the positive values of m_z) will be preferred. It follows that an average magnetic moment in the direction of the magnetic field appears (the *paramagnetic* effect). Note, however, that in states with $L = S = 0$ we have $H' = 0$. Consequently for these states the energy splittings are no longer linearly dependent on H, and we are obliged to consider quadratic terms in H, for example the perturbation

$$H_1' = \frac{e^2}{8mc^2}\sum_k (H\times r_k)^2 \tag{4.3}$$

neglected in (3.2). The energy correction to first order of perturbation theory will then be given by

$$\frac{e^2}{8mc^2}\left\langle \sum_k (H\times r_k)^2\right\rangle = \frac{e^2}{8mc^2}\sum_k \langle (Hr_k \sin\theta_k)^2\rangle.$$

To calculate $\langle (Hr\sin\theta)^2\rangle$ we use the fact that the wavefunction with $L = S = 0$ has spherical symmetry. Hence

$$\langle \sin^2\theta\rangle = 1-\langle\cos^2\theta\rangle = 1-\tfrac{1}{3} = \tfrac{2}{3}$$

and therefore the level shift for states with $L = S = 0$ will be

$$\Delta E = \frac{e^2H^2}{2mc^2}\sum_k \langle \mathbf{r}_k^2\rangle. \tag{4.4}$$

Since the magnetic moment is given by $\mathbf{m} = \dfrac{\partial \Delta E}{\partial \mathbf{H}}$, we have $\mathbf{m} = \chi\mathbf{H}$, where $\chi = -\dfrac{e^2}{6mc^2}\sum_k \mathbf{r}_k^2$. The quantity χ (< 0) is called the *diamagnetic susceptibility*, and thus the term (4.3) is responsible for the diamagnetic properties of atoms. Since c^2 appears in the denominator of the expression (4.4), diamagnetic effects can in general be observed only if paramagnetic ones are absent (which is the case if $L = S = 0$).

5. If the interaction of the electron with the magnetic field is of the same order of magnitude as the spin–orbit coupling, we have to take

$$H' = \phi(r)\mathbf{l}.\mathbf{s}+\frac{e\hbar}{2mc}H(l_z+2s_z), \quad \phi(r) = \frac{\hbar^2e^2}{2m^2c^2}\frac{1}{r^3}, \tag{5.1}$$

as the perturbation operator. In the unperturbed states, the constants of the motion are the square magnitudes and the z-projections of the orbital and of the spin angular momenta (i.e. the quantum numbers of the states are $l, m, \frac{1}{2}, \pm\frac{1}{2}$). It has been seen (cf. problem 18, Chapter VI) that the unperturbed electron states can alternatively be characterized by the quantum numbers n, l, j, m_j. Now the degeneracy of an unperturbed level $E_n^{(0)}$ is, as we know, equal to $2n^2$. However, there is no need to solve a secular equation of this order. Indeed, in the perturbed states, the z-projection of the total angular momentum is a constant of the motion. The correct zero-order wavefunctions of the perturbed states will accordingly be linear combinations of functions $\psi^{(0)}_{nljm_j}$ which have the same values of n, l, m_j, and have $j = l+\frac{1}{2}$ and $l-\frac{1}{2}$, i.e.

$$\psi^{(0)} = c_1\psi^{(0)}_{n,l,j=l+1/2,m_j}+c_2\psi^{(0)}_{n,l\,j=l-1/2,m_j} \tag{5.2}$$

or, using the results of problem 18, Chapter VI,

$$\psi^{(0)} = c_1\frac{R(r)}{\sqrt{2l+1}}\begin{pmatrix}\sqrt{l+m_j+\frac{1}{2}}\,Y_l^{m_j-1/2}\\ \sqrt{l-m_j+\frac{1}{2}}\,Y_l^{m_j+1/2}\end{pmatrix}+c_2\frac{R(r)}{\sqrt{2l+1}}\begin{pmatrix}-\sqrt{l-m_j+\frac{1}{2}}\,Y_l^{m_j-1/2}\\ \sqrt{l+m_j+\frac{1}{2}}\,Y_l^{m_j+1/2}\end{pmatrix}. \tag{5.3}$$

Let us define now the quantity

$$A = \frac{me^4}{\hbar^2}\left(\frac{e^2}{\hbar c}\right)^2\frac{1}{n^3l(l+1)(2l+1)}. \tag{5.4}$$

Then, from (5.1)–(5.3) above, and (7.6) of Chapter III, we obtain without difficulty

$$\left\langle n, l, l+\frac{1}{2}, m_j\,|H'|\,n, l, l+\frac{1}{2}, m_j\right\rangle = A\frac{1}{2}+H\mu_0m_j\left(1+\frac{1}{2l+1}\right),$$

$$\left\langle n, l, l-\frac{1}{2}, m_j\,|H'|\,n, l, l-\frac{1}{2}, m_j\right\rangle = -A\frac{l+1}{2}+H\mu_0m_j\left(1-\frac{1}{2l+1}\right),$$

$$\left\langle n, l, l-\frac{1}{2}, m_j\,|H'|\,n, l, l+\frac{1}{2}, m_j\right\rangle = \left\langle n, l, l+\frac{1}{2}, m_j\,|H'|\,n, l, l-\frac{1}{2}, m_j\right\rangle$$

$$= -\frac{H\mu_0}{2l+1}\sqrt{\left(l+\frac{1}{2}\right)^2-m_j^2},$$

$$\mu_0 = \frac{|e|\,\hbar}{2mc}.$$

Since the degeneracy is relevant only in connection with the quantum number j (as $\mathbf{j}^2$ does not commute with H'), the secular equation of the problem will be of second order only:

$$\begin{vmatrix} E_n^{(0)}+A\frac{1}{2}+H\mu_0 m_j\left(1+\frac{1}{2l+1}\right)-E, & -\frac{H\mu_0}{2l+1}\sqrt{\left(l+\frac{1}{2}\right)^2-m_j^2} \\ -\frac{H\mu_0}{2l+1}\sqrt{\left(l+\frac{1}{2}\right)^2-m_j^2}, & E_n^{(0)}-A\frac{l+1}{2}+H\mu_0 m_j\left(1-\frac{1}{2l+1}\right)-E \end{vmatrix} = 0 \quad (5.5)$$

where $E_n^{(0)}$ is the energy of the unperturbed hydrogen atom. From (1.4) it follows that $E^{(1)}_{n,\,l,\,j=l+1/2} = A\frac{l}{2}$ and $E^{(1)}_{n,\,l,\,j=l-1/2} = -A\frac{l+1}{2}$. If we now define the quantities

$$E_+ = E_n^{(0)}+E^{(1)}_{n,\,l,\,j=l+1/2}, \quad E_- = E_n^{(0)}+E^{(0)}_{n,\,l,\,j=l-1/2}, \quad (5.6)$$

then the energies of the hydrogen atom in the external magnetic field, taking into account spin–orbit coupling, i.e. the solutions E of the above secular equation, can be written in the form

$$E = \frac{1}{2}(E_+ + E_-)+H\mu_0 m_j \pm\sqrt{\frac{1}{4}(E_+ - E_-)^2+H\mu_0\frac{m_j}{2l+1}(E_+ - E_-)+\frac{1}{4}H^2\mu_0^2}. \quad (5.7)$$

Note the following special cases:

(a) For weak magnetic fields, $\mu_0 H \ll (E_+ - E_-)$, it follows from (5.7) that

$$\begin{aligned} E-E_+ &\approx +H\mu_0 m_j\frac{2l+2}{2l+1} \\ E-E_- &\approx -H\mu_0 m_j\frac{2l}{2l+1}. \end{aligned} \quad (5.8)$$

As is to be expected, the relations (5.8) are special cases of (3.15), the first for $j = l+\frac{1}{2}$, and the second for $j = l-\frac{1}{2}$, with, in each case, $s = \frac{1}{2}$.

(b) For strong magnetic fields, $\mu_0 H \gg (E_+ - E_-)$, and negligible spin–orbit coupling, $E_+ \approx E_- = \tilde{E}$, it follows immediately from (5.7) that

$$E = \tilde{E}+H\mu_0(m_j \pm \tfrac{1}{2}), \quad (5.9)$$

an expression which can also be obtained from (4.2).

6. For the whole system, the steady state equation is

$$(H_1+H_2+H_{12})\psi = E\psi \quad (6.1)$$

where H_1 and H_2 are the Hamiltonians of the one-particle systems, and H_{12} is the (symmetrical) interaction Hamiltonian: $H_{12} = H_{21}$. In zero order $\psi = \phi_{n_1}(\mathbf{r}_1)\,\phi_{n_2}(\mathbf{r}_2)$ where the $\phi_n(\mathbf{r})$ satisfy the Schrödinger equation $H_1\,\phi_n(\mathbf{r}) = E_n\,\phi_n(\mathbf{r})$.

The eigenvalue $E = E_r + E_s$ (the total energy of the system) is doubly degenerate; the corresponding eigenfunctions being $\phi_r(\mathbf{r}_1)\,\phi_s(\mathbf{r}_2)$ when particle 1 is in the rth state and particle 2 in the sth state, and $\phi_s(\mathbf{r}_1)\,\phi_r(\mathbf{r}_2)$ in the opposite case. Any linear combination of the form

$$\psi = a\,\phi_r(1)\,\phi_s(2) + b\,\phi_s(1)\,\phi_r(2), \qquad |a|^2 + |b|^2 = 1 \tag{6.2}$$

is also a normalized eigenfunction with the same energy $E = E_r + E_s$. Substituting (6.2) in (6.1) we obtain $(H_{12} - \varepsilon) = 0$, $\varepsilon = E - E_r - E_s$, and so

$$(H_{12} - \varepsilon)\,[a\,\phi_r(1)\,\phi_s(2) + b\,\phi_s(1)\,\phi_r(2)] = 0. \tag{6.3}$$

Using the notation

$$\int \phi_r^*(1)\,\phi_s^*(2) H_{12}\,\phi_r(1)\,\phi_s(2)\,d\mathbf{r}_1\,d\mathbf{r}_2 = \int \phi_s^*(1)\,\phi_r^*(2) H_{12}\,\phi_s(1)\,\phi_r(2)\,d\mathbf{r}_1\,d\mathbf{r}_2 = K,$$

and

$$\int \phi_s^*(1)\,\phi_r^*(2) H_{12}\,\phi_r(1)\,\phi_s(2)\,d\mathbf{r}_1\,d\mathbf{r}_2 = \int \phi_r^*(1)\,\phi_s^*(2) H_{12}\,\phi_s(1)\,\phi_r(2)\,d\mathbf{r}_1\,d\mathbf{r}_2 = A,$$

let us multiply (6.3) first by $\phi_r^*(1)\,\phi_s^*(2)$, and then by $\phi_s^*(1)\,\phi_r^*(2)$, in each case integrating the product over $\mathbf{r}_1$ and $\mathbf{r}_2$, and obtain

$$a(K - \varepsilon) + bA = 0, \qquad aA + b(K - \varepsilon) = 0. \tag{6.4}$$

The system of equations (6.4) has non-trivial solutions if $A^2 = (K - \varepsilon)^2$. In the case $\varepsilon = K + A$, we have $a = b$ and therefore

$$E = E_r + E_s + K + A, \qquad \psi_S = \frac{1}{\sqrt{2}}\,[\phi_r(1)\,\phi_s(2) + \phi_s(1)\,\phi_r(2)]$$

which is the symmetrical solution. In the case $\varepsilon = K - A$, we have $a = -b$ and therefore

$$E = E_r + E_s + K - A, \qquad \psi_A = \frac{1}{\sqrt{2}}\,[\phi_r(1)\,\phi_s(2) - \phi_s(1)\,\phi_r(2)]$$

which is the anti-symmetrical solution.

In the energy correction $\varepsilon = K \pm A$, the term K is the classical potential energy of the Coulomb interaction between the two charge distributions $\varrho_r(1) = \phi_r^*(1)\,\phi_r(1)$ and $\varrho_s(2) = \phi_s^*(2)\,\phi_s(2)$, while A is called the *exchange energy*, and has no classical counterpart.

By taking the two particles to be electrons moving in the Coulomb field of a nucleus of charge Ze, the above results can be seen to be applicable to *two-electron atoms* (H^-, He, Li^+, etc.)

Considering now the spins of the electrons, it follows, from the total anti-symmetry of the wavefunction, that in the state ψ_S the total spin must be zero, while in the state ψ_A it must be equal to 1 (see Table VII.1). To these two possibilities correspond the so-called *para* and *ortho* states respectively.

7. At time $t = 0$, the wavefunction of the system is

$$\psi(0) = \phi_r(1)\,\phi_s(2) = \frac{1}{\sqrt{2}}[\psi_S+\psi_A]. \tag{7.1}$$

Since ψ_s and ψ_A describe stationary states of energy $E_S = E_r+E_s+K+A$ and $E_A = E_r+E_s+$ $+K-A$ respectively, their time-dependence will be

$$\psi_S(t) = \psi_S e^{-\frac{i}{\hbar}E_S t}, \quad \psi_A(t) = \psi_A e^{-\frac{i}{\hbar}E_A t}. \tag{7.2}$$

For $t > 0$ the ave-function of the system is therefore

$$\psi(t) = \frac{1}{\sqrt{2}}[\psi_S(t)+\psi_A(t)] = \left[\phi_r(1)\,\phi_s(2)\cos\left(\frac{At}{\hbar}\right)-i\,\phi_s(1)\,\phi_r(2)\sin\left(\frac{At}{\hbar}\right)\right]e^{-\frac{i}{\hbar}(E_r+E_s+K)t}. \tag{7.3}$$

The expression (7.3) shows that if, at time $t = 0$, particle 1 is found to be in the rth state and particle 2 in the sth state, then at time $t = \pi\hbar/2A$ the occupation of the states will have been reversed.

8. It is easy to verify that the Thomas–Fermi equation is the Euler equation of the variational problem

$$\delta\int_0^\infty dx(\tfrac{1}{2}\phi'^2+\tfrac{2}{5}x^{-1/2}\phi^{5/2}) = 0. \tag{8.1}$$

The functions $\phi = e^{-\alpha x}$ satisfy the boundary conditions $\phi(0) = 1$, $\phi(\infty) = 0$. Substituting $\phi = e^{-\alpha x}$ into (8.1) and remembering that $\Gamma(\frac{1}{2}) = \sqrt{\pi}$, it follows that we require $\delta\left(\frac{\alpha}{4}+\frac{2}{5}\left(\frac{2}{5\alpha}\right)^{1/2}\sqrt{\pi}\right) = 0$, whence by differentiating with respect to α we find $\alpha = \frac{2}{5}(4\pi)^{1/2} = 0{\cdot}93$.

9. From the expression (XI.4), and by using the result of the preceding problem, we have $\Phi(r) = \frac{Ze}{r}e^{-(\alpha r/b)}$, $\alpha = 0{\cdot}93$, so that the potential energy of an electron in the atom under consideration is

$$V(r) = -\frac{Ze^2}{r}\exp\left(-1{\cdot}05Z^{1/3}\frac{r}{a}\right). \tag{9.1}$$

Let us imagine that we have removed one of the electrons from the K-shell and also a unit positive charge from the nucleus. The result is a neutral atom with $Z-1$ electrons. The potential energy of the chosen electron is, according to the Thomas–Fermi model,

$$V(r) = -\frac{e^2}{r}-\frac{(Z-1)e^2}{r}e^{-(\mu r/a)}, \quad \mu = 1{\cdot}05(Z-1)^{1/3}. \tag{9.2}$$

Thus we may write $V(r) = V_0(r)+V_1(r)$, where $V_0(r) = -\frac{Ze^2}{r}$ is the potential energy due to the Coulomb field of the nucleus, and $V_1(r) = \frac{(Z-1)e^2}{r}(1-e^{-(\mu r/a)})$ is the potential energy of the (averaged out) interactions with the other electrons.

To the lowest order of approximation (see problem 33, Chapter II) we can neglect $V_1(r)$ and obtain $E_0 = -\frac{e^2Z^2}{2a}$, $\psi_{100} = \sqrt{\frac{Z^3}{\pi a^3}}\,e^{-(\mu r/a)}$ for the energy and the wavefunction of a K-shell electron. Treating $V_1(r)$ as a perturbation, to first-order approximation we obtain for the K-shell energy,

$$E = E_0+\int \psi_{100}^2\, V_1(r)\, d\mathbf{r} = -\frac{e^2}{2a}\left\{Z^2-2Z(Z-1)\left[1-\left(1+\frac{\mu}{2Z}\right)^{-2}\right]\right\} = -\frac{e^2}{2a}Z'^2$$

and so

$$Z' = Z\sqrt{1-\frac{2(Z-1)}{Z}\left[1-\left(1+\frac{\mu}{2Z}\right)^{-2}\right]}. \tag{9.3}$$

This result is valid only for atoms with Z in the range $40 < Z < 60$, since for small Z the statistical character of the Thomas–Fermi method is inappropriate, and, for large Z, relativistic corrections to the K-shell energy are important.

10. Introducing the new variable $y = 2ae^{-\mu x}$, where $a^2 = 2mV_0/\mu^2\hbar^2$, the Schrödinger equation

$$\frac{d^2\psi}{dx^2}+\frac{2m}{\hbar^2}\left[E-V_0(1-e^{\mu x})^2\right]\psi = 0,$$

becomes

$$\frac{d^2\psi}{dy^2}+\frac{1}{y}\frac{d\psi}{dy}+\left(\frac{E-V_0}{V_0}\frac{a^2}{y}+\frac{a}{y}-\frac{1}{4}\right)\psi = 0. \tag{10.1}$$

The possible energies of the molecular oscillator are those values of E for which equation (10.1) has solutions which vanish at infinity. Looking for these solutions of the form

$$\psi = e^{-(y/2)}y^s\, u(y), \tag{10.2}$$

it follows from (10.1) that if we choose

$$s^2 = -\frac{E-V_0}{V_0}\,a^2 \tag{10.3}$$

then u must satisfy a differential equation of the type (A.62):

$$\left[y\frac{d^2}{dy^2}+(1+2s-y)\frac{d}{dy}-\left(\frac{1}{2}+s-a\right)\right]u = 0. \tag{10.4}$$

The general solution of this equation is

$$u = C_1 F(\tfrac{1}{2}+s-a,\ 1+2s;\ y)+C_2 F(\tfrac{1}{2}-s-a,\ 1-2s;\ y).$$

Taking into account the asymptotic behaviour (A.64), we see that (10.2) is a solution of the problem (i.e. it vanishes at infinity) only if

$$\tfrac{1}{2}\pm s-a = -n, \qquad n = 0, 1, 2 \ldots . \tag{10.5}$$

From this condition we have that

$$s^2 = a^2-2a(n+\tfrac{1}{2})+(n+\tfrac{1}{2})^2. \tag{10.6}$$

Noting that, for $|x| \ll 1$, (small oscillations), the Morse potential is essentially a harmonic oscillator potential:

$$V = V_0(1-e^{-\mu x})^2 \approx V_0\mu^2 x^2 = \frac{m\omega^2}{2}x^2,$$

so that

$$\mu\left(\frac{2V_0}{m}\right)^{1/2} = \omega, \tag{10.7}$$

we can use (10.3), (10.6) and (10.7) to obtain finally the required energy eigenvalues of the molecular oscillator:

$$E_n = \hbar\omega\left(n+\frac{1}{2}\right)-\frac{\mu^2\hbar^2}{2m}\left(n+\frac{1}{2}\right)^2, \qquad n = 0, 1, 2, \ldots . \tag{10.8}$$

By comparing this result with that obtained for the harmonic oscillator, we can see that, for the Morse potential, the expression for the energy contains an extra term in the square of the quantum number n, which makes the levels appear closer to one another as the energy increases, in agreement with experimental data.

11. The vibrational and the rotational energy levels of the molecule are given by the solutions of the radial equation (II.15), with the potential (11a), which are bounded at the origin and which vanish at infinity. Introducing the notation

$$\lambda = \frac{2ma^2}{\hbar^2}E, \qquad \gamma^2 = \frac{2ma^2}{\hbar^2}V_0, \tag{11.1}$$

where $m = m_1 m_2(m_1+m_2)^{-1}$ is the reduced mass of the molecule, the radial equation becomes

$$\frac{d^2R_l}{d\varrho^2}+\left[\lambda+\frac{2\gamma^2}{\varrho}-\frac{\gamma^2+l(l+1)}{\varrho^2}\right]R_l = 0. \tag{11.2}$$

Note that, as $\varrho\to\infty$, the solution has the form $R_l \to \exp\left(i\sqrt{\lambda}\,\varrho\right)$, and that, as $\varrho\to 0$, it behaves like ϱ^μ where μ is the positive root $\mu = \frac{1}{2}+\sqrt{\gamma^2+(l+\frac{1}{2})^2}$ of the characteristic equation, the negative root being excluded by condition (II.16). We accordingly look for a solution of equation (11.2) of the form

$$R_l = \varrho^{\mu} e^{i\sqrt{\lambda}\varrho} v_l(\varrho), \tag{11.3}$$

and obtain for v_l the equation

$$\varrho \frac{d^2 v_l}{d\varrho^2} + (2\mu + 2i\sqrt{\lambda}\varrho) \frac{dv_l}{d\varrho} + (2\mu i\sqrt{\lambda} + 2\gamma^2) v_l = 0. \tag{11.4}$$

By introducing the variable $z = -2i\sqrt{\lambda}\varrho$, this becomes

$$z \frac{d^2 v_l}{dz^2} + (2\mu - z) \frac{dv_l}{dz} - \left(\mu - \frac{i\gamma^2}{\sqrt{\lambda}}\right) v_l = 0. \tag{11.5}$$

The solution of this equation which is regular at the origin is, except for a constant factor, the confluent hypergeometric series (A.59):

$$v_l = F\left(\mu - i\frac{\gamma^2}{\sqrt{\lambda}},\ 2\mu;\ -2i\sqrt{\lambda}\varrho\right). \tag{11.6}$$

To study the asymptotic behaviour of F we have to distinguish two cases: $\lambda < 0$ and $\lambda > 0$.

(1) If $\lambda < 0$, let $\sqrt{\lambda} = i\beta$ say, where β is real, and then

$$R_l = \varrho^{\mu} e^{-\beta\varrho} F\left(\mu - \frac{\gamma^2}{\beta},\ 2\mu;\ 2\beta\varrho\right). \tag{11.7}$$

By (A.64), as $\varrho \to \infty$, $F \sim e^{2\beta\varrho}$, so that R_l increases indefinitely. It follows that the series must reduce to a polynomial. This will occur only if

$$\mu - \frac{\gamma^2}{\beta} = -n, \qquad n = 0, 1, 2, \ldots, \tag{11.8}$$

that is, if

$$E_{nl} = -\frac{\dfrac{\hbar^2}{2ma^2}\gamma^4}{\left[n + \dfrac{1}{2} + \sqrt{\gamma^2 + \left(l + \dfrac{1}{2}\right)^2}\right]^2}. \tag{11.9}$$

The set of all negative discrete eigenvalues E forms the vibrational and rotational spectrum of the diatomic molecule. Since for molecular spectra $\gamma \gg 1$, the denominator can be expanded in a series in powers of $1/\gamma$,

$$\left[n + \frac{1}{2} + \gamma^2 + \left(l + \frac{1}{2}\right)^2\right]^{-2}$$

$$= \frac{1}{\gamma^2}\left[1 - \frac{2\left(n + \frac{1}{2}\right)}{\gamma} - \frac{\left(l + \frac{1}{2}\right)^2}{\gamma^2} + \frac{3\left(n + \frac{1}{2}\right)^2}{\gamma^2} + \frac{3\left(n + \frac{1}{2}\right)\left(l + \frac{1}{2}\right)^2}{\gamma^3} + \ldots\right].$$

And hence

$$E_{nl} = V_0\left[-1+\frac{2}{\gamma}\left(n+\frac{1}{2}\right)+\frac{1}{\gamma^2}\left(l+\frac{1}{2}\right)^2-\frac{3}{\gamma^2}\left(n+\frac{1}{2}\right)^2-\frac{3}{\gamma^3}\left(n+\frac{1}{2}\right)\left(l+\frac{1}{2}\right)^2+\ldots\right].$$

By expanding in a series the potential V about the value $s = 1$, we obtain

$$V(\varrho) \approx \frac{V_0}{a^2}(r-a)^2-V_0,$$

whence it follows that $\omega = \left(\frac{2V}{ma^2}\right)^{1/2}$ is the classical frequency of small oscillations.

Denoting by $I = ma^2$ the moment of inertia of the molecule, we shall have

$$E_n = -\frac{1}{2}I\omega^2+\hbar\omega\left(n+\frac{1}{2}\right)+\frac{\hbar^2}{2I}\left(l+\frac{1}{2}\right)^2-\frac{3\hbar^2}{2I}\left(n+\frac{1}{2}\right)^2-\frac{3\hbar^2}{2I^2\omega}\left(n+\frac{1}{2}\right)\left(l+\frac{1}{2}\right)^2+\ldots;$$

$$n, l = 0, 1, 2 \ldots. \tag{11.10}$$

The first term of the series (11.10) is a constant. The second term gives the molecular vibrational spectrum. The third term gives the rotational spectrum; for large values of l, when $(l+\frac{1}{2})^2 \approx l(l+1)$, this is the same as for the rigid rotator. The fourth term is a correction to the vibrational spectrum (cf. the preceding problem). The higher terms represent the effect of coupling between the vibrational and the rotational motion of the molecule. The dissociation energy of the molecule is equal to minus the ground state energy E_{00}, thus:

$$E_{\text{dis}} = \frac{1}{2}I^2-\frac{1}{2}\hbar\omega+\frac{1}{4}\frac{\hbar^2}{I}+\frac{3}{16}\frac{\hbar^3}{I^2\omega}+\ldots. \tag{11.11}$$

(2) If $\lambda > 0$, let $\lambda = \beta^2$, and then

$$R_l = \varrho^\mu e^{i\beta\varrho}F\left(\mu-\frac{i\gamma^2}{\beta}, 2\mu; -2i\beta\varrho\right). \tag{11.12}$$

By (A.64), as $\varrho \to \infty$,

$$R_l \sim (-2i\beta)^{-\mu}\Gamma(2\mu)\times$$

$$\times\left[e^{-i\pi\mu}\frac{e^{-(\pi\gamma^2/2\beta)}}{\Gamma\left(\mu+i\frac{\gamma^2}{\beta}\right)}e^{i(\beta\varrho+(\gamma^2/\beta)\ln 2\beta\varrho)}+\frac{e^{-(\pi\gamma^2/2\beta)}}{\Gamma\left(\mu-i\frac{\gamma^2}{\beta}\right)}e^{-i(\beta\varrho+(\gamma^2/\beta)\ln 2\beta\varrho)}\right].$$

Thus, for positive energies, the function R_l oscillates indefinitely between bounds as $\varrho \to \infty$. Any value of E (> 0) is thus an eigenvalue of the Schrödinger equation, and the corresponding eigenfunctions describe states of the dissociated molecule.

CHAPTER XII

Relativistic Quantum Mechanics

1. Definitions, Notation, Conventions

The choice of a moment of time t and a position $\mathbf{r} = (x, y, z)$ in three-dimensional space defines a "point" or "point-event", having the coordinates $x_1 = x$, $x_2 = y$, $x_3 = z$, $x_4 = ict = ix_0$, in four-dimensional space-time. Any linear transformation of the form†

$$x_\mu \to x'_\mu = a_{\mu\nu} x_\nu\,, \tag{XII.1}$$

in which the coefficients $a_{\mu\nu}$ satisfy the orthogonality conditions

$$a_{\mu\nu} a_{\mu\varrho} = \delta_{\nu\varrho}\,, \quad a_{\nu\mu} a_{\varrho\mu} = \delta_{\nu\varrho}\,, \tag{XII.2}$$

preserves the "length" of all four-vectors, and, in particular,

$$x_\mu^2 = x_1^2 + x_2^2 + x_3^2 + x_4^2 = x_\mu'^2 = x_1'^2 + x_2'^2 + x_3'^2 + x_4'^2\,.$$

Such a transformation is called a *homogeneous Lorentz transformation*. The "homogeneous Lorentz group" consists of all such transformations, from one *inertial system* to another without change of origin.

A *four-dimensional vector* (or "four-vector") V_μ is a set of four quantities V_1, V_2, V_3, V_4, which are so defined that under a transformation of the four-dimensional coordinate frame (XII.1) they transform in the same way as do the coordinates x_μ, i.e. $V_\mu \to V'_\mu = a_{\mu\nu} V_\nu$.

A *four-dimensional tensor* (or "four-tensor") of the second rank is a set of 16 quantities $T_{\mu\nu}$, which are so defined that under a transformation of the coordinates (XII.1) they transform like products of coordinates $x_\mu x_\nu$, i.e. according to the formula

$$T_{\mu\nu} \to T'_{\mu\nu} = a_{\mu\alpha} a_{\nu\beta} T_{\alpha\beta}\,.$$

† In this chapter, all Greek indices take the values 1, 2, 3, 4, and, all Roman indices the values 1, 2, 3. The summation convention, i.e. the convention of summing over repeated "dummy" indices, is adopted, thus, e.g.,

$$x_\mu^2 \to x_\mu x_\mu = x_1^2 + x_2^2 + x_3^2 + x_4^2\,.$$

2. Elements of Relativistic Mechanics

In relativistic mechanics the energy E and the momentum $\mathbf{p}$ of a free particle of mass m and velocity $\mathbf{v}$ are given by the expressions

$$E = \frac{mc^2}{\left(1-\frac{v^2}{c^2}\right)^{1/2}}, \quad \mathbf{p} = \frac{m\mathbf{v}}{\left(1-\frac{v^2}{c^2}\right)^{1/2}}. \tag{XII.3}$$

The energy of the particle in its own frame of reference, $E = E_0 = mc^2$, is called the "rest energy" of the particle, and the energy expressed in terms of the momentum is called the *Hamiltonian*

$$H = \sqrt{c^2\mathbf{p}^2+m^2c^4}. \tag{XII.4}$$

The four-vector p_μ, with components $\left(\mathbf{p}, \frac{i}{c}H\right)$, is called the *energy-momentum four-vector*. It satisfies the following relation

$$p_\mu p_\mu + m^2c^2 = 0. \tag{XII.5}$$

From a study of the transformations of the electromagnetic field vectors $\mathbf{E}$ and $\mathbf{H}$, on passing from one inertial frame to another, it can be shown that the components of these vectors form an anti-symmetrical tensor of the second rank

$$(F_{\mu\nu}) = \begin{pmatrix} 0 & H_z & -H_y & -iE_x \\ -H_z & 0 & H_x & -iE_y \\ H_y & -H_x & 0 & -iE_z \\ iE_x & iE_y & iE_z & 0 \end{pmatrix}, \tag{XII.6}$$

called *the electromagnetic field tensor*. This can best be seen by noting that the electromagnetic field can be derived from a four-vector potential A_μ, by the relations

$$F_{\mu\nu} = \frac{\partial A_\nu}{\partial x_\mu} - \frac{\partial A_\mu}{\partial x_\nu}. \tag{XII.7}$$

The spatial components $A_{1,2,3}$ of A_μ form a three-dimensional vector $\mathbf{A}$, which is in fact the usual *vector potential* of the field. The time component of A_μ, $A_4 = iA_0$ say, is such that A_0 is the usual (real) *scalar potential* of the field. The relations (XII.7) are such that $F_{\mu\nu}$ is unchanged by the following transformation of A_μ:

$$A_\mu \to A_\mu + \frac{\partial G}{\partial x_\mu}, \tag{XII.8}$$

called a *gauge transformation of the second kind*. It follows that the components $F_{\mu\nu}$ of the electromagnetic field tensor define the four-potential A_μ except for the gradient of an arbitrary scalar function G. Physical significance can therefore be attributed only to quantities invariant under the transformation (XII.8); in particular, all electrodynamical equations have to be invariant under this transformation.

The generalized momentum and the Hamiltonian of a particle of mass m and charge e, moving in an electromagnetic field having a four-potential $A_\mu = (\mathbf{A}, iA_0)$, are given by

$$\mathbf{p} = \frac{m\mathbf{v}}{\left(1-\dfrac{v^2}{c^2}\right)^{1/2}} + \frac{e}{c}\mathbf{A}, \tag{XII.9}$$

and

$$H = eA_0 + \sqrt{c^2\left(\mathbf{p}-\frac{e}{c}\mathbf{A}\right)^2 + m^2c^4}. \tag{XII.10}$$

The four-vector p_μ with components $\left(\mathbf{p}, \dfrac{i}{c}H\right)$ is called the generalized *energy-momentum four-vector*. It satisfies the relation

$$\left(p_\mu - \frac{e}{c}A_\mu\right)^2 + m^2c^2 = 0. \tag{XII.11}$$

3. The Klein–Gordon Equation and the Dirac Equation

The Klein–Gordon equation for spinless particles can be obtained by replacing the components of the energy-momentum four-vector by the differential operators $p_\mu = -i\hbar \dfrac{\partial}{\partial x_\mu}$ in (XII.5) and (XII.11) respectively; thus the Klein–Gordon equation for a free particle is

$$(p_\mu^2 + m^2c^2)\psi = 0 \tag{XII.12}$$

and the Klein–Gordon equation for a particle with charge e in an electromagnetic field A_μ is

$$\left[\left(p_\mu - \frac{e}{c}A_\mu\right)^2 + m^2c^2\right]\psi = 0. \tag{XII.13}$$

Introducing the notation $k = mc/\hbar$ and the d'Alembert operator

$$\square = \nabla^2 - \frac{1}{c^2}\frac{\partial^2}{\partial t^2} = \frac{\partial^2}{\partial x_\mu \partial x_\mu}$$

the equations (XII.12) and (XII.13) can be written in the form

$$(\square - k^2)\psi(\mathbf{r}, t) = 0, \tag{XII.12$'$}$$

$$\left[\left(\frac{\partial}{\partial x_\mu} - \frac{ie}{\hbar c}A_\mu\right)^2 - k^2\right]\psi(\mathbf{r}, t) = 0. \tag{XII.13$'$}$$

Free particles of mass m and spin $\frac{1}{2}$ are described by the Dirac equation

$$i\hbar \frac{\partial \psi(\mathbf{r}, t)}{\partial t} = H_D \psi(\mathbf{r}, t), \tag{XII.14}$$

in which

$$H_D = -i\hbar c\boldsymbol{\alpha} \cdot \nabla + mc^2\beta = c\boldsymbol{\alpha} \cdot \mathbf{p} + mc^2\beta, \tag{XII.15}$$

and $\boldsymbol{\alpha} = (\alpha_1, \alpha_2, \alpha_3)$ and β are 4×4 Hermitian matrices which satisfy the equations

$$\begin{aligned} &\alpha_1^2 = \alpha_2^2 = \alpha_3^2 = \beta^2 = 1, \\ &\alpha_i\beta + \beta\alpha_i = 0, \quad \alpha_i\alpha_k + \alpha_k\alpha_i = 2\delta_{ik}. \end{aligned} \tag{XII.16}$$

The Dirac wavefunction ψ is a one-column matrix

$$\psi = \begin{pmatrix} \psi_1 \\ \psi_2 \\ \psi_3 \\ \psi_4 \end{pmatrix} \tag{XII.17}$$

and is called a double or bi-spinor, for reasons which will appear later. The Dirac equation (XII.14) and (XII.15) can be rewritten in four-vector notation in the form

$$(\gamma_\mu p_\mu - imc)\psi = 0, \tag{XII.18}$$

where

$$\boldsymbol{\gamma} = -i\beta\boldsymbol{\alpha}, \quad \gamma_4 = \beta \tag{XII.19}$$

are Hermitian matrices which satisfy the equation

$$\gamma_\mu\gamma_\nu + \gamma_\nu\gamma_\mu = 2\delta_{\nu\mu}. \tag{XII.20}$$

The most useful representations of $\boldsymbol{\alpha}$ and β are

(1) The Pauli representation

$$\boldsymbol{\alpha} = \begin{bmatrix} 0 & \boldsymbol{\sigma} \\ \boldsymbol{\sigma} & 0 \end{bmatrix}, \quad \beta = \begin{bmatrix} I & 0 \\ 0 & -I \end{bmatrix}, \tag{XII.21}$$

(2) The Kramers representation

$$\boldsymbol{\alpha}^K = \begin{bmatrix} \boldsymbol{\sigma} & 0 \\ 0 & -\boldsymbol{\sigma} \end{bmatrix}, \quad \beta^K = \begin{bmatrix} 0 & I \\ I & 0 \end{bmatrix}, \tag{XII.21'}$$

(3) The Majorana representation

$$\alpha_1^M = \begin{bmatrix} \sigma_1 & 0 \\ 0 & -\sigma_1 \end{bmatrix}, \quad \alpha_2^M = \begin{bmatrix} 0 & -I \\ -I & 0 \end{bmatrix}, \quad \alpha_3^M = \begin{bmatrix} \sigma_3 & 0 \\ 0 & -\sigma_3 \end{bmatrix}, \quad \beta^M = \begin{bmatrix} \sigma_2 & 0 \\ 0 & -\sigma_2 \end{bmatrix}, \tag{XII.21''}$$

where $\boldsymbol{\sigma} = (\sigma_1, \sigma_2, \sigma_3)$ is a vector having the Pauli matrices as its components.

The Dirac equation for an electron in an electromagnetic field having the four-potential $A_\mu = (\mathbf{A}, iA_0)$ is, in four-vector notation,

$$\left[\gamma_\mu\left(p_\mu - \frac{e}{c}A_\mu\right) - imc\right]\psi = 0. \tag{XII.22}$$

In the notation of (XII.12′) and (XII.13′), the equations (XII.18) and (XII.22) can be written in the form

$$\left(\gamma_\mu \frac{\partial}{\partial x_\mu} + k\right)\psi = 0, \tag{XII.18′}$$

$$\left[\gamma_\mu\left(\frac{\partial}{\partial x_\mu} - \frac{ie}{\hbar c}A_\mu\right) + k\right]\psi = 0. \tag{XII.22′}$$

Problems

1. Starting from the Klein–Gordon equation (XII.12′), deduce the equation of conservation

$$\frac{\partial \varrho}{\partial t} + \nabla \mathbf{j} = 0, \tag{1a}$$

where

$$\varrho = \mathrm{Im}\left(\frac{\hbar}{mc^2}\frac{\partial \psi^*}{\partial t}\psi\right), \quad \mathbf{j} = \mathrm{Re}\left(\frac{\hbar}{im}\psi^*\nabla\psi\right). \tag{1b}$$

Considering the plane wave solution

$$\psi(\mathbf{r}, t) = A\exp\left[\frac{i}{\hbar}(\mathbf{p}\cdot\mathbf{r} - Et)\right], \tag{1c}$$

show that the energy can have both a plus and a minus sign. What form do the expressions (1b) take in the non-relativistic limit $|\mathbf{p}| \ll mc$?

2. Show, by means of the matrices

$$\tau_1 = \begin{bmatrix} 0 & 1 \\ 1 & 0 \end{bmatrix}, \quad \tau_2 = \begin{bmatrix} 0 & -i \\ i & 0 \end{bmatrix}, \quad \tau_3 = \begin{bmatrix} 1 & 0 \\ 0 & -1 \end{bmatrix}, \quad I = \begin{bmatrix} 1 & 0 \\ 0 & 1 \end{bmatrix}, \tag{2a}$$

that the Klein–Gordon equation (XII.12′) can be put into the form of a "Schrödinger" equation

$$i\hbar\frac{\partial \psi}{\partial t} = H\psi, \tag{2b}$$

where

$$\psi = \begin{bmatrix} \phi \\ \chi \end{bmatrix}, \tag{2c}$$

$$\phi = \frac{1}{\sqrt{2}}\left(\psi - \frac{\hbar}{mc}\frac{\partial \psi}{\partial x_4}\right), \quad \chi = \frac{1}{\sqrt{2}}\left(\psi + \frac{\hbar}{mc}\frac{\partial \psi}{\partial x_4}\right), \tag{2d}$$

$$H = -\frac{\hbar^2}{2m}(\tau_3 + i\tau_2)\nabla^2 + mc^2\tau_3. \tag{2e}$$

3. Show, using the notation of the preceding problem, that the expressions (1b) can be written in the form

$$\varrho = \tfrac{1}{2}(\phi^*\phi - \chi^*\chi) = \tfrac{1}{2}\psi^+\tau_3\psi, \tag{3a}$$

$$\mathbf{j} = \frac{\hbar}{4im}[\psi^+\tau_3(\tau_3+i\tau_2)\nabla\psi - (\nabla\psi_0)^+\tau_3(\tau_3+i\tau_2)\psi]. \tag{3b}$$

4. From the results of the preceding problem, show

(1) that if A is a 2×2 matrix which does not contain x, y, z or derivatives with respect to x, y, z, then the definition (3.6) of the Hermitian conjugate A^H of A reduces to

$$A^H = \tau_3 A^+ \tau_3, \tag{4a}$$

and the unitarity condition (3.7) to

$$U^H = U^{-1}, \tag{4b}$$

and (2) that

$$\begin{aligned} &\tau_1^H = -\tau_1, \quad \tau_2^H = -\tau_2, \quad \tau_3^H = \tau_3 \\ &(i\tau_1)^H = i\tau_1, \quad (i\tau_2)^H = i\tau_2, \quad (i\tau_3)^H = -i\tau_3. \end{aligned} \tag{4c}$$

Express $\langle A\rangle_{FV}$ in terms of the solution $\psi(\mathbf{r}, t)$ of the Klein–Gordon equation (XII.12′) (see the remarks made in the answer to question 3 above).

5. By using the expressions (3.8)–(3.10), express the energy, the momentum and the angular momentum of a spinless free particle in terms of the solution $\psi(\mathbf{r}, t)$ of the Klein–Gordon equation (XII.12′).

6. From the Klein–Gordon equation (XII.13′), deduce the equation of conservation

$$\frac{\partial j_\mu}{\partial x_\mu} = 0, \tag{6a}$$

where

$$j_\mu = \frac{\hbar}{2mi}\left(\psi^*\frac{\partial\psi}{\partial x_\mu} - \frac{\partial\psi^*}{\partial x_\mu}\psi\right) - \frac{e}{mc}A_\mu\psi^*\psi = (\mathbf{j}, ic\varrho). \tag{6b}$$

7. Solve problem 2 for the Klein–Gordon equation (XII.13′) which describes the relativistic motion of charged spinless particles in an electromagnetic field A_μ. How may (6.6) and (6.7) be written in the matrix formalism?

8. Using (1c), (2c) and (2d), find the positive and negative energy solutions of the Klein–Gordon equation for a free particle in the "Schrödinger" form (2b). Normalize these solutions in the F-V metric and discuss their significance.

9. Show that, in relativistic quantum mechanics, the motion of a charged spinless particle in an electromagnetic field can be described by the equation

$$\left(\Box - k^2 + \frac{e^2A_0^2}{\hbar^2c^2} - 2\frac{ie}{\hbar c^2}A_0\frac{\partial}{\partial t}\right)\psi(\mathbf{r}, t) = 0. \tag{9a}$$

Find the corresponding stationary-state equation.

10. A system consisting of a π^- meson bound by Coulomb forces to a nucleus is called a π-mesic atom. Using equation (9.3), find the energies of the stationary states of a π-mesic atom whose nucleus has a charge $+Ze$.

11. Verify the following relations

$$\begin{aligned} \gamma_\mu\gamma_\mu &= 4, \quad \gamma_\mu\gamma_\nu\gamma_\mu = -2\gamma_\nu \\ \gamma_\mu\gamma_\nu\gamma_\lambda\gamma_\mu &= 4\delta_{\lambda\nu}, \quad \gamma_\mu\gamma_\nu\gamma_\lambda\gamma_\varrho\gamma_\mu = -2\gamma_\varrho\gamma_\lambda\gamma_\nu \end{aligned} \tag{11a}$$

$$\mathrm{Tr}\,(\gamma_\mu\gamma_\nu) = 4\delta_{\mu\nu}. \tag{11b}$$

12. Consider the following set of 16 elements†

$$\left.\begin{array}{c} I \\ \gamma_1, \quad \gamma_2, \quad \gamma_3, \quad \gamma_4 \\ i\gamma_1\gamma_3, \quad i\gamma_2\gamma_1, \quad i\gamma_3\gamma_2, \quad i\gamma_1\gamma_4, \quad i\gamma_2\gamma_4, \quad i\gamma_3\gamma_4 \\ i\gamma_1\gamma_2\gamma_3, \quad i\gamma_2\gamma_1\gamma_4, \quad i\gamma_3\gamma_2\gamma_4, \quad i\gamma_3\gamma_1\gamma_4 \\ \gamma_1\gamma_2\gamma_3\gamma_4 = \gamma_5 \end{array}\right\} \tag{12a}$$

Denoting the elements of the set (12a) by Γ_A, $A = 1, 2, \ldots, 16$, verify that they have the following properties:

(1)
$$\Gamma_A^2 = I. \tag{12b}$$

(2) The product of any two elements of the set is proportional to a third element of the set, i.e.

$$\Gamma_A\Gamma_B = a\Gamma_c, \tag{12c}$$

where $a = \pm 1$ or $\pm i$, and $a = +1$ if and only if $A = B$.

(3) If $\Gamma_A \neq I$, then Γ_B can always be found such that

$$\Gamma_B\Gamma_A\Gamma_B = -\Gamma_A. \tag{12d}$$

(4) If $\Gamma_A \neq I$, then

$$\mathrm{Tr}\,(\Gamma_A) = 0. \tag{12e}$$

13. Show that the elements Γ_A, $A = 1, 2, \ldots, 16$, of the preceding problem are linearly independent.

14. Show that if a 4×4 matrix commutes with all the matrices γ_μ, then it is a multiple of the unit matrix.

15. Show that the 4×4 unitary matrices

$$U_{P\to K} = \frac{1}{\sqrt{2}}\begin{bmatrix} I & I \\ I & -I \end{bmatrix}, \quad U_{K\to M} = \frac{e^{i\pi/4}}{\sqrt{2}}\begin{bmatrix} I & \sigma_2 \\ -\sigma_2 & I \end{bmatrix} \tag{15a}$$

† In this problem, the only assumption that need be made is that the γ_μ satisfy the commutation relations (XII.20); there is no need to assume that they are 4×4 Hermitian matrices (see the next problem).

transform the matrices, γ_μ from the Pauli representation into the Kramers representation, and from the Kramers representation into the Majorana representation, i.e. that

$$\gamma_\mu^K = U_{P\to K}\gamma_\mu U_{P\to K}^{-1} \quad \text{and} \quad \gamma_\mu^M = U_{K\to M}\gamma_\mu^K U_{K\to M}^{-1}. \tag{15b}$$

Find the unitary matrix $U_{P\to M}$ and the corresponding transformation of the two-component spinor (XII.17), which, together with the transformations (15b), leave the Dirac equation (XII.18) unchanged in form.

16. Starting from the Dirac equation (XII.18′) deduce the equation of conservation

$$\frac{\partial j_\mu}{\partial x_\mu} = 0, \tag{16a}$$

where

$$j_\mu = ic\bar{\psi}\gamma_\mu\psi, \tag{16b}$$

and

$$\bar{\psi} = \psi^+\gamma_4$$

($\bar{\psi}$ is called the "Dirac conjugate" of ψ).

17. Find the solutions of the Dirac equation (XII.14) which represent stationary states of free particles with well-defined linear momenta.

18. By using equation (V.4), find the time dependence of the position operator $\mathbf{r}$ of a free particle of spin $\frac{1}{2}$ in the Heisenberg picture. Discuss the result obtained.

19. Find a real function f such that the solution ψ of the Dirac equation (17.2) can, through the unitary transformation e^{iS}, with

$$S = -\frac{i}{2mc}\beta\boldsymbol{\alpha}\,.\,\mathbf{p}f\left(\frac{|\mathbf{p}|}{mc}\right), \tag{19a}$$

be written in the form

$$\hat{\psi} = e^{iS}\psi = \hat{\psi}_+ + \hat{\psi}_-, \tag{19b}$$

in which $\hat{\psi}_+$ and $\hat{\psi}_-$ are the eigen-spinors of the Hamiltonian

$$\hat{H}_D = e^{iS}H_D e^{-iS} \tag{19c}$$

with positive ($E = +E_p$) and negative ($E = -E_p$) eigenvalues respectively, which have the last (or, respectively, the first) two components equal to zero.

From the answer to question 17, express the bi-spinors (17.19, 17.19′) in the above representation, which is called the Foldy–Wouthuysen representation.

20. Show that, in the non-relativistic limit $|\mathbf{p}| \ll mc$, the negative energy part (19.6′) of the two-component spinor $\hat{\psi}$ is negligible compared with the positive energy part (19.6).

21. Determine the form of the Hamiltonian $\hat{H}_D$ of problem 19, and find an explicit expression for the transformation operator e^{iS} if, instead of (19.3), the following form is chosen for f:

$$f = -\frac{mc}{|\mathbf{p}|} \tan^{-1} \frac{mc}{|\mathbf{p}|}. \tag{21a}$$

22. Show that the matrix U in the relation

$$\psi'(x'_\mu) = U\psi(x_\mu), \tag{22a}$$

in which $\psi(x_\mu)$ and $\psi'(x'_\mu)$ are corresponding solutions of the Dirac equation (XII.18) in two different frames of reference related by the transformation (XII.1), satisfies the equation

$$U^{-1}\gamma_\mu U = a_{\mu\nu}\gamma_\nu. \tag{22b}$$

23. Show that, under the transformation (22a) of ψ, the Dirac conjugate function $\bar{\psi}$ transforms as follows:

$$\bar{\psi}'(x'_\mu) = \frac{a_{44}}{|a_{44}|}\bar{\psi}(x_\mu)U^{-1}. \tag{23a}$$

24. Show that, for an infinitesimal transformation

$$x'_\mu = a_{\mu\nu}x_\nu = (\delta_{\mu\nu}+\varepsilon_{\mu\nu})x_\nu \tag{24a}$$

where $\varepsilon_{\mu\nu}$ is an infinitesimal four-tensor, the matrix U introduced in problem 22 can be written in the form

$$U = 1+\tfrac{1}{4}\varepsilon_{\mu\nu}\gamma_\mu\gamma_\nu. \tag{24b}$$

25. Using the relations (22a), (22b) and (23a), show that if the orthogonal transformations (XII.1) do not change the sign of the time ($a_{44} > 0$), then the following quantities:

$$\begin{aligned} S &= \bar{\psi}\psi, \quad V_\mu = \bar{\psi}\gamma_\mu\psi, \\ T_{\mu\nu} &= \tfrac{1}{2}\bar{\psi}(\gamma_\mu\gamma_\nu-\gamma_\nu\gamma_\mu)\psi \\ A_\mu &= \bar{\psi}\gamma_\mu\gamma_5\psi, \quad P = \bar{\psi}\gamma_5\psi \end{aligned} \tag{25a}$$

transform respectively as a scalar, a four-vector, an anti-symmetrical four-tensor, an axial four-vector, and a pseudo scalar.

26. Show that the Klein–Gordon equation (XII.13) and the Dirac equation (XII.22) are invariant under a gauge transformation of the second kind (XII.8), if the following unitary transformation is made simultaneously:

$$\psi \to \psi \exp\left(\frac{ie}{\hbar c}G\right) \tag{26a}$$

This is called a "gauge transformation of the first kind".

27. Using in succession the Pauli, the Kramers and the Majorana representations for the matrices γ_μ, find the matrix elements of the "charge conjugation" operator, which transforms the solution ψ of the Dirac equation (XII.21) into the solution ψ_C of the equation obtained from (XII.21) by changing the charge e to $-e$.

Show that, if ψ describes a stationary state of positive energy

$$\psi(\mathbf{r}, t) = \psi(\mathbf{r})e^{-\frac{i}{\hbar}Et}, \quad E > 0,$$

then ψ_C describes a stationary state of negative energy $-E$. What is the relation between $\bar{\psi}$ and $\bar{\psi}_C$?

28. Find the "Schrödinger" form of the Dirac equation for a charged particle in an electromagnetic field $A_\mu = (\mathbf{A}, iA_0)$.

29. The interaction of a Dirac particle with an electromagnetic field can lead to transitions between the states with positive energy and those with negative energy. Hence, in contrast with the case of a free particle (see problem 19), the Foldy–Wouthuysen transformation,

$$\hat{\psi} = e^{iS}\psi, \tag{29a}$$

can result in a covariant separation of the contributions of these states only to some order of non-relativistic approximation.

By expanding the operator S in a power series in $1/mc^2$, find the Hamiltonian of the transformed equation

$$i\hbar \frac{\partial \hat{\psi}}{\partial t} = \hat{H}\hat{\psi} \tag{29b}$$

to an approximation which retains terms of order

(kinetic energy$/mc^2)^3$ and (kinetic energy). (field energy)$/(mc^2)^2$.

30. A free electron of energy $E = E_p$ moving in the z-direction meets a "step" in the electrostatic potential at the surface $z = 0$ (Fig. XII.1). Calculate the reflection and transmission coefficients R and T, defined in (II.7), in which, in this case, $\mathbf{j}$ is given by (16.4).

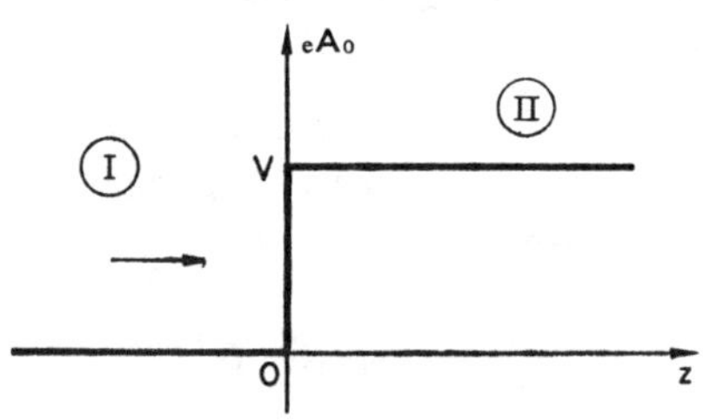

FIG. XII.1.

31. Find the values of the coefficients R and T of the preceding problem if

(1) $V = 0$; (2) $0 < V < E_p - E_0$; (3) $V = E_p - E_0$; (4) $E_p - E_0 < V < E_p + E_0$; (5) $V = E_p + E_0$; (6) $E_p + E_0 < V < \infty$; and (7) $V = \infty$.

Discuss the results by comparing them with the non-relativistic ones obtained from the Schrödinger equation (see problem 22, Chapter II).

How can the paradoxical situation (the Klein paradox†), which arises if $E_p + E_0 < V \leqslant \infty$, be explained?

32. Using the Dirac equation, find the energies of the bound states of an electron in the Coulomb potential $A_0 = |e|Z/r$ of an infinitely heavy nucleus (the problem of hydrogen-like atoms treated relativistically).

33. Using the Dirac equation (XII.22), find the energy levels of an electron in a homogeneous constant magnetic field $\mathbf{H}$. Compare the result with that of problem 34 of Chapter VI.

34. Show that Maxwell's equations in vacuo:

$$\nabla \times \mathbf{E} + \frac{1}{c}\frac{\partial \mathbf{H}}{\partial t} = 0, \quad \nabla \times \mathbf{H} - \frac{1}{c}\frac{\partial \mathbf{E}}{\partial t} = 0$$
$$\nabla . \mathbf{H} = 0, \qquad \nabla . \mathbf{E} = 0 \tag{34a}$$

can be written in the "Schrödinger" form

$$i\hbar \frac{\partial \psi}{\partial t} = H\psi, \tag{34b}$$

in which

$$\psi = \begin{bmatrix} \psi_1 \\ \psi_2 \\ \psi_3 \end{bmatrix}, \quad \psi_j = E_j + iH_j, \quad j = 1, 2, 3, \tag{34c}$$

and

$$\frac{\partial \psi_j}{\partial x_j} = 0. \tag{34d}$$

35. Find the solutions of (34b) which correspond to states of well-defined linear momentum of a free photon.

36. By choosing the z-axis of the coordinate system to be in the direction of $\mathbf{p}$, find the physical meaning of the solutions (35.9) of the eigenvalue equation (35.10) given in the answer to problem 35 above.

† O. Klein, *Zs. f. Phys.* **53**, 157 (1929).

Solutions

1. Multiplying (XII.12′) by ψ^*, and its complex conjugate by ψ, and subtracting the results, we find that

$$\psi^* \nabla^2\psi - (\nabla^2\psi^*)\,\psi - \frac{1}{c^2}\left(\psi^* \frac{\partial^2\psi}{\partial t^2} - \frac{\partial^2\psi^*}{\partial t^2}\psi\right) = 0,$$

i.e.

$$\nabla[\psi^* \nabla\psi - (\nabla\psi^*)\,\psi] - \frac{1}{c^2}\frac{\partial}{\partial t}\left(\psi^* \frac{\partial\psi}{\partial t} - \frac{\partial\psi^*}{\partial t}\psi\right) = 0. \tag{1.1}$$

Now, since

$$\frac{i\hbar}{2mc^2}\left(\psi^* \frac{\partial\psi}{\partial t} - \frac{\partial\psi^*}{\partial t}\psi\right) = \mathrm{Im}\left(\frac{\hbar}{mc^2}\psi \frac{\partial\psi^*}{\partial t}\right) = \varrho, \tag{1.2}$$

and

$$\frac{\hbar}{2mi}[\psi^* \nabla\psi - (\nabla\psi^*)\,\psi] = \mathrm{Re}\left(\frac{\hbar}{im}\psi^* \nabla\psi\right) = \mathbf{j}, \tag{1.3}$$

equation (1a) follows immediately.

On substituting the plane wave (1c) into (XII.12′), we obtain

$$\left(-\frac{\mathbf{p}^2}{\hbar^2} + \frac{E^2}{c^2\hbar^2} - k^2\right)\psi(\mathbf{r}, t) = 0,$$

whence

$$E = \pm c\sqrt{\mathbf{p}^2 + m^2c^2} = \pm E_p. \tag{1.4}$$

From the two possible signs of the energy, it can be seen that the general solution of the Klein–Gordon equation for a free particle can, owing to the linearity of the equation, be regarded as a superposition of plane waves travelling in different directions. Since the differential equation (XII.12′) is of second order in t (unlike the non-relativistic ones), a knowledge of the function $\psi(\mathbf{r}, t)$ at a given time t, without specifying also the function $[\partial\psi(\mathbf{r}, t)/\partial t]$ at the same time, is not sufficient to determine it at any later time. Furthermore, the quantity $\varrho(\mathbf{r}, t)$ need not always be positive; thus for the stationary state (1c), we have

$$\varrho = \frac{E}{mc^2}|\psi|^2 \quad \text{and} \quad \mathbf{j} = \frac{\mathbf{p}}{m}|\psi|^2. \tag{1.5}$$

Hence, unless the eigenfunctions with negative energy eigenvalues can be regarded as having no physical significance, the function $\psi(\mathbf{r}, t)$ cannot lead to the type of statistical prediction which is so successful in non-relativistic quantum mechanics. In the non-relativistic limit $|\mathbf{p}| \ll mc$, it can be seen that $E_p \approx mc^2$, $\mathbf{p}/m \approx \mathbf{v}$, and that the expressions (1.5) become the well-known non-relativistic expressions for the probability density and the probability flux:

$$\varrho \approx |\psi|^2, \quad \mathbf{j} \approx \mathbf{v}\,|\psi|^2.$$

Remarks: For a free particle, the solutions of the Klein–Gordon equation with negative energies can be ignored without loss of consistency, since the Hilbert space spanned by the solutions with positive energies is a complete one. Transitions from the positive to the negative energy states are then not possible, and the quantity ϱ given by (1.5) is always positive, so that a statistical interpretation is possible, at least in some circumstances.† The eigenfunctions of the position operator $\mathbf{r}$ are no longer δ functions, but are more complicated functions which allow the particle to be located only in a volume having linear dimensions of the same order as the Compton wavelength $\hbar/mc$ of the particle.

2. With the notation

$$\psi_4 = -\frac{1}{k}\frac{\partial \psi}{\partial x_4}, \tag{2.1}$$

the Klein–Gordon equation (XII.12′) becomes equivalent to the system of equations

$$\begin{aligned} &\frac{\partial \psi}{\partial x_4} + k\psi_4 = 0 \\ &\nabla^2\psi - k\frac{\partial \psi_4}{\partial x_4} - k^2\psi = 0. \end{aligned} \tag{2.2}$$

From (2d) we obtain

$$\psi = \frac{1}{\sqrt{2}}(\phi+\chi), \quad \psi_4 = \frac{1}{\sqrt{2}}(\phi-\chi) \tag{2.3}$$

and thus from (2.2) and (2.3) the following system of equations:

$$\begin{aligned} i\hbar\frac{\partial \phi}{\partial t} &= -\frac{\hbar^2}{2m}\nabla^2(\phi+\chi) + mc^2\phi \\ i\hbar\frac{\partial \chi}{\partial t} &= \frac{\hbar^2}{2m}\nabla^2(\phi+\chi) - mc^2\chi \end{aligned} \tag{2.4}$$

or, in matrix form,

$$i\hbar\frac{\partial}{\partial t}\begin{bmatrix}\phi\\ \chi\end{bmatrix} = -\frac{\hbar^2}{2m}\nabla^2\begin{bmatrix}1 & 1\\ -1 & -1\end{bmatrix}\begin{bmatrix}\phi\\ \chi\end{bmatrix} + mc^2\begin{bmatrix}1 & 0\\ 0 & -1\end{bmatrix}\begin{bmatrix}\phi\\ \chi\end{bmatrix}. \tag{2.5}$$

With the aid of the matrices (2a) we can write

$$\begin{bmatrix}1 & 1\\ -1 & -1\end{bmatrix} = \tau_3 + i\tau_2$$

and thus equation (2b) follows immediately. It can be seen that it is of the same form as the Schrödinger equation of non-relativistic quantum mechanics.

† R. G. Newton and E. P. Wigner, *Rev. Mod. Phys.* **21**, 400 (1949).

3. From (2.1) and (2.3) we have

$$\varrho = \operatorname{Im}\left(\frac{\hbar}{mc^2}\frac{\partial \psi^*}{\partial t}\psi\right) = \operatorname{Im}(i\psi_4^*\psi) = \frac{1}{2}[\psi_4^*\psi + (\psi_4^*\psi)^*]$$

$$= \frac{1}{4}[(\phi^*-\chi^*)(\phi+\chi)+(\phi-\chi)(\phi^*+\chi^*)] = \frac{1}{2}(\phi^*\phi-\chi^*\chi)$$

$$= \frac{1}{2}\,\boxed{\phi^*\chi^*}\begin{bmatrix}1 & 0\\ 0 & -1\end{bmatrix}\begin{bmatrix}\phi\\ \chi\end{bmatrix} = \frac{1}{2}\psi^+\tau_3\psi.$$

Similarly

$$\mathbf{j} = \operatorname{Re}\left(\frac{\hbar}{im}\psi^*\nabla\psi\right) = \frac{\hbar}{2im}(\psi^*\nabla\psi - \psi\nabla\psi^*)$$

$$= \frac{\hbar}{4im}[(\phi^*+\chi^*)\nabla(\phi+\chi)-(\phi+\chi)\nabla(\phi^*+\chi^*)]$$

$$= \frac{\hbar}{4im}\left[\boxed{\phi^*\chi^*}\begin{bmatrix}1 & 1\\ 1 & 1\end{bmatrix}\begin{bmatrix}\nabla\phi\\ \nabla\chi\end{bmatrix} - \boxed{\nabla\phi^*\,\nabla\chi^*}\begin{bmatrix}1 & 1\\ 1 & 1\end{bmatrix}\begin{bmatrix}\phi\\ \chi\end{bmatrix}\right]$$

$$= \frac{\hbar}{4im}[\psi^+\tau_3(\tau_3+i\tau_2)\nabla\psi - (\nabla\psi)^+\tau_3(\tau_3+i\tau_2)\psi].$$

Remarks: In non-relativistic quantum mechanics a metric in the space of the wavefunctions was introduced by defining a scalar product such that

$$\begin{aligned}\langle\psi_1,\psi_2\rangle &= \int \psi_1^*\psi_2\,d\mathbf{r}\\ \langle\psi,\psi\rangle &= \int |\psi|^2\,d\mathbf{r} = \int \varrho\,d\mathbf{r}.\end{aligned} \tag{3.1}$$

In this metric, called the *Schrödinger metric*, the mean value $\langle A\rangle$ and the Hermitian conjugate A^+ of an observable A are defined respectively by the expressions†

$$\langle A\rangle = \langle\psi, A\psi\rangle = \int \psi^* A\psi\,d\mathbf{r} \tag{3.2}$$

and

$$\langle\psi_1, A\psi_2\rangle = \langle A^+\psi_1, \psi_2\rangle. \tag{3.3}$$

As was shown in problem 2, the Klein–Gordon equation (XII.12′) can be put into the Schrödinger form (2b) by the use of two-component wavefunctions. Intuition suggests that it should be possible to express the energy, the linear momentum and the angular momentum of spinless particles in relativistic quantum mechanics as mean values of differential operators if a metric can be suitably defined in the space of the two-component functions. Arguing by analogy (cf. (3.1) and (3a)) let us define a metric in the space of the two-component wave-

† The mean value $\langle A\rangle$ given by (3.2) differs from that given by (III.3) by a factor equal to the norm of the wavefunction $\langle\psi,\psi\rangle$. Since the latter is a constant of the motion, the two definitions are equivalent so far as the present discussion is concerned.

functions by defining a scalar product of $\psi_1(\mathbf{r}, t)$ and $\psi_2(\mathbf{r}, t)$ as follows:

$$\langle\psi_1, \psi_2\rangle_{FV} = \tfrac{1}{2}\int \psi_1^+\tau_3\psi_2\, d\mathbf{r}. \tag{3.4}$$

The resultant metric is called the *Feshbach–Villars metric.*† In the F–V metric, the mean value $\langle A\rangle_{FV}$, and the Hermitian conjugate A^H of an operator A, are defined by the relations

$$\langle A\rangle_{FV} = \langle\psi, A\psi\rangle_{FV} = \tfrac{1}{2}\int \psi^+\tau_3 A\psi\, d\mathbf{r}, \tag{3.5}$$

and

$$\langle\psi_1, A\psi_2\rangle_{FV} = \langle A^H\psi_1, \psi_2\rangle_{FV} \tag{3.6}$$

respectively.

The operator A is said to Hermitian if

$$A^H = A \tag{3.6'}$$

and unitary if

$$\langle A\psi_1, A\psi_2\rangle_{FV} = \langle\psi_1, \psi_2\rangle_{FV}. \tag{3.7}$$

Using (3.5), we find the following expressions for the energy, the linear momentum, and the angular momentum of a spinless free particle:

$$E = \langle\psi, H\psi\rangle_{FV} \tag{3.8}$$

$$\mathbf{p} = \left\langle \psi, \frac{\hbar}{i}\nabla\psi\right\rangle_{FV} \tag{3.9}$$

$$\mathbf{l} = \left\langle \psi, \mathbf{r}\times\frac{\hbar}{i}\nabla\psi\right\rangle_{FV}. \tag{3.10}$$

The values of E, $\mathbf{p}$ and $\mathbf{l}$ given by these expressions agree with those calculated from the energy-momentum tensor of the Klein–Gordon field (see problem 5). This gives a justification of the definition (3.5).

4. Using (3.5) we can write

$$\langle\psi_1, A\psi_2\rangle_{FV} = \tfrac{1}{2}\int \psi_1^+\tau_3\psi_2\, d\mathbf{r} = \langle A^H\psi_1, \psi_2\rangle_{FV} = \tfrac{1}{2}\int \psi_1^+(A^H)^+\tau_3\psi_2\, d\mathbf{r}.$$

Since $\tau_3^+ = \tau_3$ in this problem, (4a) follows immediately. Similarly, from

$$\langle A\psi_1, A\psi_2\rangle_{FV} = \tfrac{1}{2}\int \psi_1^+ A\tau_3 A\psi_2\, d\mathbf{r} = \langle\psi_1, \psi_2\rangle_{FV} = \tfrac{1}{2}\int \psi_1^+\tau_3\psi_2\, d\mathbf{r}$$

we can obtain (4b).

The equalities (4c) are obtained immediately by using (4a) and the well known properties of the matrices τ_i:

$$\tau_i^+ = \tau_i, \quad \tau_k\tau_l + \tau_l\tau_k = 2\delta_{lk}, \quad \tau_k\tau_l = i\tau_m, \tag{4.1}$$

in which (k, l, m) stands for any cyclic permutation of the indices (1, 2, 3).

† H. Feshbach and F. Villars, *Rev. Mod. Phys.* **30**, 24 (1958).

Note that, in the F–V metric, τ_3, $i\tau_1$ and $i\tau_2$ are Hermitian, and τ_1, τ_2 and $i\tau_3$ are anti-hermitian.

Using (2c) and (2d) we have that

$$\langle A\rangle_{FV} = \frac{1}{2}\int \overline{\phi^*\chi^*}\begin{bmatrix}1 & 0\\ 0 & -1\end{bmatrix}\begin{bmatrix}A\phi\\ A\chi\end{bmatrix}d\mathbf{r} = \frac{i\hbar}{2mc^2}\int\left(\psi^* A\frac{\partial\psi}{\partial t} - \frac{\partial\psi^*}{\partial t}A\psi\right)d\mathbf{r}. \tag{4.2}$$

5. Using (3.8), (2c)–(2e) and (XII.12′) we have that

$$E = \frac{1}{2}\int \overline{\phi^*\chi^*}\begin{bmatrix}-\frac{\hbar^2}{2m}\nabla^2+mc^2 & -\frac{\hbar^2}{2m}\nabla^2\\ -\frac{\hbar^2}{2m}\nabla^2 & -\frac{\hbar^2}{2m}\nabla^2+mc^2\end{bmatrix}\begin{bmatrix}\phi\\ \chi\end{bmatrix}d\mathbf{r}$$

$$= -\frac{\hbar^2}{2mc^2}\int\left(\psi^*\frac{\partial^2\psi}{\partial t^2} - \frac{\partial\psi^*}{\partial t}\frac{\partial\psi}{\partial t}\right)d\mathbf{r}. \tag{5.1}$$

For the operators $\frac{\hbar}{i}\nabla$ and $\mathbf{r}\times\frac{\hbar}{i}\nabla$ respectively, (4.2) yields the results

$$\mathbf{p} = \frac{i\hbar}{2mc^2}\int\left[\psi^*\left(\frac{\hbar}{i}\nabla\right)\frac{\partial\psi}{\partial t} - \frac{\partial\psi^*}{\partial t}\left(\frac{\hbar}{i}\nabla\right)\psi\right]d\mathbf{r}, \tag{5.2}$$

$$\mathbf{l} = \frac{i\hbar}{2mc^2}\int\left[\psi^*\left(\mathbf{r}\times\frac{\hbar}{i}\nabla\right)\frac{\partial\psi}{\partial t} - \frac{\partial\psi^*}{\partial t}\left(\mathbf{r}\times\frac{\hbar}{i}\nabla\right)\psi\right]d\mathbf{r}. \tag{5.3}$$

The quantities E, $\mathbf{p}$ and $\mathbf{l}$ given by (5.1)–(5.3) agree with the energy, the linear momentum and the angular momentum of the Klein–Gordon field, as calculated from its energy-momentum tensor.†

Note that the operators H, $-i\hbar\nabla$ and $-i\hbar\mathbf{r}\times\nabla$ are Hermitian in the sense of (3.6′).

6. With the notation

$$D_\mu = \frac{\partial}{\partial x_\mu} - \frac{ie}{\hbar c}A_\mu, \tag{6.1}$$

the Klein–Gordon equation (XII.13′) can be written as

$$(D_\mu^2 - k^2)\psi = 0. \tag{6.2}$$

Taking the complex conjugate of (6.2), and defining

$$\bar{D}_\mu = \frac{\partial}{\partial x_\mu} + \frac{ie}{\hbar c}A_\mu \tag{6.3}$$

† See Corinaldesi-Strocchi (1963), Chapter II.

we obtain

$$(\bar{D}_\mu^2 - k^2)\psi^* = 0.$$

Following the procedure of problem 1 we have then that

$$\psi^*(D_\mu^2\psi) - (\bar{D}_\mu^2\psi^*)\psi = 0.$$

On imposing the Lorentz gauge condition

$$\frac{\partial A_\mu}{\partial x_\mu} = 0, \tag{6.4}$$

it follows that

$$\frac{\partial}{\partial x_\mu}\left(\psi^* \frac{\partial \psi}{\partial x_\mu} - \frac{\partial \psi^*}{\partial x_\mu}\psi - \frac{2ie}{\hbar c} A_\mu \psi^* \psi\right) = 0. \tag{6.5}$$

On multiplying (6.5) by $\hbar/2mi$, we obtain the continuity equation (6a), in which the components of the four-vector j_μ are given by

$$\mathbf{j} = \frac{\hbar}{2mi}[\psi^* \nabla\psi - (\nabla\psi^*)\psi] - \frac{e}{mc}\mathbf{A}\psi^*\psi = \text{Re}\left(\frac{\hbar}{mi}\psi^* \nabla\psi - \frac{e}{mc}\mathbf{A}\psi^*\psi\right), \tag{6.6}$$

$$\varrho = \frac{i\hbar}{2mc^2}\left(\psi^* \frac{\partial\psi}{\partial t} - \frac{\partial\psi^*}{\partial t}\psi\right) - \frac{e}{mc^2}A_0\psi^*\psi = \text{Im}\left(\frac{\hbar}{mc^2}\frac{\partial\psi^*}{\partial t}\psi - \frac{ie}{mc^2}A_0\psi^*\psi\right). \tag{6.7}$$

In the absence of an electromagnetic field, ($\mathbf{A} = 0$, $A_0 = 0$), (6.6) and (6.7) reduce to (1b).

7. With the notation

$$\psi_4 = -\frac{1}{k}D_4\psi, \tag{7.1}$$

$$D_4 = \frac{\partial}{\partial x_4} - \frac{ie}{\hbar c}A_4, \qquad k = \frac{mc}{\hbar},$$

the Klein–Gordon equation (XII.13′) is equivalent to the system of equations

$$\begin{aligned} D_4\psi + k\psi_4 &= 0 \\ \nabla^2\psi - kD_4\psi_4 - k^2\psi &= 0. \end{aligned} \tag{7.2}$$

Let us now take

$$\psi = \begin{bmatrix}\phi \\ \chi\end{bmatrix}, \tag{7.3}$$

where

$$\phi = \frac{1}{\sqrt{2}}(\psi + \psi_4) \quad \text{and} \quad \chi = \frac{1}{\sqrt{2}}(\psi - \psi_4). \tag{7.4}$$

Then, from (7.2) and (7.4), the following system of equations is obtained

$$i\hbar\frac{\partial\phi}{\partial t} = \frac{1}{2m}\left(\frac{\hbar}{i}\nabla-\frac{e}{c}\mathbf{A}\right)^2(\phi+\chi)+(eA_0+mc^2)\phi$$
$$i\hbar\frac{\partial\chi}{\partial t} = -\frac{1}{2m}\left(\frac{\hbar}{i}\nabla-\frac{e}{c}\mathbf{A}\right)^2(\phi+\chi)+(eA_0-mc^2)\chi, \tag{7.5}$$

which is equivalent to the matrix equation

$$i\hbar\frac{\partial\psi}{\partial t} = H\psi, \tag{7.6}$$

$$H = \frac{1}{2m}(\tau_3+i\tau_2)\left(\frac{\hbar}{i}\nabla-\frac{e}{c}\mathbf{A}\right)^2+mc^2\tau_3+eA_0I. \tag{7.7}$$

From (7.4) it follows that

$$\psi = \frac{1}{\sqrt{2}}(\phi+\chi) \quad \text{and} \quad \frac{\partial\psi}{\partial t} = \frac{mc^2}{i\hbar\sqrt{2}}\left[\phi-\chi+\frac{eA_0}{mc^2}(\phi+\chi)\right]. \tag{7.8}$$

Thus (6.6) and (6.7) can be written as

$$\mathbf{j} = \frac{\hbar}{4im}[\psi^+\tau_3(\tau_3+i\tau_2)\nabla\psi-(\nabla\psi)^+\tau_3(\tau_3+i\tau_2)\psi]-\frac{e}{2mc}\mathbf{A}\psi^+\tau_3(\tau_3+i\tau_2)\psi, \tag{7.9}$$

$$\varrho = \tfrac{1}{2}(\phi^*\phi-\chi^*\chi) = \tfrac{1}{2}\psi^+\tau_3\psi. \tag{7.10}$$

Remarks: The advantages of the matrix formalism just developed are as follows:

(a) The formal equivalence of the Klein–Gordon equation (7.6) and the Schrödinger equation offers, in some circumstances (see problems 8 and 9), the possibility of a one-particle interpretation of the results.

(b) The density (7.10) appears as the difference of two positive densities, as it must be in a theory which describes simultaneously particles with charges of both signs (for example, the π^+ and the π^- mesons). Thus $e\mathbf{j}$ can be regarded† as a net electric current and $e\varrho$ as a net density of electric charge, both expressed as the difference between the current (or density) of positively and negatively charged particles respectively.

(c) In this formalism the charge symmetry which is characteristic of the relativistic equation appears explicitly.

Thus, if we take the complex conjugates of equations (7.5):

$$i\hbar\frac{\partial\chi^*}{\partial t} = \frac{1}{2m}\left(\frac{\hbar}{i}\nabla+\frac{e}{c}\mathbf{A}\right)^2(\phi^*+\chi^*)+(mc^2-eA_0)\chi^*$$
$$i\hbar\frac{\partial\phi^*}{\partial t} = -\frac{1}{2m}\left(\frac{\hbar}{i}\nabla+\frac{e}{c}\mathbf{A}\right)^2(\phi^*+\chi^*)-(mc^2+eA_0)\phi^* \tag{7.11}$$

† W. Pauli and V. F. Weisskopf, *Helv. Phys. Acta*, **1**, 709 (1934).

we see that, if (7.3) satisfies the equation

$$i\hbar \frac{\partial \psi}{\partial t} = H(e)\psi, \tag{7.12}$$

then there exists a function

$$\psi_C = \begin{bmatrix} \chi^* \\ \phi^* \end{bmatrix} = \tau_1 \psi^* \tag{7.13}$$

obtained from ψ by the operation of "charge conjugation", which satisfies the equation

$$i\hbar \frac{\partial \psi_C}{\partial t} = H(-e)\psi_C. \tag{7.14}$$

From (7.12)–(7.14) it follows that

$$\tau_1 H^*(e)\tau_1 = -H(-e). \tag{7.15}$$

Thus, if ψ is an eigenfunction of $H(e)$ with energy E, i.e. if $H(e)\psi = E\psi$, then the conjugate function ψ_C is an eigenfunction of $H(-e)$ with energy $-E$, and conversely.

Note also that

$$\varrho_C = \tfrac{1}{2}\overset{+}{\psi_C}\tau_3\psi_C = -\varrho \quad \text{and} \quad \mathbf{j}_C = \mathbf{j}. \tag{7.16}$$

The law of conservation of electric charge makes it possible, with a proper normalization of ψ, to have

$$\int \varrho \, d\mathbf{r} = \tfrac{1}{2}\int \psi^+ \tau_3 \psi \, d\mathbf{r} = \pm 1 \tag{7.17}$$

and

$$\tfrac{1}{2}\int \overset{+}{\psi_C}\tau_3\psi_C \, d\mathbf{r} = \mp 1. \tag{7.18}$$

It is important to note that, in the limit $e \to 0$ the function ψ_C becomes a solution of (2b) which is independent of ψ (see problem 8).

Thus, in the absence of an electromagnetic field, there is a one-to-one correspondence between the "positive" and the "negative" solutions of (2b). The positive solution, normalized to $+1$ (according to the F–V metric (3.4)), describes a positive charge, the negative solution, normalized to -1, describes a negative charge.

This correspondence must remain in the presence of an electromagnetic field.

8. The solutions of (2b), corresponding to the eigenvalues (1.4), viz.,

$$E = \pm c\sqrt{\mathbf{p}^2 + m^2c^2} = \pm E_p$$

of the Hamiltonian (2e), are

$$\psi^{(+,-)} = \begin{bmatrix} \phi^{(+,-)} \\ \chi^{(+,-)} \end{bmatrix} = \begin{bmatrix} u^{(+,-)} \\ v^{(+,-)} \end{bmatrix} \exp\left[\frac{i}{\hbar}(\mathbf{p}.\mathbf{r} \mp E_p t)\right] \tag{8.1}$$

where

$$u^{(+,-)} = \frac{A}{mc^2\sqrt{2}}(mc^2 \pm E_p), \quad v^{(+,-)} = \frac{A}{mc^2\sqrt{2}}(mc^2 \mp E_p). \tag{8.2}$$

Note that, in accordance with what was said in the preceding problem, the function obtained from $\psi^{(+)}$ by charge conjugation:

$$\psi_C^{(+)} = \begin{bmatrix} \chi^{(+)*} \\ \phi^{(+)*} \end{bmatrix} = \begin{bmatrix} v^{(+)} \\ u^{(+)} \end{bmatrix} \exp\left[\frac{i}{\hbar}(-\mathbf{p}.\mathbf{r}+E_p t)\right] = \begin{bmatrix} u^{(-)} \\ v^{(-)} \end{bmatrix} \exp\left[\frac{i}{\hbar}(-\mathbf{p}.\mathbf{r}+E_p t)\right], \quad (8.3)$$

is an eigenfunction of (2b) with the negative energy eigenvalue $-E_p$ [only the square of p appears in (1.4)].

Thus, if the solution $\psi^{(+)}$ represents a particle of positive charge, $\psi^{(-)}$ will represent a particle of negative charge, and by (7.17) and (7.18), we have that

$$\tfrac{1}{2}\int \psi^{(+)+}\tau_3\psi^{(+)}\, d\mathbf{r} = +1$$

and

$$\tfrac{1}{2}\int \psi^{(-)+}\tau_3\psi^{(-)}\, d\mathbf{r} = -1.$$

If we take the normalization volume equal to unity, it follows from the above relations that we shall have $A = (mc^2/E_p)^{1/2}$ and hence

$$u^{(+)} = \frac{mc^2+E_p}{\sqrt{2mc^2E_p}}, \quad v^{(+)} = \frac{mc^2-E_p}{\sqrt{2mc^2E_p}}, \quad \frac{1}{2}\left(u^{(+)2}-v^{(+)2}\right) = +1$$

$$u^{(-)} = \frac{mc^2-E_p}{\sqrt{2mc^2E_p}}, \quad v^{(-)} = \frac{mc^2+E_p}{\sqrt{2mc^2E_p}}, \quad \frac{1}{2}\left(u^{(-)2}-v^{(-)2}\right) = -1.$$

If a particle is such that its motion is correctly described by the wavefunction ψ, then a particle whose motion is correctly described by ψ_C is said to be the "anti-particle" of the first one. Thus, if $\psi^{(+)}$ describes the free motion of a π^+ meson, the function $\psi_C^{(+)}$ will describe the free motion of a π^- meson, which is the anti-particle of the π^+. Particles and anti-particles differ not only in electric charge but in other properties also (e.g., in magnetic moment, in baryon number, etc.). Under the charge conjugation operation these quantities change their sign. Particles which have no electric charge are sometimes identical with their own anti-particles; this applies, e.g., to the π^0 meson and the photon, but not to the neutron or to the neutrino. From what has been said above it can easily be deduced that a π^0 meson can be described by a real wavefunction, a fact which is true for all particles which are identical with their anti-particles.

9. For an electrostatic field, $\mathbf{A} = 0$, and the Lorentz condition (6.4) reduces to

$$\frac{\partial A_0}{\partial t} = 0. \quad (9.1)$$

Equation (9a) then follows from (XII.13′). The equation for stationary states is obtained as usual if we substitute into (9a) the trial solution

$$\psi(\mathbf{r}, t) = \phi(\mathbf{r}) \exp\left(-\frac{i}{\hbar}Et\right). \quad (9.2)$$

After some simple calculation it is found that

$$\left[\nabla^2 - k^2 + \left(\frac{E - eA_0}{\hbar c}\right)^2\right]\phi(\mathbf{r}) = 0. \tag{9.3}$$

Remarks: From (6.7) it follows that, in the stationary state (9.2), the charge density is

$$e\varrho = e\frac{E - eA_0}{mc^2}|\phi|^2. \tag{9.4}$$

For $E > eA_0$ the sign of the charge density is the same as the sign of the charge of the particles. But for large potential energies, if $E < eA_0$, the sign of $e\varrho$ is opposite to the sign of e. Thus in strong fields the one-particle interpretation, which is valid for weak fields and for free particles, is no longer valid.

The reason why the charge density changes its sign in strong fields can be given only in the context of field theory, which takes into account the processes of pair creation and annihilation.

10. Since $m_\pi = 273 m_e$, the nuclear motion can be neglected, without serious loss of accuracy.

Thus, treating the nucleus as an infinitely heavy point, the energies of the bound states of the mesic atom are simply the energies of the bound states of a π^- meson in the Coulomb potential

$$A_0 = \frac{|e|Z}{r}. \tag{10.1}$$

Since the potential energy of the π^- meson is then $eA_0 = -\dfrac{e^2Z}{r}$, (9.3) becomes

$$\left(\nabla^2 - \frac{E_0^2 - E^2}{\hbar^2c^2} + \frac{2\alpha Ze}{\hbar c}\frac{1}{r} + \frac{\alpha^2Z^2}{r^2}\right)\phi(\mathbf{r}) = 0, \tag{10.2}$$

in which

$$E_0 = m_\pi c^2 \quad \text{and} \quad \alpha = \frac{e^2}{\hbar c} = \frac{1}{137{\cdot}2}$$

is the fine structure constant.

If we use spherical coordinates and write

$$\phi(\mathbf{r}) = \frac{R_l(r)}{r} Y_l^m(\theta, \phi),$$

the radial variable can be separated from the angular variables, and the radial function $R_l(r)$ will satisfy the equation [cf. (II.10)–(II.17)]:

$$\frac{d^2R_l}{dr^2} - \left(\frac{E_0^2 - E^2}{\hbar^2c^2} - \frac{2\alpha ZE}{\hbar c}\frac{1}{r} + \frac{l(l+1) - \alpha^2Z^2}{r^2}\right)R_l = 0. \tag{10.3}$$

By introducing the dimensionless variable

$$\xi = \beta r, \quad \text{where} \quad \beta = \frac{2}{\hbar c}(E_0^2 - E^2)^{1/2},$$

(10.3) becomes

$$\frac{d^2R_l}{d\xi^2} - \left(\frac{1}{4} - \frac{\varepsilon}{\xi} + \frac{l(l+1) - \alpha^2 Z^2}{\xi^2}\right) R_l = 0. \tag{10.4}$$

The energy eigenvalues for which, in the asymptotic region ($\xi \to \infty$), the solutions R_l remain bounded, are contained in the parameter

$$\varepsilon = \frac{\alpha Ze}{\sqrt{E_0^2 - E^2}}. \tag{10.5}$$

The asymptotic solution of (10.4) which satisfies the required condition is $R_l \sim \exp\left(-\frac{\xi}{2}\right)$.

Looking then for R_l in the form

$$R_l = e^{-\xi/2} u(\xi) \tag{10.6}$$

we obtain for $u(\xi)$ the differential equation

$$\frac{d^2u}{d\xi^2} - \frac{du}{d\xi} + \left(\frac{\varepsilon}{\xi} - \frac{l(l+1) - \alpha^2 Z^2}{\xi^2}\right) u = 0. \tag{10.7}$$

Since $\xi = 0$ is a singularity of (10.7) we put

$$u(\xi) = \xi^s \sum_{k=0}^{\infty} c_k \xi^k, \quad s > 0, \quad c_0 \neq 0. \tag{10.8}$$

Substituting (10.8) into (10.7) we find

$$\sum_{k=0}^{\infty} \{[s(s-1) - l(l+1) + \alpha^2 Z^2] c_k \xi^{k+s-2} + [2s(k+1) c_{k+1} + (\varepsilon - s) c_k] \xi^{k+s-1} + [(k+1)(k+2) c_{k+2} - (k+1) c_{k+1}] \xi^{k+s}\} = 0. \tag{10.9}$$

This equation is satisfied only if the coefficients of all powers of ξ vanish. Putting equal to zero the coefficient of the lowest power of ξ, i.e. ξ^{s-2}, and of the general term ξ^{k+s-1}, we obtain the relations

$$s(s-1) - l(l+1) + \alpha^2 Z^2 = 0, \tag{10.10}$$

$$c_{k+1} = \frac{s+k-\varepsilon}{(k+1)(2s+k)} c_k. \tag{10.11}$$

For large values of k, $c_{k+1} \approx \frac{1}{k} c_k$, and and hence for large k the series (10.8) behaves like $\exp(\xi)$ and the factor $\exp(-\xi/2)$ from (10.6) cannot ensure the correct behaviour of R_l as $\xi \to \infty$. The series must therefore reduce to a polynomial, i.e. starting from a particular

$k = n'$ say, all coefficients must be equal to zero, and hence

$$s+n'-\varepsilon = 0, \qquad n' = 0, 1, 2 \ldots \tag{10.12}$$

It is evident that s is the positive solution of (10.10) for any given value of $l = 0, 1, 2, \ldots,$ i.e. that

$$s = \tfrac{1}{2}+(l+\tfrac{1}{2})\sqrt{1-\alpha^2Z^2(l+\tfrac{1}{2})^{-2}}. \tag{10.13}$$

Finally, from (10.5), (10.12) and (10.13), the following energy eigenvalues are obtained

$$E_{n'l} = E_0\left\{1+\alpha^2Z^2\left[n'+\tfrac{1}{2}+(l+\tfrac{1}{2})\sqrt{1-\alpha^2Z^2(l+\tfrac{1}{2})^{-2}}\right]^{-2}\right\}^{-1/2}. \tag{10.14}$$

By introducing a principal quantum number

$$n = n'+l+1, \qquad n = 1, 2, 3, \ldots$$

we find for the energies of bound states of a π^- meson the following expression

$$E_{nl} = E_0\left\{1+\alpha^2Z^2\left[(n-l-\tfrac{1}{2})+(l+\tfrac{1}{2})\sqrt{1-\alpha^2Z^2(l+\tfrac{1}{2})^{-2}}\right]^{-2}\right\}^{-1/2}. \tag{10.15}$$

In order to compare this relativistic result with that obtained from the Schrödinger equation [cf. the problem of hydrogen-like atoms, (II.35)] we expand (10.15) in a series in powers of αZ up to terms in $(\alpha Z)^4$, thus:

$$E_{nl} = E_0-\frac{E_0(\alpha Z)^2}{2n^2}+\frac{E_0(\alpha Z)^4}{2n^4}\left(\frac{3}{4}-\frac{n}{l+\frac{1}{2}}\right)+\ldots \tag{10.16}$$

The first term is the rest energy of the π^- meson. The second term gives the energy spectrum of the bound states in the non-relativistic approximation [cf. (35.11) Chapter II]. The third term is a relativistic correction which depends on the orbital quantum number, owing to which the energy levels calculated in non-relativistic approximation are split into components which depend on l, thus giving a *fine structure* to the spectrum.

As an historical detail we mention here that soon after the Klein–Gordon equation was proposed in 1926, a calculation similar to the one given above was carried out, in order to try to explain the observed fine structure in the spectra of hydrogen-like atoms. The calculated results disagreed with experimental observations; thus the predicted spread in energy

$$E_{n,\,n-1}-E_{n,\,0} = \frac{m_e e^8 Z^4}{c^2\hbar^4 n^3}\,\frac{n-1}{n-\frac{1}{2}} \tag{10.17}$$

of the fine-structure splitting of the level with principal quantum number n is much larger than the experimental value.

Later the reason for this discrepancy became clear: the Klein–Gordon equation describes spinless particles, so that the $\frac{1}{2}$ spin of the electron had been ignored. A correct formula for the fine structure, which is in fact very similar to (10.15), can be obtained from the Dirac equation (see problem 32).

11. The first of the relations (11a) results directly from (XII.20). The other relations can be verified as follows:

$$\gamma_\mu\gamma_\nu\gamma_\mu = (\gamma_\mu\gamma_\nu+\gamma_\nu\gamma_\mu)\gamma_\mu-\gamma_\nu\gamma_\mu\gamma_\mu = 2\gamma_\nu-4\gamma_\nu = -2\gamma_\nu,$$

$$\gamma_\mu\gamma_\nu\gamma_\lambda\gamma_\mu = (\gamma_\mu\gamma_\nu+\gamma_\nu\gamma_\mu)\gamma_\lambda\gamma_\mu-\gamma_\nu\gamma_\mu\gamma_\lambda\gamma_\mu = 2(\delta_{\mu\nu}\gamma_\lambda\gamma_\mu+\gamma_\nu\gamma_\lambda) = 2(\gamma_\lambda\gamma_\nu+\gamma_\nu\gamma_\lambda) = 4\delta_{\lambda\nu},$$

$$\begin{aligned}\gamma_\mu\gamma_\nu\gamma_\lambda\gamma_\varrho\gamma_\mu &= (\gamma_\mu\gamma_\nu+\gamma_\nu\gamma_\mu)\gamma_\lambda\gamma_\varrho\gamma_\mu-\gamma_\nu\gamma_\mu\gamma_\lambda\gamma_\varrho\gamma_\mu = 2(\gamma_\lambda\gamma_\varrho\gamma_\nu-2\gamma_\nu\delta_{\lambda\varrho})\\ &= 2(2\gamma_\nu\delta_{\lambda\varrho}-\gamma_\varrho\gamma_\lambda\gamma_\nu-2\gamma_\nu\delta_{\lambda\varrho}) = -2\gamma_\varrho\gamma_\lambda\gamma_\nu.\end{aligned}$$

Remembering that

$$\mathrm{Tr}\,(AB) = \mathrm{Tr}\,(BA) \tag{11.1}$$

(11b) follows immediately, since

$$\mathrm{Tr}\,(\gamma_\mu\gamma_\nu) = \mathrm{Tr}\,(\gamma_\nu\gamma_\mu) = \tfrac{1}{2}\,\mathrm{Tr}\,(\gamma_\nu\gamma_\mu+\gamma_\mu\gamma_\nu) = \delta_{\mu\nu}\,\mathrm{Tr}\,(I) = 4\delta_{\mu\nu}.$$

12. Propositions (1) to (3) of this problem can be verified by direct calculation using the commutation relations (XII.20). Thus, with e.g. the notation

$$\Gamma_6 = i\gamma_1\gamma_3 \quad \text{and} \quad \Gamma_{16} = \gamma_1\gamma_2\gamma_3\gamma_4,$$

we have that

$$\begin{aligned}\Gamma_6\Gamma_{16}\Gamma_6 &= -\gamma_1\gamma_3\gamma_1\gamma_2\gamma_3\gamma_4\gamma_1\gamma_3 = +\gamma_1\gamma_1\gamma_3\gamma_1\gamma_2\gamma_3\gamma_4\gamma_3\\ &= -\gamma_3\gamma_3\gamma_1\gamma_2\gamma_3\gamma_4 = -\gamma_1\gamma_2\gamma_3\gamma_4 = -\Gamma_{16}.\end{aligned}$$

Using (11.1), the relation (12e) follows as a direct consequence of (12d):

$$-\mathrm{Tr}\,(\Gamma_A) = \mathrm{Tr}\,(\Gamma_B\Gamma_A\Gamma_B) = \mathrm{Tr}\,(\Gamma_A\Gamma_B^2) = \mathrm{Tr}\,(\Gamma_A) = 0.$$

13. We have to show first that

$$\sum_{A=1}^{16} a_A\Gamma_A = 0 \tag{13.1}$$

is true if, and only if, all the coefficients a_A, ($A = 1, 2, \ldots, 16$), are equal to zero.

By taking the trace of (13.1), and using (12e), we find that $a_1 = 0$. Similarly, by multiplying (13.1) by Γ_A and using (12b, c, e), we find that $a_A = 0$. It follows that the "hyper-complex numbers" γ_μ cannot be represented by square matrices smaller than 4×4, since there cannot existi16 linearly independent $n\times n$ matrices if $n < 4$. In fact, the γ_μ can be represented by a sut able choice of 16 linearly-independent 4×4 matrices. This representation (and all representations equivalent to it, obtained by a similarity transformation $\gamma_\mu \to U\gamma_\mu U^{-1}$) is irreducible.† Any other representation can be reduced to the form

$$\begin{bmatrix}\gamma_\mu & & 0\\ & \ddots & \\ 0 & & \gamma_\mu\end{bmatrix}.$$

† See W. Pauli (1936) and R. H. Good (1955).

It is worth mentioning that in abstract algebra any set of hyper-complex numbers which satisfy the anti-commutation relations (XII.20) is said to constitute a Clifford algebra.

14. Let X be a 4×4 matrix which commutes with all the matrices γ_μ and hence with all the Γ_A. From the linear independence of the Γ_A it follows that X can be written in the form

$$X = \sum_{A=1}^{16} x_A \Gamma_A \tag{14.1}$$

in which we have, from (12b, c, e), that

$$x_A = \tfrac{1}{4}\,\mathrm{Tr}\,(X\Gamma_A). \tag{14.2}$$

Let us write (14.1) in the form

$$X = x_B\Gamma_B + \sum_{C\neq B} x_C\Gamma_C, \quad \Gamma_B \neq I. \tag{14.3}$$

Now let Γ_A be the matrix such that, in accordance with (12d), $\Gamma_A\Gamma_B\Gamma_A = -\Gamma_B$. By hypothesis $\Gamma_A X\Gamma_A = X$, so that, if we multiply (14.3) on the left and on the right by Γ_A, we obtain

$$X = -x_B\Gamma_B + \sum_{C\neq B} x_C\Gamma_A\Gamma_C\Gamma_A = -x_B\Gamma_B + \sum_{C\neq B} (\pm 1)x_C\Gamma_C, \tag{14.4}$$

in which we have $+1$ if Γ_C and Γ_A commute, and -1 if they anti-commute. If we multiply (14.3) and (14.4) by Γ_B, and take the trace, it follows that $x_B = -x_B = 0$.

Since Γ_B could equally well have been any one of the Γ matrices (excluding I), it follows that the only non-vanishing coefficient in the expression (14.1) is that of I, a fact which completes the proof of the statement made in the problem.

15. Using (XII.21–21″) and (XII.19), this statement can be verified by elementary calculation.

The unitary matrix which transforms the matrices γ_μ from the Pauli representation to the Majorana representation is

$$U_{P\to M} = U_{K\to M}U_{P\to K} = \frac{e^{i\pi/4}}{2}\begin{bmatrix} I+\sigma_2 & I-\sigma_2 \\ I-\sigma_2 & -I-\sigma_2 \end{bmatrix}, \tag{15.1}$$

and the required transformations of the bi-spinor (XII.17) are

$$\psi_K = U_{P\to K}\psi \quad \text{and} \quad \psi_M = U_{K\to M}\psi_K = U_{P\to M}\psi \tag{15.2}$$

respectively.

16. Let us first find the equation for $\bar{\psi}$ (i.e. the conjugate Dirac equation). Taking the Hermitian conjugate of (XII.18′) we have

$$\frac{\partial\psi^+}{\partial x_i}\gamma_i - \frac{\partial\psi^+}{\partial x_4}\gamma_4 + k\psi^+ = 0.$$

Multiplying this equation on the right by γ_4, and using (XII.20) and (16c), we obtain

$$\frac{\partial\bar{\psi}}{\partial x_\mu}\gamma_\mu - k\bar{\psi} = 0. \tag{16.1}$$

Now, from (XII.18′) and (16.1), and using the same procedure as in problems 1 and 6, we can derive the equation of conservation (16a).

If $j_\mu = (\mathbf{j}, ic\varrho)$, then, in three-dimensional notation, (16a) becomes

$$\frac{\partial\varrho}{\partial t} + \nabla\cdot\mathbf{j} = 0, \tag{16.2}$$

where

$$\varrho = \psi^+\psi = \psi_\mu^*\psi_\mu, \tag{16.3}$$

$$\mathbf{j} = C\psi^+\boldsymbol{\alpha}\psi. \tag{16.4}$$

The density (16.3) is clearly positive. With the usual boundary conditions, it follows from (16.2) and Gauss's theorem that

$$\frac{\partial}{\partial t}\int \psi^+\psi \, d\mathbf{r} = 0.$$

Hence the normalization

$$\int \psi^+\psi \, d\mathbf{r} = 1 \tag{16.5}$$

remains unchanged in time and thus ϱ can be regarded as a probability density. We should mention here that, as in the case of free spinless particles (see problem 1), the probabilistic interpretation given above is only approximately valid. The centre of any wave-packet corresponding to a free particle of spin $\frac{1}{2}$, besides having a uniform motion in a straight line, performs in addition a rapid oscillatory motion ("Zitterbewegung"), whose amplitude is of the order of the particle's Compton wavelength $\hbar/mc$ (see problem 18). Using the normalization (16.5), the mean value of an operator A is given by

$$\langle A\rangle = \int \psi^+ A\psi \, d\mathbf{r} = \int \bar{\psi}\gamma_4 A\psi \, d\mathbf{r}. \tag{16.6}$$

17. For stationary states, the wavefunction is of the form

$$\psi(\mathbf{r}, t) = \psi(\mathbf{r}) \exp\left(-\frac{i}{\hbar}Et\right). \tag{17.1}$$

By substituting (17.1) into (XII.14), we obtain the equation

$$H_D\psi(\mathbf{r}) = E\psi(\mathbf{r}). \tag{17.2}$$

It is useful to express the bi-spinor $\psi(\mathbf{r})$ in the form of two spinor functions, each having two components,

$$\phi = \begin{bmatrix}\psi_1\\ \psi_2\end{bmatrix}, \quad \chi = \begin{bmatrix}\psi_3\\ \psi_4\end{bmatrix}, \tag{17.3}$$

as follows:

$$\psi(\mathbf{r}) = \begin{bmatrix} \psi_1 \\ \psi_2 \\ \psi_3 \\ \psi_4 \end{bmatrix} = \begin{bmatrix} \begin{bmatrix} \psi_1 \\ \psi_2 \end{bmatrix} \\ \begin{bmatrix} \psi_3 \\ \psi_4 \end{bmatrix} \end{bmatrix} = \begin{bmatrix} \phi \\ \chi \end{bmatrix}. \tag{17.4}$$

In states of specified linear momentum, the dependence on the momentum of the bi-spinor (17.4) is of the form

$$\psi(\mathbf{r}) = N \begin{bmatrix} u_1 \\ u_2 \\ u_3 \\ u_4 \end{bmatrix} e^{\frac{i}{\hbar}\mathbf{p}.\mathbf{r}} = N \begin{bmatrix} \begin{bmatrix} u_1 \\ u_2 \end{bmatrix} \\ \begin{bmatrix} u_3 \\ u_4 \end{bmatrix} \end{bmatrix} e^{\frac{i}{\hbar}\mathbf{p}.\mathbf{r}} = N \begin{bmatrix} u \\ v \end{bmatrix} e^{\frac{i}{\hbar}\mathbf{p}.\mathbf{r}} = \begin{bmatrix} \phi \\ \chi \end{bmatrix}, \tag{17.5}$$

where N is a normalization constant and the spinors u and v are independent of the time and of the coordinates; only the matrix operators $\boldsymbol{\sigma}$ can operate on them. Substituting (17.5) into (17.2), and defining

$$H_D = \begin{bmatrix} mc^2 & -i\hbar c\boldsymbol{\sigma}.\nabla \\ -i\hbar c\boldsymbol{\sigma}.\nabla & -mc^2 \end{bmatrix}, \tag{17.6}$$

we obtain the set of homogeneous equations

$$\begin{aligned} (mc^2-E)\,\phi+c\boldsymbol{\sigma}.\mathbf{p}\chi &= 0 \\ c\boldsymbol{\sigma}.\mathbf{p}\phi-(mc^2+E)\,\chi &= 0. \end{aligned} \tag{17.7}$$

Non-trivial solutions of these equations exist only if the determinant of the coefficients of ϕ and χ vanishes, i.e. if

$$\begin{vmatrix} mc^2-E & c\boldsymbol{\sigma}.\mathbf{p} \\ -c\boldsymbol{\sigma}.\mathbf{p} & mc^2+E \end{vmatrix} = 0. \tag{17.8}$$

This determinant can be evaluated by using the identity

$$(\boldsymbol{\sigma}.\mathbf{A})\,(\boldsymbol{\sigma}.\mathbf{B}) = \mathbf{A}.\mathbf{B}+i\boldsymbol{\sigma}.(\mathbf{A}\times\mathbf{B}) \tag{VI.9a}$$

which is valid if $[\boldsymbol{\sigma}, \mathbf{A}] = [\boldsymbol{\sigma}, \mathbf{B}] = 0$.

After some simple calculation it is found that $E = \pm E_p$, where

$$E_p = c\sqrt{\mathbf{p}^2+m^2c^2} \tag{17.9}$$

is the magnitude of the energy of the particle.

The two signs of E correspond to two types of stationary solution of the Dirac equation for a free particle: the one with $E = +E_p$ is called the *positive-energy solution* and the one with $E = -E_p$ the *negative-energy solution.*

Let us introduce now the unitary Hermitian operator

$$\Lambda = \frac{H_D}{E_p} = \frac{1}{E_p}(c\boldsymbol{\alpha}.\mathbf{p}+mc^2\beta). \tag{17.10}$$

Since $\Lambda^2 = 1$ it follows that the eigenvalues of Λ are $\lambda = E/E_p = \pm 1$. The eigenvalue $\lambda = +1$ corresponds to positive-energy solutions ($E = +E_p$), and $\lambda = -1$ to negative-energy solutions ($E = -E_p$).

Since $[H_D, H_D] = [H_D, \mathbf{p}] = [H_D, \Lambda] = 0$, it follows that, for a free particle, the energy E_p, the momentum $\mathbf{p}$, and the eigenvalues λ of the operator Λ are constants of the motion, and therefore can have well defined values simultaneously.

From the equations (17.7), the spinors ϕ and χ can be expressed in terms of each other, thus, for example,

$$\chi = \frac{c\boldsymbol{\sigma}.\mathbf{p}}{mc^2+E}\phi, \tag{17.11}$$

and hence the bi-spinor which represents a state with momentum $\mathbf{p}$ and energy λE_p can be written in the form

$$\psi_{p\lambda}(\mathbf{r}) = N\begin{bmatrix} u \\ \dfrac{c\boldsymbol{\sigma}.\mathbf{p}}{mc^2+\lambda E_p}u \end{bmatrix}\exp\left(\frac{i}{\hbar}\mathbf{p}.\mathbf{r}\right). \tag{17.12}$$

Since N is still undetermined, any arbitrary normalization condition can be applied to the spinor u, and in particular we may take

$$u^+u = |u_1|^2+|u_2|^2 = 1. \tag{17.13}$$

It can then be verified immediately that if we take

$$N = (2\pi\hbar)^{-3/2}\left(\frac{mc^2+\lambda E_p}{2\lambda E_p}\right)^{1/2}, \tag{17.14}$$

the bi-spinors $\psi_{\mathbf{p}\lambda}$ will be normalized in the sense that

$$\int \psi^+_{\mathbf{p}\lambda}\psi_{\mathbf{p}'\lambda'}\,d\mathbf{r} = \delta_{\lambda\lambda'}\delta(\mathbf{p}-\mathbf{p}'). \tag{17.15}$$

In the non-relativistic approximation, we have, for the positive-energy solutions, $E = E_p = mc^2+E'$ where $E' \ll mc^2$, and thus from (17.11) it follows that

$$\chi = \frac{c\boldsymbol{\sigma}.\mathbf{p}}{2mc^2+E'}\phi \approx \frac{\boldsymbol{\sigma}.\mathbf{p}}{2mc}\phi \ll \phi. \tag{17.16}$$

Thus, if the velocity of the particle in the frame of reference chosen is small compared with c, two of the four components of the bi-spinor $\psi_{\mathbf{p}\lambda}$ are small in comparison with the other two. For this reason ψ_1, ψ_2 are called the *large components* and ψ_3, ψ_4 the *small components*. In states with $E = -E_p$ (the negative-energy solutions), ψ_1 and ψ_2 are small and ψ_3 and ψ_4 are large.

We shall now show that, besides having different signs of E/E_p, the various stationary states for a given linear momentum of a free particle can be distinguished from one other by means of another physical quantity, namely, the projection of the spin of the particle onto its direction of motion. To show this, let us define a "helicity operator"

$$s = \frac{\mathbf{\Sigma}.\mathbf{p}}{|\mathbf{p}|}, \tag{17.16'}$$

where

$$\mathbf{\Sigma} = \frac{\hbar}{2}\begin{bmatrix} \boldsymbol{\sigma} & 0 \\ 0 & \boldsymbol{\sigma} \end{bmatrix}. \tag{17.17}$$

The operator (17.16′) commutes with the Hamiltonian (17.6), and hence the corresponding physical quantity is a constant of the motion for a free particle. Since the momentum $\mathbf{p}$ is a constant, we can choose the z-axis to lie along the direction of $\mathbf{p}$, and the helicity operator then reduces to

$$s = \Sigma_z = \frac{\hbar}{2}\begin{bmatrix} \sigma_z & 0 \\ 0 & \sigma_z \end{bmatrix}. \tag{17.17'}$$

But Σ_z is simply the spin operator of the spin-$\frac{1}{2}$ particle. The eigenfunctions $U_\pm$ of Σ_z, with eigenvalues $\pm\hbar/2$, can be written in the form of the bi-spinors (17.12) by taking

$$u_+ = \begin{bmatrix} 1 \\ 0 \end{bmatrix}, \quad u_- = \begin{bmatrix} 0 \\ 1 \end{bmatrix}. \tag{17.18}$$

In the state $u_+ = \begin{bmatrix} 1 \\ 0 \end{bmatrix}$, the spin of the particle is said to be directed along its momentum, since $\boldsymbol{\sigma}.\mathbf{p}\begin{bmatrix} 1 \\ 0 \end{bmatrix} = p\begin{bmatrix} 1 \\ 0 \end{bmatrix}$, and, in the state $u_- = \begin{bmatrix} 0 \\ 1 \end{bmatrix}$, to be directed in the opposite sense, since $\boldsymbol{\sigma}.\mathbf{p}\begin{bmatrix} 0 \\ 1 \end{bmatrix} = -p\begin{bmatrix} 0 \\ 1 \end{bmatrix}$. Thus in the states described by the spinors (17.18) the spin projection has a well-defined value. There exist, of course, states in which the spin projection has no defined value; such states are described by spinors of the form $a\begin{bmatrix} 1 \\ 0 \end{bmatrix} + b\begin{bmatrix} 0 \\ 1 \end{bmatrix}$. In conclusion, the time-independent wavefunctions, for a given momentum p directed along the z-axis, and with well-defined values of λ $(= \pm 1)$ and of Σ_z $(= \pm\hbar/2)$, can be written as

$$\psi_{p,\lambda,1/2} = \sqrt{\frac{mc^2+\lambda E_p}{2\lambda E_p}}\begin{bmatrix} 1 \\ 0 \\ \dfrac{cp}{mc^2+\lambda E_p} \\ 0 \end{bmatrix}\frac{\exp\left(\dfrac{i}{\hbar}pz\right)}{(2\pi\hbar)^{3/2}} \tag{17.19}$$

$$\psi_{p,\lambda,-1/2} = \sqrt{\frac{mc^2+\lambda E_p}{2\lambda E_p}}\begin{bmatrix} 0 \\ 1 \\ 0 \\ \dfrac{-cp}{mc^2+\lambda E_p} \end{bmatrix}\frac{\exp\left(\dfrac{i}{\hbar}pz\right)}{(2\pi\hbar)^{3/2}}. \tag{17.19'}$$

The bi-spinors (17.19, 19′) satisfy the orthonormalization condition

$$\int \psi^{+}_{(p\lambda\Sigma_z)}\psi_{(p'\lambda'\Sigma_z')}\,d\mathbf{r} = \delta_{\lambda\lambda'}\delta_{\Sigma_z\Sigma_z'}\delta(\mathbf{p}'-\mathbf{p}). \tag{17.20}$$

An arbitrary state of given λ can be written in the form

$$\psi_\lambda = \Sigma_{(\Sigma_z)} \int A(\mathbf{p})\,\psi_{(p\lambda\Sigma_z)}\,d\mathbf{p}. \tag{17.21}$$

Bearing in mind that $H_D\psi_{(p\lambda\Sigma_z)} = \lambda E_p\psi_{(p\lambda\Sigma_z)}$, it follows that

$$\Lambda\psi_\lambda = \Sigma_{(\Sigma_z)} \int \frac{A(\mathbf{p})}{E_p} H_D\psi_{p\lambda\Sigma_z}\,d\mathbf{p} = \lambda\psi_\lambda. \tag{17.22}$$

With the help of Λ, projection operators for the energy states $\pm E_p$ can be constructed thus:

$$\Gamma_\pm = \tfrac{1}{2}(1\pm\Lambda); \tag{17.23}$$

these operators act on the bi-spinors ψ_λ as follows:

$$\Gamma_+\psi_+ = \psi_+, \quad \Gamma_+\psi_- = 0, \quad \Gamma_-\psi_+ = 0, \quad \Gamma_-\psi_- = \psi_-. \tag{17.24}$$

18. Let $\mathbf{r}(t)$ be the position operator, $\boldsymbol{\alpha}(t)$ the operator (XII.21), $\mathbf{p}$ the momentum operator and H the Hamiltonian (XII.15) of a free particle of spin $\frac{1}{2}$, in the Heisenberg picture (see Chapter V).

Using the Heisenberg equation (V.3), the commutation rules $[x_k, p_l] = i\hbar\delta_{kl}$, and (XII.16), we find after some simple calculation that

$$\frac{d\mathbf{r}}{dt} = \frac{1}{i\hbar}[\mathbf{r}, H] = c\boldsymbol{\alpha} \tag{18.1}$$

$$\frac{d\boldsymbol{\alpha}}{dt} = \frac{1}{i\hbar}[\boldsymbol{\alpha}, H] = \frac{i}{\hbar}(H\boldsymbol{\alpha}+\boldsymbol{\alpha}H) - \frac{2i}{\hbar}\boldsymbol{\alpha}H = \frac{2i}{\hbar}(c\mathbf{p}-\boldsymbol{\alpha}H). \tag{18.2}$$

Since $\mathbf{p}$ and H are constants of the motion, (18.2) can easily be integrated. The result is

$$\boldsymbol{\alpha}(t) = c\frac{\mathbf{p}}{H} + \left(\boldsymbol{\alpha}(0) - c\frac{\mathbf{p}}{H}\right)\exp\left(-\frac{2i}{\hbar}Ht\right). \tag{18.3}$$

Similarly, from (18.1) and (18.3), we obtain

$$\mathbf{r}(t) = \mathbf{a} + c^2\frac{\mathbf{p}}{H}t + ic\hbar\left(\boldsymbol{\alpha}(0) - c\frac{\mathbf{p}}{H}\right)\frac{\exp\left(-\frac{2i}{\hbar}Ht\right)}{2H}, \tag{18.4}$$

in which

$$\mathbf{a} = \mathbf{r}(0) - ic\hbar\left(\boldsymbol{\alpha}(0) - c\frac{\mathbf{p}}{H}\right)\frac{1}{2H}. \tag{18.5}$$

From (18.4) (which expresses the time dependence of the position operator in the Heisenberg picture) the motion of the centroid $\langle \mathbf{r} \rangle$ of a wave-packet for a free particle with spin $\frac{1}{2}$ can be found.

By comparison with the classical motion (which is uniform and rectilinear for a free particle),

$$\mathbf{r}_{\rm cl}(t) = \mathbf{r}_{\rm cl}(0) + \left(c\frac{\mathbf{p}}{H}\right)_{\rm cl} t,$$

it can be seen that, in addition to the classical uniform motion in a straight line, the centroid of the wave-packet has a rapid oscillatory motion

$$\left\langle ic\hbar\left(\boldsymbol{\alpha}(0) - c\frac{\mathbf{p}}{H}\right)\frac{\exp\left(-\frac{2i}{\hbar}Ht\right)}{2H}\right\rangle,$$

whose amplitude and period are of the order of $\hbar/mc$ and of $\hbar/mc^2$ respectively. This oscillatory motion was called the "Zitterbewegung" by E. Schrödinger. Because of it, the localization of a free particle is possible only in a volume of linear dimensions of order $\hbar/mc$ (= the Compton wavelength of the particle). The oscillatory motion vanishes if the wave-packet is a superposition of positive-energy solutions only, i.e. is of the form (17.21) with $\lambda = +1$, or of negative-energy solutions only ($\lambda = -1$). To see this it is sufficient to show that

$$\Gamma_{\pm}\left(\boldsymbol{\alpha} - c\frac{\mathbf{p}}{H}\right)\frac{\exp\left(-\frac{2i}{\hbar}Ht\right)}{2H}\Gamma_{\pm} = 0,$$

where the $\Gamma_{\pm}$ are the projection operators (17.23). Using (18.2), (17.10), (17.23) and (17.22), it can be shown that

$$[H, \boldsymbol{\alpha}] = 2(c\mathbf{p} - \boldsymbol{\alpha}H), \quad [\Gamma_{\pm}, \boldsymbol{\alpha}] = \pm c\frac{\mathbf{p}}{E_p} \mp \boldsymbol{\alpha}\frac{H}{E_p},$$

$$H\Gamma_{\pm} = \pm E_p\Gamma_{\pm},$$

and hence that

$$0 = \Gamma_{\pm}[\Gamma_{\pm}, \boldsymbol{\alpha}]\Gamma_{\pm} = \Gamma_{\pm}\left(c\frac{\mathbf{p}}{H} - \boldsymbol{\alpha}\right)\Gamma_{\pm},$$

which verifies the above statement.

It follows that the oscillatory motion is due to interference between the positive- and the negative-energy solutions which are normally required to form a wave-packet, since neither set alone constitutes a complete set of functions.

19. Note that the Hermitian matrix S commutes with $\beta\alpha_j$ and anti-commutes with β, i.e.

$$S\beta\boldsymbol{\alpha}.\mathbf{p} = \beta\boldsymbol{\alpha}.\mathbf{p}S \quad \text{and} \quad S\beta = -\beta S.$$

Thus the Dirac Hamiltonian of a free particle becomes, in the Foldy–Wouthuysen representation,

$$\hat{H}_D = e^{iS}(c\boldsymbol{\alpha}.\mathbf{p}+\beta mc^2)\,e^{-iS} = e^{iS}\beta(c\beta\boldsymbol{\alpha}.\mathbf{p}+mc^2)\,e^{-iS} = e^{iS}\beta\, e^{-iS}\beta(c\boldsymbol{\alpha}.\mathbf{p}+\beta mc^2)$$
$$= e^{2iS}\beta^2(c\boldsymbol{\alpha}.\mathbf{p}+\beta mc^2) = e^{2iS}H_D.$$

Expanding e^{2iS} in a power series,

$$e^{2iS} = 1+\frac{1}{mc}\beta\boldsymbol{\alpha}.\mathbf{p}f+\frac{1}{2!}\frac{1}{m^2c^2}(\beta\boldsymbol{\alpha}.\mathbf{p})^2f^2+\frac{1}{3!}\frac{1}{m^3c^3}(\beta\boldsymbol{\alpha}.\mathbf{p})^3f^3+\ \dots$$

and bearing in mind that

$$(\beta\boldsymbol{\alpha}.\mathbf{p})^n = \begin{cases} \pm|\mathbf{p}|^n & \text{for } n \text{ even} \\ \pm|\mathbf{p}|^{n-1}\beta\boldsymbol{\alpha}.\mathbf{p} & \text{for } n \text{ odd,} \end{cases}$$

we obtain

$$e^{2iS} = \cos\frac{|\mathbf{p}|f}{mc}+\beta\frac{\boldsymbol{\alpha}.\mathbf{p}}{|\mathbf{p}|}\sin\frac{|\mathbf{p}|f}{mc}, \tag{19.1}$$

and hence

$$\hat{H}_D = \frac{\boldsymbol{\alpha}.\mathbf{p}}{|\mathbf{p}|}\left(c\,|\mathbf{p}|\cos\frac{|\mathbf{p}|f}{mc}-mc^2\sin\frac{|\mathbf{p}|f}{mc}\right)+\beta\left(mc^2\cos\frac{|\mathbf{p}|f}{mc}+c\,|\mathbf{p}|\sin\frac{|\mathbf{p}|f}{mc}\right). \tag{19.2}$$

The function f has to be chosen so that the coefficient of the operator $\boldsymbol{\alpha}.\mathbf{p}$, which mixes the upper and lower components of the bi-spinor (19b), vanishes. This is the case if we choose

$$f = \frac{mc}{|\mathbf{p}|}\tan^{-1}\frac{|\mathbf{p}|}{mc}. \tag{19.3}$$

Then

$$\hat{H}_D = \beta\left[mc^2\frac{mc}{\sqrt{\mathbf{p}^2+m^2c^2}}+|\mathbf{p}|\,c\frac{|\mathbf{p}|}{\sqrt{\mathbf{p}^2+m^2c^2}}\right] = \beta c\sqrt{p^2+m^2c^2} = \beta E_p. \tag{19.4}$$

Therefore in the Pauli representation (XII.21), the two upper components of the solution of the equation

$$\hat{H}_D\hat{\psi} = E_p\hat{\psi} \tag{19.5}$$

correspond to the positive energy, and the two lower ones to the negative energy, and thus the bi-spinor

$$\hat{\psi} = \begin{bmatrix}\hat{\psi}_1\\ \hat{\psi}_2\\ \hat{\psi}_3\\ \hat{\psi}_4\end{bmatrix} = \begin{bmatrix}\begin{bmatrix}\hat{\psi}_1\\ \hat{\psi}_2\end{bmatrix}\\ \begin{bmatrix}\hat{\psi}_3\\ \hat{\psi}_4\end{bmatrix}\end{bmatrix} = \begin{bmatrix}\hat{\phi}\\ \hat{\chi}\end{bmatrix}$$

can be written in the form (19b), with

$$\hat{\psi}_+ = \frac{1}{2}(I+\beta)\hat{\psi} = \begin{bmatrix} I & 0 \\ 0 & 0 \end{bmatrix}\begin{bmatrix} \hat{\phi} \\ \hat{\chi} \end{bmatrix} = \begin{bmatrix} \hat{\phi} \\ 0 \end{bmatrix}, \tag{19.6}$$

$$\hat{\psi}_- = \frac{1}{2}(I-\beta)\hat{\psi} = \begin{bmatrix} 0 & 0 \\ 0 & I \end{bmatrix}\begin{bmatrix} \hat{\phi} \\ \hat{\chi} \end{bmatrix} = \begin{bmatrix} 0 \\ \hat{\chi} \end{bmatrix}, \tag{19.6'}$$

and then

$$\hat{H}_D\hat{\psi}_+ = E_p\hat{\psi}_+, \quad \hat{H}_D\hat{\psi}_- = -E_p\hat{\psi}_-, \tag{19.7}$$

which was to be proved.

Thus, in the F–W representation, the operators $\frac{1}{2}(I\pm\beta)$ project the solutions of equation (19.5) onto the states of positive and of negative energy respectively.

The projection operators of the solution ψ of the Dirac equation (17.2) onto states of specified energy $\hat{\psi}_\pm$ are $\frac{1}{2}(I\pm\beta)e^{iS}$, since

$$\hat{\psi}_\pm = \tfrac{1}{2}(I\pm\beta)e^{iS}\psi. \tag{19.8}$$

By choosing f as in (19.3), the explicit form of the F–W transformation operator can be obtained from (19.1):

$$\begin{aligned} e^{iS} &= \cos\frac{|\mathbf{p}|f}{2mc} + \beta\frac{\boldsymbol{\alpha}\cdot\mathbf{p}}{|\mathbf{p}|}\sin\frac{|\mathbf{p}|f}{2mc} = \frac{1}{2}\sqrt{\frac{2E_p}{E_p+mc^2}}\left(1+\frac{c\beta\boldsymbol{\alpha}\cdot\mathbf{p}+mc^2}{E_p}\right) \\ &= \frac{1}{2}\sqrt{\frac{2E_p}{E_p+mc^2}}\left(1+\beta\frac{H_D}{E_p}\right). \end{aligned} \tag{19.9}$$

Using this expression and the relations (19.8), after some simple calculation we obtain expressions for the bi-spinors (17.19, 19′) in the F–W representation:

$$\hat{\psi}_{p,1,1/2} = \begin{bmatrix} 1 \\ 0 \\ 0 \\ 0 \end{bmatrix}\frac{\exp\left(\frac{i}{\hbar}pz\right)}{(2\pi\hbar)^{3/2}}, \quad \hat{\psi}_{p,1,-1/2} = \begin{bmatrix} 0 \\ 1 \\ 0 \\ 0 \end{bmatrix}\frac{\exp\left(\frac{i}{\hbar}pz\right)}{(2\pi\hbar)^{3/2}}, \tag{19.10}$$

$$\hat{\psi}_{p,-1,1/2} = \begin{bmatrix} 0 \\ 0 \\ 1 \\ 0 \end{bmatrix}\frac{\exp\left(\frac{i}{\hbar}pz\right)}{(2\pi\hbar)^{3/2}}, \quad \hat{\psi}_{p,-1,-1/2} = \begin{bmatrix} 0 \\ 0 \\ 0 \\ 1 \end{bmatrix}\frac{\exp\left(\frac{i}{\hbar}pz\right)}{(2\pi\hbar)^{3/2}}. \tag{19.10'}$$

20. In the non-relativistic limit, $E_p \approx mc^2$, and the transformation operator (19.9) becomes

$$e^{iS} \approx \frac{1}{2}\left(2+\beta\frac{\boldsymbol{\alpha}\cdot\mathbf{v}}{c}\right) \tag{20.1}$$

in which we have written $\mathbf{p} = m\mathbf{v}$.

Using (19.8) we find that

$$\hat{\psi}_- = \frac{1}{2}(I-\beta)e^{iS}\psi \approx \frac{1}{4}(I-\beta)\left(2+\beta\frac{\boldsymbol{\alpha}\cdot\mathbf{v}}{c}\right)\psi$$
$$\hat{\psi}_+ = \frac{1}{2}(I+\beta)e^{iS}\psi \approx \frac{1}{4}(I+\beta)\left(2+\beta\frac{\boldsymbol{\alpha}\cdot\mathbf{v}}{c}\right)\psi, \tag{20.2}$$

or, by (XII.21) and (17.4),

$$\hat{\psi}_- \approx -\frac{1}{2}\begin{bmatrix} 0 \\ \dfrac{\boldsymbol{\sigma}\cdot\mathbf{v}}{c}\phi+2\chi \end{bmatrix}, \quad \hat{\psi}_+ \approx \frac{1}{2}\begin{bmatrix} \dfrac{\boldsymbol{\sigma}\cdot\mathbf{v}}{c}\chi+2\phi \\ 0 \end{bmatrix}. \tag{20.3}$$

Since, in non-relativistic mechanics, the (kinetic) energy of a free particle is always positive, it follows by (17.16) that $\frac{\boldsymbol{\sigma}\cdot\mathbf{v}}{c}\phi+2\chi \ll \frac{\boldsymbol{\sigma}\cdot\mathbf{v}}{c}\chi+2\phi$ and thus, in the non-relativistic limit, $\hat{\psi}_- \ll \hat{\psi}_+$.

21. Choosing the function f to have the form (21a), the second term of the Hamiltonian (19.4) vanishes and we obtain

$$\hat{H}_D = \frac{\boldsymbol{\alpha}\cdot\mathbf{p}}{|\mathbf{p}|}\left(c|\mathbf{p}|\frac{1}{\sqrt{1+\left(\dfrac{mc}{|\mathbf{p}|}\right)^2}}+mc^2\frac{\dfrac{mc}{|\mathbf{p}|}}{\sqrt{1+\left(\dfrac{mc}{|\mathbf{p}|}\right)^2}}\right) = \frac{\boldsymbol{\alpha}\cdot\mathbf{p}}{|\mathbf{p}|}E_p. \tag{21.1}$$

In the same way as we established the expression (19.9) for e^{iS}, we find in this case that

$$e^{iS} = \frac{E_p+c|\mathbf{p}|-\beta\dfrac{\boldsymbol{\alpha}\cdot\mathbf{p}}{|\mathbf{p}|}mc^2}{\sqrt{2E_p(E_p+c|\mathbf{p}|)}}. \tag{21.2}$$

The transformation of the bi-spinor by the unitary operator (21.2) is called the Cini–Touschek transformation.†

22. By (22a) and the relation $p'_\mu = a_{\mu\nu}p_\nu$, the equation

$$(\gamma_\mu p'_\mu - imc)\,\psi'(x'_\mu) = 0 \tag{22.1}$$

becomes

$$(\gamma_\mu a_{\mu\nu}p_\nu - imc)U\psi = 0. \tag{22.2}$$

Multiplying this equation on the left by U^{-1} we find that

$$(U^{-1}\gamma_\mu U a_{\mu\nu}p_\nu - imc)\psi = 0. \tag{22.3}$$

† M. Cini and B. Touschek, *Nuovo Cimento*, **7**, 422 (1958).

Now (22.3) is the same as (XII.18) if and only if

$$U^{-1}\gamma_\mu U a_{\mu\nu} = \gamma_\nu. \tag{22.4}$$

By the orthogonality relations (XII.2), (22b) then follows from (22.4).

It can easily be shown that (22b) is valid also if $\psi(x_\mu)$ is a solution of the Dirac equation (XII.22) for a particle in an electromagnetic field.

23. Since the fourth coordinate $x_4 = ict$ is a pure imaginary quantity, the matrix U is not unitary. Of the elements of the transformation (XII.1), only a_{44} and a_{kl} $(k, l = 1, 2, 3)$ are real, while the elements a_{4k} are pure imaginary. Thus, on account of the Hermitian property of γ_μ, and of the relation (22b), we find that

$$(U^{-1}\gamma_4 U)^+ = a_{44}\gamma_4 - a_{4k}\gamma_k, \qquad k = 1, 2, 3.$$

Multiplying this expression on the right by γ_4 and remembering (XII.20), we obtain

$$(U^{-1}\gamma_4 U)^+ \gamma_4 = U^+\gamma_4(U^+)^{-1}\gamma_4 = \gamma_4 a_{4\mu}\gamma_\mu.$$

By changing the right-hand side with the help of (22b) we have

$$\gamma_4 U^+\gamma_4(U^+)^{-1}\gamma_4 = U^{-1}\gamma_4 U. \tag{23.1}$$

Since $\gamma_4 = \gamma_4^{-1}$, (23.1) can be written as

$$(\gamma_4 U^+\gamma_4)\gamma_4(\gamma_4 U^+\gamma_4)^{-1} = U^{-1}\gamma_4 U$$

and hence

$$\gamma_4 U^+\gamma_4 = \lambda U^{-1}, \tag{23.2}$$

where $\lambda = \pm 1$.

In order to decide when $\lambda = +1$ and when $\lambda = -1$, consider the identity $U^+U = U^+\gamma_4\gamma_4 U$. By transforming the right-hand side of this identity with the help of (22b) and of (23.2), we obtain

$$U^+U = \lambda\gamma_4 U^{-1}\gamma_4 U = \lambda(a_{44} + a_{4k}\gamma_4\gamma_k), \qquad k = 1, 2, 3.$$

Taking the trace of both sides, and remembering that $\mathrm{Tr}\,(\gamma_4\gamma_k) = 0$ [see (12e)], we find

$$\mathrm{Tr}\,(U^+U) = \lambda a_{44}. \tag{23.3}$$

Since $\mathrm{Tr}\,(U^+U) > 0$ always, it follows that $\lambda = +1$ for those transformations which do not change the sign of the time, i.e. which have $a_{44} > 0$, and $\lambda = -1$ for those transformations which change the sign of the time, i.e. for which $a_{44} < 0$. Thus

$$\gamma_4 U^+\gamma_4 = \frac{a_{44}}{|a_{44}|}U^{-1}. \tag{23.4}$$

Taking the Hermitian conjugate of (22a) and multiplying it on the right by γ_4, we find that

$$\bar{\psi}' = \psi^+ U^+\gamma_4 = \bar{\psi}\gamma_4 U^+\gamma_4. \tag{23.5}$$

The relation (23a) follows immediately from (23.4) and (23.5).

24. In order to keep the lengths of all four-vectors unchanged under the transformation (24a), we need, to first order in $\varepsilon_{\mu\nu}$,

$$x_\mu'^2 = (\delta_{\mu\nu}+\varepsilon_{\mu\nu})(\delta_{\mu\varrho}+\varepsilon_{\mu\varrho})x_\nu x_\varrho = x_\nu^2,$$

whence $\varepsilon_{\mu\nu}$ must be anti-symmetrical, i.e. $\varepsilon_{\mu\nu} = -\varepsilon_{\nu\mu}$. Since the transformation (24a) is an infinitesimal one, the matrix U will differ from the unit matrix by infinitesimal quantities proportional to $\varepsilon_{\mu\nu}$ only, thus

$$U = 1+M_{\mu\nu}\varepsilon_{\mu\nu}. \tag{24.1}$$

In order to determine the matrix "generators" $M_{\mu\nu}$ of the transformation U we use (22b), which, for $a_{\mu\nu} = \delta_{\mu\nu}+\varepsilon_{\mu\nu}$, leads to the equation

$$(1-M_{\lambda\nu}\varepsilon_{\lambda\nu})\gamma_\mu(1+M_{\lambda\nu}\varepsilon_{\lambda\nu}) = \gamma_\mu+\varepsilon_{\mu\nu}\gamma_\nu,$$

i.e.

$$(\gamma_\mu M_{\lambda\nu}-M_{\lambda\nu}\gamma_\mu)\varepsilon_{\lambda\nu} = \varepsilon_{\mu\nu}\gamma_\nu. \tag{24.2}$$

Remembering that

$$\varepsilon_{\mu\nu}\gamma_\nu = \varepsilon_{\lambda\nu}\delta_{\lambda\mu}\gamma_\nu = \tfrac{1}{2}\varepsilon_{\lambda\nu}(\delta_{\lambda\mu}\gamma_\nu-\delta_{\nu\mu}\gamma_\lambda),$$

(24.2) becomes

$$(\gamma_\mu M_{\lambda\nu}-M_{\lambda\nu}\gamma_\mu-\tfrac{1}{2}\delta_{\lambda\mu}\gamma_\nu+\tfrac{1}{2}\delta_{\nu\mu}\gamma_\lambda)\varepsilon_{\lambda\nu} = 0. \tag{24.3}$$

It can easily be seen that this equation will be satisfied if we take $M_{\lambda\nu} = \frac{1}{4}\gamma_\lambda\gamma_\nu$, and thus (24b) is verified.

25. Using the expressions given in the problem, we have

$$\begin{aligned}
S' &= \bar{\psi}'\psi' = \bar{\psi}U^{-1}U\psi = \bar{\psi}\psi = S,\\
V'_\mu &= \bar{\psi}'\gamma_\mu\psi' = \bar{\psi}U^{-1}\gamma_\mu U\psi = a_{\mu\nu}\bar{\psi}\gamma_\nu\psi = a_{\mu\nu}V_\nu,\\
T'_{\mu\nu} &= \tfrac{1}{2}\bar{\psi}'(\gamma_\mu\gamma_\nu-\gamma_\nu\gamma_\mu)\psi' = \tfrac{1}{2}\bar{\psi}U^{-1}\gamma_\mu UU^{-1}\gamma_\nu U\psi\\
&\quad -\tfrac{1}{2}\bar{\psi}U^{-1}\gamma_\nu UU^{-1}\gamma_\mu U\psi = \tfrac{1}{2}a_{\mu\varrho}a_{\nu\tau}\bar{\psi}\gamma_\varrho\gamma_\tau\psi\\
&\quad -\tfrac{1}{2}a_{\nu\tau}a_{\mu\varrho}\bar{\psi}\gamma_\tau\gamma_\varrho\psi = a_{\mu\varrho}a_{\nu\tau}\tfrac{1}{2}\bar{\psi}(\gamma_\varrho\gamma_\tau-\gamma_\tau\gamma_\varrho)\psi\\
&= a_{\mu\varrho}a_{\nu\tau}T_{\varrho\tau}.
\end{aligned}$$

The anti-symmetry of $T_{\mu\nu}$ is obvious. Since, by definition $\gamma_5 = \gamma_1\gamma_2\gamma_3\gamma_4$, it follows from the preceding relations that P transforms like the product of the four coordinates $x_1x_2x_3x_4$, i.e. like a four-dimensional volume. P is therefore invariant under spatial rotations, but changes sign under a mirror reflection of the three spatial coordinates, i.e. it is a "pseudo-scalar".

From what has been said above, the axial-vector character of A_μ is obvious. In view of the result of problem 13 it can easily be seen that the quantities (25a) exhaust all the independent bi-linear forms which can be constructed from $\bar{\psi}$ and ψ, in the sense that any such bi-linear form can be expressed as a linear combination of the quantities (25a).

The bi-linear covariants (25a) contain in all 16 independent components: one component each of types S and P, 4 components each of types V_μ and A_μ, and 6 components of type $T_{\mu\nu}$.

The results of this problem are still valid if ψ is a solution of the Dirac equation for a particle in an electromagnetic field.

26. Note that, under the transformation (XII.8),

$$\left(p_\mu - \frac{e}{c} A_\mu\right) e^{\frac{ie}{\hbar c} G} \psi \to e^{\frac{ie}{\hbar c} G} \left(p_\mu - \frac{e}{c} A_\mu\right) \psi,$$

whence, after some simple calculation, it follows that

$$\left(p_\mu - \frac{e}{c} A_\mu\right)^2 e^{\frac{ie}{\hbar c} G} \psi \to e^{\frac{ie}{\hbar c} G} \left(p_\mu - \frac{e}{c} A_\mu\right)^2 \psi.$$

That the Klein–Gordon equation is invariant under gauge transformations of the first and of the second kind, follows immediately. That the same is true of the Dirac equation (XII.22) can be shown in a similar way.

If we take into account the fact that the unitary transformations (26a) do not change any of the physical properties of the system under consideration (e.g., they do not affect the mean values of the observables), we can say that equations (XII.13) and (XII.22), in so far as their physical content is concerned, are invariant under a gauge transformation of the second kind.

27. The equation satisfied by ψ_C, without specifying for the moment any particular representation of the matrices γ_μ, is

$$\left[\gamma_\mu \left(p_\mu + \frac{e}{c} A_\mu\right) - imc\right] \psi_C = 0. \tag{27.1}$$

Taking the complex conjugate of (XII.22), and remembering that $\mathbf{A}^* = \mathbf{A}$, $\mathbf{p}^* = -\mathbf{p}$, $A_0^* = -A_0$ and $p_4^* = p_4$, we find that

$$\left[\boldsymbol{\gamma}^* \cdot \left(\mathbf{p} + \frac{e}{c} \mathbf{A}\right) - \gamma_4^* \left(p_4 + \frac{e}{c} A_4\right) - imc\right] \psi^* = 0. \tag{27.2}$$

If, in (27.2), we make the following transformation

$$\psi^* = C\psi_C, \quad \text{i.e.} \quad \psi_C = C^{-1}\psi^*, \tag{27.3}$$

where the matrix C satisfies the relations

$$\boldsymbol{\gamma} = C^{-1}\boldsymbol{\gamma}^* C, \quad \gamma_4 = -C^{-1}\gamma_4^* C, \tag{27.4}$$

then (27.2) will coincide with (27.1).

In the Pauli representation we have

$$C\gamma_1 = -\gamma_1 C, \quad C\gamma_2 = \gamma_2 C, \quad C\gamma_3 = -\gamma_3 C, \quad C\gamma_4 = -\gamma_4 C$$

and thus the matrix of the charge conjugation operator in this representation is $C = \gamma_2$. In the Kramers representation we find, in a similar manner, that $C = \gamma_2^K$.

Since in the Majorana representation $C\gamma_\mu = \gamma_\mu C$, $(\mu = 1, 2, 3, 4)$, the matrix C, in accordance with problem 14, is a multiple of the unit matrix. By convention one takes $C = 1$, so that

$$\psi_C^M = \psi^{M*}. \tag{27.5}$$

For the second part of the problem we have

$$i\hbar\frac{\partial\psi_C}{\partial t} = C^{-1}\left(i\hbar\frac{\partial\psi^*}{\partial t}\right) = -C^{-1}\left(i\hbar\frac{\partial\psi}{\partial t}\right)^* = EC^{-1}\psi^* = E\psi_C.$$

In conclusion, if the function ψ describes a state of a certain charged particle, then the function ψ_C obtained from ψ by charge conjugation describes a state of a particle with the same mass and spin, but with a charge and a magnetic moment of opposite sign (cf. the electron and the positron).

As in (16.1), the equation satisfied by $\bar{\psi}$ in the presence of an electromagnetic field is

$$\left(p_\mu - \frac{e}{c}A_\mu\right)\bar{\psi}\gamma_\mu + imc\bar{\psi} = 0. \tag{27.6}$$

By carrying out the calculations described in the first part of the problem, we obtain, in analogy with (27.3),

$$\bar{\psi}_C = \bar{\psi}_C^*. \tag{27.7}$$

28. Let us write the Dirac equation (XII.22) in the form

$$\left[\boldsymbol{\gamma}\cdot\left(\mathbf{p} - \frac{e}{c}\mathbf{A}\right) + \gamma_4\left(-\frac{i\hbar}{ic}\frac{\partial}{\partial t} + \frac{e}{ic}A_0\right) - imc\right]\psi = 0.$$

Multiplying on the left by $ic\gamma_4$ and using (XII.16), we obtain the Dirac equation for a particle in an electromagnetic field in "Schrödinger" form:

$$i\hbar\frac{\partial\psi}{\partial t} = \left[c\boldsymbol{\alpha}\cdot\left(\mathbf{p} - \frac{e}{c}\mathbf{A}\right) + mc^2\beta + eA_0\right]\psi. \tag{28.1}$$

If the Hamiltonian

$$H = c\boldsymbol{\alpha}\cdot\left(\mathbf{p} - \frac{e}{c}\mathbf{A}\right) + mc^2\beta + eA_0$$

is time independent, (28.1) has stationary solutions of the form (17.1). The equation for the stationary states is then

$$H\psi(\mathbf{r}) = E\psi(\mathbf{r}), \tag{28.2}$$

i.e.

$$H_D\psi(\mathbf{r}) = (E + e\boldsymbol{\alpha}\cdot\mathbf{A} - eA_0)\psi(\mathbf{r}), \tag{28.3}$$

where H_D is the free particle Hamiltonian (XII.15).

29. Let us write the Hamiltonian (28.2) in the form

$$H = \beta mc^2 + X + Y \tag{29.1}$$

where

$$X = eA_0, \quad Y = c\boldsymbol{\alpha}\cdot\left(\mathbf{p} - \frac{e}{c}\mathbf{A}\right) \tag{29.2}$$

with the properties

$$X\beta - \beta X = 0, \quad Y\beta + \beta Y = 0. \tag{29.3}$$

From (28.1) and (29a) we have that

$$i\hbar\frac{\partial}{\partial t}e^{-iS}\hat{\psi} = H\psi = He^{-iS}\hat{\psi} = e^{-iS}\left(i\hbar\frac{\partial\hat{\psi}}{\partial t}\right) + \left(i\hbar\frac{\partial}{\partial t}e^{-iS}\right)\hat{\psi},$$

whence

$$i\hbar\frac{\partial\hat{\psi}}{\partial t} = \left[e^{iS}\left(H - i\hbar\frac{\partial}{\partial t}\right)e^{-iS}\right]\hat{\psi} \tag{29.4}$$

and thus

$$\hat{H} = e^{iS}\left(H - i\hbar\frac{\partial}{\partial t}\right)e^{-iS}. \tag{29.5}$$

Since S is of order $1/mc^2$, to the required order of approximation $[H = 0(1/m^3c^6)]$, we have by (27a) and (25a) of Chapter I,

$$\begin{aligned}\hat{H} = {} & H + i[S, H] - \frac{1}{2}[S, [S, H]] - \frac{i}{6}[S, [S, [S, H]]] \\ & + \frac{1}{24}\left[S, \left[S, [S[S, \beta mc^2]]\right]\right] - \hbar\dot{S} - \frac{i\hbar}{2}[S, \dot{S}] + \frac{\hbar}{6}[S, [S, \dot{S}]].\end{aligned} \tag{29.6}$$

In order to find the operator S, we retain first only the terms in (29.6) of zero and first order in mc^2, i.e. we take

$$\hat{H}_0 = \beta mc^2 + X + Y + i[S, \beta]mc^2. \tag{29.7}$$

In order not to have the two upper components of the bi-spinor $\hat{\psi}$ mixed (by $\hat{H}_0$) with the two lower components, the off-diagonal matrix elements of $\hat{H}_0$ have to be identically zero. This is the case if, in (29.7), we have

$$Y + imc^2[S, \beta] = 0,$$

whence it follows that, to this order of approximation, $S = S_1$, say, where

$$S_1 = -\frac{i\beta Y}{2mc^2}. \tag{29.8}$$

With this expression for S, we find for the commutators (29.6):

$$i[S_1, H] = -Y + \frac{\beta}{2mc^2}[Y, X] + \frac{\beta Y^2}{mc^2}$$

$$-\frac{1}{2}[S_1, [S_1, H]] = -\frac{\beta Y^2}{2mc^2} - \frac{1}{8m^2c^4}[Y, [Y, X]] - \frac{Y^3}{2m^2c^4}$$

$$-\frac{i}{6}[S_1, [S_1, [S_1, H]]] = \frac{Y^3}{6m^2c^4} - \frac{\beta Y^4}{6m^3c^6}$$

$$\frac{1}{24}\left[S_1, \left[S_1, \left[S_1[S_1, \beta mc^2]\right]\right]\right] = \frac{\beta Y^4}{24m^3c^6}$$

$$-\hbar\dot{S}_1 = i\hbar\frac{\beta Y}{2mc^2}, \quad -\frac{i\hbar}{2}[S_1, S_1] = -\frac{i\hbar}{8m^2c^4}[Y, \dot{Y}]$$

$$\frac{\hbar}{6}[S_1, [S_1, \dot{S}_1]] = -\frac{i\hbar\beta}{48m^3c^6}[Y, [Y, \dot{Y}]].$$

The last double commutator can be neglected to the approximation used [its order of magnitude is (field energy/m^3c^6)].

Thus, the Hamiltonian (29.6) becomes

$$\begin{aligned}\hat{H}_1 &= \beta\left(mc^2 + \frac{Y^2}{2mc^2} - \frac{Y^4}{8m^3c^6}\right) + X - \frac{1}{8m^2c^4}[Y, [Y, X]] \\ &\quad - \frac{i\hbar}{8m^2c^4}[Y, \dot{Y}] + \frac{\beta}{2mc^2}[Y, X] - \frac{Y^3}{3m^3c^4} + \frac{i\hbar\beta\dot{Y}}{2mc^2} \\ &= \beta mc^2 + X_1 + Y_1\end{aligned} \tag{29.9}$$

where

$$X_1 = \beta\left(\frac{Y^2}{2mc^2} - \frac{Y^4}{8m^3c^6}\right) + X - \frac{1}{8m^2c^4}[Y, [Y, X]] - \frac{i\hbar}{8m^2c^4}[Y, \dot{Y}],$$

$$Y_1 = \frac{\beta}{2mc^2}[Y, X] - \frac{Y^3}{3m^2c^4} + \frac{i\hbar\beta\dot{Y}}{2mc^2}.$$

Note that Y_1 is of order $1/mc^2$. By repeating the above procedure we can remove the off-diagonal term Y_1 from (29.9) if a second F–W transformation of the same form as (29.8) is carried out, with

$$S_2 = -\frac{i\beta}{2mc^2}Y_1 = -\frac{i\beta}{2mc^2}\left(\frac{\beta}{2mc^2}[Y, X] - \frac{Y^3}{3m^2c^4} + \frac{i\hbar\beta\dot{Y}}{2mc^2}\right). \tag{29.10}$$

Further, we have

$$\hat{H}_2 = e^{iS_2}\left(\hat{H}_1 - i\hbar\frac{\partial}{\partial t}\right)e^{-iS_2} = \beta mc^2 + X_1 + \frac{\beta}{2mc^2}[Y_1, X_1] + \frac{i\hbar\beta\dot{Y}_1}{2mc^2} = \beta mc^2 + X_1 + Y_2, \tag{29.11}$$

where Y_2 is of order $1/m^2c^4$.

Finally, by means of a third transformation, with

$$S_3 = -\frac{i\beta}{2mc^2}Y_2 \tag{29.12}$$

the off-diagonal term is pushed beyond the limits of the approximation required, and, after a calculation similar to the preceding one, we obtain

$$\hat{H}_3 = e^{iS_3}\left(\hat{H}_2 - i\hbar\frac{\partial}{\partial t}\right)e^{-iS_3} = \beta mc^2 + X_1$$
$$= \beta\left(mc^2 + \frac{Y^2}{2mc^2} - \frac{Y^4}{8m^3c^6}\right) + X - \frac{1}{8m^2c^4}[Y, [Y, X]] - \frac{i\hbar}{8m^2c^4}[Y, \dot{Y}]. \tag{29.13}$$

From (XII.6) and (XII.7), it follows that

$$\mathbf{H} = \nabla\times\mathbf{A}, \quad \mathbf{E} = -\nabla A_0 - \frac{1}{c}\frac{\partial\mathbf{A}}{\partial t}$$

and thus, within the limits of our approximation, we have

$$\frac{Y^2}{2mc^2} = \frac{\left[c\boldsymbol{\alpha}.\left(\mathbf{p} - \frac{e}{c}\mathbf{A}\right)\right]^2}{2mc^2} = \frac{\left(\mathbf{p} - \frac{e}{c}\mathbf{A}\right)^2}{2m} - \frac{e\hbar}{2mc}\boldsymbol{\sigma}.\mathbf{H},$$

$$\frac{1}{8m^2c^4}([Y, X] - i\hbar\dot{Y}) = \frac{ie\hbar}{8m^2c^3}\boldsymbol{\alpha}.\left(-\nabla A_0 - \frac{\partial\mathbf{A}}{\partial t}\right) = \frac{ie\hbar}{8m^2c^3}\boldsymbol{\alpha}.\mathbf{E},$$

$$\left[Y, \frac{ie\hbar}{8m^2c^3}\boldsymbol{\alpha}.\mathbf{E}\right] = \frac{ie\hbar}{8m^2c^2}[\boldsymbol{\alpha}.\mathbf{p}, \boldsymbol{\alpha}.\mathbf{E}]$$
$$= \frac{e\hbar^2}{8m^2c^2}\nabla.\mathbf{E} + \frac{ie\hbar}{8m^2c^2}\boldsymbol{\sigma}.(\nabla\times\mathbf{E}) + \frac{e\hbar}{4m^2c^2}\boldsymbol{\sigma}.(\nabla\times\mathbf{p}).$$

Thus, finally, the Hamiltonian, to the required approximation, is

$$\hat{H}_3 = \beta\left(mc^2 + \frac{\left(\mathbf{p} - \frac{e}{c}\mathbf{A}\right)^2}{2m} - \frac{p^4}{8m^3c^2}\right) + eA_0 - \frac{e\hbar\beta}{2mc}\boldsymbol{\sigma}.\mathbf{H}$$
$$- \frac{ie\hbar^2}{8m^2c^2}\boldsymbol{\sigma}.(\nabla\times\mathbf{E}) - \frac{e\hbar}{8m^2c^2}\boldsymbol{\sigma}.(\mathbf{E}\times\mathbf{p}) - \frac{e\hbar^2}{8m^2c^2}\nabla.\mathbf{E}. \tag{29.14}$$

The first term (in brackets) is the series expansion of $c\sqrt{\left(\mathbf{p}-\frac{e}{c}\mathbf{A}\right)^2+m^2c^2}$ to the required approximation, and includes purely relativistic corrections. The second and third terms are, respectively, the electrostatic and the magnetic dipole potential energies. The following two terms are spin–orbit interaction energies; thus, for a static potential with spherical symmetry ($\mathbf{A} = \mathbf{A}(r)$, $A_0 = V(r)$), we have

$$\nabla\times\mathbf{E} = 0, \quad \boldsymbol{\sigma}.(\mathbf{E}\times\mathbf{p}) = -\frac{1}{r}\frac{dV}{dr}\boldsymbol{\sigma}.(\mathbf{r}\times\mathbf{p}) = -\frac{1}{r}\frac{dV}{dr}\boldsymbol{\sigma}.\mathbf{l},$$

and hence

$$H_{\text{spin-orbit}} = \frac{e\hbar}{4m^2c^2}\frac{1}{r}\frac{dV}{dr}\boldsymbol{\sigma}.\mathbf{l}. \tag{29.15}$$

The last term in (29.14) is known as the *Darwin term*. Since $\nabla.\mathbf{E} = 4\pi\varrho$, it vanishes except inside the charge distribution which produces the field.

30. In order to solve this problem we have to find the solution of equation (28.3) with $\mathbf{A} = 0$, i.e. of

$$H_D\psi(\mathbf{r}) = (E-eA_0)\,\psi(\mathbf{r}) \tag{28.3$'$}$$

with energy $E = E_p$, and with

$$eA_0 = \begin{cases} 0 & \text{for} \quad z < 0 \\ V & \text{for} \quad z > 0. \end{cases}$$

It can be seen immediately that the required solutions will have the form (17.19, 19′), with the difference that in region II the energy E_p is replaced by E_p-V, and the momentum $p = \frac{1}{c}\sqrt{E_p^2-E_0^2}$ by

$$q = \frac{1}{c}\sqrt{(E_p-V)^2-E_0^2}, \qquad E_0 = mc^2. \tag{30.1}$$

Since the spin orientation is a constant of the motion, i.e. $[H, \Sigma_z] = 0$, it is sufficient to use for the calculation either $\psi_{p,\,1,\,1/2}$ or $\psi_{p,\,1,\,-1/2}$, as both give the same result. We shall suppose that the spin of the incident particle is directed along its momentum. The required solution of (28.3′) is then, in region I, a superposition of an incident wave of amplitude A and of a reflected wave of amplitude B, of the form

$$\psi_I(z) = A\psi_{p,\,1,\,1/2}+B\psi_{-p,\,1,\,1/2} = A\begin{bmatrix} 1 \\ 0 \\ \dfrac{cp}{E_0+E_p} \\ 0 \end{bmatrix} e^{\frac{i}{\hbar}pz} + B\begin{bmatrix} 1 \\ 0 \\ \dfrac{-cp}{E_0+E_p} \\ 0 \end{bmatrix} e^{-\frac{i}{\hbar}pz} \tag{30.2}$$

and, in region II, a wave transmitted through the potential step

$$\psi_{II}(z) = D\psi_{q,1,1/2} = D\begin{bmatrix} 1 \\ 0 \\ \dfrac{cq}{E_0+E_p-V} \\ 0 \end{bmatrix} e^{\frac{i}{\hbar}qz}. \tag{30.3}$$

The coefficients R and T are given by

$$R = \frac{|\mathbf{j}_B|}{|\mathbf{j}_A|} \quad \text{and} \quad T = \frac{|\mathbf{j}_D|}{|\mathbf{j}_A|} \tag{30.4}$$

where, in accordance with (16.4),

$$\begin{aligned} \mathbf{j}_A &= C\,|A|^2\,\psi^+_{p,1,1/2}\alpha\psi_{p,1,1/2}, \quad \mathbf{j}_B = C\,|B|^2\,\psi^+_{-p,1,1/2}\alpha\psi_{-p,1,1/2} \\ \mathbf{j}_D &= C\,|D|^2\,\psi^+_{q,1,1/2}\alpha\psi_{q,1,1/2}. \end{aligned} \tag{30.5}$$

Clearly

$$(\mathbf{j}_A)_x = (\mathbf{j}_B)_x = (\mathbf{j}_D)_x = (\mathbf{j}_A)_y = (\mathbf{j}_B)_y = (\mathbf{j}_D)_y = 0.$$

So that

$$|\mathbf{j}_A| = |(\mathbf{j}_A)_z| \quad \text{where}$$

$$|(\mathbf{j}_A)_z| = C\,|A|^2 \left[1, 0, \frac{cp}{E_0+E_p}, 0\right] \begin{bmatrix} 0 & 0 & 1 & 0 \\ 0 & 0 & 0 & -1 \\ 1 & 0 & 0 & 0 \\ 0 & -1 & 0 & 0 \end{bmatrix} \begin{bmatrix} 1 \\ 0 \\ \dfrac{cp}{E_0+E_p} \\ 0 \end{bmatrix} = \frac{2c^2p}{E_0+E_p}\,|A|^2. \tag{30.6}$$

Similarly,

$$|(\mathbf{j}_B)_z| = -\frac{2c^2p}{E_0+E_p}\,|B|^2, \quad |(\mathbf{j}_D)_z| = \frac{2c^2q}{E_0+E_p-V}\,|D|^2, \tag{30.7}$$

and hence

$$R = \left|\frac{B}{A}\right|^2, \quad T = \left|\frac{q}{p}\,\frac{E_0+E_p}{E_0+E_p-V}\right| \cdot \left|\frac{D}{A}\right|^2. \tag{30.8}$$

The ratios B/A and D/A are determined from the equations

$$\begin{aligned} A+B &= D \\ A-B &= \frac{q}{p}\,\frac{E_0+E_p}{E_0+E_p-V}\,D \end{aligned} \tag{30.9}$$

obtained from the continuity condition $\psi_I(0) = \psi_{II}(0)$.

Using the notation

$$a = \left(\frac{E_p-E_0}{E_p+E_0}\right)^{1/2}, \quad b = \left(\frac{E_p-E_0-V}{E_p+E_0-V}\right)^{1/2}, \tag{30.10}$$

the equations (30.9) become simply

$$A+B = D$$
$$A-B = \frac{b}{a} D. \tag{30.11}$$

Thus, after some elementary calculation, we obtain the result that

$$R = \left|\frac{a-b}{a+b}\right|^2 \quad \text{and} \quad T = \frac{4\,|ab|}{|a+b|^2}. \tag{30.12}$$

31. The results of the calculation are given in Table XII.1. The dependence of R and of T on V is shown in Fig. XII.2.

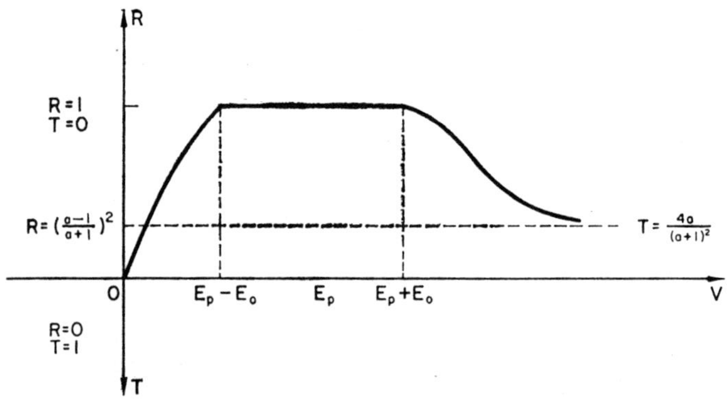

FIG. XII.2

TABLE XII.1

	V	b	R	T	$R+T$
1	0	a	0	1	1
2	$0 < V < E_p - E_0$	$0 < b < 1$	$R = \left(\frac{a-b}{a+b}\right)^2 < 1$	$T = \frac{4ab}{(a+b)^2} < 1$	1
3	$E_p - E_0$	0	1	0	1
4	$E_p - E_0 < V < E_p + E_0$	$b = ib_0$ b_0 – real	1	$\lim_{z\to\infty} (\mathbf{j}_D)_z = 0$	1
5	$E_p + E_0$	∞	1	0	1
6	$E_p + E_0 < V < \infty$	$b > 1$	$R = \left(\frac{a-b}{a+b}\right)^2 < 1$	$T = \frac{4ab}{(a+b)^2} < 1$	1
7	∞	1	$R = \left(\frac{a-1}{a+1}\right)^2 < 1$	$T = \frac{4a}{(a+1)^2} < 1$	1

DISCUSSION OF THE SPECIAL CASES

(1) For $V = 0$ there is evidently no reflection.

(2) In this case the kinetic energy of the incident electron, $E_{\text{kin}} = E_p - E_0$, is greater than the height of the potential step. In agreement with the non-relativistic result, the electron may pass through the potential step, or may be reflected off it with a non-vanishing probability.

(3) The kinetic energy of the electron is equal to the step height. As in the non-relativistic case, reflection is certain, and $T = 0$.

(4) The potential energy is greater than the kinetic energy, but does not exceed its value plus twice the rest energy. In this energy range, which is approximately 2 MeV wide, there is total reflection of the electron, which, unlike cases 3 and 5, is similar to the phenomenon of total reflection in wave optics: the electron enters region II, and returns to region I after travelling a finite distance in region II. For $z > 0$ the probability density decreases exponentially with z, as in the non-relativistic case, thus

$$\varrho = \psi_{II}^{+}\psi_{II} = |D|^2 \left(1 + \frac{c^2 |q|^2}{(E_0 + E_p - V)^2}\right) \exp\left(-\frac{2}{\hbar c}\sqrt{(V - E_p + E_0)(E_p + E_0 - V)}\, z\right).$$

Note that whereas, in the non-relativistic case, total reflection occurs whenever the potential energy is greater than the kinetic energy of the incident particle, i.e. whenever $V > E_{\text{kin}}$, in the relativistic case it does not occur if $V > E_{\text{kin}} + 2E_0 = E_p + E_0$. The value $V = E_p + E_0$ (case 5) is the largest one for which total reflection still occurs.

(6), (7) In view of the above remarks, cases 6 and 7 show a very interesting situation; the incident electron, although it does not have a kinetic energy high enough to surmount the potential step, nevertheless enters region II with a non-vanishing probability. This is so even if $V = \infty$, the probability being then $T = 4a(a+1)^{-2} \neq 0$. This surprising result is called Klein's paradox.

We shall show that, like the "Zitterbewegung", it is a natural and necessary consequence of the existence of negative energy states.

Consider Figs. XII.3 and XII.4. On the left-hand side of each figure the two zones $(-\infty, -E_0)$ and $(E_0, +\infty)$ of electron energy are shown hatched. They are separated by a region of width $2E_0 \approx 2$ MeV (see problem 17). The energy of the incident (free) electron, $E = E_p$, is necessarily in the range $(E_0, +\infty)$. On the right-hand side of the figures the same zones are shown, but displaced by an amount V, where $V > E_p + E_0$ and $E_p - E_0 < V < E_p + E_0$, in order to represent the state of affairs in cases 6 and 4, respectively.

Owing to the displacement $V > E_p + E_0$ (see Fig. XII.3) the zone $(-\infty, -E_0)$ of the left-hand side is raised up on the right-hand side so that it overlaps the zone $(E_0, +\infty)$ in which the energy $E = E_p$ of the incident electron lies. Since the passage of the electron from the region $z < 0$ into the region $z > 0$ takes place at constant total energy, the final state of

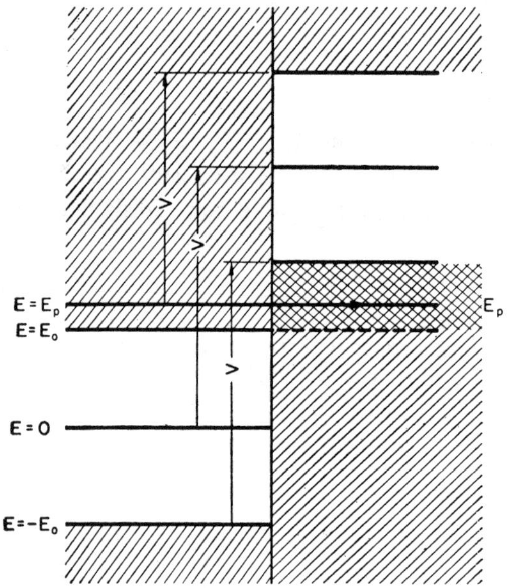

FIG. XII.3

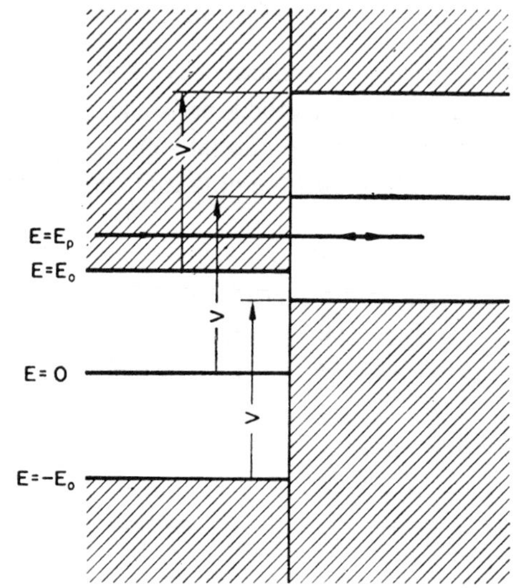

FIG. XII.4

the electron is in the lower energy zone, i.e. is one of the negative energy states according to Dirac's theory. This explains why the electron can penetrate through a potential step of height $V > E_p + E_0$.

The situation is very different if $E_p - E_0 < V < E_p + E_0$ (see Fig. XII.4). In this case the energy E_p extends into the forbidden region on the right-hand side, and hence the electron, after a finite penetration into region II, returns to region I, i.e. it undergoes total reflection.

32. This problem consists in determining the values of E in the equation

$$H_D\psi(\mathbf{r}) = \big(E - eA_0(r)\big)\,\psi(\mathbf{r}) \tag{28.3'}$$

which correspond to the bound states ($E \leqslant E_0 = mc^2$) of the electron in the potential energy well $eA_0 = -e^2Z/r$. Since the potential energy has spherical symmetry, we shall solve the equation

$$H\psi(\mathbf{r}) \equiv \left(c\boldsymbol{\alpha}\,.\,\mathbf{p} + E_0\beta - \frac{e^2Z}{r}\right)\psi(\mathbf{r}) = E(\mathbf{r}) \tag{32.1}$$

in spherical coordinates.

Using the identity (VI.9a), we can write

$$(\boldsymbol{\sigma}\,.\,\mathbf{r})(\boldsymbol{\sigma}\,.\,\mathbf{l}) = (\boldsymbol{\sigma}\,.\,\mathbf{r})\big(\boldsymbol{\sigma}\,.\,(\mathbf{r}\times\mathbf{p})\big) = \big(\mathbf{r}\,.\,(\mathbf{r}\times\mathbf{p})\big) + i\boldsymbol{\sigma}\,.\,\big(\mathbf{r}\times(\mathbf{r}\times\mathbf{p})\big) = i[(\boldsymbol{\sigma}\,.\,\mathbf{r})(\mathbf{r}\,.\,\mathbf{p}) - r^2(\boldsymbol{\sigma}\,.\,\mathbf{p})],$$

and hence

$$(\boldsymbol{\sigma}\,.\,\mathbf{p}) = \frac{(\boldsymbol{\sigma}\,.\,\mathbf{r})}{r^2}\,[(\mathbf{r}\,.\,\mathbf{p}) + i(\boldsymbol{\sigma}\,.\,\mathbf{l})].$$

Thus

$$(\boldsymbol{\alpha}\,.\,\mathbf{p}) = \begin{bmatrix} 0 & (\boldsymbol{\sigma}\,.\,\mathbf{p}) \\ (\boldsymbol{\sigma}\,.\,\mathbf{p}) & 0 \end{bmatrix} = \begin{bmatrix} 0 & \dfrac{(\boldsymbol{\sigma}\,.\,\mathbf{r})}{r} \\ \dfrac{(\boldsymbol{\sigma}\,.\,\mathbf{r})}{r} & 0 \end{bmatrix} \frac{(\mathbf{r}\,.\,\mathbf{p})-i\hbar+i(\boldsymbol{\sigma}\,.\,\mathbf{l})+i\hbar}{r},$$

or, if we introduce the "radial momentum operator"

$$p_r = \frac{(\mathbf{r}\,.\,\mathbf{p})-i\hbar}{r} = -i\hbar\left(\frac{\partial}{\partial r}+\frac{1}{r}\right) = -i\hbar\frac{1}{r}\frac{\partial}{\partial r}r \tag{II.9}$$

and use the notation

$$\alpha_r = \begin{bmatrix} 0 & \dfrac{(\boldsymbol{\sigma}\,.\,\mathbf{r})}{r} \\ \dfrac{(\boldsymbol{\sigma}\,.\,\mathbf{r})}{r} & 0 \end{bmatrix} \tag{32.2}$$

we obtain

$$(\boldsymbol{\alpha}\,.\,\mathbf{p}) = \alpha_r\left(p_r+i\frac{(\boldsymbol{\sigma}\,.\,\mathbf{l})+\hbar}{r}\right) \tag{32.3}$$

We define now an operator K through the relation

$$\hbar K = p(\boldsymbol{\Sigma}\,.\,\mathbf{l})+\hbar \tag{32.4}$$

where

$$\boldsymbol{\Sigma} = \begin{bmatrix} \boldsymbol{\sigma} & 0 \\ 0 & \boldsymbol{\sigma} \end{bmatrix}. \tag{32.5}$$

The Hamiltonian then becomes

$$H = c\alpha_r p_r + \frac{i\hbar c}{r}\alpha_r\beta K + E_0\beta - \frac{e^2 Z}{r}. \tag{32.6}$$

Since $[K, \beta] = [K, \alpha_r] = [K, p_r] = 0$, it follows that $[H, K] = 0$. Using again the identity (VI.9a), as well as the commutator equation $\mathbf{l}\times\mathbf{l} = i\hbar\mathbf{l}$, we have that

$$\begin{aligned}\hbar^2K^2 &= (\boldsymbol{\Sigma}\,.\,\mathbf{l})(\boldsymbol{\Sigma}\,.\,\mathbf{l})+2\hbar(\boldsymbol{\Sigma}\,.\,\mathbf{l})+\hbar^2 = \mathbf{l}^2+i\boldsymbol{\Sigma}\,.\,(\mathbf{l}\times\mathbf{l})+2\hbar(\boldsymbol{\Sigma}\,.\,\mathbf{l})+\hbar^2 \\ &= \left(\mathbf{l}+\frac{1}{2}\hbar\boldsymbol{\Sigma}\right)^2+\frac{\hbar^2}{4} = \mathbf{j}^2+\frac{\hbar^2}{4},\end{aligned}$$

where $\mathbf{j} = \mathbf{l}+\frac{1}{2}\hbar\boldsymbol{\Sigma} = \mathbf{l}+\mathbf{s}$ is the total angular momentum operator. The eigenvalues of the operator $\mathbf{j}^2$ are of the form $\hbar^2 j(j+1)$, where $j = l\pm\frac{1}{2}$, except for the s-state ($l = 0$), in which j can have only the value $\frac{1}{2}$. The operator $\hbar^2K^2$ is therefore a constant of the motion, with eigenvalues $\hbar^2k^2$, where $k^2 = j(j+1)+\frac{1}{4} = (j+\frac{1}{2})^2$, and hence

$$k = \pm(j+\tfrac{1}{2}) = \pm 1, \pm 2, \pm 3, \ldots . \tag{32.7}$$

We are interested in states of the electron which have a well-defined value of the total angular momentum, i.e. a well-defined value of k. The energies of these states, in accordance with (32.6), are the eigenvalues of the equation

$$\left(c\alpha_r p_r + \frac{i\hbar c}{r}\alpha_r \beta k + E_0\beta - \frac{e^2 Z}{r}\right)\psi(r) = E\psi(r), \tag{32.8}$$

in which $\psi(r)$ denotes the radial part of $\psi(\mathbf{r})$.

Since $\alpha_r^2 = \beta^2 = 1$ and $\alpha_r\beta + \beta\alpha_r = 0$, it is possible to restrict ourselves to a 2×2 representation, for example

$$\alpha_r = \begin{bmatrix} 0 & -i \\ i & 0 \end{bmatrix} \quad \text{and} \quad \beta = \begin{bmatrix} 1 & 0 \\ 0 & -1 \end{bmatrix}.$$

The angular and the spin parts of the wavefunction will then be determined from the condition that $\psi(\mathbf{r})$ be an eigenfunction of K (since $[H, K] = 0$). The radial part, which we shall write in the form

$$\psi(r) = \frac{1}{r}\begin{bmatrix} F(r) \\ G(r) \end{bmatrix} \tag{32.9}$$

is determined by the system of differential equations

$$\begin{aligned} (\hbar c)^{-1}\left(E - E_0 + \frac{e^2 Z}{r}\right)F + \frac{dG}{dr} + \frac{k}{r}G &= 0 \\ (\hbar c)^{-1}\left(E + E_0 + \frac{e^2 Z}{r}\right)G - \frac{dF}{dr} + \frac{k}{r}F &= 0 \end{aligned} \tag{32.10}$$

obtained from equation (32.8).

With the notation

$$B = \frac{E + E_0}{\hbar c}, \quad A = \frac{E_0 - E}{\hbar c}, \quad D = \sqrt{AB}, \tag{32.11}$$

and the dimensionless variable $\varrho = Dr$, the equations (32.10) become

$$\begin{aligned} \left(\frac{A}{D} - \frac{Z\alpha}{\varrho}\right)F - \left(\frac{d}{d\varrho} + \frac{k}{\varrho}\right)G &= 0 \\ \left(\frac{B}{D} + \frac{Z\alpha}{\varrho}\right)G - \left(\frac{d}{d\varrho} - \frac{k}{\varrho}\right)F &= 0, \end{aligned} \tag{32.12}$$

where $\alpha = e^2/\hbar c$ is the fine structure constant.

We shall look for solutions in the form of series

$$\begin{aligned} F(\varrho) &= e^{-\varrho}\sum_{\nu=0} a_\nu \varrho^{s+\nu}, \\ G(\varrho) &= e^{-\varrho}\sum_{\nu=0} b_\nu \varrho^{s+\nu}. \end{aligned} \tag{32.13}$$

Substituting (32.13) into (32.12), and equating to zero the coefficients of $\varrho^{s+\nu-1}$, we obtain the relations

$$\left.\begin{aligned}(k+s)\,b_0+Z\alpha a_0 &= 0\\ Z\alpha b_0+(k-s)\,a_0 &= 0\end{aligned}\right\} \tag{32.14}$$

and, for $\nu > 0$,

$$\left.\begin{aligned}\frac{A}{D}\,a_{\nu-1}-Z\alpha a_\nu-(s+\nu+k)\,b_\nu+b_{\nu-1} &= 0\\ \frac{B}{D}\,b_{\nu-1}-Z\alpha b_\nu-(s+\nu-k)\,a_\nu+a_{\nu-1} &= 0\end{aligned}\right\} \tag{32.15}$$

From (32.14) we obtain the equation $k^2-s^2-Z^2\alpha^2 = 0$, whose positive root is

$$s = (k^2-Z^2\alpha^2)^{1/2}. \tag{32.16}$$

The negative root must be excluded, since it would make the functions F and G singular at the origin.

Multiplying the first equation (32.15) by D and the second by A, and subtracting the results, we obtain

$$a_\nu\left(-\sqrt{\frac{B}{A}}\,Z\alpha+s+\nu-k\right) = b_\nu\left(\sqrt{\frac{B}{A}}\,(s+\nu+k)+Z\alpha\right). \tag{32.17}$$

As $\nu \to \infty$, it follows from (32.17) and (32.15) that

$$a_\nu \approx \sqrt{\frac{B}{A}}\,b_\nu \tag{32.17'}$$

$$\left.\begin{aligned}\frac{A}{D}\,a_{\nu-1}-Z\alpha a_\nu-\nu b_\nu+b_{\nu-1} &\approx 0\\ \frac{B}{D}\,b_{\nu-1}+Z\alpha b_\nu-\nu a_\nu+a_{\nu-1} &\approx 0\end{aligned}\right\} \tag{32.15'}$$

From these equations we obtain immediately the result that, as $\nu \to \infty$,

$$a_\nu \approx \frac{2}{\nu}\,a_{\nu-1} \quad \text{and} \quad b_\nu \approx \frac{2}{\nu}\,b_{\nu-1}.$$

This means that for large values of ν the two series (32.13) both behave like $e^{2\varrho}$. Thus, solutions which satisfy the condition at infinity can be obtained only by interrupting the series F and G, i.e. reducing them to polynomials.

Let us suppose that this happens for $\nu = N$, so that $a_{\nu+1} = b_{\nu+1} = 0$ for all $\nu \geqslant N$. Then, by (32.15),

$$\sqrt{A}\,a_N = -\sqrt{B}\,b_N. \tag{32.18}$$

Putting $\nu = N$ in (32.17), and using (32.18), we find that

$$\frac{B-A}{D} Z\alpha = 2(s+N),$$

whence, by (32.11) and (32.16),

$$E = E_0\{1+\alpha^2Z^2[N+|k|\sqrt{1-\alpha^2Z^2k^{-2}}]^{-2}\}^{-1/2}. \tag{32.19}$$

If we define a principal quantum number

$$n = N+|k| = N+j+\tfrac{1}{2}$$

(32.19) becomes

$$E_{nj} = E_0\{1+\alpha^2Z^2[(n-j-\tfrac{1}{2})+(j+\tfrac{1}{2})\sqrt{1-\alpha^2Z^2(j+\tfrac{1}{2})^{-2}}]^{-2}\}^{-1/2}. \tag{32.20}$$

This expression is very similar to the expression (10.15) obtained from the Klein–Gordon equation for π-mesic atoms. The difference arises from the fact that while the Klein–Gordon equation describes a spinless particle, and hence has $\mathbf{j} = \mathbf{l}$, the Dirac equation refers to a spin-$\frac{1}{2}$ particle, with $\mathbf{j} = \mathbf{l}+\mathbf{s}$. Expanding (32.20) in a series in powers of αZ, we obtain

$$E_{nj} = E_0-\frac{E_0(\alpha Z)^2}{2n^2}+\frac{E_0(\alpha Z)^4}{2n^4}\left(\frac{3}{4}-\frac{n}{j+\frac{1}{2}}\right)+\ \ldots . \tag{32.21}$$

The first term is the rest energy of the electron. The second term gives the energy spectrum of the bound states in non-relativistic approximation (see (33.12) Chapter II). The third term is a relativistic correction which takes into account the spin of the electron and leads to a fine structure in the spectra of hydrogen-like atoms, in very good agreement with experimental observations.

The total spread in energy

$$E_{n,\, n-1/2}-E_{n,\, 1/2} = \frac{m_e e^8 Z^4}{c^2\hbar^4 n^3}\,\frac{n-1}{2n} \tag{32.22}$$

of the component levels which form the fine structure of the states of given n is much smaller than that previously obtained without taking into account the electron spin (see 10.17).

An important feature of (32.20) is that it contains only n and j. We thus expect the states $2s_{1/2}$ ($n = 2, l = 0, j = \frac{1}{2}$) and $2p_{1/2}$ ($n = 2, l = 1, j = l-\frac{1}{2} = \frac{1}{2}$) to have the same energy.

This is true to a very good approximation only. In fact, as the measurements of Lamb and Retherford (1947) have shown, the $2s_{1/2}$ level is very slightly higher than the $2p_{1/2}$. The frequency difference corresponding to the transition $2s_{1/2} \leftrightarrow 2p_{1/2}$ is $1{\cdot}05777\times10^{-9}\text{s}^{-1}$; this is of the order of a millionth part of the usual Balmer frequencies. This relative displacement of the two levels, called the *Lamb shift*, which is contrary to the result (32.20), can be explained only by means of quantum electrodynamics, in which the electromagnetic field is treated as a quantized system.

33. From the Dirac equation (XII.22), using the substitution

$$\psi(\mathbf{r}, t) = \psi(\mathbf{r}) \exp\left(-\frac{i}{\hbar} Et\right) \tag{33.1}$$

the following equation for the stationary states is obtained:

$$\left[c\boldsymbol{\alpha}\cdot\left(\mathbf{p}-\frac{e}{c}\,\mathbf{A}\right)+mc^2\beta\right]\psi = E\psi. \tag{33.2}$$

If we write the bi-spinor ψ in the form

$$\psi = \begin{bmatrix}\phi\\ \chi\end{bmatrix}$$

and use (XII.21), the following system of equations follows from (33.2):

$$\begin{aligned} c\boldsymbol{\sigma}\cdot\left(\mathbf{p}-\frac{e}{c}\mathbf{A}\right)\phi &= (E+mc^2)\chi \\ c\boldsymbol{\sigma}\cdot\left(\mathbf{p}-\frac{e}{c}\mathbf{A}\right)\chi &= (E-mc^2)\phi. \end{aligned} \tag{33.3}$$

By choosing the z-axis to be in the direction of the field $\mathbf{H}$, and the corresponding vector potential to be

$$A_0 = A_y = A_z = 0, \quad A_x = -Hy,$$

and by eliminating the spinor χ, we find that

$$[c^2\mathbf{p}^2-e^2H^2y^2-ecH(\hbar\sigma_z-2yp_x)]\phi = (E^2-m^2c^4)\phi. \tag{33.4}$$

The operator on the left-hand side of this equation does not contain explicitly the coordinates x and z. Hence, the operators p_x and p_z commute with the Hamiltonian, i.e. the x and the z components of the (generalized) momentum are constants of the motion. Accordingly we look for a solution ϕ of the form

$$\phi = e^{\frac{i}{\hbar}(p_x x+p_z z)} f(y). \tag{33.5}$$

The eigenvalues p_x and p_z can take all values from $-\infty$ to $+\infty$. The momentum p_z is related to the velocity v_z through the relation $p_z = mv_z(1-v^2/c^2)^{-1/2}$ (see XII.9). Thus the velocity of the particle in the direction of the field can have any value; in other words, the motion along the z-axis, as in the non-relativistic case, is not quantized.

By substituting (33.5) into (33.4) the following equation for the spinor f is obtained:

$$\left[-c^2\hbar^2\frac{d^2}{dy^2}+(eHy+cp_x)^2-ec\hbar H\sigma_z\right]f = (E^2-m^2c^4-p_z^2c^2)f.$$

Introducing the dimensionless variable

$$\xi = \sqrt{\frac{|e|\,H}{\hbar c}}\left(y+\frac{cp_x}{eH}\right),$$

the following differential equation is found

$$\left(\frac{d^2}{d\xi^2}-\xi^2+a-\sigma_z\right)f = 0, \tag{33.6}$$

where

$$a = \frac{E^2-m^2c^4-p_z^2c^2}{c\hbar\,|e|\,H}. \tag{33.7}$$

The spinor f can be chosen to be an eigenfunction of σ_z, i.e.

$$\sigma_z f = 2sf \quad \text{where} \quad s = \pm\tfrac{1}{2}.$$

This means that for $s = +\frac{1}{2}$ we have

$$f = \begin{bmatrix} f_{1/2} \\ 0 \end{bmatrix}$$

and, for $$s = -\frac{1}{2}, \quad f = \begin{bmatrix} 0 \\ f_{-1/2} \end{bmatrix}.$$

Thus, for f_s, $\left(s = \pm\frac{1}{2}\right)$, we obtain the equation

$$\left[\frac{d^2}{d\xi^2}-\xi^2+(a-2s)\right]f_s = 0. \tag{33.8}$$

This equation is formally the same as the Schrödinger equation of a linear oscillator which oscillates with classical frequency $\omega = |e|H/mc$ about the point $y = -cp_x/eH$. Hence we can deduce that the constant $\frac{\hbar\omega}{2}(a-2s)$, which plays the role of the oscillator energy, can take only the values $\left(n+\frac{1}{2}\right)\hbar\omega$, with $n = 0, 1, 2, \ldots$. Remembering (33.7), the following expression for the energy levels of an electron in a homogeneous constant magnetic field is then obtained:

$$\frac{E^2-m^2c^4}{2mc^2} = \omega\hbar\left(n+\frac{1}{2}\right)+\frac{p_z^2}{2m}+2\mu_0 sH \tag{33.9}$$

where $\mu_0 = |e|\hbar/2mc$ is the Bohr magneton.

The corresponding eigenfunctions are

$$f_s = Ne^{-\xi^2/2}H_n(\xi), \tag{33.10}$$

where N is a normalization constant and $H_n(\xi)$ is the nth Hermite polynomial.

The spinor χ can be expressed in terms of ϕ as follows:

$$\chi = \frac{C}{E+mc^2}\left[2sp_z - \sqrt{\frac{|e|\hbar H}{c}}\left(\xi\sigma_x + i\sigma_y\frac{\partial}{\partial\xi}\right)\right]\phi. \tag{33.11}$$

It can be seen that, in the non-relativistic limit, the results obtained above are the same as those obtained in problem 34 of Chapter VI.

34. Using the formula valid in any right-handed system of Cartesian axes,

$$(\mathbf{A}\times\mathbf{B})_j = \varepsilon_{jkl}A_kA_l; \qquad j, k, l = 1, 2, 3,$$

where ε_{jkl} is the Levi–Civita symbol (see problem 3, Chapter VI), Maxwell's equations *in vacuo* (34a) can be written in the form

$$\left.\begin{aligned} &\varepsilon_{jkl}\frac{\partial E_l}{\partial x_k}+\frac{1}{c}\frac{\partial H_j}{\partial t}=0, \qquad \varepsilon_{jkl}\frac{\partial H_l}{\partial x_k}-\frac{1}{c}\frac{\partial E_j}{\partial t}=0\\ &\qquad\frac{\partial H_j}{\partial x_j}=0, \qquad \frac{\partial E_j}{\partial x_j}=0.\end{aligned}\right\} \tag{34.1}$$

Introducing a complex vector defined as follows

$$\psi_j = E_j + iH_j \tag{34.2}$$

the equations (34.1) become

$$\varepsilon_{jkl}\frac{\partial\psi_l}{\partial x_k}-\frac{i}{c}\frac{\partial\psi_j}{\partial t}=0 \tag{34.3}$$

$$\frac{\partial\psi_j}{\partial x_j}=0. \tag{34.3$'$}$$

Now, from (34.3),

$$\frac{i}{c}\frac{\partial}{\partial t}\left(\frac{\partial\psi_j}{\partial x_j}\right)=\varepsilon_{jkl}\frac{\partial^2\psi_l}{\partial x_k\partial x_j}=0,$$

i.e. $\dfrac{\partial\psi_j}{\partial x_j}$ is independent of t. We can therefore dispense with (34.3′) and use (34.3) only, provided we assume as an initial condition that the divergence of the solutions, $\dfrac{\partial\psi_j}{\partial x_j}$, vanishes at $t = 0$.

If we introduce the one-column matrix (34c) and the 3×3 matrices

$$s_1 = \begin{bmatrix}0 & 0 & 0\\ 0 & 0 & -i\\ 0 & i & 0\end{bmatrix}, \quad s_2 = \begin{bmatrix}0 & 0 & i\\ 0 & 0 & 0\\ -i & 0 & 0\end{bmatrix}, \quad s_3 = \begin{bmatrix}0 & -i & 0\\ i & 0 & 0\\ 0 & 0 & 0\end{bmatrix}, \tag{34.4}$$

defined by the relations

$$(s_k)_{jl} = i\varepsilon_{jkl}, \tag{34.4'}$$

and further define

$$p_k = -i\hbar \frac{\partial}{\partial x_k}, \tag{34.5}$$

then (34.3) can be written in the form

$$i\hbar \frac{\partial \psi_j}{\partial t} = c(s_k)_{jl} p_k \psi_l. \tag{34.6}$$

But this is just the j component of the (matrix) equation (34b), with $H = cs_k p_k$.

By means of dimensional considerations and formal analogies with the Schrödinger, the Klein–Gordon and the Dirac equations, the operators p_k and H can be shown to have some of the properties expected of the momentum operator and the Hamiltonian respectively of a free photon. They are not, however, equivalent to momentum and energy observables. The attempt to interpret ψ as a "photon wave-function" also leads to certain difficulties.† These are resolved in the second quantization formalism of quantum electrodynamics.

35. The required solutions of (34b), i.e. of (34.6), will have the form

$$\psi_j = u_j \exp\left[\frac{i}{\hbar}(\mathbf{p}\,.\,\mathbf{r} - Et)\right], \tag{35.1}$$

in which

$$(u_j) = \begin{bmatrix} u_1 \\ u_2 \\ u_3 \end{bmatrix}, \tag{35.2}$$

and the components u_j are independent of time and of position. Only the matrix operators s_k can operate on them.

By substituting (35.1) into (34.6), the following set of homogeneous algebraic equations is obtained:

$$\begin{aligned} Eu_1 + icp_3u_2 - icp_2u_3 &= 0 \\ -icp_3u_1 + Eu_2 + icp_1u_3 &= 0 \\ icp_2u_1 - icp_1u_2 + Eu_3 &= 0. \end{aligned} \tag{35.3}$$

For solutions to exist, the determinant of the coefficients must vanish. Since this condition is satisfied for three values of E, viz., $E = 0, \pm cp$; where $p = +\sqrt{\mathbf{p}^2}$, it follows that the eigenvalue equation

$$Hu = Eu \tag{35.4}$$

† See R. H. Good (1957).

has three linearly independent solutions, u_0, u_+ and u_-, say, with the eigenvalues $E = 0$, $+cp$ and $-cp$ respectively.

We can define more completely the solutions of (35.3) by imposing the normalization conditions

$$u_0^+u_0 = u_+^+u_+ = u_-^+u_- = 1. \tag{35.5}$$

For $E = 0$, the equations (35.3) become

$$\begin{aligned} p_3u_2 - p_2u_3 &= 0 \\ p_1u_3 - p_3u_1 &= 0 \\ p_2u_1 - p_1u_2 &= 0 \end{aligned}$$

whence it follows that $u_2 = \dfrac{p_2}{p_1}u_1$, $u_3 = \dfrac{p_3}{p_1}u_1$. (For simplicity, u_1, u_2 and u_3 are written here in place of $(u_0)_1$, $(u_0)_2$ and $(u_0)_3$.)

Using the normalization condition (35.5) we obtain

$$1 = |u_1|^2+|u_2|^2+|u_3|^2 = \frac{p_1^2+p_2^2+p_3^2}{p_1^2}|u_1|^2 = \frac{p^2}{p_1^2}|u_1|^2.$$

Hence, with the exception of a phase factor, we have

$$u_1 = \frac{p_1}{p}, \quad u_2 = \frac{p_2}{p}, \quad u_3 = \frac{p_3}{p}, \quad \text{i.e.} \quad u_0 = \frac{1}{p}\begin{bmatrix} p_1 \\ p_2 \\ p_3 \end{bmatrix}.$$

For $E = +cp$, we have

$$\begin{aligned} pu_1 + ip_3u_2 - ip_2u_3 &= 0 \\ -ip_3u_1 + pu_2 + ip_1u_3 &= 0 \\ ip_2u_1 - ip_1u_2 + pu_3 &= 0. \end{aligned}$$

From the first equation multiplied by p and the second one multiplied by $-ip_3$ we find that

$$u_1 = \frac{ip_2p - p_1p_3}{p_1^2+p_2^2}u_3.$$

Similarly, by multiplying the first one by ip_3 and the second by p, we find that

$$u_2 = -\frac{p_2p_3 + ip_1p}{p_1^2+p_2^2}u_3.$$

Using the normalization condition (35.5), we then obtain

$$u_3 = \frac{p_1^2+p_2^2}{\sqrt{2p^2(p_1^2+p_2^2)}}$$

and thus, finally,

$$u_+ = \frac{1}{\sqrt{2p^2(p_1^2+p_2^2)}}\begin{bmatrix} ipp_2-p_1p_3 \\ -ipp_1-p_2p_3 \\ p_1^2+p_2^2 \end{bmatrix}.$$

A similar expression can be found for u_-, and the complete result is then:

$$u_0 = \frac{1}{p}\begin{bmatrix} p_1 \\ p_2 \\ p_3 \end{bmatrix}, \quad u_\pm = \frac{1}{\sqrt{2p^2(p_1^2+p_2^2)}}\begin{bmatrix} \pm ipp_2-p_1p_3 \\ \mp ipp_1-p_2p_3 \\ p_1^2+p_2^2 \end{bmatrix}. \tag{35.6}$$

A simple calculation shows that the initial condition (34.3′), which, for u, takes the form

$$u_j p_j = 0, \tag{35.7}$$

is not satisfied by the zero-energy solution u_0, which, for this reason, must be eliminated.

The solutions $u_\pm$ satisfy (34.3′) identically and are called the "transverse" solutions. It can also be verified easily that

$$u_\pm^\dagger u_\mp = 0. \tag{35.8}$$

Thus a complete set of eigenfunctions for pure radiation fields is given by

$$\begin{aligned} \psi_+(\mathbf{p}, \mathbf{r}, t) &= (2\pi\hbar)^{-3/2}u_+ \exp\left[\frac{i}{\hbar}(\mathbf{p}.\mathbf{r}-cpt)\right] \\ \psi_-(\mathbf{p}, \mathbf{r}, t) &= (2\pi\hbar)^{-3/2}u_- \exp\left[\frac{i}{\hbar}(\mathbf{p}.\mathbf{r}+cpt)\right], \end{aligned} \tag{35.9}$$

where $\psi_\pm$ are the solutions of the eigenvalue equation

$$H\psi_\pm = \pm cp\psi_\pm \tag{35.10}$$

and satisfy the orthonormalization conditions

$$\begin{aligned} \int \psi_\pm^\dagger(\mathbf{p})\,\psi_\mp(\mathbf{p}')\,d\mathbf{r} &= 0 \\ \int \psi_\pm^\dagger(\mathbf{p})\,\psi_\pm(\mathbf{p}')\,d\mathbf{r} &= \delta(\mathbf{p}-\mathbf{p}'). \end{aligned} \tag{35.11}$$

36. In this problem, $\mathbf{p} = (0, 0, p)$. Putting $p_1 = 0$ and taking the limit $p_2 \to 0$, we find from (35.6) and (35.9) that

$$\psi_\pm = (2\pi\hbar)^{-3/2}\,2^{-1/2}\begin{bmatrix} \pm i \\ -1 \\ 0 \end{bmatrix} \exp\left[\frac{i}{\hbar}p(z\mp ct)\right]. \tag{36.1}$$

But since

$$\psi_{\pm;j} = E_{\pm;j}+iH_{\pm;j}, \qquad j = 1, 2, 3,$$

we find by separating the real and the imaginary parts that

$$E_{\pm;1} = \mp a \sin\left[\frac{p}{\hbar}(z \mp ct)\right], \quad H_{\pm;1} = \pm a \cos\left[\frac{p}{\hbar}(z \mp ct)\right],$$

$$E_{\pm;2} = -a \cos\left[\frac{p}{\hbar}(z \mp ct)\right], \quad H_{\pm;2} = -a \sin\left[\frac{p}{\hbar}(z \mp ct)\right],$$

$$E_{\pm;3} = 0, \qquad H_{\pm;3} = 0,$$

where $a = (2\pi\hbar)^{-3/2}2^{-1/2}$.

From the phase $\frac{p}{\hbar}(z \mp ct)$ of these plane waves it can be seen that the positive (or negative) eigenvalue solution corresponds to a wave which travels in the positive (or negative) direction of $\mathbf{p}$.

Since $E^2_{\pm;1}+E^2_{\pm;2} = H^2_{\pm;1}+H^2_{\pm;2} = a^2$, it follows that the tips of the transverse vectors $\mathbf{E}_+ \equiv (E_{+;1}, E_{+;2})$; $\mathbf{E}_- \equiv (E_{-;1}, E_{-;2})$; $\mathbf{H}_+ \equiv (H_{+;1}, H_{+;2})$ and $\mathbf{H}_- \equiv (H_{-;1}, H_{-;2})$ lie on a circle of radius a in the xy plane.

Denoting the phases by

$$\phi_+ = \frac{p}{\hbar}(z-ct), \quad \phi_- = \frac{p}{\hbar}(z+ct), \tag{36.2}$$

it can be seen from Tables XII.2, XII.3, and Figs. XII.5, XII.6, that the solution with the positive eigenvalue represents a plane wave whose circular polarization is right-handed with respect to the direction of the momentum vector, and that the solution with the negative eigenvalue represents similarly a wave whose circular polarization is left-handed.

TABLE XII.2

ϕ_+	$E_{+;1}$	$E_{+;2}$	$H_{+;1}$	$H_{+;2}$
$0 < \phi_+ < \frac{\pi}{2}$	−	−	+	−
$\frac{\pi}{2} < \phi_+ < \pi$	−	+	−	−
$\pi < \phi_+ < \frac{3\pi}{2}$	+	+	−	+
$\frac{3\pi}{2} < \phi_+ < 2\pi$	+	−	+	+

TABLE XII.3

ϕ_-	$E_{-;1}$	$E_{-;2}$	$H_{-;1}$	$H_{-;2}$
$0 < \phi_- < \frac{\pi}{2}$	+	−	−	−
$\frac{\pi}{2} < \phi_- < \pi$	+	+	+	−
$\pi < \phi_- < \frac{3\pi}{2}$	−	+	+	+
$\frac{3\pi}{2} < \phi_- \quad 2\pi$	−	−	−	+

The directions of propagation and of circular polarization of the waves $\psi_\pm$ which correspond to the eigenvalues $E = \pm cp$ are shown in Figs. XII.7 and XII.8.

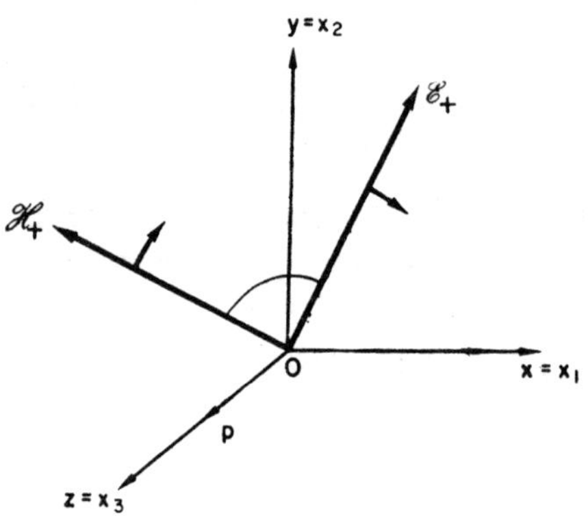

FIG. XII.5.

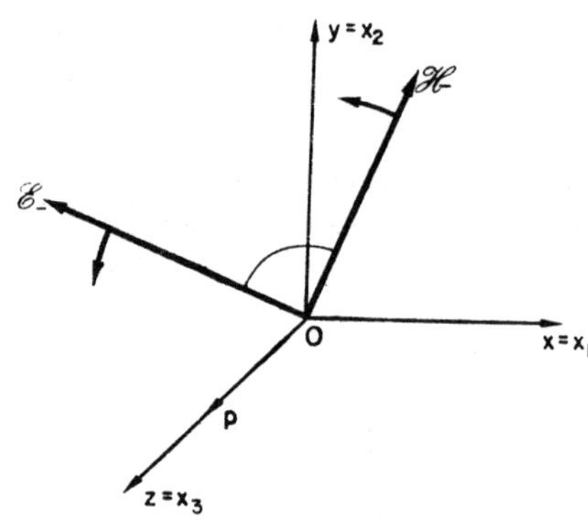

FIG. XII.6.

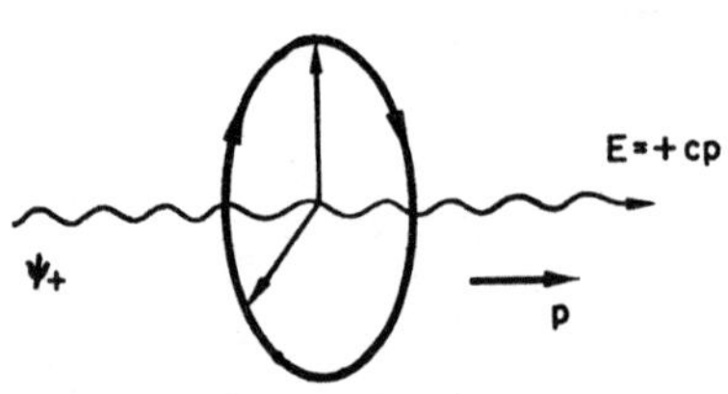

FIG. XII.7.

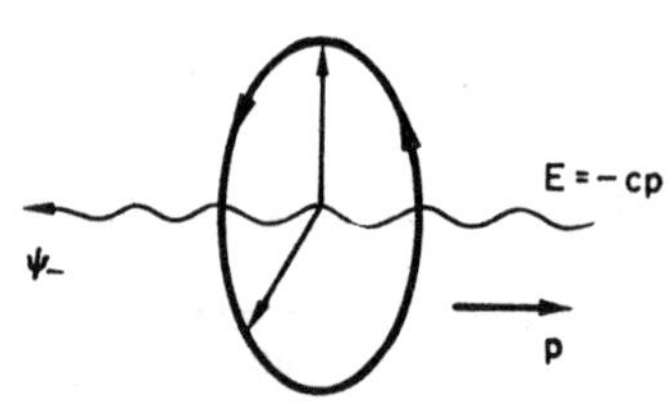

FIG. XII.8.

As can be seen from (36.2), the wave ψ_-, which corresponds to the eigenvalue $-cp$, is the same as the wave which corresponds to the eigenvalue $+cp$, except that it travels in the opposite direction. We can therefore say that any state of a free photon of momentum $\mathbf{p}$ and energy $+cp$ is represented in general by a linear superposition of two plane waves, with right-handed and left-handed circular polarization respectively with respect to the direction of the momentum vector, which latter coincides in direction with the direction of propagation. In particular, any state of linear polarization is a superposition of the two states of opposite circular polarization.

APPENDIX

Certain Functions used in Quantum Mechanics

1. Hermite Polynomials

Definition:

$$H_n(z) = (-1)^n e^{z^2}\left(\frac{d^n}{dz^n} e^{-z^2}\right), \quad n = 0, 1, 2, \ldots \tag{1}$$

Differential equation:

$$\left(\frac{d^2}{dz^2} - 2z\frac{d}{dz} + 2n\right) H_n(z) = 0. \tag{2}$$

Generating function:

$$\exp(-s^2 + 2sz) = \sum_{n=0} \frac{s^n}{n!} H_n(z). \tag{3}$$

Recurrence relations:

$$\frac{d}{dz} H_n = 2nH_{n-1} \tag{4}$$

$$\left(2z - \frac{d}{dz}\right) H_n = H_{n-1} \tag{5}$$

$$2zH_n = H_{n+1} + 2nH_{n-1}. \tag{6}$$

The first few Hermite polynomials:

$$\begin{aligned} H_0 &= 1 & H_1 &= 2z \\ H_2 &= 4z^2 - 2 & H_3 &= 8z^3 - 12z \\ H_4 &= 16z^4 - 48z^2 + 12 & H_5 &= 32z^5 - 160z^3 + 120z. \end{aligned} \tag{7}$$

2. Legendre Polynomials and the Associated Legendre Functions

Definition:

Legendre polynomials:

$$P_l(u) = \frac{1}{2^l l!} \frac{d^l}{du^l} (u^2 - 1)^l, \quad l = 0, 1, 2, \ldots. \tag{8}$$

Legendre functions:

$$P_l^m, \quad l = 0, 1, 2, \ldots; \qquad m = 0, 1, 2, \ldots, l,$$

$$P_l^m(u) = (1-u^2)^{m/2} \frac{d^m}{du^m} P_l(u) = \frac{(1-u^2)^{m/2}}{2^l l!} \frac{d^{l+m}}{du^{l+m}} (u^2-1)^l, \qquad -1 \leqslant u \leqslant 1. \tag{9}$$

Particular cases:

$$m = l, \qquad P_l^l(u) = (2l-1)!!\,(1-u^2)^{1/2} \tag{10}$$

$$m = 0, \qquad P_l^{(0)}(u) = P_l(u). \tag{11}$$

Differential equation:

$$\left[(1-u^2)\frac{d^2}{du^2} - 2u\frac{d}{du} + l(l+1) - \frac{m^2}{1-u^2}\right] P_l^m = 0. \tag{12}$$

Generating functions:

$$\frac{1}{\sqrt{1-2tr+t^2}} = \sum_{l=0}^{\infty} t^l P_l(u), \qquad |t| < 1, \tag{13}$$

$$(2m-1)!!\,(1-u^2)^{m/2} \frac{t^m}{(1-2tu+t^2)^{m+1/2}} = \sum_{l=m}^{\infty} t^l P_l^m(u), \qquad |t| < 1. \tag{14}$$

Orthonormality relation:

$$\int_{-1}^{+1} P_k^m P_l^m \, du = \frac{2}{2l+1} \frac{(l+m)!}{(l-m)!} \delta_{kl}. \tag{15}$$

Recurrence relations:

$$(2l+1)uP_l^m = (l+1-m)P_{l+1}^m + (l+m)P_{l-1}^m \tag{16}$$

$$(1-u^2)\frac{d}{du} P_l^m = -luP_l^m + (l+m)P_{l-1}^m = (l+1)uP_l^m - (l+1-m)P_{l+1}^m. \tag{17}$$

[N.B. The convention $P_{-1} = 0$ is used above.]

Particular values:

$$P_l(1) = 1, \quad P_l(-1) = (-1)^l \tag{18}$$

$$P_l^m(1) = P_l^m(-1) = 0, \quad m \neq 0 \tag{19}$$

$$P_l^m(0) = \begin{cases} (-1)^p \dfrac{(2p+2m)!}{2^l p!(p+m)!} & \text{if} \quad l-m = 2p \\ 0 & \text{if} \quad l-m = 2p+1 \end{cases} \tag{20}$$

The first few Legendre polynomials:

$$\begin{aligned} &P_0 = 1, \quad P_1 = u, \quad P_2 = \tfrac{1}{2}(3u^2-1) \\ &P_3 = \tfrac{1}{2}(5u^3-3u), \quad P_4 = \tfrac{1}{8}(35u^4-30u^2+3) \end{aligned} \tag{21}$$

3. Spherical Harmonics

Definition:

$$Y_l^m(\theta, \phi) = (-1)^m \left[\frac{2l+1}{4\pi}\frac{(l-m)!}{(l+m)!}\right]^{1/2} P_l^m(\cos\theta)e^{im\phi}, \quad m \geqslant 0 \tag{22}$$

$$Y_l^{-m}(\theta, \phi) = (-1)^m Y_l^{m*}(\theta, \phi) \tag{23}$$

$$l = 0, 1, 2, \ldots; \qquad m = -l, -l+1, \ldots, +l.$$

Y_l^m consists of a polynomial of degree $(l-|m|)$ and parity $(-1)^{l-m}$ in $\cos\theta$, multiplied by $\sin^{|m|}\theta\, e^{im\phi}$.

Particular cases:

$$m = 0, \quad Y_l^0 = \sqrt{\frac{2l+1}{4\pi}}\, P_l(\cos\theta) \tag{24}$$

$$m = l, \quad Y_l^l = (-1)^l \left[\frac{2l+1}{4\pi}\frac{(2l)!}{2^{2l}(l!)^2}\right]^{1/2} \sin^l\theta e^{il\phi}. \tag{25}$$

Differential equation:

$$\left[\frac{1}{\sin\theta}\frac{\partial}{\partial\theta}\left(\sin\theta\frac{\partial}{\partial\theta}\right)+\frac{1}{\sin^2\theta}\frac{\partial^2}{\partial\phi^2}+l(l+1)\right] Y_l^m(\theta, \phi) = 0. \tag{26}$$

Orthonormality relation:

$$\int Y_l^{m*} Y_{l'}^{m'}\, d\Omega = \int_0^{2\pi} d\phi \int_0^{\pi} \sin\theta\, d\theta Y_l^{m*}(\theta, \phi) Y_{l'}^{m'}(\theta, \phi) = \delta_{ll'}\delta_{mm'}. \tag{27}$$

Recurrence relations:

$$\begin{aligned} \cos\theta Y_l^m &= \left[\frac{(l+1+m)(l+1-m)}{(2l+1)(2l+3)}\right]^{1/2} Y_{l+1}^m + \left[\frac{(l+m)(l-m)}{(2l+1)(2l-1)}\right]^{1/2} Y_{l-1}^m \\ \sin\theta Y_l^m &= \left\{-\left[\frac{(l+1-m)(l+2-m)}{(2l+1)(2l+3)}\right]^{1/2} Y_{l+1}^{m-1} + \left[\frac{(l+m)(l+m-1)}{(2l+1)(2l-1)}\right]^{1/2} Y_{l-1}^{m-1}\right\} e^{i\phi}. \end{aligned} \tag{28}$$

The first few spherical harmonics:

$$\begin{gathered} Y_0^0 = \frac{1}{\sqrt{4\pi}}, \quad Y_1^0 = \sqrt{\frac{3}{4\pi}}\cos\theta, \quad Y_2^0 = \sqrt{\frac{5}{16\pi}}(3\cos^2\theta - 1) \\ Y_3^0 = \sqrt{\frac{7}{16\pi}}(5\cos^3\theta - 3\cos\theta) \\ Y_1^1 = -\sqrt{\frac{3}{8\pi}}\sin\theta e^{i\phi}, \quad Y_2^1 = -\sqrt{\frac{15}{8\pi}}\sin\theta\cos\theta e^{i\phi} \\ Y_3^1 = -\sqrt{\frac{21}{64\pi}}\sin\theta(5\cos^2\theta - 1)e^{i\phi} \end{gathered} \tag{29}$$

$$Y_2^2 = \sqrt{\frac{15}{32\pi}} \sin^2\theta e^{2i\phi}, \quad Y_3^2 = \sqrt{\frac{105}{32\pi}} \sin^2\theta \cos\theta e^{2i\phi}$$

$$Y_3^3 = -\sqrt{\frac{35}{64\pi}} \sin^3\theta e^{3i\phi}.$$

The addition theorem for spherical harmonics:

$$\frac{2l+1}{4\pi} P_l(\cos\alpha) = \sum_{m=-l}^{+l} Y_l^{m*}(\theta_1, \phi_1) Y_l^m(\theta_2, \phi_2) \tag{30}$$

where α is the angle between the directions (θ_1, ϕ_1) and (θ_2, ϕ_2).

4. Laguerre Polynomials

Definition:

$$\left.\begin{aligned} L_n^0 &= e^z \frac{d^n}{dz^n}(e^{-z}z^n) \\ L_n^k &= (-1)^k \frac{d^k}{dz^k} L_{n+k}^0 \end{aligned}\right\} \quad k, n = 0, 1, 2, \ldots \tag{31}$$

Differential equation:

$$\left[z\frac{d^2}{dz^2} + (k+1-z)\frac{d}{dz} + n\right] L_n^k = 0. \tag{32}$$

Generating function:

$$\frac{e^{-zt/(1-t)}}{(1-t)^{k+1}} = \sum_{n=0}^{\infty} \frac{t^n}{(n+k)!} L_n^k(z), \qquad |t| < 1. \tag{33}$$

Orthonormality relation:

$$\int_0^\infty e^{-z} z^k L_n^k L_m^k \, dz = \frac{[(n+k)!]^3}{n!} \delta_{nm}. \tag{34}$$

5. The Gamma Function

Definition:

$$\Gamma(z) = \int_0^\infty e^{-t} t^{z-1} \, dt, \qquad \mathrm{Re}\,(z) > 0. \tag{35}$$

For $-(n+1) < \mathrm{Re}\,(z) \leq -n$, $z \neq 0, -1, -2, \ldots$ the Γ function is defined by means of the recurrence formula

$$\Gamma(z) = \frac{\Gamma(z+1)}{z} = \frac{\Gamma(z+n+1)}{z(z+1)\ldots(z+n)}. \tag{36}$$

[Note that the points $z = 0, -1, -2, \ldots$ are simple poles of $\Gamma(z)$.]
Functional relations:

$$\Gamma(z+1) = z\Gamma(z)$$

$$\Gamma(z)\Gamma(1-z) = \frac{\pi}{\sin \pi z} \tag{37}$$

Asymptotic behaviour: As $|z| \to \infty$,

$$\Gamma(z) \sim e^{(z-1/2)\ln z - z + (1/2)\ln 2\pi}, \quad |\arg z| \leqslant \pi - \delta, \quad 0 < \delta \ll 1. \tag{38}$$

For real x, as $x \to \infty$,

$$\Gamma(x) \sim \sqrt{2\pi}\, x^{x-1/2} e^{-x}.$$

Particular values: If n is a positive integer,

$$\Gamma(n+1) = n!$$

$$\Gamma\left(n+\frac{1}{2}\right) = \frac{(2n-1)!!}{2^n}\sqrt{\pi}, \tag{39}$$

If β is any real number,

$$|\Gamma(1+i\beta)|^2 = \frac{\pi\beta}{\sinh \pi\beta}. \tag{40}$$

6. Bessel Functions of the First Kind

Definition:

$$J_\nu(z) = \sum_{k=0}^{\infty} \frac{(-1)^k \left(\frac{z}{2}\right)^{\nu+2k}}{\Gamma(k+1)\Gamma(k+\nu+1)} \tag{41}$$

where ν is (in general) a complex number, and, in the complex z-plane, a cut $(-\infty, 0)$ is made.†

Differential equation:

$$\left[\frac{d^2}{dz^2} + \frac{1}{z}\frac{d}{dz} + \left(1 - \frac{\nu^2}{z^2}\right)\right] J_\nu(z) = 0. \tag{42}$$

Particular cases:

$$J_n(z) = \sum_{k=0}^{\infty} \frac{(-1)^k \left(\frac{z}{2}\right)^{n+2k}}{k!\,(k+n)!}, \quad n = 0, 1, 2, \ldots \tag{43}$$

$$J_{-n}(z) = (-1)^n J_n(z). \tag{44}$$

† The restriction thus imposed on z, $|\arg z| < \pi$, is necessary to ensure the uniformity of z^ν. If ν is an integer, the cut becomes unnecessary.

7. Spherical Bessel Functions

Notation:

$j_l(\varrho)$ = the spherical Bessel functions, proper,

$n_l(\varrho)$ = the Neumann functions,

$h_l^{(+)}(\varrho)$ = the Hankel functions of the first kind,

$h_l^{(-)}(\varrho)$ = the Hankel functions of the second kind.

Definitions:

$$j_l(\varrho) = \left(\frac{\pi}{2\varrho}\right)^{1/2} J_{l+1/2}(\varrho), \quad n_l(\varrho) = (-1)^l \left(\frac{\pi}{2\varrho}\right)^{1/2} J_{-l-1/2}(\varrho)$$
$$h_l^{(\pm)}(\varrho) = n_l(\varrho) \pm i j_l(\varrho). \tag{45}$$

Differential equation:

$$\left[\frac{d^2}{d\varrho^2} + \frac{2}{\varrho}\frac{d}{d\varrho} + 1 - \frac{l(l+1)}{\varrho^2}\right] f_l = 0. \tag{46}$$

The solution which is regular at the origin is $f_l = j_l$.
Solutions which are irregular at the origin are $f_l = n_l$ and $h_l^{(\pm)}$.

The first few spherical Bessel functions:

$$_0 = \frac{\sin\varrho}{\varrho}, \quad j_1 = \frac{\sin\varrho}{\varrho^2} - \frac{\cos\varrho}{\varrho}$$
$$n_0 = \frac{\cos\varrho}{\varrho}, \quad n_1 = \frac{\cos\varrho}{\varrho^2} + \frac{\sin\varrho}{\varrho}. \tag{47}$$

Asymptotic behaviour: for $\varrho \gg l(l+1)$,

$$j_l(\varrho) \sim \frac{1}{\varrho}\sin\left(\varrho - \frac{l\pi}{2}\right), \quad n_l(\varrho) \sim \frac{1}{\varrho}\cos\left(\varrho - \frac{l\pi}{2}\right)$$
$$h_l^{(\pm)}(\varrho) \sim \frac{1}{\varrho}\exp\left[\pm i\left(\varrho - \frac{l\pi}{2}\right)\right]\left[1 \pm i\frac{l(l+1)}{2\varrho} - \dots\right]. \tag{48}$$

Behaviour near the origin: for small ϱ,

$$j_l(\varrho) = \frac{\varrho^l}{(2l+1)!!}\left[1 - \frac{\varrho^2}{2(2l+3)} + \dots\right]$$
$$n_l(\varrho) = \frac{(2l+1)!!}{2l+1}\left(\frac{1}{\varrho}\right)^{l+1}\left[1 + \frac{\varrho^2}{2(2l-1)} + \dots\right] \tag{49}$$

where $(2l+1)!! = (2l+1)(2l-1)(2l-3)\dots 3.1$.

The expansion formula for a plane wave:

$$e^{ikz} = \sum_{l=0}^{\infty} (2l+1)\, i^l j_l(kr)\, P_l(\cos\theta), \qquad z = r\cos\theta \tag{50}$$

$$e^{i\mathbf{k}.\mathbf{r}} = 4\pi \sum_{l=0}^{\infty} \sum_{m=-l}^{+l} i^l j_l(kr)\, Y_l^{m*}(\theta_k, \phi_k)\, Y_l^m(\theta_r, \phi_r) \tag{50'}$$

where (θ_k, ϕ_k) and (θ_r, ϕ_r) are the polar angles of the vectors $\mathbf{k}$ and $\mathbf{r}$ respectively.

8. The Hypergeometric Function

Definition:

$$F(a, b, c; z) = 1 + \frac{ab}{c}\frac{z}{1!} + \frac{a(a+1)\,b(b+1)}{c(c+1)}\frac{z^2}{2!} + \frac{a(a+1)(a+2)\,b(b+1)(b+2)}{c(c+1)(c+2)}\frac{z^3}{3!} + \dots \tag{51}$$

The series is convergent at all points inside the unit circle $|z| < 1$.

Differential equation:

$$\left[z(1-z)\frac{d^2}{dz^2} + [c-(a+b+1)z]\frac{d}{dz} - ab\right]u = 0. \tag{52}$$

For $c \neq -n$, $(n = 0, 1, 2, \dots)$, $F(a, b, c; z)$ is the solution of this equation which satisfies the condition $u(0) = 1$.

The general solution of the differential equation (52) is then given by

$$u = C_1 F(a, b, c; z) + C_2 z^{1-c} F(a+1-c, b+1-c, 2-c; z) \tag{53}$$

for $|z| < 1$ and $c \neq 0, \pm 1, \pm 2, \dots$.

The analytic continuation to points outside the unit circle, i.e. for $|z| \geqslant 1$, with a cut $(+1, \infty)$,† is given by

$$F(a, b, c; z) = \frac{\Gamma(c)\Gamma(c-a-b)}{\Gamma(c-a)\Gamma(c-b)} F(a, b, a+b-c+1; 1-z) + \frac{\Gamma(c)\Gamma(a+b-c)}{\Gamma(a)\Gamma(b)} (1-z)^{c-a-b} F(c-a, c-b, c-a-b+1; 1-z), \tag{54}$$

$$F(a, b, c; z) = \frac{\Gamma(c)\Gamma(b-a)}{\Gamma(b)\Gamma(c-a)} (-z)^{-a} F\left(a, a-c+1, a-b+1; \frac{1}{z}\right) + \frac{\Gamma(c)\Gamma(a-b)}{\Gamma(a)\Gamma(c-b)} (-z)^{-b} F\left(b, b-c+1, b-a+1; \frac{1}{z}\right). \tag{55}$$

† i.e. for $|\arg(-z)| < \pi$.

Asymptotic behaviour: as $|z| \to \infty$,

$$F(a, b, c; z) \sim \frac{\Gamma(c)\,\Gamma(b-a)}{\Gamma(b)\,\Gamma(c-a)}(-z)^{-a} + \frac{\Gamma(c)\,\Gamma(a-b)}{\Gamma(a)\,\Gamma(c-b)}(-z)^{-b}. \tag{56}$$

Special case:

For $a = -n$ or $b = -n$, $(n = 0, 1, 2, \ldots)$, $F(a, b, c; z)$ reduces to a polynomial of degree n (Jacobi polynomial).

Relation to Legendre functions:

$$P_l^m(z) = \frac{\Gamma(l+m+1)}{\Gamma(l-m+1)} \cdot \frac{(z^2-1)^{m/2}}{2^m \Gamma(1+m)} F\left(m-l,\, m+l+1,\, m+1, \frac{1-z}{2}\right),$$
$$|\arg(z \pm 1)| < \pi. \tag{57}$$

Relation to Legendre polynomials:

$$P_l(z) = F\left(-l, l+1, 1; \frac{1-z}{2}\right). \tag{58}$$

9. The Confluent Hypergeometric Function

Definition:

$$F(a, c; z) = 1 + \frac{a}{c}\frac{z}{1!} + \frac{a(a+1)}{c(c+1)}\frac{z^2}{2!} + \frac{a(a+1)(a+2)}{c(c+1)(c+2)}\frac{z^3}{3!} + \ldots\,. \tag{59}$$

The series is convergent in the whole of the complex plane.

For $c = -n$, $(n = 0, 1, 2, \ldots)$, $F(a, c; z)$ is defined to be

$$\lim_{c \to -n} \frac{F(a, c; z)}{\Gamma(c)} = \frac{\Gamma(a+n+1)}{\Gamma(a)} \frac{z^{n+1}}{(n+1)!} F(a+n+1, n+2; z). \tag{60}$$

Relation to the hypergeometric series:

$$F(a, c; z) = \lim_{b \to \infty} F\left(a, b, c; \frac{z}{b}\right). \tag{61}$$

Differential equation:

$$\left[z\frac{d^2}{dz^2} + (c-z)\frac{d}{dz} - a\right]u = 0. \tag{62}$$

General solution of the equation (62):

$$u = C_1 F(a, c; z) + C_2 z^{1-c} F(a-c+1, 2-c; z). \tag{63}$$

Asymptotic behaviour: as $|z| \to \infty$,

$$F(a, c; z) \sim e^{-i\pi a}\frac{\Gamma(c)}{\Gamma(c-a)} z^{-a} + \frac{\Gamma(c)}{\Gamma(a)} e^z z^{a-c} \tag{64}$$

provided $a \neq -n$, $(n = 0, 1, 2, \ldots)$, and the z-plane is cut along the positive imaginary axis.

Relation to Laguerre polynomials:

$$L_n^k(z) = \frac{[(n+k)!]^2}{n!\,k!} F(-n, k+1; z). \tag{65}$$

Relation to Hermite polynomials:

$$H_n(z) = \begin{cases} (-1)^p \dfrac{(2p)!}{p!} F\left(-p, \dfrac{1}{2}; z^2\right) & \text{for} \quad n = 2p \\ (-1)^p\, 2\, \dfrac{(2p+1)!}{p!} zF\left(-p, \dfrac{3}{2}; z^2\right) & \text{for} \quad n = 2p+1. \end{cases} \tag{66}$$

References

CHAPTER I

1. *The Abstract Hilbert Space*

1. J. von Neumann, *Mathematical Foundations of Quantum Mechanics*, Princeton University Press, Princeton, 1955: Chapter 11.
2. P. Roman, *Advanced Quantum Theory*, Addison-Wesley, Reading, Massachusetts, 1965: Appendix 1.
3. G. L. Trigg, *Quantum Mechanics*, D. Van Nostrand Co., Inc., Princeton, 1969: Chapter 3, § 3 and § 4.
4. P. Dennery and A. Krzywicki, *Mathematics for Physicists*, Harper, New York, 1967.

2. *The Hilbert Space of Ket Vectors. Operators in the Ket-Space*

1. P. A. M. Dirac, *The Principles of Quantum Mechanics*, Clarendon Press, Oxford, 1958: Chapter I and Chapter II.
2. A. Messiah, *Quantum Mechanics*, Interscience, New York, 1961, Vol. 1: Chapter VII, § 1–§ 15.
3. P. Roman, op. cit.: Chapter 1, § 1 and § 2; Appendix 1, sec. 3a.
4. G. L. Trigg, op. cit.: Chapter 4, § 1, § 3 and § 4.
5. K. Gottfried, *Quantum Mechanics*, W. A. Benjamin, New York, 1966, Vol. I: § 23.
6. J. von Neumann, op. cit.: Chapter II.

3. *The Matrix Representation of Vectors and Operators*

1. P. A. M. Dirac, op. cit.: Chapter III.
2. A. Messiah, op. cit., Vol. I, Chapter VII, § 16–§ 22 and Chapter VIII, § 15–§ 19.
3. P. Roman, op. cit.: Appendix 1, § 2.
4. A. S. Davydov, *Quantum Mechanics*, Pergamon Press, Oxford, 1965: § 27–§ 29.

4. *The Hilbert Space of Wave Functions*

1. A. Messiah, op. cit., Vol. I: Chapter V and Chapter VIII, § 15–§ 19.
2. J. von Neumann, op. cit.: Chapter III, § 1–§ 5 and Chapter IV.
3. K. Gottfried, op. cit.: § 7.
4. A. S. Davydov, op. cit.: § 1–§ 10.

CHAPTER II

1. *The Schrödinger Equation and the Schrödinger Eigenvalue-Problem. The Statistical Meaning of the Wave Function*

1. A. Messiah, *Quantum Mechanics*, Interscience, New York, 1961, Vol. I: Chapter II, § 9–§ 17 and Chapter IV, § 1–§ 3.
2. L. I. Schiff, *Quantum Mechanics*, McGraw-Hill, New York, 1965: § 6–§ 8 and § 10.
3. L. D. Landau and E. M. Lifshitz, *Quantum Mechanics*, Pergamon Press, Oxford, 1958: § 2, § 3, § 17, § 18 and § 20.

4. K. Gottfried, *Quantum Mechanics*, W. A. Benjamin, New York, 1966: § 4, § 5 and § 7.
5. A. S. Davydov, *Quantum Mechanics*, Pergamon Press, Oxford, 1965: § 4, § 15, § 16 and § 21.
6. J. von Neumann, *Mathematical Foundations of Quantum Mechanics*, Princeton University Press, Princeton, 1955: Chapter III, § 1–§ 5.

2. *One-Dimensional Quantum Systems. Bound States and Unbound States (The Transmission of Particles through Potential Barriers)*

1. A. Messiah, op. cit., Vol. 1, Chapter III and Chapter IV, § 3 and § 4.
2. L. I. Schiff, op. cit.: § 9 and § 17.
3. L. D. Landau and E. M. Lifshitz, op. cit.: § 19, § 21, § 22, § 25 and § 27.
4. A. S. Davydov, op. cit.: § 26 and § 32.

3. *The Harmonic Oscillator*

1. L. I. Schiff, op. cit.: § 13.
2. L. D. Landau and E. M. Lifshitz, op. cit.: § 23.
3. A. S. Davydov, op. cit.: § 33.
4. K. Gottfried, op. cit., Vol. I: § 31.
5. G. L. Trigg, *Quantum Mechanics*, D. Van Nostrand Co., Inc., Princeton, 1965: Chapter 6, § 1.
6. P. A. M. Dirac, *The Principles of Quantum Mechanics*, Clarendon Press, Oxford, 1958: § 34.
7. A. Messiah, op. cit., Vol. I: Chapter XII.

4. *Motion in a Central Field*

1. A. Messiah, op. cit., Vol. I: Chapter IX, § 1–§ 10.
2. P. A. M. Dirac, op. cit.: § 38.
3. L. I. Schiff, op. cit.: § 15 and § 17.
4. L. D. Landau and E. M. Lifshitz, op. cit.: § 32–§ 35.
5. G. L. Trigg, op. cit.: Chapter 6, § 3, Sec. 2, 3 and 4.
6. A. S. Davydov, op. cit.: § 34–§ 37.

5. *The Two Body Problem and the Hydrogen Atom*

1. A. Messiah, op. cit., Vol. I: Chapter IX, § 11–§ 13 and Chapter XI, § 1–§ 6.
2. L. I. Schiff, op. cit.: § 16.
3. P. A. M. Dirac, op. cit.: § 39.
4. L. Gottfried, op. cit.: § 9 and § 18.
5. G. L. Trigg, op. cit.: Chapter 6, § 3, Sec. 1 and 4.
6. L. D. Landau and E. M. Lifshitz, op. cit.: § 36 and § 37.
7. A. S. Davydov, op. cit.: § 38.
8. H. A. Bethe and E. E. Salpeter, *Quantum Mechanics of One- and Two-Electron Atoms*, Springer, Berlin, 1957: Chapter I, § 1–§ 6.

CHAPTER III

1. *The Mean Values of Dynamical Variable Constants of Motion*

1. A. Messiah, *Quantum Mechanics*, Interscience, New York, 1961: Vol. I, Chapter IV, § 5 and § 6, Chapter V, § 3, § 4, § 19 and § 20.
2. L. I. Schiff, *Quantum Mechanics*, McGraw-Hill, New York, 1955: § 7 and § 23.
3. P. Roman, *Advanced Quantum Theory*, Addison-Wesley, Reading, Massachusetts, 1965: Chapter 1, § 3.
4. A. S. Davydov, *Quantum Mechanics*, Pergamon Press, Oxford, 1965: § 6, § 7, § 17 and § 19.
5. K. Gottfried, *Quantum Mechanics*, W. A. Benjamin, New York, 1966, Vol. 1: § 4 and § 6.
6. J. von Neumann, *Mathematical Foundations of Quantum Mechanics*, Princeton University Press, Princeton, 1955: Chapter IV.

2. *The Coordinate–Momentum Uncertainty Relations, Complementarity and Causality*

1. A. MESSIAH, op. cit., Vol. I: Chapter IV, § 7–§ 9 and § 12–§ 19.
2. L. I. SCHIFF, op. cit.: § 3–§ 5.
3. K. GOTTFRIED, op. cit.: § 1 and § 3.
4. L. D. LANDAU and E. M. LIFSHITZ: *Quantum Mechanics*, Addison-Wesley, Reading, Massachusetts, 1958: § 1 and § 16.
5. W. HEISENBERG, *The Physical Principles of the Quantum Theory*, Chicago University Press, 1930.

3. *The Time–Energy Uncertainty Relation*

1. A. MESSIAH, op. cit., Vol. I: Chapter IV, § 10 and § 11, Chapter VIII, § 13 and Chapter X, § 16.
2. L. D. LANDAU and E. M. LIFSHITZ, op. cit.: § 44.

4. *The Angular Momentum–Angle Variable Uncertainty Relation*

1. D. JUDGE, *Phys. Lett.* **5,** 189 (1963).
2. D. JUDGE, *Nuovo Cimento,* **31,** 322 (1964).
3. P. CARRUTHERS and M. M. NIETO, *Rev. Mod. Phys.*, **40,** 411 (1968).

5. *General Uncertainty Relations*

1. A. MESSIAH, op. cit., Vol. I: Chapter VIII, § 4.
2. K. GOTTFRIED, op. cit., Vol. I: § 24.
3. P. ROMAN, op. cit.: Chapter 1, § 4, sec. 4a.
4. A. S. DAVYDOV, op. cit.: § 13.
5. J. VON NEUMANN, op. cit.: Chapter III, § 4.

CHAPTER IV

1. *The W. K. B. Approximation in the One-Dimensional Case*

1. A. MESSIAH, *Quantum Mechanics*, Interscience, New York, 1961, Vol. I, Chapter VI, § 6–§ 11.
2. L. I. SCHIFF, *Quantum Mechanics*, McGraw-Hill, New York, 1955: § 28.
3. L. D. LANDAU and E. M. LIFSHITZ, *Quantum Mechanics*, Pergamon Press, Oxford, 1958: § 76–§ 78, § 50–§ 53.
4. A. S. DAVYDOV, *Quantum Mechanics*, Pergamon Press, Oxford, 1965: § 23–§ 26.
5. K. GOTTFRIED, *Quantum Mechanics*, W. A. Benjamin, New York, 1966: Vol. 1, § 8.
6. G. L. TRIGG, *Quantum Mechanics*, D. Van Nostrand Co., Inc., Princeton, Chapter 6: § 4, sec. 3.
7. R. E. LANGER, *Phys. Rev.* **51,** 669 (1937).
8. W. H. FURRY, *Phys. Rev.* **71,** 360 (1947).
9. P. MORSE and H. FESHBACH, *Mathematical Methods of Theoretical Physics*, McGraw-Hill, New York, 1953, p. 1092.
10. N. FROMAN and P. O. FROMAN, *J.W.K.B. Approximation*, North-Holland Publishing Company, Amsterdam, 1965 (Micromonograph).

2. *The W.K.B. Approximation for the Radial Schrödinger Equation*

1. L. D. LANDAU and E. M. LIFSHITZ, op. cit.: § 49.
2. N. FROMAN and P. O. FROMAN, op. cit.: Chapter 11.
3. M. L. GOLDBERGER and K. M. WATSON, *Collision Theory*, John Wiley & Sons, New York, 1969: Chapter 6, § 7.

CHAPTER V

1. *The "Pictures" of Quantum Mechanics*

1. A. MESSIAH, *Quantum Mechanics*, Interscience, New York, 1961, Vol. 1: Chapter VIII, § 8–§ 14.
2. P. A. M. DIRAC, *The Principles of Quantum Mechanics*, Clarendon Press, Oxford, 1958: § 27 and § 28.

References

3. P. ROMAN, *Advanced Quantum Theory*, Addison-Wesley, Reading, Massachusetts, 1965: Chapter 1, § 5.
4. G. L. TRIGG, *Quantum Mechanics*, D. Van Nostrand Co., Inc., Princeton, 1964: Chapter 5, § 1 and § 2.
5. K. GOTTFRIED, *Quantum Mechanics*, W. A. Benjamin, New York, 1966: § 28.

2. *Particular Representations of Quantum Mechanics*

1. A. MESSIAH, op. cit., Vol. 1: Chapter VIII, § 15–§ 19.
2. P. A. M. DIRAC, op. cit.: § 22 and § 23.
3. G. L. TRIGG, op. cit.: Chapter 5, § 3.
4. A. S. DAVYDOV, *Quantum Mechanics*, Pergamon Press, Oxford, 1965: § 27–§ 29.

3. *The Density Operator*

1. A. MESSIAH, op. cit., Vol. 1: Chapter VIII, § 20–§ 25.
2. P. ROMAN, op. cit.: Chapter 1, § 8.
3. P. A. M. DIRAC, op. cit.: § 33.
4. K. GOTTFRIED, op., cit., Vol. 1: § 20.
5. A. S. DAVYDOV, op. cit.: § 14 and § 22.
6. L. D. LANDAU and E. M. LIFSHITZ, *Quantum Mechanics*, Pergamon Press, Oxford, 1958: § 14.
7. V. FANO, *Rev. Mod. Phys.*, **29,** 74 (1957).
8. D. TER HAAR, *Rep. Progr. Phys.*, **24,** 304 (1961).

CHAPTER VI

1. *Algebraic Properties of the Angular Momentum and the Eigenvalue Problem*

1. A. MESSIAH, *Quantum Mechanics*, Interscience, New York, 1961, Vol. II: Chapter XIII, § 1–§ 7.
2. A. S. DAVYDOV, *Quantum Mechanics*, Pergamon Press, Oxford, 1965: § 40 and § 44.
3. P. A. M. DIRAC, *The Principles of Quantum Mechanics*, Clarendon Press, Oxford, 1958: § 36.
4. G. L. TRIGG, *Quantum Mechanics*, D. Van Nostrand Co., Inc., Princeton, 1967: Chapter 6, § 2, sec. 1.

2. *Angular Momentum and Rotations of Coordinate Axes*

1. A. MESSIAH, op. cit., Vol. II: Chapter XIII, § 10–§ 17.
2. K. GOTTFRIED, *Quantum Mechanics*, W. A. Benjamin, New York, 1966: § 32, § 35, § 37.
3. P. ROMAN, *Advanced Quantum Theory*, Addison-Wesley, Reading, Massachusetts, 1965: Chapter 5, § 3, Sec. c.
4. A. S. DAVYDOV, op. cit.: § 43.

3. *The Addition of Angular Momenta*

1. A. MESSIAH, op. cit., Vol. II: Chapter XIII, § 24–§ 29.
2. K. GOTTFRIED, op. cit.: § 25.
3. G. L. TRIGG, op. cit.: Chapter 6, § 2, Sec. 2.
4. A. S. DAVYDOV, op. cit.: § 41 and § 42.
5. L. D. LANDAU and E. M. LIFSHITZ, *Quantum Mechanics*, Pergamon Press, Oxford, 1958: Chapter XIV.
6. W. MILLER, JR. *Symmetry Group and their Applications*, Academic Press, New York, 1972: Chapter 7, § 7 and § 8.

4. *Irreducible Tensor Operators. The Wigner–Eckart Theorem*

1. A. MESSIAH, op. cit., Vol. II: Chapter XIII, § 30–§ 33.
2. P. ROMAN, op. cit.: Chapter 6, § 2.
3. K. GOTTFRIED, op. cit.: § 36.
4. G. L. TRIGG, op. cit.: Chapter 6, § 2, Sec. 3.
5. W. MILLER, JR., op. cit.: Chapter 7, § 10.
6. E. D. WIGNER, *Group Theory and its Applications to the Quantum Mechanics of Atomic Spectra*, Academic Press, 1959.

5. *The Spin*

1. A. MESSIAH, op. cit., Vol. II: Chapter XIII, § 18–§ 23.
2. L. I. SCHIFF, *Quantum Mechanics*, McGraw-Hill, New York, 1965: § 33.
3. K. GOTTFRIED, op. cit.: § 33.
4. L. D. LANDAU and E. M. LIFSHITZ, op. cit.: Chapter VIII.
5. P. A. M. DIRAC, op. cit.: §37.
6. N. MOTT and I. SNEDDON, *Wave Mechanics and its Application*, Clarendon Press, Oxford, 1948: Chapter 4.

6. *The Orbital Angular Momentum*

1. A. MESSIAH, op. cit., Vol. I: Chapter V, § 18, Chapter IX, § 3, Vol. II, Chapter XIII, § 8 and § 9.
2. L. I. SCHIFF, op. cit.: § 17 and § 24.
3. L. D. LANDAU and E. M. LIFSHITZ, op. cit.: Chapter IV.
4. K. GOTTFRIED, op. cit.: Vol. 1, § 10.
5. P. A. M. DIRAC, op. cit.: § 35.
6. A. S. DAVYDOV, op. cit.: § 7 and § 8 and Appendix B.

CHAPTER VII

1. *Symmetry and Anti-Symmetry of State Vectors; Spin and Statistics*

1. A. MESSIAH, *Quantum Mechanics*, Interscience, New York, 1961, Vol. II: Chapter XIV, § 1–§ 8.
2. P. ROMAN, *Advanced Quantum Theory*, Addison-Wesley, Reading, Massachusetts, 1965: Chapter 1, § 6.
3. P. A. M. DIRAC, *The Principles of Quantum Mechanics*, Clarendon Press, Oxford, 1958: Chapter IX.
4. L. I. SCHIFF, *Quantum Mechanics*, McGraw-Hill, New York, 1955: § 32.
5. K. GOTTFRIED, *Quantum Mechanics*, W. A. Benjamin, New York, 1966, Vol. I: § 43.
6. A. S. DAVYDOV, *Quantum Mechanics*, Pergamon Press, Oxford, 1965: § 86–§ 88.
7. L. D. LANDAU and E. M. LIFSHITZ, *Quantum Mechanics*, Pergamon Press, Oxford, 1958: § 61–§ 63.
8. G. L. TRIGG, *Quantum Mechanics*, D. Van Nostrand Co., Inc., Princeton, 1967: Chapter 7, § 1–§ 3.
9. H. A. BETHE, *Intermediate Quantum Mechanics*, W A. Benjamin, New York, 1967: Chapter 3.

2. *Isotopic Spin*

1. A. MESSIAH, op. cit., Vol. II: Chapter XIV, § 14 and § 15.
2. P. ROMAN, op. cit.: Chapter 5, § 3, Sections f and ff.
3. L. D. LANDAU and E. M. LIFSHITZ, op. cit.: § 115.
4. G. L. TRIGG, op. cit.: Chapter 7, § 4.

3. *Second Quantization*

1. A. MESSIAH, op. cit., Vol. II: Chapter XXI, § 1–§ 8.
2. P. ROMAN, op. cit.: Chapter 1, § 7.
3. P. A. M. DIRAC, op. cit.: Chapter X.
4. K. GOTTFRIED, op. cit.: § 50–§ 53.
5. L. I. SCHIFF, op. cit.: Chapter XIII.
6. L. D. LANDAU and E. M. LIFSHITZ, op. cit.: § 64 and § 65.
7. A. S. DAVYDOV, op. cit.: Chapter XIV and Chapter XV.
8. G. L. TRIGG, op. cit.: Chapter 13 and Chapter 14.
9. H. A. BETHE, op. cit.: Chapter X.

CHAPTER VIII

1. *Stationary Perturbations*

1. A. MESSIAH, *Quantum Mechanics*, Interscience, New York, 1961, Vol. II: Chapter XVI.
2. L. I. SCHIFF, *Quantum Mechanics*, McGraw-Hill, New York, 1955: § 25.

3. P. A. M. DIRAC, *The Principles of Quantum Mechanics*, Clarendon Press, Oxford, 1958: § 42, § 43 and § 46.
4. H. A. BETHE, *Intermediate Quantum Mechanics*, W. A. Benjamin, New York, 1964, Chapter 1 and Chapter 4.
5. N. MOTT and I. SNEDDON, *Wave Mechanics and its Applications*, Clarendon Press, Oxford, 1948: § 17 and § 18.
6. K. GOTTFRIED, *Quantum Mechanics*, W. A. Benjamin, New York, 1966, Vol. I: § 44–§ 47 and § 49.
7. G. L. TRIGG, *Quantum Mechanics*, D. Van Nostrand Co., Inc., Princeton, 1964: Chapter 6, § 4, Sec. 1.
8. A. S. DAVYDOV, *Quantum Mechanics*, Pergamon Press, Oxford, 1965: § 47–§ 50.
9. L. D. LANDAU and E. M. LIFSHITZ, *Quantum Mechanics*, Pergamon Press, Oxford. 1958: § 38 and § 39.

2. *Variational Method*

1. A. MESSIAH, op. cit., Vol. II: Chapter XVII, § 1-§ 6.
2. H. A. BETHE, op. cit.: Chapter 1 and Chapter 5.
3. K. GOTTFRIED, op. cit.: § 48.
4. L. I. SCHIFF, op. cit.: § 27.
5. A. S. DAVYDOV, op. cit.: § 51.
6. G. L. TRIGG, op. cit.: Chapter 61, § 4, sec. 2.

CHAPTER IX

1. *Time-Dependent Perturbations*

1. A. MESSIAH, *Quantum Mechanics*, Interscience, New York, 1961, Vol. II: Chapter XVII.
2. P. ROMAN, *Advanced Quantum Theory*, Addison-Wesley, Reading, Massachusetts, 1965: Chapter 3, § 5.
3. P. A. M. DIRAC, *The Principles of Quantum Mechanics*, Clarendon Press, Oxford, 1958: § 44 and § 45.
4. L. I. SCHIFF, *Quantum Mechanics*, McGraw-Hill, New York, 1955: § 29.
5. K. GOTTFRIED, *Quantum Mechanics*, W. A. Benjamin, New York, 1966, Vol. I: § 58.
6. G. L. TRIGG, *Quantum Mechanics*, D. Van Nostrand Co., Inc., Princeton, 1964: Chapter 9, § 1.
7. A. S. DAVYDOV, *Quantum Mechanics*, Pergamon Press, Oxford, 1965: § 74–§ 77.
8. L. D. LANDAU and E. M. LIFSHITZ, *Quantum Mechanics*, Pergamon Press, Oxford, 1958: § 40–§ 43.
9. W. HEITLER, *The Quantum Theory of Radiation*, Clarendon Press, Oxford, 1954: § 14–§ 16.

2. *The Pure Radiation Field*

1. A. MESSIAH, op. cit., Vol. II: Chapter XXI, § 17–§ 29.
2. P. A. M. DIRAC, op. cit.: § 74–§ 77.
3. L. I. SCHIFF, op. cit.: § 48.
4. K. GOTTFRIED, op. cit.: Chapter VIII.
5. A. S. DAVYDOV, op. cit.: § 134.
6. G. L. TRIGG, op. cit.: Chapter 14, § 2, sec. 4.
7. W. HEITLER, op. cit.: Chapter I and Chapter II.

3. *The Interaction of Radiation with Atomic Systems*

1. A. MESSIAH, op. cit., Vol. II: Chapter XXI, § 30–§ 32.
2. P. A. M. DIRAC, op. cit.: § 63 and § 67.
3. L. I. SCHIFF, op. cit.: § 35, § 36, § 37, § 49 and § 50.
4. A. S. DAVYDOV, op. cit.: § 78–§ 85.
5. K. GOTTFRIED, op. cit.: § 54 and § 56.
6. G. L. TRIGG, op. cit.: Chapter 9, § 3, sec. 2.
7. H. A. BETHE, *Intermediate Quantum Mechanics*, W. A. Benjamin, New York, 1964. Chapters 12, 13, 14 and 21.
8. N. MOTT and I. SNEDDON, *Wave Mechanics and its Applications*, Clarendon Press, Oxford, 1948: Chapter 10.

References

CHAPTER X

1. *Scattering Amplitudes and Scattering Cross-sections. The Integral Equation of Potential Scattering*

1. A. Messiah, *Quantum Mechanics*, Interscience, New York, 1961, Vol. I: Chapter X, § 1–§ 7; Vol. II, Chapter XIX, § 1–§ 7.
2. P. Roman, *Advanced Quantum Theory*, Addison-Wesley, Reading, Massachusetts, 1965: Chapter 3, § fl and § 2.
3. L. I. Schiff, *Quantum Mechanics*, McGraw-Hill, New York, 1955: § 18.
4. K. Gottfried, *Quantum Mechanics*, W. A. Benjamin, New York, 1966, Vol. I: § 12 and § 13.
5. G. L. Trigg, *Quantum Mechanics*, D. Van Nostrand Co., Inc., Princeton, 1967: Chapter 8, § 3, Sections 1–3, and § 5.
6. A. S. Davydov, *Quantum Mechanics*, Pergamon Press, Oxford, 1965: § 95–§ 97, § 101 and § 102.
7. L. D. Landau and E. M. Lifshitz, *Quantum Mechanics*, Pergamon Press, Oxford, 1958: § 124, § 125, § 135 and § 137.
8. R. G. Newton, *Scattering Theory of Waves and Particles*, McGraw-Hill, New York, 1966: Chapter 8, § 1.
9. M. L. Goldberger and K. M. Watson, *Collision Theory*, John Wiley & Sons, New York, 1964: Chapter 3, § 1–§ 4.

2. *The Method of Partial Waves*

1. A. Messiah, op. cit., Vol. I: Chapter X, § 8–§ 20; Vol. II, Chapter XIX, § 9.
2. P. Roman, op. cit.: Chapter 3, § 3, Sections a–d5.
3. L. I. Schiff, op. cit.: § 19.
4. G. L. Trigg, op. cit.: Chapter 8, § 3, Sec. 4.
5. K. Gottfried, op. cit., Vol. 1: § 14–§ 16 and § 49.
6. A. S. Davydov, op. cit.: § 98 and § 99.
7. L. D. Landau and E. M. Lifshitz, op. cit.: § 122, § 123, § 126–§ 132, § 136 and § 138.
8. R. G. Newton, op. cit.: Chapter 11, § 1 and § 2 and Chapter 18.
9. M. L. Goldberger and K. M. Watson, op. cit.: Chapter 6, § 1–§ 4, § 6 and § 7.

3. *Coulomb Scattering*

1. A. Messiah, op. cit., Vol. I: Chapter XI, § 7–§ 11.
2. L. I. Schiff, op. cit.: § 20.
3. K. Gottfried, op. cit.: § 17.
4. G. L. Trigg, op. cit.: Chapter 8, § 4, sec. 4.
5. A. S. Davydov, op. cit.: § 100.
6. L. D. Landau and E. M. Lifshitz, op. cit.: § 133 and § 134.
7. R. G. Newton, op. cit.: Chapter 14, § 6.
8. M. L. Goldberger and K. M. Watson, op. cit.: Chapter 6, § 4.

4. *Inelastic Scattering*

1. L. D. Landau and E. M. Lifshitz, op. cit.: Chapter XVIII.
2. L. I. Schiff, op. cit.: § 30 and § 37.
3. P. Roman, op. cit.: Chapter 3, § 3, sec. d6.
4. G. L. Trigg, op. cit.: Chapter 8, § 3, sec. 4; Chapter 9, § 2.
5. K. Gottfried, op. cit.: § 57.
6. A. S. Davydov, op. cit.: § 103–§ 107.
7. A. Messiah, op. cit., Vol. II: Chapter XIX, § 16–§ 26.
8. R. G. Newton, op. cit.: Chapter 16 and Chapter 17.
9. M. L. Goldberger and K. M. Watson, op. cit.: Chapter 4, § 1.

5. *Jost Functions and Regge Poles*

1. P. Roman, op. cit.: Chapter 3, § 4, sec. f.
2. M. L. Goldberger and K. M. Watson, op. cit.: Chapter 6, § 5.

3. V. DE ALFARO and T. REGGE, *Potential Scattering*, North Holland Publ. Co., Amsterdam, 1965.
4. R. G. NEWTON, op. cit.: Chapter 12 and Chapter 13.
5. R. G. NEWTON, *The Complex j-Plane*, W. A. Benjamin, New York, 1967.
6. S. FRAUTSCHI, *Regge Poles*, W. A. Benjamin, New York, 1963.

6. *The Lippmann–Schwinger Equations*

1. P. ROMAN, op. cit.: Chapter 4, § 2.
2. K. GOTTFRIED, op. cit.: § 56.
3. M. L. GOLDBERGER and K. M. WATSON, op. cit.: Chapter 5, § 3.
4. R. G. NEWTON, op. cit. (1966): Chapter 7.

CHAPTER XI

1. *Atoms* (*Generalities*)

1. L. D. LANDAU and E. M. LIFSHITZ, *Quantum Mechanics*, Pergamon Press, Oxford, 1958: § 66–§ 68 and § 73–§ 77.
2. L. I. SCHIFF, *Quantum Mechanics*, McGraw-Hill, New York, 1955: § 38 and § 39.
3. A. S. DAVYDOV, *Quantum Mechanics*, Pergamon Press, Oxford, 1965: § 92 and § 93.
4. H. A. BETHE, *Intermediate Quantum Mechanics*, W. A. Benjamin, New York, 1964: Chapter 10.
5. N. MOTT and I. SNEDDON, *Wave Mechanics and its Application*, Clarendon Press, Oxford, 1948: § 27–§ 29.

2. *Molecules* (*Generalities*)

1. L. D. LANDAU and E. M. LIFSHITZ, op. cit.: Chapter XI and Chapter XIII.
2. L. I. SCHIFF, op. cit.: § 40.
3. A. S. DAVYDOV, op. cit.: Chapter XIII.
4. H. A. BETHE, op. cit.: Chapter 11.
5. A. MESSIAH, *Quantum Mechanics*, Interscience, New York, 1961: Vol. II, Chapter XVIII, § 12–§ 17.
6. G. L. TRIGG, *Quantum Mechanics*, D. Van Nostrand Co., Inc., Princeton, 1967: Chapter 7, § 3, sec. 3.
7. N. MOTT and I. SNEDDON, op. cit.: Chapter 7.

3. *Spin-Orbit Coupling*

1. L. D. LANDAU and E. M. LIFSHITZ, op. cit.: § 72.
2. L. I. SCHIFF, op. cit.: § 38 and § 44.
3. A. S. DAVYDOV, op. cit.: § 66.
4. G. L. TRIGG, op. cit.: Chapter 11, § 2, sec. 3.
5. H. A. BETHE, op. cit.: Chapter 10.
6. K. GOTTFRIED, *Quantum Mechanics*, W. A. Benjamin, New York, 1966, Vol. I, § 46.
7. P. ROMAN, *Advanced Quantum Theory*, Addison-Wesley, Reading, Massachusetts, 1965: Chapter 6, § 3, Sec. a3.

4. *The Self-Consistent Field Approximation*

1. A. MESSIAH, op. cit., Vol. II: Chapter XVIII, § 7–§ 11.
2. H. A. BETHE, op. cit.: Chapter 6.
3. L. D. LANDAU and E. M. LIFSHITZ, op. cit.: § 69.
4. L. I. SCHIFF, op. cit.: § 38.
5. A. S. DAVYDOV, op. cit.: § 90.
6. N. MOTT and I. SNEDDON, op. cit.: § 28.

5. *The Statistical Model of Thomas and Fermi*

1. A. MESSIAH, op. cit., Vol. II: Chapter XIV, § 13.
2. P. GOMBAS, *Rev. Mod. Phys.*, **35**, 512 (1963) and P. GOMBAS, *Die statistiche Theorie des Atoms und ihre Anwendungen*, Springer, Wien, 1949.
3. L. D. LANDAU and E. M. LIFSHITZ, op. cit.: § 70.

4. L. I. SCHIFF, op. cit.: § 38.
5. A. S. DAVYDOV, op. cit.: § 91.
6. H. A. BETHE, op. cit.: Chapter 7.

CHAPTER XII

1. *Elements of Relativistic Mechanics and of Relativistic Electrodynamics. The Lorentz Group*

1. L. D. LANDAU and E. M. LIFSHITZ, *Classical Theory of Fields*, Addison-Wesley, Reading, Massachusetts, 1951: § 1–§ 9, § 15–§ 18, § 23–§ 24.
2. S. S. SCHWEBER, *An Introduction to Relativistic Quantum Field Theory*, Row, Peterson, Evanston, Illinois, 1961: Chapter 2.
3. C. MØLLER, *The Theory of Relativity*, Oxford University Press, 1952.

2. *The Klein–Gordon Equation*

1. J. D. BJORKEN and S. D. DRELL, *Relativistic Quantum Mechanics*, McGraw-Hill, New York, 1967: Chapter 1, § 2.
2. E. CORINALDESI and F. STROCCHI, *Relativistic Wave Mechanics*, North-Holland Publ. Co., Amsterdam, 1963.
3. P. ROMAN, *Advanced Quantum Theory*, Addison-Wesley, Reading, Massachusetts, 1965: Chapter 2, § 1.
4. L. I. SCHIFF, *Quantum Mechanics*, McGraw-Hill, New York, 1965: § 42.
5. A. S. DAVYDOV, *Quantum Mechanics*, Pergamon Press, Oxford, 1965: § 54–§ 59.
6. H. A. BETHE, *Intermediate Quantum Mechanics*, W. A. Benjamin, New York, 1964: Chapter 16.
7. S. S. SCHWEBER, op. cit.: Chapter 3.
8. A. MESSIAH, *Quantum Mechanics*, Interscience, New York, 1961, Vol. II: Chapter XX, § 5.
9. G. L. TRIGG, *Quantum Mechanics*, D. Van Nostrand Co., Inc., Princeton, 1964: Chapter 10, § 1.
10. N. MOTT and I. SNEDDON, *Wave Mechanics and its Applications*, Clarendon Press, Oxford, 1949: § 51.
11. H. FESHBACH and F. VILLARS, *Rev. Mod. Phys.* **30,** 27 (1958).

3. *The Dirac Equation Generalities*

1. P. A. M. DIRAC, *The Principles of Quantum Mechanics*, Clarendon Press, Oxford, 1958: Chapter XI.
2. J. D. BJORKEN and S. D. DRELL, op. cit.: Chapter 1, § 3 and § 4; Chapter 2, 3 and 5.
3. E. CORINALDESI and F. STROCCHI, op. cit.
4. P. ROMAN, op. cit.: Chapter 2, § 2, Sections 2a–2e and Appendix 3, § 2.
5. L. I. SCHIFF, op. cit.: § 43.
6. A. MESSIAH, op. cit., Vol. II: Chapter XX, § 11–§ 24, § 28–§ 30 and § 36–§ 39.
7. A. S. DAVYDOV, op. cit.: § 60, § 61, § 63–§ 68, § 69, § 72 and § 73.
8. G. L. TRIGG, op. cit.: Chapter 10, § 2; Chapter 11, § 1, § 2, Sections 1 and 2; § 3.
9. H. A. BETHE, op. cit.: Chapter 17 and Chapter 18.
10. S. S. SCHWEBER, op. cit.: Chapter 4, § 1–§ 5, § 7 and § 8.
11. N. MOTT and I. SNEDDON, op. cit.: § 52–§ 58.
12. W. PAULI, in *Encyclopedia of Physics*, Edited by S. FLÜGGE, Vol. V, Part 1, Springer, Berlin, 1958: § 18–§ 24.

4. *The Dirac Algebra*

1. P. ROMAN, op. cit.: Appendix 3, § 1.
2. A. MESSIAH, op. cit., Vol. II: Chapter XX, § 10.
3. S. S. SCHWEBER, op. cit.: Chapter 4, § 2.
4. H. A. BETHE, op. cit.: Chapter 17.
5. R. H. GOOD, JR., *Rev. Mod. Phys.*, **27,** 187 (1955).
6. W. PAULI, *Ann. de l'Institut Henri Poincaré*, **6,** 137 (1936).

5. *The Foldy-Wouthuysen Transformation*

1. L. L. FOLDY and S. A. WOUTHUYSEN, *Phys. Rev.* **78,** 29 (1950).

References

2. J. D. BJORKEN and S. D. DRELL, op. cit.: Chapter 4.
3. E. CORINALDESI and F. STROCCHI, op. cit.
4. P. ROMAN, op. cit.: Chapter 2, § 2, sec. 2f.
5. A. S. DAVYDOV, op. cit.: § 62 and § 67.
6. A. MESSIAH, op. cit., Vol. II: Chapter XX, § 31–§ 33.
7. G. L. TRIGG, op. cit.: Chapter 11, § 2, sec. 3.
8. S. S. SCHWEBER, op. cit.: Chapter 4, § 6.

6. *The Dirac Particle in a Central Field. The Relativistic Hydrogen Atom*

1. P. A. M. DIRAC, op. cit.: § 72.
2. J. D. BJORKEN and S. D. DRELL, op. cit.: Chapter 4, § 4.
3. E. CORINALDESI and F. STROCCHI, op. cit.
4. A. MESSIAH, op. cit.: Vol. II, Chapter XX, § 25–§ 27.
5. L. I. SCHIFF, op. cit.: § 44.
6. A. S. DAVYDOV, op. cit.: § 70 and § 71.
7. G. L. TRIGG, op. cit.: Chapter 12.
8. N. MOTT and I. SNEDDON, op. cit.: § 55.
9. H. A. BETHE and E. E. SALPETER, *Quantum Mechanics of One- and Two-Electron Atoms*, Springer, Berlin, 1957: § 10–§ 14, § 17.

7. *The Quantum Mechanics of the Photon*

1. B. KURSUNOGLU, *Modern Quantum Theory*, W. H. Freeman, San Francisco, 1962: Chapter II.
2. R. H. GOOD, JR., *Phys. Rev.* **105,** 1914 (1957).
3. W. J. ARCHIBALD, *Can. J. Phys.* **33,** 565 (1955).
4. A. I. AKHIEZER and V. B. BERESTETSKI, *Quantum Electrodynamics*, Moscow, 1969 (3rd edition): § 2 and § 3.

Other Problem Books in Quantum Mechanics

1. D. TER HAAR, *Problems in Quantum Mechanics*, Academic Press, New York, 1975.
2. S. FLÜGGE, *Practical Quantum Mechanics*, Vol. I and Vol. II, Springer, Berlin.
3. V. I. KOGAN and V. M. GALITSKI, *Problems in Quantum Mechanics*, Prentice Hall, 1963.
4. I. I. GOLDMAN and V. D. KRIVCHENKOV, *Sbornik Zadach po Kvantovoi Mekhanike*, Moscow, 1956.

Index

Index

Index

OTHER TITLES IN THE SERIES IN NATURAL PHILOSOPHY

OTHER TITLES IN THE SERIES

Vol. 52. Rousseau and Mathieu—Problems in Optics
Vol. 53. Bowler—Nuclear Physics
Vol. 54. Pomraning—The Equations of Radiation Hydrodynamics
Vol. 55. Belinfante—A Survey of Hidden Variables Theories
Vol. 56. Scheibe—The Logical Analysis of Quantum Mechanics
Vol. 57. Robinson—Macroscopic Electromagnetism
Vol. 58. Gombas and Kisdi—Wave Mechanics and its Applications
Vol. 59. Kaplan and Tsytovich—Plasma Astrophysics
Vol. 60. Kovacs and Zsoldos—Dislocations and Plastic Deformation
Vol. 61. Auvray and Fourrier—Problems in Electronics
Vol. 62. Mathieu—Optics
Vol. 63. Atwater—Introduction to General Relativity
Vol. 64. Muller—Quantum Mechanics: A Physical World Picture
Vol. 65. Bilenky—Introduction to Feynman Diagrams
Vol. 66. Vodar and Romand—Some Aspects of Vacuum Ultraviolet Radiation Physics
Vol. 67. Willet—Gas Lasers: Population Inversion Mechanisms
Vol. 68. Akheizer *et al.*—Plasma Electrodynamics, Volume 1—Linear Theory
Vol. 69. Glasby—The Nebular Variables
Vol. 70. Bialynicki-Birula—Quantum Electrodynamics
Vol. 71. Karpman—Non-linear Waves in Dispersive Media
Vol. 72. Cracknell—Magnetism in Crystalline Materials
Vol. 73. Pathria—The Theory of Relativity
Vol. 74. Sitenko and Tartakovskii—Lectures on the Theory of the Nucleus
Vol. 75. Belinfante—Measurement and Time Reversal in Objective Quantum Theory
Vol. 76. Sobolev—Light Scattering in Planetary Atmospheres
Vol. 77. Novakovic—The Pseudo-spin Method in Magnetism and Ferroelectricity
Vol. 78. Novozhilov—Introduction to Elementary Particle Theory
Vol. 79. Busch and Schade—Lectures on Solid State Physics
Vol. 80. Akheizer *et al.*—Plasma Electrodynamics, Volume 2
Vol. 81. Soloviev—Theory of Complex Nuclei
Vol. 82. Taylor—Mechanics: Classical and Quantum
Vol. 83. Srinivasan and Parthasathy—Some Statistical Applications in X-Ray Crystallography
Vol. 84. Rogers—A Short Course in Cloud Physics
Vol. 85. Ainsworth—Mechanisms of Speech Recognition
Vol. 86. Bowler—Gravitation and Relativity
Vol. 87. Klinger—Problems of Linear Electron (Polaron) Transport Theory in Semiconductors
Vol. 88. Weiland and Wilhelmson—Coherrent Non-Linear Interaction of Waves in Plasmas
Vol. 89. Pacholczyk—Radio Galaxies
Vol. 90. Elgaroy—Solar Noise Storms
Vol. 91. Heine—Group Theory in Quantum Mechanics
Vol. 92. Ter Haar—Lectures on Selected Topics in Statistical Mechanics
Vol. 93. Bass and Fuks—Wave Scattering from Statistically Rough Surfaces
Vol. 94. Cherrington—Gaseous Electronics and Gas Lasers
Vol. 95. Sahade and Wood—Interacting Binary Stars
Vol. 96. Rogers—A Short Course in Cloud Physics (2nd Edition)
Vol. 97. Reddish—Stellar Formation
Vol. 98. Patashinskii and Pokrovskii—Fluctuation Theory of Phase Transitions
Vol. 99. Ginzburg—Theoretical Physics and Astrophysics
Vol. 100. Constantinescu—Distributions and their Appications in Physics
Vol. 101. Gurzadyan—Flare Stars
Vol. 102. Lominadze—Cyclotron Waves in Plasma
Vol. 103. Alkemade—Metal Vapours in Flames
Vol. 104. Akheizer and Peletminsky—Methods of Statistical Physics
Vol. 105. Klimontovich—Kinetic Theory of Non-Ideal Gases and Non-Ideal Plasmas
Vol. 106. Ginzburg—Waynflete Lectures on Physics
Vol. 107. Sitenko—Fluctuations and Non-linear Interactions of Waves in Plasmas
Vol. 108. Sinai—Theory of Phase Transitions: Rigorous Results
Vol. 109. Davydov—Biology and Quantum Mechanics
Vol. 110. Demianski—Relativistic Astrophysics